SUPERSYMMETRY, SUPERGRAVITY AND SUPERSTRINGS '86

SUPERSYMMETRY, SUPERGRAVITY, SUPERSTRINGS '86

Proceedings of the Trieste Spring School
7–15 April 1986

Edited by

B. de Wit
P. Fayet
M. Grisaru

World Scientific

Published by

World Scientific Publishing Co Pte Ltd
P. O. Box 128, Farrer Road, Singapore 9128.

Library of Congress Cataloging-in-Publication data is available.

SUPERSYMMETRY, SUPERGRAVITY AND SUPERSTRINGS '86

ISBN 9971-50-144-9

Printed in Singapore by Kim Hup Lee Printing Co. Pte. Ltd.

PREFACE

These are the Proceedings of the 1986 Spring School on Supersymmetry, Supergravity and Superstrings which was held at the International Centre for Theoretical Physics at Trieste, Italy, from 7-15 April.

The lectures at the School, the fourth in the series, reflect the shift that has taken place since the previous one, as Strings and Superstrings have come to take centre stage with supersymmetry one of the main tools and ten-dimensional supergravity the "phenomenological" output of these theories. The topics covered ranged from aspects of two-dimensional supersymmetry and supergravity, reflecting the world-sheet aspects of strings, through the four-dimensional phenomenology of the new particles predicted by supersymmetric theories, and on to the higher-dimensional worlds where strings are supposed to live. Introductory lectures on strings were followed by detailed discussion of the Riemann surface approach to string loops, of the (Kac-Moody) algebra of vertex operators, and of the covariant string field theory approach. As in the past years, the School was followed by a three-day Workshop on related topics.

We are grateful to the local organizer, Professor R. Iengo, for the crucial role he played and the general help he provided. We also wish to thank the staff of the Centre for their assistance. Finally, it is a pleasure to thank the Director of the Centre, Professor Abdus Salam, for his support and hospitality.

The Editors

CONTENTS

Spring School And Workshop On Supersymmetry Supergravity
And Superstrings
I.C.T.P. 7 – 18 April 1986 Trieste

INTRODUCTION

"Supergravity is dead. Long live supergravity in the context of superstrings." This seemed to be the motto of the Fourth Spring School on Supergravity and Supersymmetry which was held at the International Centre for Theoretical Physics at Trieste between 7-15 April 1986.

The School was attended by 250 exceptionally enthusiastic young Physicists and Mathematicians.

The lecture courses from L. Alvarez-Gaumé, M. Green, T. Banks, P. West, D. Olive, M. Grisaru, P. van Nieuwenhuizen and H. Nilles dealt respectively with problems of strings in their first quantized aspects, spin structures, Riemann surfaces, second quantized strings, string algebras, lattices and vertices, as well as problems of compactification in anomalies associated with quantum strings. The only lectures devoted to non-string subjects were those of P. Fayet, who gave a beautiful introduction to supersymmetric theories and their phenomenology.

The enthusiasm of the participants as well as of the lecturers, was so intense that we have decided to have a repeat of this Spring School devoted entirely to superstrings next April in 1987.

The Spring School was followed by the traditional Workshop on related subjects during 16-18 April 1986. Both the School and the Workshop were directed by B. de Wit, P. Fayet, M. Grisaru, R. Iengo, and E. Sezgin to whom we owe profound appreciation.

Abdus Salam

N=1 AND N=2 SUPERSYMMETRIC THEORIES OF PARTICLES

Pierre Fayet

Laboratoire de Physique Théorique de l'Ecole Normale Supérieure
24 rue Lhomond , 75231 Paris Cedex O5
FRANCE

ABSTRACT

We discuss ordinary (N=1) supersymmetric theories of particles, and how to search for the new particles predicted. We then discuss extended (N=2) supersymmetric grand-unified theories, and their formulation in a 6 dimensional spacetime.

1. N=1 SUPERSYMMETRIC THEORIES OF PARTICLES

Supersymmetric theories are invariant under a set of transformations which change the spins of particles by 1/2 unit, turning bosons into fermions, and conversely. These supersymmetry transformations are generated by a self-conjugate 4-component Majorana spin-1/2 operator Q, which satisfies the algebra[1)]

$$\begin{aligned} \left[Q , P^\mu \right] &= 0 \\ \left\{ Q , \overline{Q} \right\} &= -2 \gamma_\mu P^\mu \end{aligned} \qquad (1)$$

(For review articles on supersymmetry and supergravity, see e.g. Refs. 2-9).)

For many years supersymmetry was considered by the vast majority of physicists as a beautiful mathematical structure, very interesting from the point of view of quantum field theory, but of no relevance to particle physics :

one could not exhibit even a single pair of known particles of different spins that could be directly related under supersymmetry. The supersymmetry algebra (1) appeared, formally, as a possible extension for the Lorentz and Poincaré algebras underlying the theory of relativity - but the Laws of Nature did not seem to be supersymmetric !

However Nature may be supersymmetric, if one postulates the existence of a whole class of new particles - gluinos and photinos, winos and zinos, spin-0 leptons and quarks, etc. - as yet unobserved, which would be the superpartners of the ordinary particles[10]. Before discussing the properties of the many new particles predicted, it is useful to recall briefly the reasons for introducing them.

1.1 Why Do We Need To Introduce Superpartners ?

Could one use supersymmetry to relate the spin-1 photon with a massless spin-1/2 neutrino - and, at the same time, their charged electroweak partners, the spin-1 $W^{\pm}$ and the spin-1/2 electrons $e^{\pm}$? This is possible in principle[11], but leads to several difficulties when the other particles are considered. We have to take into account the other leptons (ν_μ , μ^- ; ν_τ ; τ^-), and quarks. Of course one could attempt to use N=2 or even N=4 supersymmetry to relate the photon with 2 or even 4 neutrinos (this was actually one of the initial motivations for N=2 supersymmetry[12]).But even in that case, we would still need new bosonic partners for quarks (these partners cannot be the eight neutral spin-1 gluons). Preserving a symmetry of treatment between leptons and quarks leads one to associate all of them with new bosonic partners.

There are other reasons for not associating the W^- with the electron e^-. This would require part of the Dirac field of the electron to transform, like the W^-, as the lower member of an electroweak triplet ($I_3 = -1$). We know that this cannot be the case : e_L^- and e_R^- should have $I_3 = -1/2$ and 0, respectively.

In the same spirit we might still attempt to relate a left-handed fermionic doublet $\begin{pmatrix} \nu \\ \ell^- \end{pmatrix}_L$ with a doublet of spin-0 fields $\begin{pmatrix} \varphi^0 \\ \varphi^- \end{pmatrix}$, which would trigger the electroweak breaking. But the translation of φ^0 would result in a charged Dirac field $\ell_L^- + \lambda_R^-$ (λ^- denoting what we now call a gaugino, with $I_3 = -1$)[11]. Such a field, even if light, cannot, nowadays, be interpreted as describing a charged lepton e^-, μ^- or τ^-. Moreover the field ν_L, originally intended to represent a massless or light neutrino, would acquire a mass by combining with the gaugino $\lambda_{Z\,R}$ associated with the Z. This is, actually, how the particles now called "winos" and "zinos" - both mixtures of "gauginos" and "higgsinos" - entered in supersymmetric electroweak theories. The fermionic partner of the photon (the gaugino field λ_γ) which could not finally be identified with any of the known neutrinos, was called "photonic neutrino", subsequently shortened into "photino". Similarly, the spin-1/2 partners of the gluons were called "gluinos", etc.[10].

If supersymmetry relates spin-1 with spin-1/2 particles, or spin-1/2 with spin-0 particles, it does not relate directly any of the known particles together. This was the first apparent failure of supersymmetry. Despite that, Nature may still be supersymmetric, if we postulate the existence of new particles, as yet unobserved, which would be the superpartners of the ordinary ones.

We just mentioned the winos, zinos, photinos and gluinos, which are the spin-1/2 superpartners of the $W^\pm$, Z, γ and gluons, respectively. But what about the superpartners of leptons and quarks ?

Relating leptons and quarks with new spin-1 particles (still an open possibility, especially in the framework of extended supersymmetry) would require a very large gauge group ; and, therefore, a large number of new gauge bosons, presumably very heavy, as well as many additional spin-0 bosons, associated with the spontaneous breaking of this large gauge group. The simplest and most economic possibility, which does not require a very large extension of the gauge group, consists in relating all leptons and quarks with new spin-0 particles, often called "sleptons" and "squarks".

The hypothesis of the existence of these new particles was taken more and more seriously, due to the remarkable properties of supersymmetric theories, briefly summarized below :

i) the relations they provide between massive gauge bosons and Higgs bosons ;

ii) their relations with gravitation (supergravity) ;

iii) the improved convergence properties of supersymmetric quantum field theories, leading in particular to the possibility of a solution for the hierarchy problem ;

iv) and, finally, their relations with the fashionable superstring theories.

1.2 The "Supersymmetric Standard Model"

Since every known particle gets associated with a superpartner, one could think that making a theory supersymmetric simply consists in duplicating all particles, by an appropriate use of the prefix "s" or the suffix "ino". If suitably constructed, however, supersymmetric theories can provide us with new relations between two classes of particles which are already present in ordinary gauge theories, namely massive spin-1 gauge bosons and spin-0 Higgs bosons. The latter appear (together of course with spin-1/2 inos) as spin-0 partners of spin-1 gauge bosons, in massive multiplets of supersymmetry[10,11,13]. This has important consequences in simple N=1 supersymmetric theories, and even more in extended supersymmetric theories, especially when grand-unification is considered (cf. section 3).

In particular, supersymmetric theories of particles include a charged Higgs boson $w^{\pm}$ associated with the $W^{\pm}$, and a neutral one z, associated with the Z. They would be degenerated in mass with the $W^{\pm}$ and Z, respectively, if supersymmetry were unbroken. After spontaneous breaking of supersymmetry the masses are no longer equal in general, but they still satisfy mass relations, like :

$$\begin{cases} m^2 \text{ (charged Higgs boson } w^{\pm}) = m_W^2 + \Delta \\ m^2 \text{ (neutral Higgs boson } z) = m_Z^2 + \Delta \end{cases} \qquad (2)$$

Δ depends on the supersymmetry breaking mechanism considered, and is usually positive (e.g. $\Delta = 4\, m_{3/2}^2$) for gravity-induced supersymmetry breaking.

This association between gauge and Higgs bosons is not a trivial property. Gluons are associated with gluinos, which are also color octets. Quarks are associated with spin-0 quarks, which are also color triplets. But the $W^{\pm}$ and Z are associated with Higgs bosons $w^{\pm}$ and z, having different electroweak gauge transformation properties. For example, the $W^{\pm}$ transforms as a member of an electroweak triplet, while the associated Higgs boson $w^{\pm}$ transforms as a member of an electroweak doublet (i.e. $I_3(W^{\pm}) = \pm 1$ differs from $I_3(w^{\pm}) = \pm 1/2$). Moreover the gauge and Higgs bosons have very different couplings to leptons and quarks, proportional to g and $g \dfrac{m_{fermion}}{m_W}$, respectively.

Despite that, these gauge and Higgs bosons appear as related under supersymmetry ! This relation may even be made explicit using the superfield formalism of N=1 supersymmetry. Higgs bosons may then be described by spin-0 components of massive gauge superfields $W^{\pm}(x, \theta, \bar{\theta})$, $Z(x, \theta, \bar{\theta})$, which also describe the spin-1 gauge bosons $W^{\pm}$ and Z. (In that formalism, however, the Lagrangian density is not polynomial.) More precisely, we can write the following expansions for massive gauge superfields[6,13) :

$$\begin{cases} W^{\pm}(x,\theta,\bar\theta) = \dfrac{1}{m_W}\, w^{\pm}(x) + \dots - \theta\sigma_\mu\bar\theta\, W^{\mu\pm}(x) + \dots \\ Z\,(x,\theta,\bar\theta) = \dfrac{1}{m_Z}\, z\,(x) - \dots - \theta\sigma_\mu\bar\theta\, Z^{\mu}(x) + \dots \end{cases} \tag{3}$$

In these expansions we have omitted for simplicity linear or trilinear terms in the anticommuting Grassman coordinates θ , $\bar\theta$. Their coefficients involve spinors χ (subcanonical) and λ (canonical), respectively. $m\,\chi$ and λ correspond to what we usually call higgsinos and gauginos, respectively. The gauge superfield mass terms may be expressed in terms of

$$\begin{cases} m_W^2 \int W^{+}(x,\theta,\bar\theta)\, W^{-}(x,\theta,\bar\theta)\; d^4\theta \\ \dfrac{1}{2}\, m_Z^2 \int Z^2(x,\theta,\bar\theta)\; d^4\theta \end{cases} \tag{4}$$

In the presence of the mass terms (4) the gauge superfields $W^{\pm}$ and Z acquire additional degrees of freedom. Some of their component fields (previously auxiliary, and which could be gauged away) now become promoted to the role of physical propagating fields describing spin-0 Higgs bosons and spin-1/2 higgsinos. Expressions (4) include, in particular, kinetic energy terms for Higgs bosons and higgsinos ; mass terms for both gauge and Higgs bosons ; and additional mass terms mixing the gauginos ($\lambda_W^{\pm}$, λ_Z) with the higgsinos ($m_W\,\chi_W^{\pm}$, $m\ \chi_Z$), despite their different electroweak properties (see Ref. 6) for more details).After diagonalization of the fermion mass matrix we get, as long as supersymmetry is unbroken :

- 2 charged Dirac winos

$$\tilde{W}^-_{1,2} = (\gamma_5) \frac{\lambda^- \pm m_W \chi^-}{\sqrt{2}} \qquad \text{(mass } m_W\text{)} \qquad (5)$$

and

- 2 neutral Majorana zinos

$$\tilde{Z}_{1,2} = (\gamma_5) \frac{\lambda_Z \pm m_Z \chi_Z}{\sqrt{2}} \qquad \text{(mass } m_Z\text{)} \qquad (6)$$

Winos and zinos are all gaugino/higgsino mixtures, associated under supersymmetry with both the massive spin-1 gauge bosons $W^\pm$ and Z, and the massive spin-0 Higgs bosons $w^\pm$ and z, in irreducible multiplets of supersymmetry. In the presence of supersymmetry breaking, the mass matrices of winos and zinos should be rediagonalized ; there is often one wino lighter than the W, and one zino lighter than the Z, but this is not necessary.

The electroweak symmetry breaking, in a supersymmetric theory, makes use of two doublet chiral Higgs superfields $\begin{pmatrix} S^0 \\ S^- \end{pmatrix}$ and $\begin{pmatrix} T^0 \\ T^- \end{pmatrix}$, left-handed and right-handed, respectively, so as to provide the necessary degrees of freedom for inos and Higgs bosons. (With only one doublet Higgs superfield, one would remain with a massless two-component charged Dirac fermion.) The first doublet is also responsible for the masses of charge −1/3 quarks and charged leptons ; the second one, for the masses of charge 2/3 quarks. The three chiral superfields S^-, T^- and $\frac{S^0 - T^{0\dagger}}{\sqrt{2}}$ can be gauged away, while the gauge superfields $W^\pm$ and Z acquire masses. The remaining uneaten neutral chiral Higgs superfield

$$H^0 = \frac{S^0 + T^{0\dagger}}{\sqrt{2}} = \frac{h^0 - i\, h'^0}{2} + \theta\, \tilde{h}^0 + \cdots \qquad (7)$$

describes a Majorana spin-1/2 higgsino $\tilde{h}^{\circ}$, and two neutral spin-0 Higgs bosons, h° (scalar) and h'° (pseudoscalar), with dilatonlike and axionlike couplings to leptons and quarks, respectively. Altogether we get, for the supersymmetric extension of the standard model, the following minimal particle content, represented in Table 1[10].

Table 1 : Minimal particle content of a N=1 supersymmetric theory, after the spontaneous breaking of the electroweak symmetry. Supersymmetry relates the massive spin-1 gauge bosons $W^{\pm}$ and Z to the spin-0 Higgs bosons $w^{\pm}$ and z. (If, in addition, an extra U(1) group is gauged, there is another neutral gauge boson, the U, which acquires a mass while the pseudoscalar Higgs boson h'° is gauged away.)

Spin 1	Spin 1/2	Spin 0	
gluons g photon γ	gluinos $\tilde{g}$ photino $\tilde{\gamma}$		
$W^{\pm}$ Z	2 (Dirac) winos $\tilde{W}_i^{\pm}$ 2 (Majorana) zinos $\tilde{Z}_i$	$w^{\pm}$ z	Higgs bosons
	1 (Majorana) higgsino $\tilde{h}^{\circ}$	h° (standard) h'° (pseudoscalar)	Higgs bosons
	leptons ℓ quarks q	spin-0 leptons $\tilde{\ell}_L$, $\tilde{\ell}_R$ spin-0 quarks $\tilde{q}_L$, $\tilde{q}_R$	

Table 1 is essentially model-independent, although the mass spectrum of the various particles involved depends on the symmetry breaking mechanisms considered. We refer the reader to earlier lectures at the Trieste School in 1984 for a discussion of different mechanisms of spontaneous breaking of global or local supersymmetry - and of their consequences for the mass spectrum of the new particles predicted[6)]. See also, more generally, the review articles of Refs. 2-10).

In the model of Ref. 10) supersymmetry was spontaneously broken by means of an extension of the gauge group to SU(3)xSU(2)xU(1)xU(1). This requires the existence of a new neutral gauge boson U, with a predominantly axial coupling to leptons and quarks, related by supersymmetry to the spin-1/2 goldstino generated by the spontaneous breaking of the global supersymmetry. The pseudoscalar h'° which appears in Table 1 is then a would-be Goldstone boson, gauged away while the U boson acquires a mass.

Models of this type, in which an extra U(1) is used to trigger spontaneous supersymmetry breaking, have been very useful as prototypes, allowing one to study the phenomenological properties of the new particles. They are now disfavoured, for several reasons : i) they tend to predict massless or light gluinos, as well as relatively light selectrons $\lesssim$ 40 GeV/c^2 ; ii) extra Higgs singlets should be introduced, in addition to the two Higgs doublets, in order to avoid unacceptable effects of the extra U(1) gauge boson U ; iii) it is hard to obtain an anomaly-free model while keeping an acceptable mass spectrum ; this requires additional fields, such as mirror lepton and quark fields and additional Higgs doublets, for example.

Nowadays one prefers to use a mechanism of gravity-induced supersymmetry breaking which relies on an "hidden sector" of the supergravity theory (see for example Ref. 5), and references therein). Then the gauge particle of supersymmetry - the spin-3/2 gravitino - acquires a large mass $m_{3/2}$

by the super-Higgs mechanism. One can generate large masses for all spin-0 leptons and quarks at the classical level, without having to introduce an extra U(1) group. When particles coupled with gravitational strength are disregarded, one gets a globally supersymmetric theory with additional soft-breaking[14] terms of dimensions $\leqslant$ 3, which are responsible for the mass splittings between bosons and fermions in the multiplets of supersymmetry.

1.3 R-Symmetry And The Interactions Of The New Particles

Let us consider the minimal content of a supersymmetric theory, as given in Table 1. At this stage the theory has an unbroken continuous R-symmetry, defined as follows[10) :

$$V(x,\theta,\bar\theta) \longrightarrow V(x,\theta e^{-i\alpha},\bar\theta e^{i\alpha}) \qquad (8)$$

for gauge superfields,

$$\begin{cases} S(x,\theta,\bar\theta) \quad \text{left-handed} \longrightarrow S(x,\theta e^{-i\alpha},\bar\theta e^{i\alpha}) \\ T(x,\theta,\bar\theta) \quad \text{right-handed} \longrightarrow T(x,\theta e^{-i\alpha},\bar\theta e^{i\alpha}) \end{cases} \qquad (9)$$

for the chiral Higgs superfields responsible for gauge symmetry breaking, and

$$\begin{cases} S(x,\theta,\bar\theta) \quad \text{left-handed} \rightarrow e^{i\alpha} S(x,\theta e^{-i\alpha},\bar\theta e^{i\alpha}) \\ T(x,\theta,\bar\theta) \quad \text{right-handed} \rightarrow e^{-i\alpha} T(x,\theta e^{-i\alpha},\bar\theta e^{i\alpha}) \end{cases} \qquad (10)$$

for the chiral superfields describing matter (leptons and quarks).

This corresponds to the conservation of an <u>additive</u> quantum number :

$$\begin{cases} R = 0 \quad \text{for ordinary particles} \\ \qquad\quad \text{(gauge bosons, Higgs bosons, leptons and quarks)} \\ R = +1\text{ , or }-1 \quad \text{for their superpartners} \\ \qquad\quad \text{(inos, spin-0 leptons and quarks)} \end{cases} \qquad (11)$$

In particular R-symmetry acts in a chiral way on the self-conjugate Majorana spinors λ representing the gluinos $\tilde{g}$ or the photino $\tilde{\gamma}$:

$$\lambda \longrightarrow e^{\gamma_5 \alpha} \lambda \tag{12}$$

Such a continuous R-symmetry, if preserved, will in principle prevent the gluinos and photino from acquiring a mass, even if supersymmetry has been spontaneously broken. This seems experimentally excluded : massless or light gluinos could combine with quarks, antiquarks and gluons, and lead to relatively light hadronic states called R-hadrons[15) ; those would decay into ordinary hadrons by emitting photinos, carrying away part of the energy-momentum. Since no such events have been detected, we need some sort of R-symmetry breaking, so that gluinos can acquire a sizeable mass. The introduction of supergravity - with a massive spin-3/2 gravitino, usually rather heavy - does necessarily lead to violations of the R-symmetry, of the order of the gravitino mass $m_{3/2}$. We shall then get massive gluinos and photinos, either at the tree approximation, or as a result of radiative corrections.

This breaking of the continuous R-symmetry group due to the coupling to supergravity does preserve, however, a discrete subgroup of R-parity transformations. This corresponds to a <u>multiplicatively conserved</u> quantum number called R-parity, $R_p = (-1)^R$. It follows from eqs.(11) that

$$\begin{cases} R_p = + \ : & \text{for ordinary particles} \\ & \text{(gauge bosons, Higgs bosons, leptons and quarks)} \\ R_p = - \ : & \text{for their superpartners} \\ & \text{(inos, spin-0 leptons and quarks)} \end{cases} \tag{13}$$

This is equivalent to the definition[15) :

$$\text{R-parity} \quad R_p = (-1)^R = (-1)^{3B + L + 2\,\text{Spin}} \tag{14}$$

The latter formula illustrates that the conservation of baryon and lepton numbers (or simply of their difference B - L) necessarily implies R-parity conservation. A contrario, a breaking of R-parity would tend to generate unwanted B or L violating effects (unless some sort of fine-tuning of the parameters is performed).

Therefore we shall only discuss the "standard phenomenology" of supersymmetry, in which R-parity is conserved. As a result, the new R-odd particles may only be pair-produced. Most of them will be unstable, excepted the lightest one, which should be absolutely stable. Which one is the lightest, stable, R-odd particle does depend on the mass spectrum, and is therefore model-dependent. Most of the time one assumes that it is the photino, but it is not necessary. (It might be, also, a spin-0 t or b quark, or a wino or a zino, or a higgsino, etc.).

2. EXPERIMENTAL SEARCHES FOR THE NEW PARTICLES

As we just said, the new particles can only be pair-produced, owing to R-parity conservation. Most of them will be unstable, excepted the lightest one, here assumed to be the photino. The pair production of superpartners should then lead, ultimately, to the production of a pair of photinos. Photinos have very small interaction cross-sections with matter[16)]. Although quasi-invisible, they carry away energy-momentum. "Missing energy" and "missing momentum" are fundamental characteristics of the production of the new particles, and could then be the signature of supersymmetry. This is how most experiments performed to date have been looking for evidence for supersymmetry[4-10)].

2.1 Spin-0 Leptons And Photinos

Spin-0 leptons may be pair-produced in e^+e^- annihilations. Assuming the photino to be the "lightest susy particle", one can search for the process[17)]

$$e^+e^- \longrightarrow \tilde{\ell}^+\tilde{\ell}^- \longrightarrow \underbrace{\ell^+\ell^-}_{\text{Non coplanar lepton pair}} + \underbrace{(\tilde{\gamma}\,\tilde{\gamma})}_{\text{2 unobserved photinos}} \qquad (15)$$

which leads to a non-coplanar lepton pair, with at least about 1/2 of the energy missing, in average (see Fig. 1).

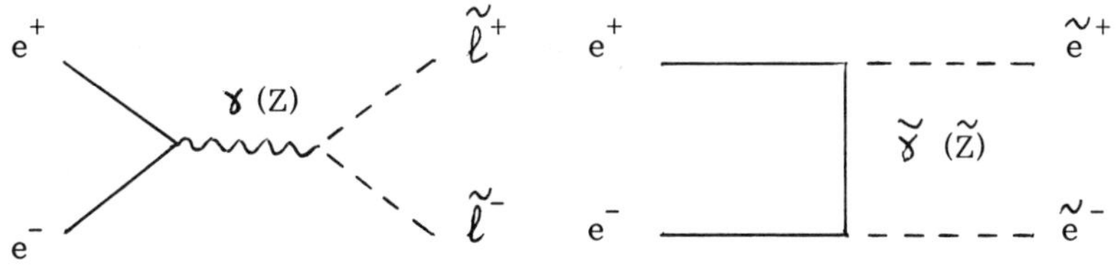

Figure 1 : Pair production of spin-0 leptons in e^+e^- annihilations.

Searches performed at PETRA (and also at PEP) have set the following lower bounds on spin-0 lepton masses [18)]

$$\begin{cases} m\ (\tilde{e}\ ) > 22\ \text{GeV}/c^2 \\ m\ (\tilde{\mu}\ ) > 21\ \text{GeV}/c^2 \\ m\ (\tilde{\tau}\ ) > 18\ \text{GeV}/c^2 \end{cases} \qquad (16)$$

assuming for simplicity that the two charged spin-0 leptons in each family are almost degenerated in mass, and that the photino is relatively light. (If $m_{\tilde{\gamma}}$ is close to $m_{\tilde{\ell}}$, the lepton pair produced does not carry much energy and tends to escape detection, so that the constraints get weaker).

The limits (16) are close to the maximum beam energy available at PETRA.

But it is also possible to search for the single production of a spin-0 electron, in association with an electron and a photino[17,19] :

$$e^+e^- \longrightarrow e \quad \tilde{e} \; \tilde{\gamma} \rightarrow (e) \quad e \, (\tilde{\gamma} \; \tilde{\gamma} \;) \tag{17}$$

This raises the lower limit on selectron masses to nearly 30 GeV/c^2, if the photino is light[20].

Much better limits may be obtained by searching for the radiative production of a pair of photinos, according to the reaction

$$e^+e^- \longrightarrow \gamma \left(\tilde{\gamma} \; \tilde{\gamma} \right) \tag{18}$$

induced by spin-0 electron exchanges[21] (see Fig. 2).

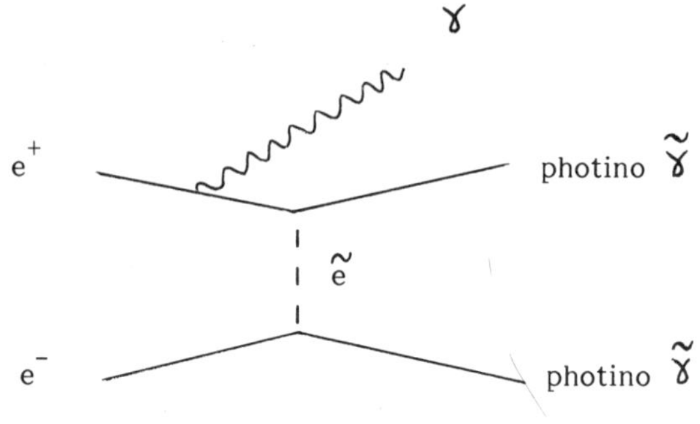

Figure 2 : Radiative production of a photino pair in e^+e^- annihilation.

The experimental signal is the production of a single photon, with missing energy-momentum carried away by the two unobserved photinos. These experiments have now reached such a high degree of precision that one should already start taking into account the concurrent reaction[22] $e^+e^- \longrightarrow \gamma \, \nu \, \bar{\nu}$.

Present results from the ASP and MAC experiments at PEP[23] imply

$$m(\tilde{e}) \;>\; 51 \text{ GeV/c}^2 \qquad (\text{if } m_{\tilde{\gamma}} \text{ small}) \tag{19}$$

If the photino is heavier, the lower limit on the spin-0 electron mass gets smaller : e.g., for a 10 GeV/c^2 photino, $m_{\tilde{\gamma}} > 33$ GeV/c^2.

2.2 Spin-0 Quarks And Gluinos

Gluinos and squarks, which are strongly interacting particles, may be copiously produced in strong interaction reactions, like pp or p$\bar{\text{p}}$ collisions. They should then decay, according to the reactions :

$$\tilde{q} \longrightarrow q\tilde{g} \;,\quad \tilde{q} \longrightarrow q\tilde{\gamma} \;,\quad \tilde{g} \longrightarrow q\bar{q}\tilde{\gamma} \;, \tag{20}$$

which ultimately lead to the production of photinos in the final state.

One can search for light gluinos by looking for the reinteraction with matter of the photinos produced by gluino decays, in a beam dump experiment[15,24]. Since the reinteraction cross-section of photinos behaves roughly like[16]

$$\sigma \;(\text{photino + matter}) \;\sim\; \frac{\alpha\,\alpha_s}{m_{\tilde{q}}^4}\, s \tag{21}$$

the limits on the pair production cross-sections of short-lived gluinos - and therefore on their masses - depend on the spin-0 quark masses $m_{\tilde{q}}$. The results of beam dump experiments[25] imply, for example,

$$m_{\tilde{g}} \gtrsim 3 \text{ GeV/c}^2 \qquad \text{if} \qquad m_{\tilde{q}} < 200 \text{ GeV/c}^2 \tag{22}$$

If spin-0 quarks are heavier, gluinos tend to have longer lifetimes. Lower bounds on their masses - at least of the order of 2 GeV/c^2 - may then be derived from searches for quasistable particles (see e.g. Ref. 26), and references therein), or from recent experiments searching for the decay $\Upsilon \rightarrow g\tilde{g}\tilde{g}$ [27,28] at PETRA, or for relatively long-lived R-hadrons that could be produced using a 300 GeV/c π^- beam at the CERN SPS[29].

Another way to search for the new particles is to look for the missing energy, or missing momentum, carried away by the emitted photinos. This method was first used in 1978 to search for light short-lived gluinos in a fixed target experiment performed at Fermilab with a 400 GeV/c proton beam[15]. This experiment[30], performed at a center of mass energy $\sqrt{s} \simeq 27$ GeV, was only sensitive to rather small values of gluino masses $\lesssim 2$ GeV/c^2. Today, $p\bar{p}$ colliders, which have much higher center of mass energies ($\sqrt{s} \sim$ 600 GeV at CERN), provide a much more efficient way to search for the missing energy-momentum that could signal the production of gluinos or squarks.

Many authors have studied these processes, and in particular the so-called monojet (and multijet) events to which they could lead[31]. Such events could arise, for example, from the pair production of gluinos

$$p\bar{p} \longrightarrow \tilde{g}\,\tilde{g} \longrightarrow \tilde{\gamma}\, q\,\bar{q} + \tilde{\gamma}\, q\,\bar{q} \tag{23}$$

or from similar processes involving squarks.

A variety of processes within the standard model can also lead to monojets : for example the production of $W^{\pm} \rightarrow \tau^{\pm}\nu$, with the τ decaying into ν + jet ; or the production of gluon + Z ($\rightarrow \nu\bar{\nu}$), etc. Presently there does not seem to be any significant excess of monojets compared to what is expected from the standard model alone[32]. From UA1 data at the CERN collider one may be able to extract the following limits (preliminary, to be confirmed) :

$$\begin{cases} m(\tilde{g}) > 60 \text{ GeV}/c^2 \\ m(\tilde{q}) > 70 \text{ GeV}/c^2 \end{cases} \tag{24}$$

except for a small allowed window corresponding approximately to m ($\tilde{g}$) $\sim$

3 to 5 GeV/c^2, with m ($\tilde{q}$) $\gtrsim$ 100 GeV/c^2 [31,32].

2.3 Winos And Zinos

Present experiments imply that the superpartners of light particles cannot be too light, i.e. that the "supergap" cannot be too small. This opens the possibility that, conversely, some of the superpartners of heavy particles - like winos and zinos - may be relatively light, and accessible to present experiments.

If supersymmetry were unbroken the $W^{\pm}$ would be degenerated in mass with two Dirac spin-1/2 winos, and the Z with two Majorana spin-1/2 zinos, as indicated in Table 1. The actual mass spectrum for these particles - which are mixtures of gaugino and higgsino components - depends on the supersymmetry breaking.

In models with extra U(1) breaking the two winos are given by

$$\begin{cases} \text{Wino}_1^- = \text{higgsino}_L^- + \text{gaugino}_R^- \\ \text{Wino}_2^- = \text{gaugino}_L^- + \text{higgsino}_R^- \end{cases} \tag{25}$$

They satisfy the mass relation

$$m^2(\text{wino}_1) + m^2(\text{wino}_2) = 2\, m_W^2 \tag{26}$$

which implies that one wino is lighter than the W, while the other is heavier[10)]

This may or may not remain true with gravity-induced supersymmetry breaking, for which the winos are mixed differently[33,34)]. In the absence of direct gaugino mass terms there is always one wino (or zino) lighter than the $W^{\pm}$ (or Z), while the other is heavier. Otherwise, it is also possible that both winos are heavier than the $W^{\pm}$, and both zinos heavier than the Z.

To illustrate this, let us consider as a particular example a simple model with gravity-induced supersymmetry breaking[35)], in which the masses of spin-0 leptons and quarks are given by

$$m_{\tilde{\ell},\tilde{q}} = \left| m_{3/2} \pm m_{\ell,q} \right| \tag{27}$$

and the masses of the Higgs bosons $w^{\pm}$ and z associated with the $W^{\pm}$ and Z, by

$$\begin{cases} m^2(w^{\pm}) = m_W^2 + 4\,m_{3/2}^2 \\ m^2(z) = m_Z^2 + 4\,m_{3/2}^2 \end{cases} \tag{28}$$

(up to radiative correction effects).

With the two Higgs doublets acquiring equal vacuum expectation values, we get the following 2x2 mass matrices for winos and zinos (in a higgsino/gaugino basis) :

$$\begin{bmatrix} -m_{3/2} & m_W \\ m_W & m_\lambda \end{bmatrix} , \quad \begin{bmatrix} -m_{3/2} & m_Z \\ m_Z & m_\lambda \end{bmatrix} , \tag{29}$$

in which m_λ parametrizes direct gravity-induced gaugino mass terms (which may or may not be present).

In the absence of such terms we find

$$\begin{cases} m(\text{winos}) = \left(m_W^2 + \frac{1}{4} m_{3/2}^2 \right)^{1/2} \pm \frac{1}{2} m_{3/2} \\ m(\text{zinos}) = \left(m_Z^2 + \frac{1}{4} m_{3/2}^2 \right)^{1/2} \pm \frac{1}{2} m_{3/2} \end{cases} \tag{30}$$

There is necessarily a wino lighter than the W, and a zino lighter than the Z.

On the other hand, with $m_\lambda = - m_{3/2}$, or $+ m_{3/2}$, we would get[34)]

$$m\,(\text{gluinos}) = m\,(\text{photino}) = m_{3/2} \tag{31}$$

and

$$\begin{cases} m\,(\text{winos}) = |\, m_W \pm m_{3/2} \,| \\ m\,(\text{zinos}) = |\, m_Z \pm m_{3/2} \,| \end{cases} \tag{32}$$

or, alternately

$$\begin{cases} m\,(\text{winos}) = (\, m_W^{\,2} + m_{3/2}^{\,2})^{1/2} \\ m\,(\text{zinos}) = (\, m_Z^{\,2} + m_{3/2}^{\,2})^{1/2} \end{cases} \tag{33}$$

(More precisely, formulas (28,31,33) can be obtained in the framework of N=2 extended supersymmetric theories, as we shall discuss in section 3 ; these formulas will be interpreted in a 5 or 6 dimensional spacetime, in which inos carry momentum $\pm\ m_{3/2}$,and Higgs bosons $\pm\ 2\ m_{3/2}$, along the extra compact dimensions[36].)

A relatively light wino could be pair-produced in e^+e^- annihilations :

$$e^+e^- \longrightarrow \tilde{W}^+\ \tilde{W}^- \tag{34}$$

It could then decay, according for example to the reactions

$$\begin{cases} \tilde{W}^{\pm} \longrightarrow \tilde{\gamma}\ \ell^{\pm}\ \overset{(-)}{\nu} \\ \tilde{W}^{\pm} \longrightarrow \tilde{\gamma}\ q\ \bar{q}' \end{cases} \tag{35}$$

induced by $W^{\pm}$ (or spin-0 lepton or quark) exchanges. Such a wino would look somewhat like an additional charged lepton (see Fig. 3). Present PETRA limits[37] imply that

$$m\,(\text{winos}) \gtrsim 22\ \text{GeV}/c^2 \tag{36}$$

One may also search for the process

$$e^+e^- \longrightarrow \tilde{Z}\,(\text{unstable})\ \tilde{\gamma} \tag{37}$$

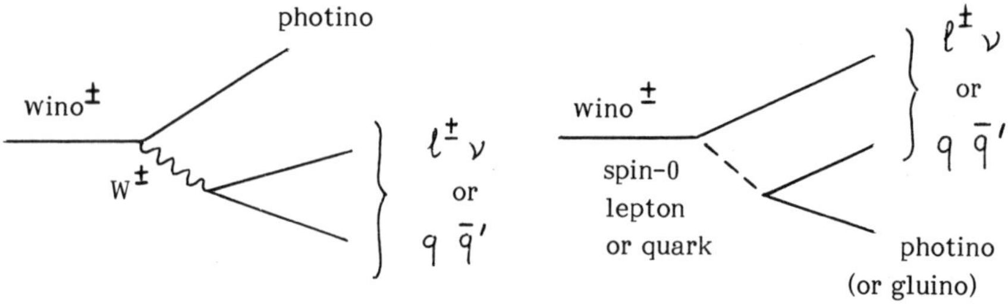

Figure 3 : Possible decay modes of winos.

induced by spin-0 electron exchanges. This leads to combined constraints on the $\tilde{\gamma}$, $\tilde{Z}$ and $\tilde{e}$ masses[37]. However in practice these constraints are not as restrictive as they look : the masses of the various particles are usually correlated in such a way that if $\tilde{\gamma}$ and $\tilde{Z}$ are light enough (so that $e^+e^- \rightarrow \tilde{Z}\,\tilde{\gamma}$ is kinematically allowed at present energies) the $\tilde{e}$'s are rather heavy, and the expected production cross section(37)is rather small (see for example the mass formulas (27-32)).

If sufficiently light, inos could also be produced in $W^{\pm}$ and Z decays, according for example to the reactions[7,9,33,34] :

$$\begin{cases} W^{\pm} \longrightarrow \tilde{W}^{\pm}\,\tilde{\gamma}\ ,\ \tilde{W}^{\pm}\,\tilde{Z}\ ,\ \tilde{W}^{\pm}\,\tilde{h} \\ Z \longrightarrow \tilde{W}^{+}\,\tilde{W}^{-},\ \tilde{Z}\ \ \tilde{h}\ ,\ \ldots \end{cases} \tag{38}$$

They would then decay (see e.g. Fig. 3) :

$$\text{ino} \longrightarrow \text{lighter ino} + q\bar{q}'\ (\text{or } \ell\,\bar{\ell}'\,) \tag{39}$$

leading to final states with missing energy-momentum, and leptons or quark jets.

3. N=2 SUPERSYMMETRIC GRAND-UNIFIED THEORIES AND EXTRA SPACE DIMENSIONS

3.1 Why N=2 Supersymmetry ?

Simple (N=1) supersymmetric theories cannot be considered as completely satisfactory from a theoretical point of view, especially when grand-unification[38] is introduced. We shall now discuss theories which are invariant under an extended algebra involving N=2 supersymmetry generators Q^i.

What are the motivations for such a construction[36, 39-41] ?

i) to realize a much larger association between massive gauge bosons and Higgs bosons, so as to find a place for the many Higgses of supersymmetric GUTs :

$$\begin{matrix} \text{1 massive spin-1} & \text{N=2 supersymmetry} & \underline{1 \text{ or } 5} \text{ spin-0} \\ \text{gauge boson} & \longleftrightarrow & \text{Higgs bosons} \end{matrix} \tag{40}$$

ii) to reduce the arbitrariness of N=1 theories : the superpotential of N=1 theories becomes fixed by N=2 supersymmetry ; the direct gaugino masses m_λ are also fixed, and equal to the gravitino mass $m_{3/2}$. N=2 supersymmetric GUTs will depend on a very small number of arbitrary parameters, e.g. e, m_X, m_W and $m_{3/2}$, only, for an N=2 SU(5) GUT, at least as long as leptons and quarks are not considered ;

iii) the possibility of determining algebraically the grand-unification mass m_X, from the value of the central charge Z[42] which appears in the N=2 algebra, and is related to the weak hypercharge operator Y[40] :

$$\{ Q^i, \bar{Q}^j \} = -2 \not{P} \, \delta^{ij} + 2 \, Z \, \varepsilon^{ij} \tag{41}$$

with

$$| Z (X^{\pm 4/3}) | \equiv m_X \tag{42}$$

iv) the possibility of expressing these theories in a 6 (or maybe 10) dimensional spacetime, in such a way that m_X , $m_{3/2}$ originate from momenta carried by the GUT or SUSY particles along extra compact dimensions,and may be computed, ultimately, in terms of the sizes of these compact dimensions.

N=2 theories have the following general features :

i) existence of larger multiplets and therefore of a new set of gravitinos, photinos, gluinos, winos and zinos, etc. ;

ii) existence of spin-0 photons and spin-0 gluons ; and of a spin-1 "graviphoton";

iii) existence of mirror leptons and quarks having V+A charged current weak interactions ;

iv) existence of additional relations between spin-1 gauge bosons and spin-0 Higgs bosons.

The motivation for extended supersymmetry is not apparent at this stage ; indeed N=1 supersymmetric theories of weak, electromagnetic and strong interactions, whose minimal content is given in Table 1, are quite appealing. The situation, however, changes when one considers grand-unification : supersymmetric grand-unified theories require a rather large number of spin-0 Higgs bosons. The Higgs sector gets quite complicated, and we no longer have a simple classification, as in Table 1.

Even with only a $\underline{24}$, a $\underline{5}$ and a $\underline{\bar{5}}$ chiral Higgs superfields - certainly a minimal choice for a N=1 supersymmetric SU(5) theory - we get <u>three</u> Higgs bosons of $\pm$ 1 unit charge, one of them only being related to the $W^{\pm}$. N=2 supersymmetry will introduce two additional Higgs bosons of $\pm$ 1 unit charge, but, ultimately, all <u>five</u> charged Higgses get related with the $W^{\pm}$ [40]:

$$W^{\pm} \longleftrightarrow \text{5 charged Higgs bosons} \qquad (43)$$

and, similarly,

$$Z \longleftrightarrow \text{5 neutral Higgs bosons} \qquad (44)$$

N=2 supersymmetry also leads us to introduce a second octet of gluinos, and a second photino. One may of course question the necessity of introducing them, especially since ordinary gluinos and photinos have not been observed yet.

But the second octet of gluinos ($\hat{g}$) and the second photino ($\hat{\gamma}$) are not really new particles. They are already present, although in a hidden way, in a grand-unified theory with only a simple (N=1) supersymmetry. The complex spin-0 Higgs field (such as the $\underline{24}$ of SU(5)) which breaks spontaneously the GUT symmetry describes an octet and a singlet of spin-0 particles, which will be interpreted later as spin-0 gluons and spin-0 photons, respectively. Their fermionic partners are a second octet of Majorana fermions ("paragluinos" $\hat{g}$), similar to the gluinos $\tilde{g}$, and a singlet one (" paraphotino" $\hat{\gamma}$), similar to the photino $\tilde{\gamma}$.

Requiring N=2 extended supersymmetry means, in particular, that there is no essential difference between the by-now familiar octet of gluinos, and the second octet of colored fermions which appears as a consequence of grand-unification. Altogether we get two octets of gluinos, two photinos, as well as a complex octet of spin-0 gluons, and a complex spin-0 photon. This is summarized in Fig. 4.

3.2 The "N=2 Supersymmetric Standard Model"[40)]

The extended supersymmetry algebra reads :

$$\left\{ Q^1 \; , \; \bar{Q}^1 \right\} = \left\{ Q^2 \; , \; \bar{Q}^2 \right\} = -2 \not{P} \qquad (45)$$

as in formula (1). In addition, the two supersymmetry generators Q^1 and Q^2 also satisfy an anticommutation relation

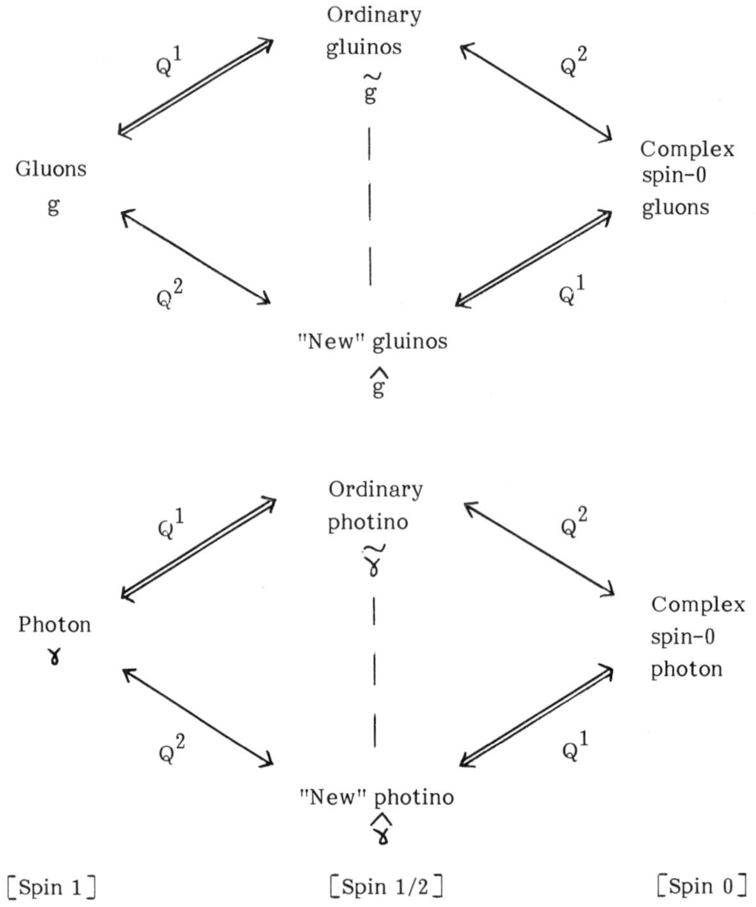

Figure 4 : Relations between the gluons, the photon, and their spin-1/2 and spin-0 partners, in an N=2 extended supersymmetric theory. Q^1 denotes the action of the first supersymmetry generator, Q^2 the action of the second one. Spin-0 gluons and spin-0 photons are described by the complex adjoint Higgs field (e.g. a 24 of SU(5)) which breaks spontaneously the GUT symmetry. (They will appear, subsequently, as the 5th and 6th components of the gluon and photon 6-vector fields, in a 6-dimensional spacetime.)

$$\left\{ Q^1 , \bar{Q}^2 \right\} = - \left\{ Q^2 , \bar{Q}^1 \right\} = 2 Z \qquad (46)$$

Z is a spin-0 symmetry generator called a central charge[42], which has the dimension of a mass. Its explicit expression includes a spontaneously generated part[40] involving neutral uncolored grand-unification symmetry generators, such as the weak hypercharge $Y = 2 (Q - T_3)$.

With SU(5) as the grand-unification group we find :

$$\left\{ Q^1 , \bar{Q}^2 \right\} = - \left\{ Q^2 , \bar{Q}^1 \right\} = 2 \left[\begin{matrix} \text{Global symmetry} \\ \text{generator} \end{matrix} -(3/5) m_X \; Y \right] \qquad (47)$$

The central charge Z vanishes for the $W^{\pm}$ and Z, γ and gluons, but equals

$$Z = \mp m_X \qquad (48)$$

for the GUT bosons $X^{\pm 4/3}$ and $Y^{\pm 1/3}$ (which have weak hypercharges Y = ± 5/3). We also find the mass relation

$$m_Y{}^2 = m_X{}^2 + m_W{}^2 \qquad (49)$$

In N=2 extended supersymmetry the minimal particle content gets increased compared with that given in Table 1. We give in Table 2 the minimal particle content of a N=2 supersymmetric SU(5) theory, after gauge symmetry breaking. As a consequence of extended supersymmetry, almost every Higgs boson now appears as a superpartner of a gauge boson. (This association becomes complete if an extra U(1) is gauged.) We also give in Table 3 the corresponding table relative to a N=2 supersymmetric O(10) theory, constructed using the same principles.

Note the existence of 3 different types of massive gauge hypermultiplets, with different field contents[39] :

- type I (like the $W^{\pm}$ and Z) : they describe 1 massive gauge boson, 4 spin-1/2

Table 2 : Minimal particle content of a N=2 supersymmetric SU(5) grand-unified theory. The spontaneous breaking $SU(5) \rightarrow SU(3)_{QCD} \times U(1)_{QED}$ is induced by a complex adjoint $\underline{24}$, and 4 quintuplets ($\underline{5}$ and $\underline{\bar{5}}$) of spin-0 Higgs fields.

Spin 1	Spin 1/2	Spin 0
$X^{\pm 4/3}$	2x3 Dirac xinos	3 charged Higgs bosons
$Y^{\pm 1/3}$	4x3 Dirac yinos	5x3 charged Higgs bosons
$W^{\pm}$	4 Dirac winos	5 charged Higgs bosons
Z	4 Majorana zinos	5 neutral Higgs bosons
γ	2 Majorana photinos	2 spin-0 photons
gluons	2x8 Majorana gluinos	2x8 spin-0 gluons
	2 neutral Majorana higgsinos	4 neutral Higgs bosons
	leptons and quarks + mirror partners	spin-0 leptons and quarks + mirror partners
+ gravitation multiplet : 1 spin-2 graviton , 2 spin-3/2 gravitinos, 1 spin-1 "graviphoton"		

Table 3 : Particle content of a N=2 supersymmetric O(10) grand-unified theory. The spontaneous breaking O(10) → $SU(3)_{QCD} \times U(1)_{QED}$ is induced by a complex adjoint 45, and 4 hexadecuplets (16 and $\overline{16}$) of spin-0 Higgs fields.

Spin 1	Spin 1/2	Spin 0
$X^{\pm 4/3}$	2x3 Dirac xinos	3 charged Higgs bosons
$Y^{\pm 1/3}$	4x3 Dirac yinos	5x3 charged Higgs bosons
$W^{\pm}$	4 Dirac winos	5 charged Higgs bosons
Z	4 Majorana zinos	5 neutral Higgs bosons
γ	2 Majorana photinos	2 spin-0 photons
gluons	2x8 Majorana gluinos	2x8 spin-0 gluons
$X'^{\pm 2/3}$	4x3 Dirac xinos '	5x3 charged Higgs bosons
$Y'^{\mp 1/3}$	4x3 Dirac yinos '	5x3 charged Higgs bosons
$X_s^{\pm 2/3}$	4x3 Dirac xinos_s	5x3 charged Higgs bosons
$W'^{\pm}$	4 Dirac winos'	5 charged Higgs bosons
Z'	4 Majorana zinos '	5 neutral Higgs bosons
	2 neutral Majorana higgsinos	4 neutral Higgs bosons
	2 neutral Majorana higgsinos	4 neutral Higgs bosons
	leptons and quarks + mirror partners	spin-0 leptons and quarks + mirror partners
+ gravitation multiplet : 1 spin-2 graviton , 2 spin-3/2 gravitinos , 1 spin-1 "graviphoton"		

inos and 5 spin-0 Higgs bosons ; and carry no central charge :

$$\mathcal{M} > |Z| = 0 \tag{50}$$

- type II (like the $X^{\pm 4/3}$) : they describe 1 massive gauge boson, 2 spin-1/2 inos and 1 spin-0 Higgs boson, all complex ; they carry a non-vanishing value of the central charge Z, and verify :

$$\mathcal{M} = |Z| > 0 \tag{51}$$

- type III (like the $Y^{\pm 1/3}$) : they describe 1 massive gauge boson, 4 spin-1/2 inos and 5 spin-0 Higgs bosons, all complex ; they carry a non-vanishing value of the central charge Z, and verify :

$$\mathcal{M} > |Z| > 0 \tag{52}$$

The actual mass spectrum of the new particles will depend on the way in which the symmetry breaking is performed. If the supersymmetry breaking is induced by gravitation or by dimensional reduction, one can get the following kind of mass spectrum[36)] :

$$m_{\text{gluinos}} = m_{\text{photino}} = m_{3/2} \tag{53}$$

$$\left\{ \begin{aligned} &m\,(\text{winos}) = (m_W^{\;2} + m_{3/2}^{\;2})^{1/2} \\ &m\,(\text{zinos}) = (m_Z^{\;2} + m_{3/2}^{\;2})^{1/2} \end{aligned} \right. \tag{54}$$

and

$$\left\{ \begin{aligned} &m\,(\text{charged Higgs bosons}) = m_W, \text{ or } (m_W^{\;2} + 4\, m_{3/2}^{\;2})^{1/2} \\ &m\,(\text{neutral Higgs bosons}) = m_Z, \text{ or } (m_Z^{\;2} + 4\, m_{3/2}^{\;2})^{1/2} \end{aligned} \right. \tag{55}$$

(up to radiative correction effects), as will be discussed in subsection 3.4. A crucial question is, of course, whether both N=2 supersymmetry generators break simultaneously, or does one have a sequential breaking N=2 $\longrightarrow$ N=1 $\longrightarrow$

N=0 ? In the second case we would no longer expect mirror leptons and quarks, and spin-0 photons and gluons, to be present at relatively low energies.

3.3 Proton Stability In N=2 Supersymmetry GUTs

A consequence of the appearance of the weak hypercharge operator Y in the anticommutation relation of the two different supersymmetry generators (eq. (47)) is the existence of mass splittings $\simeq m_X$ in all multiplets of the grand-unification group, i.e. in lepton and quark multiplets as well as in the gauge boson multiplet.

As a result the grand-unification symmetry associates light leptons with heavy quarks of mass $\simeq m_X$ (or $2\, m_X$), and conversely. It is therefore necessary to perform a replication of representations, in order to describe every single family of quarks and leptons[43].

It follows that the $X^{\pm 4/3}$ and $Y^{\pm 1/3}$ gauge bosons (as well as their Higgs partners) do not couple directly light quarks to light leptons. The usual diagrams responsible for the standard proton decay mode $p \rightarrow \pi^0 e^+$ does not exist ! (cf. Fig. 5). Actually in minimal N=2 SUSY GUTs, the proton tends to be totally stable.

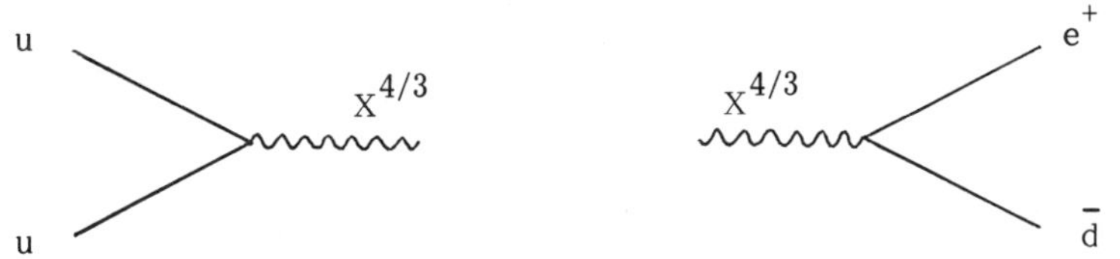

Figure 5 : Forbidden couplings in N=2 SUSY GUTs : the $X^{\pm 4/3}$ and $Y^{\pm 1/3}$ gauge bosons do not couple directly to light leptons and light quarks.

3.4 Supersymmetric GUTs In A 6 Dimensional Spacetime

An additional interest of extended supersymmetric theories is that they can be formulated in a 6 dimensional spacetime[36, 39-41, 44]. The two photinos appearing in Fig. 4, or Tables 2 and 3, originate from a single Weyl (chiral) spinor in 6 dimensions :

$$\begin{array}{ccc} \text{1 Weyl photino} & \longrightarrow & \text{2 Majorana photinos} \\ \text{in 6 dim.} & & \text{in 4 dim.} \end{array} \qquad (56)$$

$$\begin{array}{ccc} \text{1 Weyl gluino octet} & \longrightarrow & \text{2 Majorana gluino octets} \\ \text{in 6 dim.} & & \text{in 4 dim.} \end{array} \qquad (57)$$

Weyl spinors in 6 dimensions also describe, at the same time, ordinary leptons and quarks as well as their mirror partners.

The $W^{\pm}$ and Z masses are already present in the 6 dimensional spacetime, (i.e., they can be generated in a 6 d Poincaré invariant way). In 6 dimensions the $X^{\pm 4/3}$ is still massless, while the $Y^{\pm 1/3}$ has the same mass as the $W^{\pm}$:

$$\begin{cases} m^{6d}\,(X^{\pm 4/3}) = m^{6d}\,(\gamma) = m^{6d}\,(\text{gluons}) = 0 \\ m^{6d}\,(Y^{\pm 1/3}) = m^{6d}\,(W^{\pm}) = m_W \\ m^{6d}\,(Z) = m_W \big/ \cos\theta \end{cases} \qquad (58)$$

This corresponds to the existence, in the 6 dimensional spacetime, of a manifest electrostrong symmetry $SU(4) \supset SU(3)_{QCD} \times U(1)_{QED}$, which relates the photon with the eight colored gluons. It survives the "electroweak breaking" $SU(5) \rightarrow SU(4)$ (or $O(10) \rightarrow SU(4)$, etc.), which generates a mass for the $W^{\pm}$ and Z bosons.

The grand-unification mass in 4 dimensions appears as the result of the dimensional reduction from 6 (or 5) to 4 dimensions. The central charge Z, which is present in the N=2 supersymmetry algebra (47), gets replaced by the fifth component of the covariant momentum

$$Z \rightsquigarrow \mathcal{P}^5 = -i\,\mathcal{D}^5 \tag{59}$$

The grand-unification gauge bosons $X^{\pm 4/3}$ and $Y^{\pm 1/3}$ now carry covariant momentum $|\mathcal{P}^5| = m_X$ along the compact fifth dimension.

The 4 dimensional mass spectrum is given by the general formula :

$$\mathcal{M}^2_{4d} = \mathcal{M}^2_{6d} + (\mathcal{P}^5)^2 + (\mathcal{P}^6)^2 \tag{60}$$

which implies in particular

$$m_Y^2 = m_W^2 + m_X^2 \tag{61}$$

thereby providing us with a simple physical interpretation for this N=2 mass formula[40].

In a similar way we can also use the R-invariance of the 6-dimensional theory to generate mass-splittings between bosons and fermions in the 4-dimensional spacetime. The definitions (8-12) of R-invariance may be extended to 6 dimensional theories[36]. The 6d R-quantum numbersof the various fields are as follows :

$$\begin{cases} R = 0 \text{ for gauge bosons} \\ R = \pm 1 \text{ for inos (most of them gaugino/higgsino mixtures)} \\ R = 0, \text{ or } \pm 2 \text{ for Higgs bosons} \end{cases} \tag{62}$$

$$\begin{cases} R = 0 \text{ for leptons and quarks} \\ R = \pm 1 \text{ for spin-0 leptons and quarks} \end{cases} \tag{63}$$

As a result one can arran ge in such a way that inos, and spin-0 leptons or quarks, carry momentum $\pm\, m_{3/2}$ along the compact sixth dimension - for example - while Higgs bosons carry momentum 0 or $\pm\, 2\, m_{3/2}$. The mass formula (60) does then lead to the 4d mass spectrum :

$$m\,(\text{photino}) = m\,(\text{gluinos}) = m_{3/2}$$

$$\begin{cases} m\,(\text{winos}) = m_W^2 + m_{3/2}^2 \\ m\,(\text{zinos}) = m_Z^2 + m_{3/2}^2 \end{cases}$$

$$\begin{cases} m^2\,(\text{charged Higgs bosons}) = m_W^2, \text{ or } (M_W^2 + 4\,m_{3/2}^2) \\ m^2\,(\text{neutral Higgs bosons}) = m_Z^2, \text{ or } (M_Z^2 + 4\,m_{3/2}^2) \end{cases}$$

$$m_{\tilde{\ell}} \simeq m_{\tilde{q}} \simeq m_{3/2}$$

$$\text{etc.} \qquad (64)$$

in which we have disregarded the effects of radiative corrections, and of lepton and quark masses (leptons and quarks still remain massless at this stage).

Up to now m_X and $m_{3/2}$ were arbitrary parameters. One can also relate them, in various possible ways, to the lengths of the extra compact dimensions, L_5 and L_6 [36,40] ; for example :

$$m_X = \frac{\pi \hbar}{L_5\, c}, \qquad m_{3/2} = \frac{\pi \hbar}{L_6\, c} \qquad (65)$$

Although they are probably unrealistically simple (especially since the space formed by the extra compact dimensions is likely to be more complicated than a flat torus), these formulas illustrate how both the grand-unification and supersymmetry breakings may appear, in 4 dimensions, as an effect of extra compact dimensions of spacetime ; and how some of the fundamental scales of particle physics may be computed in terms of geometric parameters associated with this compact space.

In ordinary supersymmetric theories superpartners are usually expected to show up at a mass scale $\sim m_W$, or in any case $\lesssim O(\mathrm{TeV}/c^2)$. Extended supersymmetric theories, and higher dimensional theories, however, do also suggest less conventional possibilities. While both m_X and $m_{3/2}$ could be very large - in which case supersymmetry would only become apparent at very high energies - it is also possible that $m_{3/2}$ remains of order $\sim m_W$. In that case a relation like $m_{3/2} \sim \frac{\hbar}{Lc}$ would then indicate that both supersymmetry and extra dimensions might show up at future colliders.

REFERENCES

1) Yu A. Gol'fand and E.P. Likhtman, J.E.T.P. Lett. 13, 323 (1971) ; D.V. Volkov and V.P. Akulov, Phys.Letters 46B, 109 (1973) ; J. Wess and B. Zumino, Nucl.Phys. B70, 39 (1974).

2) P. Fayet and S. Ferrara, Phys. Reports 32C, 249 (1977).

3) P. van Nieuwenhuizen, Phys. Reports 68C, 189 (1981).

4) J. Ellis et al., Phys. Reports 105, 1 (1984).

5) H.P. Nilles, Phys. Reports 110, 1 (1984).

6) P. Fayet, Proc. of the 1984 Trieste Spring School on Supersymmetry and Supergravity (World Scientific, Singapour) 114 (1984).

7) H. Haber and G. Kane, Phys. Reports 117, 75 (1985).

8) M. Sohnius, Phys. Reports 128, 39 (1985).

9) Proc. of the 13th SLAC Institute on Particle Physics (1985).

10) P. Fayet, Phys. Letters 64B, 159 (1976) ; 69B, 489 (1977) ; "Unification of the Fundamental Particle Interactions", Proc. Europhysics Study Conf., Erice, 1980, eds. S. Ferrara, J. Ellis and P. van Nieuwenhuizen (Plenum,

N.Y.), 587 (1980).

11) P. Fayet, Nucl.Phys. B90, 104 (1975).

12) P. Fayet, Nucl.Phys. B113, 135 (1976).

13) P. Fayet, Nucl.Phys. B237, 367 (1984).

14) L. Girardello and M. Grisaru, Nucl.Phys. B194, 65 (1982).

15) G.R. Farrar and P. Fayet, Phys. Letters 76B, 575 (1978) ; 79B, 442 (1978).

16) P. Fayet, Phys. Letters 86B, 272 (1979).

17) G.R. Farrar and P. Fayet, Phys. Letters 89B, 191 (1980).

18) CELLO collaboration, Phys. Letters 114B, 287 (1982) ;

JADE " , " 152B, 385 (1985) ; 152B, 392 (1985) ;

MARK J " , " 152B, 439 (1985) ;

TASSO " , " 117B, 365 (1982) ;

See also S.L. Wu, preprint DESY 86-007.

19) M.K. Gaillard, L. Hall and I. Hinchliffe, Phys. Letters 116B, 279 (1982).

20) CELLO collaboration, contribution submitted to the 23rd Int. Conf. on High-Energy Physics at Berkeley (U.S.A.) , (1986).

21) P. Fayet, Phys. Letters 117B, 460 (1982) ;

J. Ellis and J.S. Hagelin, Phys. Letters 122B, 303 (1983).

22) E. Ma and J. Okada, Phys.Rev.Lett. 41, 287 (1978) ;

K.J.F. Gaemers, R. Gastmans and F.M. Renard, Phys.Rev. D19, 1605 (1979).

23) E. Fernandez et al., Phys.Rev.Lett. 54, 1118 (1985) ;

G. Bartha et al., Phys.Rev.Lett. 56, 685 (1986) ;

H. Küster, Proc. of the 21st Rencontre de Moriond at Les Arcs (France), (1986) ;

N. Jonker, Proc. of the 21st Rencontre de Moriond at Les Arcs (France), (1986).

24) G. Kane and J. Léveillé, Phys. Letters 112B, 227 (1982).

25) CHARM collaboration, Phys. Letters 121B, 429 (1983) ;
R.C. Ball et al., Phys.Rev.Lett. 53, 1314 (1984) ;
W.A. 66 collaboration, Phys. Letters 160B, 212 (1985).

26) S. Dawson, E. Eichten and C. Quigg, Phys.Rev. D31, 1581 (1985).

27) B.A. Campbell, J. Ellis and S. Rudaz, Nucl.Phys. B198, 1 (1982).

28) ARGUS collaboration, Phys. Letters 167B, 360 (1986).

29) NA 3 collaboration, Z. Phys. C31, 21 (1986).

30) J.P. Dishaw et al., Phys. Letters 85B, 142 (1979).

31) See, for example :
R.M. Barnett, Proc. of the 1985 SLAC Summer Institute on Particle Physics, 95 ;
J. Ellis, Proc. of the 1985 Int. Symposium on Lepton and Photon Interactions at High Energies (Kyoto, Japan) 849 ; and references therein.

32) R. Batley, talk given at the 6th Int. Conf. on $p\bar{p}$ Physics (Aachen, 1986).

33) S. Weinberg, Phys.Rev.Lett. 50, 387 (1983) ;
R. Arnowitt, A.H. Chamseddine and P. Nath, Phys.Rev.Lett. 50, 232 (1983).

34) P. Fayet, Phys. Letters 125B, 178 (1983) ; 133B, 363 (1983).

35) E. Cremmer, P. Fayet and L. Girardello, Phys. Letters 122B, 41 (1983).

36) P. Fayet, Nucl.Phys. B263, 649 (1986), in particular the Appendix.

37) S. Komamiya, Proc. of the 1985 Int. Symposium on Lepton and Photon Interactions at High Energies (Kyoto, Japan) 611 ;
S.L. Wu, preprint DESY 86-007 ;
H. Küster, Proc. of the 21st Rencontre de Moriond at Les Arcs (France), (1986) ; and references therein.

38) H. Georgi and S.L. Glashow, Phys.Rev.Lett. 32, 438 (1974) ;

H. Georgi, H.R. Quinn and S. Weinberg, Phys.Rev.Lett. 33, 451 (1974) ;
P. Langacker, Phys. Reports 72, 185 (1981) ; and references therein.

39) P. Fayet, Nucl. Phys. B149, 137 (1979).

40) P. Fayet, Nucl. Phys. B246, 89 (1984) ; Phys. Letters 146B, 41 (1984).

41) P. Fayet, Phys. Letters 159B, 121 (1985).

42) R. Haag, J.T. Łopuszański and M. Sohnius, Nucl. Phys. B88, 257 (1975).

43) P. Fayet, Phys. Letters 153B, 397 (1985).

44) F. Gliozzi, J. Scherk and D. Olive, Nucl. Phys. B122, 253 (1977) ;
L. Brink, J. Scherk and J.H. Schwarz, Nucl. Phys. B121, 77 (1977).

SUPERGRAVITY AND THE LOW-ENERGY LIMIT OF SUPERSTRING THEORIES

H.P. Nilles
CERN - Geneva

ABSTRACT

The structure of low-energy supergravity models is reviewed and the connection to a possible low-energy limit of superstring theories is investigated. While some encouraging relations can be found, many questions remain open. These include an understanding of the magnitude of gauge coupling constants and the origin of the compactification scale. Attempts to approximate the superstring theories at low energies are described.

1. INTRODUCTION

The subjects of my lectures are a review of the supersymmetric extension of the SU(3) × SU(2) × U(1) standard model and the possible connection between such a model and the low-energy field theory derived from superstrings. My first two lectures were entirely devoted to the construction of the supersymmetric standard model and the description of its phenomenological properties. Since this part already exists in written form[1)], I shall here only briefly describe the structure of such a "low-energy supergravity model" in order to compare it with the field theories obtained from the string. This is done in Section 2. A complementary discussion can be found in the lectures of Pierre Fayet at this school. The second two lectures were devoted to the low-energy limit of superstring theories. This is still an open subject, and I cannot present to you a solution of all the problems one encounters in its discussion. Since there is not yet a straightforward derivation of this low-energy limit, there have been many attempts to study this question from different points of view, and I can only discuss some special aspects of this approach[2]. I shall exclusively discuss the case in which N = 1 supergravity is present in the compactified four-dimensional theory; not only for the reason that this supersymmetry might allow us to understand the mass scales of fundamental scalar particles, but also because this theory is conceptually much simpler than the non-supersymmetric case. In Section 3, I shall briefly discuss the d = 10 field theory and the question of compactification and some model-independent constraints. I will then discuss the possible zero modes in a resulting four-dimensional field theory. This will allow us to have a rather model-independent first look at the possible d = 4 field theory. In Section 5 I discuss the breakdown of supersymmetry via gaugino condensates, still at a qualitative level. To compute the scalar potential of such a theory, we will be forced to make some approximations for the metric of the compactified dimensions. Such an approximation will be discussed in Section 6. Here we will then realize that a lot of unanswered questions still remain. In the last two chapters, I shall discuss a complementary review of these questions concerning the classical symmetries of string theories and their breakdown in the string-loop expansion. Most of the material presented here is contained in three papers written in collaboration with J.-P. Derendinger and L. Ibañez, while the last part contains still-unpublished results[3)-5)].

2. THE STRUCTURE OF "LOW-ENERGY SUPERGRAVITY MODELS"

These models have attracted quite some attention in past years. They are motivated by a theoretical disease of fundamental scalar particles, their masses being quadratically divergent in perturbation theory. This implies that the masses of fundamental scalar particles are arbitrary input parameters in such a model, which in turn does not allow us to understand the magnitude of the breakdown scale of weak interactions. Moreover, the existence of a large scale M_P due to gravity makes it impossible to understand why M_W should be so small compared to M_P. One solution of this problem is achieved by supersymmetry, in which the magnitude of scalar masses could be understood in terms of the breakdown scale of supersymmetry, where the presence of supersymmetric partners supplies a physical cut-off for the quadratic divergence in the mass of scalar particles. In the attempts to turn this approach into a phenomenologically-acceptable model, certain constraints, like the absence of too-fast proton decay or the absence of flavour-changing neutral currents, have to be satisfied. This forces us into a situation with a rather large supersymmetry breakdown scale $M_S \sim 10^{11}$ GeV, with a breakdown mechanism somewhat distant from the usual supermultiplets containing quarks and leptons. As a result, the final model consists of two sectors: an observable and a hidden sector. The observable sector consists of the quarks, leptons and gauge bosons including their supersymmetric partners, as well as (at least) two Higgs supermultiplets. The content of the hidden sector is not yet specified. In the original proposal it consisted of a gauge supermultiplet of a new gauge group[6)]. The only rôle of the hidden sector is to break supersymmetry which could be achieved by gaugino condensation. Meanwhile, other choices of a hidden sector have also been proposed, but the mechanism of gaugino condensation will become important again in the framework of superstrings. The hidden sector (as the name already indicates) is only very weakly coupled to the observable sector, and thus the effect of SUSY breakdown on the latter is screened. In that sense it is possible to have a SUSY-breakdown scale M_S of 10^{11} GeV in the hidden sector, whereas the splitting of the masses of supermultiplets in the observable sector (e.g., the electron and its scalar partners) is as small as a few hundred GeV. Of course, one has to have a very weak interaction between the two sectors. There exists such a force in Nature: gravity. This is one of the reasons that the construction of such a model requires local supersymmetry, i.e., supergravity. The gauge particle of this local symmetry is a spin-3/2 particle called the gravitino (the partner of the graviton). In the case of spontaneously-broken symmetry, a Super-Higgs effect occurs, and the gravitino becomes massive:

$$m_{3/2} = M_S^2/\sqrt{3}\,M_P \tag{1}$$

where $M_P = 2 \times 10^{18}$ GeV is the Planck mass. The magnitude of supersymmetry breakdown in the observable sector is of the order of $m_{3/2}$, i.e., in the TeV range if we choose $M_S \sim 10^{11}$ GeV. In the flat limit ($M_P \to \infty$, $m_{3/2}$ fixed) the observable sector becomes globally supersymmetric with certain soft breaking parameters, gaugino masses, masses of scalar partners of quarks and leptons, and trilinear scalar self-couplings with scale $m_{3/2}$. Through radiative corrections, these parameters induce a breakdown of SU(2) × U(1), the order of magnitude given by $gm_{3/2}$, where g is a typical coupling constant. Observe that such an approach only makes sense if $m_{3/2}$ is small compared to the Planck mass.

A phenomenologically-acceptable supersymmetric extension of the standard model thus necessarily leads to a theory containing gravitational interactions. This is quite a desirable situation, indicating a road to a unified theory of all interactions, including gravity. However, the incorporation of gravity at the level of N = 1 supergravity in four dimensions (d = 4) is theoretically not without problems, given the unrenormalizability of such a theory. At the moment, this can at most be regarded as an effective approximation to a more complete and theoretically-satisfactory theory still to be found. This belief is supported by the existence of the hidden sector whose properties remain a mystery within the described model, and can only be understood (if at all) in a more complete theory. One way to proceed would be a speculation about the presence of N ⩾ 1 extended supersymmetry in d = 4, but these models face a so-called "chirality problem". Only with N = 1 or N = 0 can we have chiral fermions in d = 4, a property necessary for understanding the existence of light fermions, like quarks and leptons, which can only receive a mass after the breakdown of SU(2) × U(1). The same chirality problem persists in higher-dimensional models based on the approach of Kaluza and Klein if one wants to explain all gauge interactions in d = 4 via gravitation in higher dimensions. Superstring models, however, necessarily contain gauge interactions in d > 4 and give some hope of avoiding this problem; this is why we concentrate here on these models.

3. LOW-ENERGY LIMIT OF SUPERSTRING THEORIES IN d = 10

Superstring theories have been discussed in various lectures during this school. They typically require d = 10 and come with gauge groups $E_8 \times E_8$ or O(32). They are candidates for finite theories, including gravity, and therefore unify all known interactions. Being defined in d = 10, some compactification of the six extra dimensions would be required to make contact with phenomenology. This process is at the moment not understood at all; one has to make crude approximations and then check for consistency a posteriori. One well-defined starting point for such an approach is the theory in the so-called zero slope limit, i.e., the d = 10 field theory of the massless string states. For the known superstring theories this is N = 1 supergravity in d = 10 coupled to pure $E_8 \times E_8$ or O(32) gauge multiplets. The spectrum of this theory is given by the supergravity multiplet (g_{MN}, $\psi_{M\alpha}$, B_{MN}, λ_α, ϕ) where M,N = 0, ..., 9 are world indices and α is a Majorana-Weyl spinor index, as well as the gauge multiplet (A_M^A, χ_α^A) where A = 1, ..., 496 labels the adjoint representation of $E_8 \times E_8$ or O(32). In the Type I theory, these correspond to the massless closed (open) string states respectively. The action of such a theory, including up to two derivatives, is unique and given by[7]:

$$e_{10}^{-1}\mathcal{L} = -\frac{1}{2}R - \frac{i}{2}\bar{\psi}_M \Gamma^{MNP} D_N(\omega)\psi_P + \frac{9}{16}\left(\frac{\partial_M\varphi}{\varphi}\right)^2 +$$

$$+ \frac{3}{4}\varphi^{-3/2} H_{MNP} H^{MNP} + \frac{i}{2}\bar{\lambda}\Gamma^M D_M(\omega)\lambda + \frac{3\sqrt{2}}{8}\bar{\psi}_M\left(\frac{\partial\!\!\!/\varphi}{\varphi}\right)\Gamma^M\lambda$$

$$-\frac{\sqrt{2}}{16}\varphi^{-3/4} H_{MNP}\left(i\bar{\psi}_Q \Gamma^{QMNPR}\psi_R + 6i\bar{\psi}^M\Gamma^N\psi^P\right.$$
$$\left.+\sqrt{2}\,\bar{\psi}_Q\Gamma^{MNP}\Gamma^Q\lambda - i\bar{\chi}\Gamma^{MNP}\chi\right) -$$
$$-\frac{1}{4}\varphi^{-3/4} F_{MN}F^{MN} + \frac{i}{2}\bar{\chi}\Gamma^M D_M(\omega)\chi -$$
$$-\frac{i}{4}\varphi^{-3/8}\left(\bar{\chi}\Gamma^M\Gamma^{NP}F_{NP}\right)\left(\psi_M + \frac{i\sqrt{2}}{12}\Gamma_M\lambda\right)$$
$$+ \text{four fermion interactions} \tag{2}$$

where Γ denote Dirac matrices in d = 10 and

$$F^A_{MN} = \frac{1}{2}\partial_{[M}A^A_{N]} + f^{ABC}A^B_M A^C_N \tag{3}$$

(written for short as $F = dA+A^2$) denotes the gauge field strength. Supersymmetry requires the field strength H_{MNP} of the antisymmetric tensor field B_{MN} not just to be the curl of B, but

$$H_{MNP} = \partial_{[M}B_{NP]} + \omega^{YM}_{MNP} \tag{4}$$

where the Chern-Simons term is given by

$$\omega^{YM} = Tr\left(AF - \frac{2}{3}A^3\right) \tag{5}$$

i.e., B_{NP} has to transform non-trivially under the $E_8 \times E_8$ [or O(32)] gauge transformations. This theory as it stands has gravitational anomalies and is too naive an approximation to the anomaly-free superstring theory. The absence of anomalies can be simulated[8)] by adding an additional term to (4):

$$H = dB + \omega^{YM} - \omega^{L} \tag{6}$$

with

$$\omega^{L} = Tr\,(\omega R - \tfrac{2}{3}\omega^{3}) \tag{7}$$

where ω_M^{ab} is the spin connection. ω contains a derivative, thus ω^L contains three and appears squared in the action. This term is purely bosonic and for a supersymmetric action requires additional terms which up to now are only partially known[9),10)]. The action in (2) thus requires further terms in order to be an adequate low-energy limit of string theory. The action (2) was derived by truncating all heavy string states. For a better approximation they should be integrated out, leaving a low-energy theory with higher derivatives and terms in a higher order in α' (the slope parameter). These terms appear in what is usually called "σ-model perturbation theory", not to be confused with the string loop expansion, which, at least in the heterotic case, is an expansion in g, the gauge coupling constant. This expansion in powers of α' is classical at the string level. There might also be world-sheet non-perturbative effects that play a rôle at this classical level. Looking at (2), one might wonder what g (the gauge coupling constant) is. Observe that the gauge fields have non-minimal gauge kinetic terms. Here g is not an input parameter, but g will be determined dynamically

$$\frac{1}{g^2} = \langle \varphi^{-3/4} \rangle \tag{8}$$

consistent with the expectations in the string theory. We have to be aware of the fact that the coupling constant as determined by the naive approximation might be different from that determined by the string theory. Also, the problem that we have only one naive field theory but [at least in the case of O(32)] two different string theories cannot be resolved in this context. This approximation is probably only useful in defining the important interactions at low energies. In order to ask more fundamental questions, like the determination of the fundamental coupling constants, the approximation probably has to be improved. This can already be seen when we discuss compactification. One possible way is to compactify on a six-torus T^6, leading to N = 4 supergravity in d = 4, which does not resemble known d = 4 phenomenology. One might therefore ask the question for more non-trivial compactifications (still postponing the question of why these should be more likely than the trivial ones). Defining $\phi = (3/4)\log\varphi$ and neglecting fermionic terms, the equation of motion for ϕ is:

$$\Box\phi = \exp(-\phi)\left[F_{MN}^2 + \exp(-\phi)H_{MNP}^2\right] \tag{9}$$

Integrating $\Box\phi$ over a compact manifold without boundary leads to a vanishing result. The right-hand side is positive definite and therefore has to vanish. This implies trivial compactification unless $\phi \to \infty$, which is outside the validity of our approximation[11]. The addition of ω^L in H does not change the situation, but this term requires supersymmetric completion which necessitates the presence of R^2 terms. They actually appear in the Euler combination[9),10)]

$$-\exp(-\phi)\left[R_{MNPQ}^2 - 4R_{MN}^2 + R^2\right] \tag{10}$$

on the right-hand side of (9), ensuring the absence of ghosts. With these terms from the α' expansion, non-trivial compactification is possible: R^2 can be compensated by F^2, and this implies a breakdown of gauge symmetries in the presence of compactification. Notice, however, that the scale of compactification is not yet fixed. There exists an independent argument confirming this result. For the H field to be well defined, the integral of the curl of H over a compact manifold without boundary should vanish[12]:

$$\int_{C_4} dH = \int_{C_4} [Tr\, F \wedge F - Tr\, R \wedge R] = 0 \tag{11}$$

leading to a compensation of F and R in extra dimensions. These results are very encouraging. If E $\times$ E_8 or O(32) were to remain unbroken in d = 4, they would not be able to lead to chiral fermions. The discussed constraints involve integrated quantities and could have various solutions. Only the simplest possibility - a vanishing integrand - has been studied so far[13]. It implies a direct identification of F and R. The spin connection ω_m^{ab} (m = 4, ..., 9; a,b = 1, ..., 6) can be viewed as a gauge field of an O(6) subgroup of the Lorentz group O(9,1), identified with A_m^A in an O(6) subgroup of $E_8 \times E_8$ or O(32) in order to fulfil the constraints. The question of a remaining supersymmetry in d = 4 is related to the holonomy group of the compact manifold, which in turn is a subgroup of O(6). I shall not explain this relation here in detail, but just give a heuristic argument. The gravitino ψ_M^α transforms like a 4 of O(6). N = 1 supersymmetry will be present in d = 4 if the decomposition of the 4 with respect to the holonomy group contains exactly one singlet. If there are more singlets, one will have extended supersymmetries, e.g., in the case of the torus the holonomy group is trivial and 4 = 1+1+1+1, resulting in N = 4 supersymmetry. The simplest choice for N = 1 is to have SU(3) holonomy, which leads to 4 = 1+3 and 6 = 3+$\bar{3}$, and is used in the Calabi-Yau approach. But there are certainly more possibilities, even with discrete subgroups of SU(3) corresponding to certain orbifolds[14]. For simplicity, I shall here assume SU(3) holonomy, although there seem to be some problems with Ricci flat Kähler manifolds of SU(3) holonomy from the viewpoint of string theory. With this identification of ω and A at least an SU(3) subgroup of O(32) or $E_8 \times E_8$ will break down during compactification. In the case of O(32), this will lead to $O(26) \times U(1)$ with possible zero modes in the decomposition of the adjoint of O(32), giving exclusively real representations of O(26). Based on this argument, one usually concludes that O(32) will not lead to a phenomenologically successful model, although not all possibilities have yet been studied. The situation in the case of $E_8 \times E_8$ looks much better. A decomposition of the adjoint of E_8 with respect to $E_6 \times SU(3)$ leads to $248 = (78,1) + (27,3) + (\overline{27},\overline{3}) + (1,8)$ and contains chiral representations. Moreover, E_6 is one of the more successful candidates for a grand unified gauge group with a family of quarks and leptons in 27, the number of these zero modes being defined by topological properties of the compact manifold. Here is then a common starting point for "superstring-inspired models" involving a further breakdown of E_6, renormalization group analysis of coupling constants, intermediate scale breaking, possibilities of additional U(1)'s at low energies and the question of Yukawa couplings, where one seems to need more input to explain neutrino masses and the absence of proton decay. I have not the time to discuss this here [this has already been reviewed[2)]], but will concentrate on questions which are less model-dependent.

4. A FIRST LOOK AT THE POSSIBLE THEORY IN d = 4

We have first to discuss the possible zero modes. Let us define indices $M = (\mu,m)$ ($\mu = 0, \ldots, 3$; $m = 4, \ldots, 9$) and start with the metric

$$g_{MN} = \left(\begin{array}{c|c} g_6^{-1/2}\,\hat{g}_{\mu\nu} & \\ \hline & g_{mn} \end{array} \right) \tag{12}$$

where g_6 = det g_{mn} is used to redefine $g_{\mu\nu}$ in order to have usual kinetic terms for the graviton. The integral over extra dimensions

$$\int d^6y \sqrt{-g_6} = R_c^6 \sim \frac{1}{M_c^6} \tag{13}$$

defines the average radius of compactification. Defining $g_{mn} = \exp(\sigma)\hat{g}_{mn}$, one can then normalize $\int d^6y\sqrt{-\hat{g}_6} = M_P^{-6}$ and $\exp(\sigma)$ defines the radius of compactification in units of the Planck length. Depending on the topological properties of the manifold, g_{mn} gives rise to zero modes that are scalars in d = 4 (we will not discuss off-diagonal pieces in g_{MN} like $g_{\mu m}$ that give rise to gauge bosons depending on the isometries of the manifold). g_{mn} corresponds to a symmetric tensor of O(6) with respect to the SU(3) subgroup discussed earlier; we have $21 = 1+8+6+\bar{6}$. With the notation $m = (i,\bar{j})$, the latter correspond to modes of $g_{i\bar{j}}$, g_{ij}, $g_{\bar{i}\bar{j}}$, while σ is the singlet.

Turning to the gravitino ψ_M^α, we can view α as an eight-dimensional index which transforms as a 4 of O(6) and a Weyl spinor of O(3,1). ψ_μ^α corresponds to spin-3/2 particles in d = 4 with $N_{max} = 4$ as already discussed. ψ_m^α can give rise to spin-$\frac{1}{2}$ zero modes. To obtain canonical kinetic terms for the gravitino, as in the case of the metric, a rescaling

$$\tilde{\psi}_\mu = \exp(-3\sigma/4)\,\psi_\mu \tag{14}$$

is required.

The antisymmetric tensor field B_{MN} could give rise to $B_{\mu\nu}$, $B_{m\nu}$ and B_{mn} (corresponding to the Betti numbers b_0, b_1 and b_2). A zero mode from $B_{\mu\nu}$ corresponds to one pseudoscalar degree of freedom θ defined through a duality transformation

$$H_{\mu\nu\rho}\epsilon^{\mu\nu\rho\sigma} = \varphi^{3/2}\exp(-6\sigma)\partial^{\sigma}\theta + \cdots \tag{15}$$

$B_{m\nu}$ could give rise to extra gauge bosons which (although possibly interesting) we cannot discuss here. B_{mn} will again correspond to pseudoscalars in d = 4. A decomposition with respect to SU(3) gives $15 = 1+3+\bar{3}+8$ with the singlet corresponding to the "trace" $\eta = \epsilon^{mn}B_{mn}$ and $B_{\bar{i}\bar{j}}$, B_{ij} and $B_{i\bar{j}}$ corresponding to 3, $\bar{3}$, and 8 respectively. All these modes appear in the action only through the field strength H implying derivative couplings, i.e., they show axion-like behaviour[12]. From the λ,ϕ members of the supergravity multiplet, we expect additional spin-½ (0) particles in d = 4.

The discussion of the zero modes of A_M^A involves some complication because of the identification of ω_m^{ab} and A_m^A in an SU(3) subgroup. A_μ^A will, of course, give rise to gauge bosons in the adjoint representations of the unbroken gauge group, e.g., A = 1, ..., 78 for E_6. A_m^A will give rise to scalars in d = 4, and we are mostly interested in those transforming as 27 (or $\overline{27}$) under E_6. Let us therefore write A = (a,i) or $(\bar{a},\bar{i})$, a = 1, ..., 27. The states $C^b = A_{\bar{i}}^{b,i}$ and $B^{\bar{b}} = A_i^{\bar{b},\bar{i}}$ then transform as 27, $\overline{27}$ with respect to E_6 and are singlets under the diagonal subgroup SU(3) of the product of SU(3) $\subset$ O(6) and SU(3) $\subset$ E_8. These bosons will have supersymmetric partners from the zero modes of χ_α^A. The number of the possible zero modes is of course entirely defined by the topological properties of the manifold under consideration[13].

We can now have a first look at the possible interactions of these zero modes in d = 4 starting from the d = 10 action given in (2). Of course, in general we expect here not only the influence of topological properties, but also the explicit form of the metric of the compact manifold will become important. Nonetheless we will be able to obtain some non-trivial results that are rather independent of the special form of the metric. We will do that exclusively in the framework of N = 1 supergravity in d = 4, firstly because of the reasons given in Section 2, and secondly because this theory is simpler than the non-supersymmetric case.

N = 1 supergravity in d = 4 [with action including terms up to two derivatives[15)]; for higher derivatives see Ref. 10)] is defined through two functions of the chiral superfields ϕ_i. The first is an analytic function $f(\phi_i)$ defining the gauge kinetic terms $f(\phi_i)W^{\alpha}W_{\alpha}$. In a component language, f appears in many places, but it can be extracted most efficiently from

$$\mathrm{Re}\, f(\varphi_i) F_{\mu\nu} F^{\mu\nu} + \mathrm{Im}\, f(\varphi_i)\, \epsilon_{\mu\nu\rho\sigma} F^{\mu\nu} F^{\rho\sigma} \tag{15}$$

where φ_i denotes the (complex) scalar component of ϕ_i. The second is the so-called Kähler potential

$$G(\phi_i, \phi_i^*) = K(\phi_i, \phi_i^*) + \log |W(\phi_i)|^2 \tag{16}$$

Unlike f, G is not analytic and contains the left-handed chiral superfields along with their complex conjugates. The second term in (16) contains the analytic function $W(\phi_i)$: the superpotential. The action in component form usually contains G in complicated form; the scalar kinetic terms, e.g., are

$$G_i{}^j (\partial_\mu \varphi^i)(\partial^\mu \varphi_j^*)\,; \qquad G_i{}^j \equiv \frac{\partial^2 G}{\partial \varphi^i \partial \varphi_j^*} \tag{17}$$

whereas the scalar potential is given by

$$V = \exp(G)[G_k (G^{-1})^k_\ell G^\ell - 3] \tag{18}$$

which makes it difficult to extract G once an action is given in component form. There is only one term which allows a rather simple identification of G, and this is a term involving the gravitino

$$e_4 \exp(G/2) \bar{\psi}_\mu \gamma^{\mu\nu} \gamma_5 \psi_\nu \tag{19}$$

which will later be used extensively after the correct redefinitions of the gravitino in d = 4.

Let us now consider the action in d = 10 in order to learn something about the possible action in d = 4. We start with the gauge kinetic term

$$e_{10} \varphi^{-3/4} F_{MN} F^{MN} \tag{20}$$

Since we are interested in the $F^2_{\mu\nu}$ part, we write

$$e_4 e_6 \varphi^{-3/4} F_{\mu\nu} F_{\rho\sigma} g^{\mu\rho} g^{\nu\sigma} \tag{21}$$

where, with the definitions given earlier, we would like to extract f from

$$\hat{e}_4 \, \mathrm{Re} f \, F_{\mu\nu} F^{\mu\nu} \tag{22}$$

with $\hat{e}_4 = (\det \hat{g}_{\mu\nu})^{\frac{1}{2}} = \exp(6\sigma) e_4$, and indices are contracted with the "hatted" metric. Integrating the extra six dimensions with the normalization given in (13) using $M_P \equiv 1$, we obtain

$$Re S \equiv Re f = \varphi^{-3/4} \exp(3\sigma) \tag{23}$$

as the real part of the scalar component of a chiral superfield denoted by S[16]. This is a rather amazing result. Remember that at no point in the derivation did we have to know something about the metric of the compact six-dimensional space, so this constitutes a rather model-independent result. Observe that f is usually non-trivial, that its vacuum expectation value (vev) will determine the gauge coupling constant, and that the couplings of E_8 (or E_6) and E_8' coincide.

Let us now discuss the imaginary part of f, to be extracted from $F_{\mu\nu}F_{\rho\sigma}\varepsilon^{\mu\nu\rho\sigma}$. The relevant degree of freedom comes from $B_{\mu\nu}$ as discussed earlier. $B_{\mu\nu}$ couples only through its field strength $H_{\mu\nu\rho}$ and has therefore only derivative couplings. Taking the relevant terms in the d = 10 action and integrating the extra dimensions, we obtain

$$\varphi^{-3/2} \exp(6\sigma) H_{\mu\nu\rho} H^{\mu\nu\rho} + H_{\mu\nu\rho} O^{\mu\nu\rho} \tag{24}$$

where $O^{\mu\nu\rho}$ contains fermion bilinears. H has to satisfy a constraint (neglecting R^2-terms for the moment)

$$\partial_{[\mu} H_{\nu\rho\sigma]} = - Tr\, F_{[\mu\nu} F_{\rho\sigma]} \tag{25}$$

which we take into account by adding a Lagrange multiplier

$$\theta \epsilon^{\mu\nu\rho\sigma} (\partial_\mu H_{\nu\rho\sigma} + Tr\, F_{\mu\nu} F_{\rho\sigma}) \tag{26}$$

Next we eliminate H via the equations of motion and arrive at an action containing the terms

$$\varphi^{3/2} \exp(-6\sigma)(\partial_\mu \theta)^2 + \theta \epsilon^{\mu\nu\rho\sigma} Tr(F_{\mu\nu} F_{\rho\sigma}) \tag{27}$$

which tells us that Imf = θ, and for the scalar component of S we obtain

$$S = \varphi^{-3/4} \exp(-3\sigma) + i\theta \tag{28}$$

as a mixture of g_{MN} and B_{MN} zero modes. The fermionic partner is a combination of ψ_m and λ zero modes which we will not discuss here in detail. Observe that θ couples only with derivatives except for the last term in (27), and that the d = 4 action has a Peccei-Quinn-like symmetry under shifts of θ by a real constant, thus θ couples like an axion. Let me stress again that all these statements about the action and the form of (28) are model-independent and could be derived without explicit knowledge of the metric.

Unfortunately, the situation changes once we try to extract the Kähler potential. As already indicated, the term to investigate is the d = 4 "gravitino mass term" (19). The extraction of this term is rather complicated due to several redefinitions of the gravitino field. A general form has been given in Ref. 4), and we will not repeat the derivation here. Many of the terms appearing there depend explicitly on the metric and spin-connection of the six-dimensional compact space. A model-independent statement can only be made about the structure of the superpotential, because it is an analytic function in the chiral superfields. Symbolically the "gravitino mass term" is obtained as

$$\exp(G/2) = \varphi^{-3/4} \exp(-3\sigma) \Gamma^{mnp} H_{mnp} \tag{29}$$

and from (16) we can try to read off the superpotential. $W(\phi_i)$ is defined to be an analytic function in the chiral superfields and should not contain derivatives. A first inspection of (29) therefore suggests that a possible candidate for a superpotential is the A^3 term contained in the Yang-Mills Chern-Simons term (5) included in H. This then gives rise to a trilinear superpotential involving the C and B fields defined earlier[3),16)]. At the moment it is not clear whether these are the only possible terms in the superpotential, although this seems to be likely. Observe that, for example, the superfield S as defined in (28) cannot appear in the superpotential, since its pseudoscalar component has only derivative couplings. We will come back to these points later. In any case, a more detailed discussion of the Kähler potential requires more information (or approximations) about the d = 6 metric. Before we tackle this topic, let me first present a discussion about supersymmetry breakdown in d = 4.

5. GAUGINO CONDENSATION AND SUPERSYMMETRY BREAKDOWN

N = 1 supergravity in d = 4 still needs the incorporation of supersymmetry breakdown at a scale small compared to the Planck mass. For the phenomenological reasons mentioned earlier, this should appear in a hidden sector only coupled gravitationally to the observable sector. Superstring models now miraculously contain such a hidden sector, the sector that contains the particles transforming non-trivially under the second E_8. Notice that the observable sector (for definiteness called the E_6 sector) only couples gravitationally to the E_8' sector (there are no particles that transform non-trivially both under E_6 and E_8'). Moreover, the E_8' sector contains in d = 10 a pure super-Yang-Mills multiplet, suggesting a possible breakdown of supersymmetry via gaugino condensates[3]. This breakdown has already been discussed in the framework of supergravity models, both at the level of an effective Lagrangian[6] and at the level of the complete classical action[17]. Assume asymptotically-free gauge interactions (here E_8' or a subgroup thereof) with a scale

$$\Lambda = \mu \exp\left(-1/b_0 g^2(\mu)\right) \tag{30}$$

which is renormalization-group invariant at the level of the one-loop β-function. In analogy to QCD, which leads to $q\bar{q}$ condensates, we will here assume that the gauge fermions condense at a scale

$$\langle \chi\chi \rangle = \Lambda^3 \tag{31}$$

As long as Λ is small compared to M_P, we assume that gravity will not qualitatively disturb this dynamical mechanism. The question whether such a condensate breaks supersymmetry can be studied by investigating the supersymmetry transformation laws of the fermionic fields of the theory. The non-derivative terms in these transformations will give us the auxiliary fields that serve as order parameters for supersymmetry breakdown. The relevant objects here are the auxiliary fields of the chiral superfields

$$F_K = \exp(G/2)G_K - \frac{1}{4} f_K (\chi\chi) + \cdots \tag{32}$$

where f is the gauge kinetic function discussed earlier and f_K is its derivative with respect to ϕ_K. A necessary condition for the breakdown of supersymmetry via gaugino condensates is therefore a non-trivial f-function[17]. This condition is fulfilled in the framework of superstring-inspired models, since we have seen in the last section that f = S in a rather model-independent way. Whether this is also sufficient for the breakdown of supersymmetry can only be checked by minimizing the potential

$$V = F_\kappa (G^{-1})^\kappa_\ell F^\ell - 3\exp(G) \tag{33}$$

since the different terms in (32) might cancel at the minimum. But let us for the moment assume that only the second term in (32) receives a vev. Since $f_s = 1$ in units of M_P, we find a supersymmetry breakdown scale

$$\langle F_s \rangle = M_S^2 \approx \Lambda^3/M_P \tag{34}$$

and a scale of $\Lambda \sim 10^{13}$ GeV would lead to a gravitino mass in the TeV range. Once we understand why Λ is five orders of magnitude smaller than M_P, we shall understand why $m_{3/2}/M_P \sim 10^{-15}$. Λ now depends on the E_8' gauge coupling and the spectrum of low-energy modes. Identifying g_6 with g_8 would in many circumstances lead to too large a value for Λ, and one might speculate that E_8' should break during compactification. We shall, however, see later that the equality of g_6 and g_8 seems to be only an artifact of the classical approximation, which is not true in the full theory. Thus the shadow E_8' sector of the superstring takes the rôle of the hidden sector of supergravity models and might explain the smallness of $m_{3/2}$ compared to M_P. But how does this breakdown of SUSY in the hidden sector influence the observable sector? In general, we would expect gaugino masses (m_0), scalar masses ($\tilde{m}$) and the trilinear couplings (Am) to be of the order of magnitude of $m_{3/2}$. A naive inspection shows that this might also be true here. Gaugino masses in the observable sector are in general given by

$$m_0 = f_\kappa (G^{-1})^\kappa_\ell F^\ell \tag{35}$$

where f is the gauge kinetic function of the observable sector. With $F^{\ell} = (1/4)f^{\ell}\langle\chi\chi\rangle$ we would therefore obtain $m_0 = m_{3/2}$. In the same way we would obtain under these circumstances A = 1 and scalar masses of order $m_{3/2}$. To make a definite statement we have to watch out for possible cancellations, which can only be studied once we have a better knowledge of the Kähler potential, a question which we want to discuss now.

6. REDUCTION AND TRUNCATION

A first approximation for G (that might simulate an orbifold approximation of interest in this context) is obtained through reduction and truncation[16)]. One first compactifies the d = 10 theory on a six-torus T^6. The resulting theory is N = 4 supersymmetric in d = 4. From this theory one truncates unwanted states, to obtain an N = 1 theory. From the gauge singlet sector one keeps only those states that transform as singlets under an $SU(3) \subset O(6)$ of the Lorentz group. Since ψ_μ^α transforms as a 4 of O(6) and thus as 1+3 under SU(3), we remain with one gravitino. As already explained in Section 4, there are only a few gauge singlets that survive this truncation. For the bosonic modes we have φ, σ from the metric as well as θ and η from the antisymmetric tensor. For the gauge non-singlet fields one has to remember the identification of spin-connection and gauge fields. Here one keeps those states which are singlets under the diagonal subgroup of the product of $SU(3) \subset O(6)$ and $SU(3) \subset E_8$. This leaves us with one 27 of E_6 in this case, corresponding to $C^b = A_{\bar{i}}^{b,i}$ (b = 1, ..., 27, cf. Section 4). With this well-defined procedure based on simple reduction on T^6, the component Lagrangian in d = 4 can be deduced. From this we can immediately read off f = S and $W = d_{abc}C^aC^bC^c$, which should not be surprising. Moreover, from the "gravitino mass term" formula (29) one obtains[4)]

$$G = \log\left(e^{-6\sigma}\varphi^{-3/2}\right) + \log|W|^2 \tag{36}$$

The components φ and σ are not separately the lowest components of chiral superfields. One combination $S = \varphi^{-3/4}\exp(3\sigma) + i\theta$ has already been defined earlier. To define the other combination, the information from (36) is not enough. The charged fields C do not yet appear in the first term of G in (36) and the correct definition of the superfields has yet to be found. This can be done, for example, by using the scalar kinetic terms[16]. It leads to a second superfield in which φ,σ and the C-modes mix

$$T = \exp(\sigma)\varphi^{3/4} + |C_a|^2 + i\eta \tag{37}$$

where η is the mode from $\varepsilon^{mn}B_{mn}$ as discussed earlier, and the Kähler potential from (36) thus reads

$$G = -\log(S+S^*) - 3\log(T+T^* - 2|C|^2) + \log|W|^2 \tag{38}$$

a form already previously mentioned in the framework of supergravity models[18]. The scalar potential derived from this G-function has some remarkable properties

$$V = \frac{1}{16 s t_c^3}\left[|W|^2 + \frac{t_c}{3}|W'|^2\right] + D^2\text{-terms} \tag{39}$$

where $s = \mathrm{Re}\, S$ and $t_c = \mathrm{Re}\, T - |C_a|^2 = t - |C_a|^2$ and W' is the derivative of W with respect to the C-field. The potential is positive definite ($t_c > 0$ is required by the kinetic terms) and has a minimum with vanishing vacuum energy $V = 0$. This minimum is obtained at $W = W' = 0$ independent of the values of s and t. This implies that at this level the gauge coupling constant and the radius of compactification is not yet fixed. The theory has classical symmetries which allow shifts of the values of s and t, as well as Peccei-Quinn symmetries corresponding to shifts in θ and η. This, of course, makes the use of this approximation as an effective low-energy limit of the superstring very problematic. Certain crucial parameters, like the value of the gauge coupling constant and the scale of compactification, which we believe to be dynamically determined in the full string theory, are not yet fixed. To determine these quantities in the truncated theory does not necessarily lead to the same results that would be obtained in the full theory. Actual calculations of radiative corrections in the truncated theory have been performed at the one-loop level. It was found that the resulting potential is unbounded from below with some vev running to infinity[19),20)]. This result should actually not be too surprising. The model was obtained by truncating states with mass of the order of the compactification scale M_c, which in a certain way corresponds to the limit $M_c \to \infty$. In the truncated theory this scale is determined by the vev of $\exp(-\sigma)$ and if the approach is consistent the only logical possibilities are either M_c undetermined or $M_c \to \infty$. These questions can only be solved once we know more about the theory of the massive states or the full string dynamics[21)]. The model as it stands should be regarded as an approximation for the possible interactions of the zero modes, rather than a tool to determine the fundamental dynamical quantities of string theories.

This becomes even more apparent when we include the concept of gaugino condensation within this framework[22)]. Since the gauge coupling constant is not determined, Λ in (30) is also unknown. Using (33) and (38), we get for the potential

$$V = \frac{1}{16 s t_c^3}\left[\,|W - 2(st_c)^{3/2}(\chi\chi)|^2 + \frac{t_c}{3}|W'|^2\right] \tag{40}$$

where $(\chi\chi)$ depends on g^2 through $\exp(-S/b_0)$. The potential is still positive definite and has a minimum at $V = 0$ which is still degenerate. Now the minimum need not necessarily imply $W = W' = 0$, but we could have a non-trivial vev of W. Given fixed $\langle W\rangle \neq 0$ by some unknown mechanism (not to be understood in the truncated theory), such as a vev of dB in H or a slight mismatch of the vevs of the Chern-Simons terms[23], the value of the gauge coupling constant would be fixed. In order to minimize the potential, the theory slides to a coupling constant which, through (30), gives a value of the condensate that exactly cancels the contribution of W. In other words, this means that the dilaton S slides to a value that cancels the vacuum energy in the same way as an axion slides to cancel a possible θ-parameter of a gauge theory [observe that $\exp(-S/b_0)$ contains both s and θ]. Although we do not yet understand the magnitude of supersymmetry breakdown, this mechanism to ensure $E_{vac} = 0$ after SUSY breakdown appears very attractive. Actually, I have still to convince you that supersymmetry is broken, since in (40) a certain cancellation of $\langle W\rangle$ and $\langle\chi\chi\rangle$ appears. It actually tells us that the auxiliary field F_S of the S-superfield vanishes in the vacuum. Nonetheless, here F_T requires a non-vanishing vev once $\langle W\rangle \neq 0$, and supersymmetry is broken

$$F_T = \exp(G/2)\, G_T \neq 0 \tag{41}$$

Another question is the breakdown of SUSY as it is felt in the observable sector, and here things look different. For example, the gaugino masses are given by $m_0 = f_K (G^{-1})^K_{\ \ell} F^\ell$, and only f_S is different from zero. In this case we therefore obtain $m_0 = 0$, and the same is true for the scalar masses and the trilinear coupling A. Thus the observable sector remains supersymmetric. The fact that scalar masses remain zero in this case has brought people to the idea of constructing models in which $m_{3/2}$ is as big as M_P but still having small SUSY breakdown parameters in the observable sector, a situation which is not likely to be meaningful once radiative corrections are fully considered. As already mentioned, the full corrections at one loop in the truncated theory lead to disaster[19),20)]. Nonetheless, the naive approximation discussed in this section has some attractive properties as well as these obvious diseases. The question remains whether these diseases are a result of the special approximation or a more general property of classical considerations.

7. CLASSICAL SYMMETRIES

A more complete picture of the classical approximation can be obtained by a study of the symmetries of the field theory Lagrangian. In this chapter, we will present such a discussion in the spirit of Ref. 24).

We turn first to the Peccei-Quinn-like symmetries originating from the B_{MN} modes which couple only with derivatives. This implies that the superfields S and T in d = 4 which contain these modes as pseudoscalars cannot appear in the superpotential, at least in any order of perturbation theory. In a more general framework than that discussed previously, not only one T-field but several could appear. For our purpose it is, however, sufficient to discuss only one mode; the generalization to more than one is trivial. The symmetries of the classical action are

$$S \longrightarrow S + i\alpha \; ; \quad T \longrightarrow T + i\beta \tag{42}$$

with α,β being real constants. This implies that the exact value of the imaginary parts of S and T must be unphysical. As a consequence, the functions G and the real part of f should not depend on the imaginary part of S and T. Observe that the imaginary part of f which multiplies $F\tilde{F}$ can very well depend on these fields.

Inspecting the d = 10 action given in (2), we can identify two more classical symmetries. They are

$$g_{MN} \rightarrow t\, g_{MN} \qquad \lambda \rightarrow t^{-1/4}\lambda$$

$$\varphi \rightarrow t^{-4/3}\varphi \qquad \chi \rightarrow t^{-1/4}\chi$$

$$\psi_a \rightarrow t^{-1/4}\psi_a \qquad \mathcal{L}_{10} \rightarrow t^4\mathcal{L}_{10} \tag{43}$$

where a is a flat index, and

$$\varphi \rightarrow r^{2/3}\varphi \qquad \chi \rightarrow r^{-1}\chi$$

$$\psi_M \rightarrow r^{-1}\psi_M \qquad K_{10} \rightarrow r$$

$$\lambda \rightarrow r^{-1}\lambda \qquad \mathcal{L}_{10} \rightarrow r^{-2}\mathcal{L}_{10} \tag{44}$$

where K_{10} is the gravitational coupling in the d = 10 theory. Observe that under both transformations the action is multiplied by an overall factor, implying that these symmetries are just classical and expected to be broken in string perturbation theory. These are symmetries of the field-theory Lagrangian; is there any connection to string theory? There is. The classical symmetries can be understood as a rescaling of K_{10} into the string tension (keeping the relation $g_{10} = \sqrt{T_s}\, K_{10}$ in the heterotic string), and it is understood that in string theory g_{10} is not an input parameter but the choice of a vacuum. For the d = 4 theory, this implies that classically the gauge coupling constant and the scale of compactification are not yet fixed. We have therefore (as already previously explained) a theory of the zero modes which in itself cannot solve these fundamental questions. A solution has to come from higher modes or non-classical or non-perturbative effects. In the d = 4 action, the symmetries correspond to

$$\begin{aligned} \hat{g}_{\mu\nu} &\to t^4 \hat{g}_{\mu\nu} & \qquad \tilde{\psi}_\mu &\to t\,\tilde{\psi}_\mu \\ S &\to t^4 S & \chi &\to t^{-1}\chi \\ T &\to T & \lambda &\to t^{-1}\lambda \end{aligned} \tag{45}$$

with $\mathcal{L}_4 \to t^4 \mathcal{L}_4$; $\tilde{\psi}_\mu$ is the rescaled gravitino and

$$\begin{aligned} S &\to r^{-1/2} S & \qquad C^a &\to r^{1/4} C^a \\ T &\to r^{1/2} T \end{aligned} \tag{46}$$

with $\mathcal{L}_4 \to r^{-\frac{1}{2}} \mathcal{L}_4$. These classical symmetries imply the existence of flat directions in the potential leading to degenerate vacua in which T and S are not determined. Sometimes these symmetries are discussed as a part of SU(1,1) × SU(n,1) symmetries present at the level of the scalar kinetic terms[18)]. These bigger symmetries seem to be as relevant as the SU(n) symmetries of the kinetic Lagrangian in a usual theory of complex scalar fields. They are broken if the string theory provides a superpotential, and only (42), (45) and (46) remain.

We can now study the implications of these symmetries for the classical form of f and G. To discuss f, we consider the term $\hat{e}^4 \mathrm{Re} f F^2_{\mu\nu}$ and deduce that

$$Re f \longrightarrow t^4 r^{-1/2} Re f \tag{47}$$

which is the transformation behaviour of S. Including the restrictions from the Peccei-Quinn symmetries (independence of Re f on Im S,T), we deduce

$$f = S \tag{48}$$

which should not be too big a surprise, since we had already deduced this result earlier in a model-independent way.

Again, the discussion of G is more complicated because G is not an analytic function of the chiral superfields. Investigating the gravitino mass term, we obtain

$$\exp(G/2) \longrightarrow t^{-2} r^{1/4} \exp(G/2) \tag{49}$$

Writing according to (16)

$$\exp(G) = |W|^2 \exp(K) \tag{50}$$

we observe that W has to be homogeneous of degree n. There are no restrictions on n from the symmetries, but we have earlier discussed an argument that the most likely choice is n = 3. To present the general form of G, we restrict ourselves to the modes discussed in the last chapter. Generalizations are conceptually trivial but technically more complicated. We assume that W is just a function of the C^a's, and arrive at

$$G = -\log(S+S^*) - n\log(T+T^*+g(C,C^*)) + \log|W|^2 + \bar{K}\left(\frac{C}{\sqrt{T+T^*}}, \frac{C^*}{\sqrt{T+T^*}}\right) \quad (51)$$

where $g(C,C^*)$ should scale like T, i.e., CC^*. Actually, the functions g and $\bar{K}$ can be transformed into each other, giving the general form of a supergravity potential with flat directions[25]. We also see that (51) is quite close to (38) found in the simple approximation, and the classical symmetries restrict the form of f and G drastically. But it should again be stressed that these symmetries have to be broken before we can understand the magnitude of the gauge coupling constant and the scale of compactification. There are many sources of such a possible breakdown: higher terms in σ-model perturbation theory including non-perturbative effects such as world-sheet instantons[26)-28)], even at the classical level. These effects might also break the non-renormalization of the superpotential usually expected in perturbation theory [S,T might appear exponentially in the superpotential[28]]. These results seem to depend strongly, however, on the chosen compactification scheme. Another source is given by new terms in the string-loop expansion which we will discuss in the next section.

8. BEYOND THE CLASSICAL LEVEL

The Peccei-Quinn-like symmetries (42) are believed to survive this loop expansion in any finite order of perturbation theory and can only be broken by non-perturbative effects like gauge or world-sheet instantons. (Actually, the symmetry related to S will not be disturbed by world-sheet non-perturbative effects.) For the other classical symmetries the situation is different. Classically the action is scaled and the symmetries are broken by quantum effects. In the heterotic string this loop expansion is governed by the coupling constant g, which in turn is defined through a vev of the dilaton fields. This will allow us to construct a definite loop expansion in the dilaton fields and still give us restrictions on how the classical symmetries are broken by loop effects. But before we discuss the loop expansion in more general terms, let us examine some aspects at the one-loop level. We can do that because of the mechanism of anomaly cancellation in the d = 10 field theory. Green and Schwarz have observed that the cancellation of anomalies requires certain new local counterterms with definite finite coefficients in the one-loop effective action to cancel the gauge non-invariance of present non-local terms. In general, such terms appear with infinite coefficients, but the possible symmetry of the effective action forces us to renormalize the theory in such a way that these gauge-variant local counterterms have a well-defined finite coefficient. An example of such a term is

$$\epsilon\, \epsilon^{MNPQRSTUVW} B_{MN}\, Tr(F_{PQ}F_{RS})\, Tr(F_{TU}F_{VW}) \tag{52}$$

where $\varepsilon = 1/720(2\pi)^5$. While this gives rise to many new interaction terms in the d = 4 theory, one possible manifestation seems to be of particular importance. Replacing one of the TrF^2 terms by their vev in extra dimensions, one arrives at

$$\eta\, \epsilon^{\mu\nu\rho\sigma}\, Tr(F_{\mu\nu}F_{\rho\sigma}) \tag{53}$$

η is the imaginary part of T, and unlike in the classical case it now (in addition to θ) couples to $F\tilde{F}$. Observe that (53) is gauge-invariant, while (52) is not, but is required by the absence of anomalies in d = 10. This shows that the remnants of such terms originate in ten dimensions, and are one of the few places where we could in principle see whether we live in higher dimensions. (53) suggests that not only θ, but also η, couples like an axion[12),4),29)]. To make sure that this does not lead just to a redefinition of θ at the one-loop level, all anomaly cancellation terms have to be considered. Doing this and satisfying $TrF^2 = TrR^2$ in extra dimensions, one arrives at the result that η couples differently to E_6 and E_8':

$$\epsilon \eta [(F\tilde{F})_8 - (F\tilde{F})_6] \tag{54}$$

while

$$\theta [(F\tilde{F})_8 + (F\tilde{F})_6] \tag{55}$$

This fact has interesting consequences, some of which we will now list.

a) The second axion is a candidate to solve the strong CP problem of QCD in the observable sector. One axion (like θ alone) would not be sufficient, because it is used to adjust the θ-angle of E_8' and becomes massive, of order $m_{3/2}$.

b) Supersymmetry requires the same behaviour of the real parts of S and T as that of the imaginary part; i.e., ReS and ReT couple differently to E_6 and E_8'. Since the vevs of these fields define the gauge coupling constants, g_6 and g_8' need no longer be equal. This might have consequences for the condensation scale of E_8'.

c) There exist now two axion dilaton pairs, and this might generalize the relaxation of the cosmological constant to the observable sector in the same way as it appears in the hidden sector [compare Eq. (40)].

d) Imposition of supersymmetry also requires new terms in the Kähler potential at the one-loop level[5]. We will discuss this later.

e) As expected, these effects at the one-loop level lead to an induced breakdown of supersymmetry in the observable sector once it is broken in the hidden sector. Remember our discussion in Section 6, where the observable sector remained supersymmetric. Gaugino masses are given by

$$m_0 = F_T f_T + F_S f_S \tag{56}$$

and vanish because $F_S = f_T = 0$. But here we now have $f = S+\varepsilon T$, and f_T no longer vanishes. As a result, non-trivial gaugino masses (and also non-trivial scalar masses and A-parameters) of order $\varepsilon m_{3/2}$ are transmitted to the observable sector[5]. This shows again that in a theory with $m_{3/2}$ of the order of the Planck mass, no sign of supersymmetry can survive in the TeV region.

According to the classical symmetries, the new terms scale as

$$\mathcal{L}_{1\text{-loop}} \longrightarrow t^0 r^{-1/2} \mathcal{L}_{1\text{-loop}} \tag{57}$$

consistent with the expectation of a g^2-expansion. New counterterms at the n-loop level would therefore scale as

$$\mathcal{L}_{n\text{-loop}} \longrightarrow t^{4(1-n)} r^{2n-1/2} \mathcal{L}_{n\text{-loop}} \tag{58}$$

We can now try to extract the restrictions on f and G within this framework. Since, following (47), f scales like the action we write:

$$f = \sum_{n=0}^{\infty} f_n \tag{59}$$

with

$$f_n \longrightarrow t^{4(1-n)} r^{2n-1/2} f_n \tag{60}$$

We know that $f_0 = S$. f_1 cannot contain S since this is the only field transforming non-trivially under the t-symmetry, leaving $f_1 = T+C^2$. For

$n \geqslant 2$, the analyticity of f and the fact that Re f should be independent of Im S,T forces f_n to vanish[30]. Thus we have a non-renormalization theorem for f beyond the one-loop level and

$$f = S + \epsilon (T + \alpha C^2) \tag{61}$$

Observe that the presence of supersymmetry allows us to obtain such a restrictive result from broken symmetries, since supersymmetry forces f to be analytic in the chiral superfields. This is not in contradiction with the logarithmic variation of gauge coupling constants defined through the vev of f, since this involves a discussion of the potential where non-analytic pieces appear.

G can be examined in a similar way. At the n-th level, we obtain

$$\exp(G/2) \longrightarrow t^{-4n-2} r^{n+1/4} \exp(G/2) \tag{62}$$

and this leaves us with

$$\begin{aligned} G = & -\log(S+S^*) - 3\log(T+T^*+g(C,C^*)) \\ & + \log|W|^2 + 2\log\left(1+\sum_{n=1}^{\infty} a_n \epsilon^n \left(\frac{S+S^*}{T+T^*}\right)^n\right) + \\ & + \bar{K}\left(\frac{C}{\sqrt{T+T^*}}, \frac{C^*}{\sqrt{T+T^*}}\right) \end{aligned} \tag{63}$$

where we expect the non-renormalization theorem for W to hold in any finite order of perturbation theory, and the a_n are unknown coefficients. A discussion of the resulting scalar potential is very complicated and has not yet been attempted. As long as supersymmetry remains unbroken (due to the non-renormalization theorems), we expect this potential still to have flat directions at $E_{vac} = 0$, but with redefined S and T fields. However, once a breakdown of supersymmetry and the corresponding gaugino bilinears are included, this potential might fix

$\langle t \rangle$ as well as $\langle s \rangle$, and one could come closer to a useful low-energy effective theory of the superstring.

REFERENCES

1) Nilles, H.P., Lectures given at Mt. Sorak, Korea (1984), in "Supersymmetry", ed. H.S. Song, Seoul Nat. Univ. Press (1984); for a complete review, see:
Nilles, H.P., Physics Reports 110, 1 (1984).

2) There exist some reviews on the phenomenology of superstring models which contain comprehensive lists of references to the original papers:
Segrè, G.C., Univ. of Pennsylvania report (1985),
Ibañez, L.E., CERN preprint TH.4308/85 (1985);
Ellis, J., CERN preprint 4255/85 (1985).

3) Derendinger, J.-P., Ibañez, L.E. and Nilles, H.P., Phys. Lett. 155B, 65 (1985).

4) Derendinger, J.-P., Ibañez, L.E. and Nilles, H.P., Nucl. Phys. B267, 365 (1986).

5) Ibañez, L.E. and Nilles, H.P., Phys. Lett. 169B, 354 (1986).

6) Nilles, H.P., Phys. Lett. 115B, 193 (1982); Nucl. Phys. B217, 366 (1983).

7) Chapline, G.F. and Manton, N.S., Phys. Lett. 120B, 105 (1983).

8) Green, M.B. and Schwarz, J., Phys. Lett. 149B, 117 (1984).

9) Romans, L.J. and Warner, N.P., Caltech preprint Calt-68-1291 (1985).

10) Cecotti, S., Ferrara, S., Girardello, L., Porrati, M. and Pasquinucci, A., Preprint UCLA/85/TEP/24 (1985).

11) Freedman, D.Z., Gibbons, G.W. and West, P.C., Phys. Lett. 124B, 491 (1983).

12) Witten, E., Phys. Lett. 153B, 243 (1983).

13) Candelas, P., Horowitz, G., Strominger, A. and Witten, E., Nucl. Phys. B258, 151 (1985).

14) Dixon, L., Harvey, J., Vafa, C. and Witten, E., Nucl. Phys. B261, 678 (1985).

15) Cremmer, E., Ferrara, S., Girardello, L. and Van Proeyen, A., Nucl. Phys. B212, 413 (1983).

16) Witten, E., Phys. Lett. 155B, 151 (1985).

17) Ferrara, S., Girardello, L. and Nilles, H.P., Phys. Lett. 125B, 457 (1983).

18) For a review, see:
Ellis, J., in Ref. 2), and:
Dragon, N., Ellwanger, U. and Schmidt, M., Heidelberg preprint HD-THEP-86-3 (1986).

19) Ahn, Y.J., Breit, J.D. and Segrè, G., University of Pennsylvania preprint (1986).

20) Quiros, M., CERN preprint TH.4363/86 (1986).

21) I thank F. Del Aguila for a discussion on this point.

22) Dine, M., Rohm, R., Seiberg, N. and Witten, E., Phys. Lett. 156B, 55 (1985).

23) Rohm, R. and Witten, E., Princeton preprint (1985).

24) Burgess, C.P., Font, A. and Quevedo, F., University of Texas preprint UTTG-31-85 (1985).

25) Barbieri, R., Cremmer, E., and Ferrara, S., CERN preprint TH.4177/85 (1985).

26) Wen, X.G. and Witten, E., Phys. Lett. 166B, 397 (1986).

27) Ellis, J., Gomez, C., Nanopoulos, D.V. and Quiros, M., CERN preprint TH.4395/86 (1986).

28) Dine, M., Seiberg, N., Wen, X. and Witten, E., Princeton preprint (1986).

29) Choi, K. and Kim, J.E., Phys. Lett. 165B, 71 (1985).

30) Del Aguila, F., Ibañez, L.E. and Nilles, H.P., CERN preprint, in preparation.

SUPERGRAVITY IN TEN SPACE-TIME DIMENSIONS

B. de Wit

Institute for Theoretical Physics
Princetonplein 5, P.O. Box 80.006
3508 TA Utrecht, The Netherlands

1. INTRODUCTION

It is a valuable tradition of this school to start with series of introductory lectures on supersymmetry and supergravity. In the past these lectures dealt primarily with 4-dimensional supersymmetry and supergravity as well as with generic aspects of supergravity theories, relevant irrespective of the number of space-time dimensions. As this material has been extensively recorded in the proceedings of previous schools [1,2] we will not to repeat it here. Motivated by superstring considerations, we will instead concentrate on supergravity in 10 space-time dimensions (for an introduction to superstring theory, see M. Green's lectures in this volume and references quoted therein).

In section 2 we first derive the various representations of the 10-dimensional supersymmetry algebra that are relevant for supersymmetric Yang-Mills and supergravity theories. From the existence of two inequivalent N=1 gravitino supermultiplets, we argue that there are two inequivalent N=2 supergravity theories in 10 dimensions. Free field equations which describe the physical states contained in the various supermultiplets are discussed in section 3. It is then straightforward to write down supersymmetric actions (or sets of supersymmetric field equations) at the linearized level. The interaction terms and the corresponding nonlinear modifications of the transformation rules can in principle be found by iteration. We do not consider this iterative procedure in any detail, but discuss a number of specific topics in N=1,2 supergravity in section 4. We devote particular attention to the superconformal features of Einstein-Yang-Mills supergravity and their implications for the off-shell structure of this theory, and to the SU(1,1) invariance of chiral N=2 supergravity.

2. STRUCTURE OF d=10 SUPERGRAVITY MULTIPLETS

In d=2 mod 8 dimensional Minkowski space spinors can be both real and chiral (in Euclidean space this is so for d=0 mod 8). In d=10 we may therefore simultaneously impose the conditions

$$\Gamma^{(11)}\lambda = \pm\lambda \, , \tag{2.1}$$

$$C^{-1}\,\bar{\lambda}^T = \lambda \, , \tag{2.2}$$

where

$$\Gamma^{(11)} \equiv i\,\Gamma_1\,\Gamma_2\,\Gamma_3 \cdots \Gamma_{10}$$

is the analogue of γ_5 in 4 dimensions, and C is the d=10 charge conjugation matrix, satisfying

$$C\,\Gamma_\mu\,C^{-1} = -\,\Gamma_\mu{}^T \, , \tag{2.3}$$

$$C\,\Gamma^{(11)}\,C^{-1} = -\,\Gamma^{(11)T} \, . \tag{2.4}$$

The Γ-matrices generate the d=10 Clifford algebra, i.e.

$$\Gamma_\mu\,\Gamma_\nu + \Gamma_\nu\,\Gamma_\mu = 2\delta_{\mu\nu}\mathbb{1} \, . \tag{2.5}$$

A d=10 spinor, on which (2.5) is realized, has $2^5 = 32$ components, which can be restricted to 16 real components by means of the chirality condition (2.1) and the Majorana condition (2.2). Such spinors are called Majorana-Weyl spinors.

The smallest Poincaré supersymmetry algebra in d=10 dimensions is therefore based on Majorana-Weyl generators Q_α, satisfying

$$\{Q_\alpha, Q_\beta^\dagger\} = -2i\,(\not{P}\,\Gamma_{10})_{\alpha\beta}\;, \tag{2.6}$$

where we select the negative chirality components of Q (i.e. $\Gamma^{(11)}\,Q = -Q$) . For supergravity we are interested in massless representations, so that the energy-momentum vector P_μ is lightlike. Choosing the spatial components of P_μ in the 9-th direction yields

$$P_\mu = (o,o,\ldots,\omega,i\omega)\quad,\quad(\omega > o) \tag{2.7}$$

which is invariant under SO(8) rotations of the transverse momenta. Because $P^2 = 0$, it is easy to see that $\not{P}\Gamma_{10}$ is proportional to a projection operator,

$$(\not{P}\Gamma_{10})(\not{P}\Gamma_{10}) = -\,P^2 + 2i\omega\not{P}\Gamma_{10} = 2i\omega(\not{P}\Gamma_{10})\;. \tag{2.8}$$

Using (2.7), we find

$$-i\not{P}\Gamma_{10} = \omega\,(\mathbb{1} - i\Gamma_9\,\Gamma_{10})\;. \tag{2.9}$$

Furthermore

$$\begin{aligned} i\Gamma_9\Gamma_{10} &= (\Gamma_1\Gamma_2\ldots\Gamma_8)\;\Gamma^{(11)} \\ &= \Gamma^{(9)}\Gamma^{(11)} = \Gamma^{(11)}\Gamma^{(9)}\;, \end{aligned} \tag{2.10}$$

where $\Gamma^{(9)}$ is the analogue of γ_5 for the d=8 Clifford algebra associated with the transverse momenta,

$$\Gamma^{(9)} \equiv \Gamma_1\Gamma_2\ldots\Gamma_8\;. \tag{2.11}$$

In the chiral subspace where $\Gamma^{(11)}Q = -\,Q$, we may drop $\Gamma^{(11)}$, so that

$$-i\not{P}\Gamma_{10} = \omega\big(\mathbb{1} + \Gamma^{(9)}\big)\;, \tag{2.12}$$

and likewise

$$-i\Gamma_{10}\not{p} = \omega\left(\mathbb{1} - \Gamma^{(9)}\right) , \qquad (2.13)$$

which are thus proportional to the chiral projectors in d=8 spinor space.

We are interested in representations of the supersymmetry algebra that contain only states of positive norm. Therefore, the right-hand side of (2.6) implies, via (2.12), that all charges Q with negative d=8 chirality ($\Gamma^{(9)}Q = -Q$) should vanish. Hence, the Q's can be restricted to charges with $\Gamma^{(9)}Q = Q$, and in this subspace (2.6) acquires the form

$$\begin{aligned} \{Q_\alpha, Q_\beta^\dagger\} &= 2\omega\, \delta_{\alpha\beta} , \\ \Gamma^{(11)}Q = -\,Q \,, &\qquad \Gamma^{(9)}Q = Q \,. \end{aligned} \qquad (2.14)$$

Because Q has now been restricted to 8 components (and is just an SO(8) Majorana-Weyl spinor) (2.14) defines an 8-dimensional Clifford algebra, which has a unique 16-dimensional representation. Consequently massless supersymmetry representations must decompose into 16-dimensional representations, which in turn consist of two 8-dimensional SO(8) representations. As is well-known SO(8) representations appear in a three-fold variety (triality). With the exception of certain representations, such as the adjoint 28-dimensinal one, the three types of representations are inequivalent, and are distinguished by labels s, v and c (see, for instance [3]). We shall denote the representation according to which Q transforms (the positive chirality representation) as $\mathbf{8}_v$; the 16-dimensional representation of (2.14) then decomposes into the $\mathbf{8}_s$ and $\mathbf{8}_c$ representations. Observe that this is consistent with the generic situation for Clifford algebra representations: conventionally one starts from 2n Clifford algebra generators (e.g. gamma matrices) transforming under the defining representation of SO(2n), which act on 2n-dimensional spinors transforming according to the two chiral spinor representations of SO(2n).

The smallest massless supermultiplet has now been constructed, and consists of 8 fermionic and 8 bosonic states, which we assign to the $\mathbf{8}_c$ and $\mathbf{8}_s$ representation, respectively. This is just the supersymmetric Yang-Mills multiplet in 10 dimensions [4], whose field-theoretic description will be discussed in the next section.

Before constructing the supermultiplets that are relevant for d=10 supergravity, let us first discuss some other properties of SO(8) representations. One way to distinguish the inequivalent representations is to investigate how they decompose into representations of an SO(7) subgroup. Each of the 8-dimensional representations leaves a different SO(7) subgroup of SO(8) invariant, called the isotropy group, Therefore there is an SO(7) subgroup under which the $\mathbf{8}_s$ representation branches into

$$\mathbf{8}_s \to \mathbf{7} + \mathbf{1} \; . \tag{2.15a}$$

Under this SO(7) the other two 8-dimensional representations branch into

$$\mathbf{8}_v \to \mathbf{8} \quad , \quad \mathbf{8}_c \to \mathbf{8} \; . \tag{2.15b}$$

where $\mathbf{8}$ is the spinor representation of SO(7). Corresponding branching rules for the 28-, 35- and 56-dimensional representations are

$$\begin{aligned} \mathbf{28} &\to \mathbf{7} + \mathbf{21} \\ \mathbf{35}_v &\to \mathbf{35} \\ \mathbf{35}_s &\to \mathbf{1} + \mathbf{7} + \mathbf{21} \\ \mathbf{35}_c &\to \mathbf{35} \\ \mathbf{56}_v &\to \mathbf{8} + \mathbf{48} \\ \mathbf{56}_s &\to \mathbf{21} + \mathbf{35} \\ \mathbf{56}_c &\to \mathbf{8} + \mathbf{48} \end{aligned} \tag{2.16}$$

In order to obtain the supersymmetry representations relevant for supergravity we consider tensor products of the smallest supermultiplet consisting of $\mathbf{8}_s + \mathbf{8}_c$, with one of the 8-dimensional representations. There are thus three different possibilities, each leading to a 128-dimensional supermultiplet. Using the multiplication rules for SO(8) representations,

$$\begin{aligned}
\mathbf{8}_v \times \mathbf{8}_v &= \mathbf{1} + \mathbf{28} + \mathbf{35}_v \\
\mathbf{8}_s \times \mathbf{8}_s &= \mathbf{1} + \mathbf{28} + \mathbf{35}_s \\
\mathbf{8}_c \times \mathbf{8}_c &= \mathbf{1} + \mathbf{28} + \mathbf{35}_c \\
\mathbf{8}_v \times \mathbf{8}_s &= \mathbf{8}_c + \mathbf{56}_c \\
\mathbf{8}_s \times \mathbf{8}_c &= \mathbf{8}_v + \mathbf{56}_v \\
\mathbf{8}_c \times \mathbf{8}_v &= \mathbf{8}_s + \mathbf{56}_s
\end{aligned} \tag{2.17}$$

it is straightforward to obtain the three new multiplets. Multiplying $\mathbf{8}_s$ with $\mathbf{8}_s + \mathbf{8}_c$ yields $\mathbf{8}_s \times \mathbf{8}_s$ bosonic and $\mathbf{8}_s \times \mathbf{8}_c$ fermionic states, and leads to the SO(8) decomposition shown in table 1. This supermultiplet contains the representation $\mathbf{35}_s$, which can be associated with the states of the graviton in d=10 dimensions (the field-theoretic description that we will present in section 3 will further clarify this assignment). Therefore this supermultiplet will be called the <u>graviton multiplet</u>. Multiplication with $\mathbf{8}_c$ or $\mathbf{8}_v$ goes in the same fashion, except that we will associate the $\mathbf{8}_c$ and $\mathbf{8}_v$ representations with fermionic quantities (note that these are the representations to which the fermion states of the Yang-Mills multiplet and the supersymmetry charges are assigned). Consequently, we interchange the boson and fermion assignments in these products. Multiplication with $\mathbf{8}_c$ then leads to $\mathbf{8}_c \times \mathbf{8}_c$ bosonic and $\mathbf{8}_c \times \mathbf{8}_s$ fermionic states, whereas multiplication with $\mathbf{8}_v$ gives $\mathbf{8}_v \times \mathbf{8}_c$ bosonic and $\mathbf{8}_v \times \mathbf{8}_s$ fermionic states. These supermultiplets contain fermions transforming according to the $\mathbf{56}_v$ and $\mathbf{56}_c$ representations, respectively, which can be associated with gravitino states, but no graviton states as those transform in the $\mathbf{35}_s$ representation. Therefore these two supermultiplets are called <u>gravitino multiplets</u>, and we have thus established the existence of two inequivalent gravitino multiplets. The explicit SO(8) decompositions of the Yang-Mills, graviton and gravitino supermultiplets are shown in table 1.

supermultiplet	bosons	fermions
Yang-Mills multiplet	$\mathbf{8}_s$	$\mathbf{8}_c$
graviton multiplet	$\mathbf{1} + \mathbf{28} + \mathbf{35}_s$	$\mathbf{8}_v + \mathbf{56}_v$
gravitino multiplet	$\mathbf{1} + \mathbf{28} + \mathbf{35}_c$	$\mathbf{8}_v + \mathbf{56}_v$
gravitino multiplet	$\mathbf{8}_s + \mathbf{56}_s$	$\mathbf{8}_c + \mathbf{56}_c$

Table 1: Massless N=1 supermultiplets in ten space-time dimensions containing 8+8 or 64+64 bosonic and fermionic degrees of freedom.

By combining a graviton and a gravitino multiplet it is possible to construct an N=2 supermultiplet of 128 + 128 bosonic and fermionic states. However, since there are two inequivalent gravitino multiplets, there will also be two inequivalent N=2 supermultiplets containing the states corresponding to a graviton and 2 gravitini. The content of these two multiplets has been shown in table 2. According to the construction presented above, one N=2 supermultiplet may be beviewed as the tensor product of two identical supermultiplets (namely $\mathbf{8}_s$+$\mathbf{8}_c$). Such a multiplet follows if one starts from a supersymmetry algebra based on two Majorana-Weyl spinor charges Q with the same chirality. One can perform rotations between these spinor charges which leave the supersymmetry algebra unaffected, and this feature should result in a certain degeneracy of some of the states of the supermultiplet. Indeed, the explicit SO(8) decomposition in table 2 shows such a degeneracy for the states assigned to the **1**, **28**, $\mathbf{8}_v$ and $\mathbf{56}_v$ representations. The theory based on this multiplet is chiral N=2 supergravity [5] (sometimes called type 2b supergravity).

The second supermultiplet in table 2 may be viewed as the tensor product of a ($\mathbf{8}_s$+$\mathbf{8}_c$) supermultiplet with a second supermultiplet ($\mathbf{8}_s$+$\mathbf{8}_v$). The fermionic states appear then with both chiralities. Such a multiplet can be derived from a supersymmetry algebra based on two Majorana-Weyl spinor charges Q, but now with opposite chirality. The theory based on this multiplet is nonchiral N=2 supergravity (some

chiral N=2 supermultiplet

$$(8_s + 8_c) \times (8_s + 8_c)$$

bosons	fermions
$1 + 1 + 28 + 28 + 35_s + 35_c$	$8_v + 8_v + 56_v + 56_v$

nonchiral N=2 supermultiplet

$$(8_s + 8_v) \times (8_s + 8_c)$$

bosons	fermions
$1 + 28 + 35_s + 8_s + 56_s$	$8_v + 8_c + 56_v + 56_c$

Table 2: Chiral and nonchiral N=2 supergravity multiplets in ten space-time dimensions.

times called type 2a supergravity), which can be obtained by a straightforward reduction of d=11 supergravity [6]. The latter follows from the fact that two d=10 Majorana-Weyl spinors with opposite chirality can be combined into a single d=11 Majorana spinor.

3. FIELD REPRESENTATIONS

In this section we consider a variety of field representations satisfying certain massless field equations, and analyze the representation content of the corresponding plane wave solutions. These field representations can be used to give a field-theoretic treatment of the supergravity multiplets derived in the previous section.

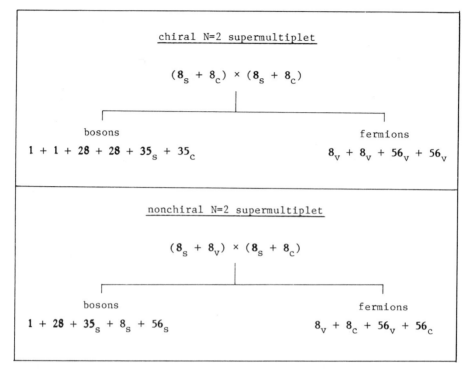

3.1. Vector gauge field

A vector gauge field $A_\mu(q)$ in the momentum representation can be decomposed into the following set of linearly independent vectors

$$q_\mu = (\vec{q}, iq_o) \quad , \quad \bar{q}_\mu = (\vec{q}, -iq_o) \ ,$$

$$\varepsilon^i_\mu = (\vec{\varepsilon}^i, 0) \quad , \quad i=1,2,\ldots,8, \quad \vec{\varepsilon}^i \cdot \vec{q} = 0 \quad , \quad \vec{\varepsilon}^i \cdot \vec{\varepsilon}^j = \delta^{ij} \ . \tag{3.1}$$

Introducing 10 coefficient functions $a_i(q)$, $b(q)$ and $c(q)$, we may write

$$A_\mu(q) = a_i(q)\, \varepsilon^i_\mu + b(q)\, \bar{q}_\mu + c(q)\, q_\mu \ . \tag{3.2}$$

We expect the coefficient function $c(q)$ to have no physical signifiance, as it can be changed by a gauge transformation on $A_\mu(q)$. Indeed, $c(q)$ does not appear in the field strength corresponding to (3.2), which is proportional to

$$F_{\mu\nu}(q) \propto a_i(q)\, \varepsilon^i_{[\mu}\, q_{\nu]} + b(q)\, \bar{q}_{[\mu}\, q_{\nu]} \ . \tag{3.3}$$

The Maxwell equation, $\partial^\mu F_{\mu\nu} = 0$, now implies

$$q^\mu\, F_{\mu\nu}(q) \propto q^2\, a_i(q)\, \varepsilon^i_\nu + b(q)\, \left(q^2\, \bar{q}_\nu - q\cdot\bar{q}\, q_\nu\right) = 0 \ , \tag{3.4}$$

from which we conclude that $b(q) = 0$ and $a_i(q) \neq 0$ provided that $q^2 = 0$ (note that $q\cdot\bar{q} > 0$). Consequently a vector gauge field that satisfies the Maxwell equation leads to 8 massless plane wave solutions with transverse polarization vectors ε^i_μ. These solutions transform under an 8-dimensional representation of the transverse rotation group SO(8). We denote this representation by $\mathbf{8}_s$.

3.2. Chiral spinors

Consider a chiral spinor $u(q)$ satisfying the massless Dirac equation, i.e.

$$\not{q}\ u(q) = 0\ ,$$

$$\Gamma^{(11)}\ u(q) = \pm\ u(q)\ . \tag{3.5}$$

The Dirac equation implies that $q^2 = 0$. Using the same manipulations as those leading to (2.13) it is easy to deduce that the spinor must satisfy

$$-\ i\Gamma_{10}\ \not{q}\ u(q) = \omega\left(1 + \Gamma^{(9)}\Gamma^{(11)}\right)\ u(q) = 0\ ,$$

or

$$\Gamma^{(9)}\ u(q) = \mp\ u(q)\ . \tag{3.6}$$

Hence the spinor u(q) is reduced to an 8-dimensional spinor, transforming under SO(8) according to the representation $\mathbf{8}_c$ (if $\Gamma^{(11)}\ u(q) = +\ u(q)$) or $\mathbf{8}_v$ (if $\Gamma^{(11)}\ u(q) = -\ u(q)$).

3.3. Graviton fields

The linearized Einstein equation for $g_{\mu\nu} = \delta_{\mu\nu} + \kappa h_{\mu\nu}$ implies that

$$R_{\mu\nu} \propto \partial^2 h_{\mu\nu} + \partial_\mu \partial_\nu h - \partial_\mu \partial_\nu h_{\rho\nu} - \partial_\nu \partial_\rho h_{\rho\mu} = 0\ , \tag{3.7}$$

where $h \equiv h_{\mu\mu}$ and $R_{\mu\nu}$ is the Ricci tensor. Again we may decompose $h_{\mu\nu}$ in the momentum representation, using the vectors (3.1):

$$h_{\mu\nu}(q) = a_{ij}(q)\ \varepsilon^i_\mu\ \varepsilon^j_\nu + b_i(q)\ \varepsilon^i_{(\mu}\ \bar{q}_{\nu)} + c(q)\ \bar{q}_\mu\ \bar{q}_\nu$$

$$+\ d_i(q)\ \varepsilon^i_{(\mu}\ q_{\nu)} + e(q)\ q_{(\mu}\ \bar{q}_{\nu)} + f(q)\ q_\mu\ q_\nu\ , \tag{3.8}$$

where a_{ij} is a symmetric tensor in 8 dimensions. Inserting this decomposition into (3.7) gives

$$R_{\mu\nu}(q) \propto q^2\big(a_{ij}(q)\, \varepsilon^i_\mu\, \varepsilon^j_\nu + b_i(q)\, \varepsilon^i_{(\mu}\, \bar{q}_{\nu)} + c(q)\, \bar{q}_\mu\, \bar{q}_\nu\big)$$

$$+ \big(a_{ij}(q) + c(q)\, q^2\big)\, q_\mu q_\nu - q.\bar{q}\, b_i(q)\, \varepsilon^i_{(\mu}\, q_{\nu)}$$

$$- 2q.\bar{q}\, c(q)\, q_{(\mu}\, \bar{q}_{\nu)}$$

$$= 0. \qquad (3.9)$$

The degrees of freedom corresponding to the coefficient functions $d_i(q)$, $e(q)$ and $f(q)$ have no physical significance as they are subject to general coordinate transformations whose linearized form is $\delta h_{\mu\nu} = \partial_\mu \xi_\nu + \partial_\nu \xi_\mu$. Since the Ricci tensor is covariant under general coordinate transformations these gauge degrees of freedom are not present in (3.9).

From (3.9) one concludes that plane wave solutions must satisfy

$$q^2\, a_{ij}(q) = 0 \quad , \quad a_{ii}(q) = b_i(q) = c(q) = 0 \ , \qquad (3.10)$$

so that the solutions are massless and characterized by a traceless symmetric tensor $a_{ij}(q)$. This implies that the graviton has $\frac{1}{2}d(d-3)$ degrees of freedom in d space-time dimensions. When d=3 there are thus no dynamic degrees of freedom associated with gravitational field as is well-known (for instance, see [7]); when d=2 one finds the same result, except that in this case (3.7) does not apply and one should use $R_{\mu\nu} = \frac{1}{2}g_{\mu\nu}R$. For d=10 there are 35 degrees of freedom, which transform under transverse rotations as a symmetric traceless tensor. This defines the SO(8) representation $\mathbf{35}_s$.

3.4. Gravitino fields

The gravitino field equation is

$$\Gamma^\mu\, (\partial_\mu \psi_\nu - \partial_\nu \psi_\mu) = 0 \ . \qquad (3.11)$$

Decomposing the gravitino field as

$$\psi_\mu(q) = u_i(q)\ \varepsilon^i_\mu + v(q)\ \bar{q}_\mu + w(q)\ q_\mu\ , \tag{3.12}$$

where the coefficient functions $u_i(q)$, $v(q)$ and $w(q)$ are spinors, the field equation takes the form

$$\not{q}\ u_i(q)\ \varepsilon^i_\nu - \left(\not{\varepsilon}^i\ u_i(q) - \bar{\not{q}}\ v(q)\right)\ q_\nu - \not{q}\ v(q)\ \bar{q}_\nu = 0\ , \tag{3.13}$$

where $\not{\varepsilon}^i \equiv \varepsilon^i_\mu\ \Gamma^\mu$. The spinor $w(q)$, which is subject to gauge transformations $\delta\psi_\mu = \partial_\mu\varepsilon$, is not determined by the gauge invariant field equation (3.11), and has disappeared from (3.13). The remaining spinors $u_i(q)$ and $v(q)$ satisfy

$$\not{q}\ u_i(q) = 0\quad ,\quad \not{q}\ v(q) = 0\quad ,\quad \bar{\not{q}}\ v(q) = \not{\varepsilon}^i\ u_i(q)\ . \tag{3.14}$$

Multiplying the last equation with $\not{q}$ and using the first two equations and $\varepsilon^i.q = 0$, one derives $q.\bar{q}\ v(q) = 0$; hence $v(q) = 0$, so that we are left with two equations for $u_i(q)$,

$$\not{q}\ u_i(q) = 0\ ,\ \not{\varepsilon}^i\ u_i(q) = 0\ . \tag{3.15}$$

Let us now assume d=10 space-time dimensions and a chiral gravitino field

$$\Gamma^{(11)}\ \psi_\mu = \pm\ \psi_\mu\ . \tag{3.16}$$

According to the result of subsection 3.2, $u_i(q)$ now transforms as a chiral vector-spinor, which constitutes a tensor product $\mathbf{8}_s \times \mathbf{8}_c$ ($\mathbf{8}_s \times \mathbf{8}_v$) for positive (negative) chirality spinors. According to the multiplication rules (2.17) this product decomposes into $\mathbf{8}_v + \mathbf{56}_v$ or $\mathbf{8}_c + \mathbf{56}_c$. However, the second equation in (3.15), which is SO(8) covariant, imposes eight conditions thus suppressing the $\mathbf{8}_v$ or $\mathbf{8}_c$ representation. Consequently chiral gravitini transform according to the $\mathbf{56}_v$ or $\mathbf{56}_c$ representation of SO(8) depending on whether ψ_μ has positive or negative chirality, respectively.

3.5. Antisymmetric tensor gauge fields

Antisymmetric tensor gauge fields have field strength tensors which are antisymmetric and of rank p. They satisfy field equations and Bianchi identities, generalizations of the Maxwell equations, which read

$$\partial_{[\nu}F_{\mu_1\dots\mu_p]} = 0 \,, \tag{3.17}$$

$$\partial_\nu F^{\nu\mu_1\dots\mu_{p-1}} = 0 \,. \tag{3.18}$$

A trivial example is the case p=1, which describes an ordinary scalar field: the solution of (3.17) yields $F_\mu = \partial_\mu\phi$, and (3.18) implies the Klein-Gordon equation on ϕ. Another example is the case of a vector gauge field discussed in subsection 3.1, which corresponds to p=2.

There are two ways of dealing with (3.17) and (3.18). One is to first solve (3.17) by substituting

$$F_{\mu_1\dots\mu_p} = p\,\partial_{[\mu_1}A_{\mu_2\dots\mu_p]} \,, \tag{3.19}$$

where $A_{\mu_1\dots\mu_{p-1}}$ is an antisymmetric tensor gauge field of rank p-1 subject to gauge transformations

$$\delta A_{\mu_1\dots\mu_{p-1}} = \partial_{[\mu_1}\xi_{\mu_2\dots\mu_{p-1}]} \,. \tag{3.20}$$

The equation of motion (3.18) is then a generalization of the Maxwell equation, i.e.

$$\partial^2 A_{\mu_1\dots\mu_{p-1}} + (-)^{p-1}\,(p-1)\,\partial_{[\mu_1}\partial^\rho A_{\mu_2\dots\mu_{p-1}]\rho} = 0 \,. \tag{3.21}$$

The alternative is to first solve (3.18) by substituting

$$F_{\mu_1\dots\mu_p} = \frac{i}{(d-p-1)!}\,\varepsilon_{\mu_1\dots\mu_p\rho\nu_1\dots\nu_{d-p-1}}\,\partial^\rho B^{\nu_1\dots\nu_{d-p-1}} \,, \tag{3.22}$$

where $B_{\nu_1\dots\nu_{d-p-1}}$ is an antisymmetric gauge field of rank d-p-1, subject to gauge transformations similar to (3.20). This procedure coincides with the previous one, but now based on the dual field strength defined by

$$F_{\mu_1\ldots\mu_p} = \frac{i}{(d-p)!}\,\varepsilon_{\mu_1\ldots\mu_p\nu_1\ldots\nu_{d-p}}\,\tilde{F}^{\nu_1\ldots\nu_{d-p}}\ , \tag{3.23}$$

so that

$$\tilde{F}_{\nu_1\ldots\nu_{d-p}} = (d-p)\,\partial_{[\nu_1}B_{\nu_2\ldots\nu_{d-p}]}\ . \tag{3.24}$$

For $\tilde{F}$ the two equations (3.17) and (3.18) are interchanged, so that the solution in terms of $B_{\nu_1\ldots\nu_{d-p-1}}$ is the dual formulation of the one in terms of $A_{\mu_1\ldots\mu_{p-1}}$.

By using a so-called "first-order" formulation it is possible to have a lagrangian that encompasses both descriptions. Namely one chooses the lagrangian

$$L = \frac{1}{p!}\,F^{\mu_1\ldots\mu_p}\left(F_{\mu_1\ldots\mu_p} - 2p\,\partial_{\mu_1}A_{\mu_2\ldots\mu_p}\right)\ , \tag{3.25}$$

where $F_{\mu_1\ldots\mu_p}$ is an unconstrained field, and $A_{\mu_1\ldots\mu_{p-1}}$ is the Lagrange multiplier for (3.18). One may now first solve the field equation for $F_{\mu_1\ldots\mu_p}$, which leads to (3.19), and substitute this result into the lagrangian which then yields the field equation (3.18). Alternatively one may solve the field equation for $A_{\mu_1\ldots\mu_{p-1}}$, which leads to (3.22), and obtain a lagrangian in terms of $B_{\mu_1\ldots\mu_{d-p-1}}$. Observe that one could also have started from a lagrangian containing $B_{\mu_1\ldots\mu_{d-p-1}}$ as a Lagrange multiplier to impose (3.18), with the same results.

Although both the descriptions in terms of a rank-(p-1) and a rank-(d-p-1) tensor gauge field lead to the same physical degrees of freedom, as those are entirely determined by (3.17) and (3.18), they differ in the number of field components: an rank-n antisymmetric tensor field subject to gauge transformations of type (3.20) and (3.22) contains

$$\binom{d}{n} - \binom{d-1}{n-1} = \binom{d-1}{n} \tag{3.26}$$

independent components, which is not the same number for n=p-1 as for n=d-p-1. This observation is relevant in the context of off-shell formulations of supersymmetric field theories, where the balance between the number of bosonic and fermionic field components is crucial. Note also that (3.26) is the dimension of the rank-n antisymmetric tensor representation of SO(d-1).

Let us now examine the plane wave solutions of (3.17) and (3.18). As before we start from a decomposition of $F_{\mu_1 \ldots \mu_p}$ in the momentum representation, employing the vectors (3.1)

$$\begin{aligned} F_{\mu_1 \ldots \mu_p}(q) \propto\ & a_{i_1 \ldots i_p}\, \varepsilon^{i_1}_{\mu_1} \ldots \varepsilon^{i_p}_{\mu_p} \\ & + \varepsilon^{i_1}_{[\mu_1} \ldots \varepsilon^{i_{p-1}}_{\mu_{p-1}} \left(\bar{q}_{\mu_p]}\, b_{i_1 \ldots i_{p-1}}(q) + q_{\mu_p]}\, c_{i_1 \ldots i_{p-1}}(q)\right) \\ & + d_{i_1 \ldots i_{p-2}}(q)\, \varepsilon^{i_1}_{[\mu_1} \ldots \varepsilon^{i_{p-2}}_{\mu_{p-2}}\, \bar{q}_{\mu_{p-1}}\, q_{\mu_p]} \ . \end{aligned} \tag{3.27}$$

Imposing (3.17) yields

$$\begin{aligned} q_{[\nu} F_{\mu_1 \ldots \mu_p]}(q) \propto\ & a_{i_1 \ldots i_p}(q)\, \varepsilon^{i_1}_{[\mu_1} \ldots \varepsilon^{i_p}_{\mu_{p-1}}\, q_{\nu]} \\ & + b_{i_1 \ldots i_{p-1}}(q)\, \varepsilon^{i_1}_{[\mu_1} \ldots \varepsilon^{i_{p-1}}_{\mu_{p-1}}\, \bar{q}_{\mu_p}\, q_{\nu]} \\ & = 0 \ , \end{aligned}$$

or

$$a_{i_1 \ldots i_p}(q) = b_{i_1 \ldots i_{p-1}}(q) = 0 \ . \tag{3.28}$$

Similarly, (3.18) leads to

$$d_{i_1 \ldots i_{p-2}}(q) = 0 \quad , \quad q^2\, c_{i_1 \ldots i_{p-1}}(q) = 0 \ , \tag{3.29}$$

p	number of solutions	SO(8) assignment
1	1	**1**
2	8	$\mathbf{8}_s$
3	28	**28**
4	56	$\mathbf{56}_s$
5	70	$\mathbf{35}_v + \mathbf{35}_c$

Table 3: Solutions described by p-rank antisymmetric field strenghts satisfying (3.17) and (3.18) in 10 space-time dimensions. The case p=5 is described in terms of a 4-rank 8-dimensional tensor, which decomposes into a selfdual and a antiselfdual part, leading to the $\mathbf{35}_v$ and $\mathbf{35}_c$ representations. Hence these degrees of freedom transform in a different manner than those of the graviton, which transform according to the inequivalent $\mathbf{35}_s$ representation. By imposing a duality condition (cf. 3.30) one can restrict the 70 degrees of freedom to a single irreducible representation.

so that the solutions are massless and characterized by an antisymmetric (p-1)-th rank tensor in the transverse (d-2)-dimensional space; hence the number of solutions equals $\binom{d-2}{p-1}$. In d=10 dimensions it is now easy to find the SO(8) assignment of these solutions, by verifying the branching into SO(7) representations. Table 3 presents the results for p=1,...,5 (note that the higher p-values are related to the lower ones by duality (cf. 3.14)).

If d=2 mod 4 and $p=\frac{1}{2}d$ it is possible to restrict the tensor $F_{\mu_1\ldots\mu_p}$ to be selfdual or antiselfdual, viz.

$$F_{\mu_1\ldots\mu_p} = \frac{\pm i}{p!}\,\varepsilon_{\mu_1\ldots\mu_{2p}}\,F^{\mu_{p+1}\ldots\mu_{2p}} \,. \tag{3.30}$$

For such tensors (3.17) and (3.18) are no longer independent. Imposing one of these equations the duality condition (3.30) induces a corresponding (d-2)-dimensional duality condition (but now in the space of transverse momenta, which is Euclidean) on the coefficients

$c_{i_1 \ldots i_{p-1}}$ in (3.27),

$$c_{i_1 \ldots i_{p-1}} = \mp \frac{1}{(p-1)!} \varepsilon_{i_1 \ldots i_{p-2}} c^{i_p \ldots i_{p-2}} \tag{3.31}$$

Consequently the number of independent solutions associated with the antisymmetric tensor is reduced by a factor 2. Note that for tensors satisfying (3.30) the lagrangian (3.25) cannot be written down as the square of a selfdual tensor vanishes in Minkowski space.

4. SUPERGRAVITY IN 10 DIMENSIONS

After the previous results it is straightforward to set up free field theories for the supermultiplets shown in tables 1 and 2. The fermions of these multiplets are described by Majorana-Weyl spinor and Rarita-Schwinger fields. Supersymmetric Yang-Mills theory requires positive chirality Majorana-Weyl fields, one for each gauge group generator. The gravitini of chiral N=2 supergravity are described in terms of two Majorana-Weyl Rarita-Schwinger fields with positive chirality, while those of nonchiral N=2 supergravity are described by Rarita-Schwinger fields of both chiralities. The remaining fermion fields of chiral N=2 supergravity are two Majorana-Weyl spinor fields of negative chirality; those of nonchiral N=2 supergravity are spinor fields of both chiralities.

The bosonic sector of supersymmetric Yang-Mills theory is described by vector gauge fields. The graviton in the supergravity theories has the usual description in terms of a zehnbein field, yielding gravitational degrees of freedom in the $\mathbf{35}_s$ representation. The bosonic degrees of freedom corresponding to the **28** and $\mathbf{56}_s$ representations can be described in terms of rank-2 and rank-3 tensor gauge fields, the singlet representations in terms of scalar fields, and the $\mathbf{8}_s$ representations in terms of a vector gauge field; finally the $\mathbf{35}_c$ representation of chiral N=2 supergravity can be associated with a selfdual rank-5 antisymmetric field strength (cf. table 3).

From the free field theories one can deduce the linearized supersymmetry transformations for the fields (for the physical states these transformations follow from the supersymmetry algebra). At this stage the theories are invariant under rigid supersymmetry and the Poincaré transformations of d=10 space-time. When extending these invariances to local supersymmetry and invariance under general coordinate transformations, interactions become necessary, which induce corresponding nonlinear modifications of the supersymmetry transformation rules. The nonlinear terms in the field equations (or the lagrangian) and the transformation rules can be determined by means of an iterative procedure. In order to facilitate this procedure one first covariantizes the free-field results with respect to general coordinate transformations, as invariance under the latter is implied by local supersymmetry. Subsequently one determines the coupling of the gravitino fields to the non-gravitational fields by requiring that the terms proportional to the derivative of the supersymmetry transformation parameter cancel. Obviously this coupling takes the form $\bar{\psi}_\mu J^\mu$, where ψ_μ denotes the gravitino field (the gauge field of supersymmetry) and the spinor current J_μ coincides with the supersymmetry Noether current. One then proceeds and determines the variation of J_μ in the Noether-coupling term, whose cancellation requires the addition of new interaction terms as well as nonlinear modifications of the supersymmetry transformation rules. It is at this point that one may discover that a locally supersymmetric theory is not possible. When no such difficulty is encountered dimensional arguments often ensure that the iterative procedure can be completed in a finite number of steps. We will not discuss these laborious iterative constructions of supergravity in any detail, but concentrate on a number of more specific topics. All of the various supergravity theories contain the standard Einstein and Rarita-Schwinger equations, but they differ in the chirality assignments of the gravitini, and in the remaining fields.

4.1. Nonchiral N=2 supergravity

Table 4 presents the field content of nonchiral N=2 supergravity, and indicates how these fields combine into the fields of d=11 supergravity [6]: the gravitino field Ψ_M, the elfbein field $E_M{}^A$ and a 3-rank

fermions	$\psi_\mu^+ + \psi_\mu^-$	144 + 144	$\mathbf{56}_v + \mathbf{56}_c$	Ψ_M
	$\lambda^- + \lambda^+$	16 + 16	$\mathbf{8}_v + \mathbf{8}_c$	
bosons	$e_\mu^a + A_\mu$	45 + 9	$\mathbf{35}_s + \mathbf{8}_s$	$E_M{}^A$
	ϕ	1	$\mathbf{1}$	
	$A_{\mu\nu} + A_{\mu\nu\rho}$	36 + 84	$\mathbf{28} + \mathbf{56}_s$	A_{MNP}

Table 4: Fields of nonchiral N=2 supergravity. The first column shows the fields decomposed into the two N=1 submultiplets. The superscript $\pm$ on the fermion fields indicates their chirality. The second column gives the corresponding numbers of field components (after subtraction of gauge degrees of freedom), and the third column the SO(8) representations of the corresponding massless states. The last column exhibits how the fields combine into those of d=11 supergravity.

antisymmetric tensor gauge field A_{MNP}. Rather than analyzing nonchiral supergravity one usually prefers to consider full d=11 supergravity, or one restricts attention to the truncated N=1 theory, whose field content can also be read off from table 4 (see, however, [14]).

4.2. N=1 Einstein-Yang-Mills supergravity

The interesting feature of N=1 supergravity is that it can be coupled to supersymmetric Yang-Mills theory. The easiest way to obtain the lagrangian and transformation rules of N=1 supergravity is by truncation from d=11 supergravity [8,9]. In the coupling to Yang-Mills theory new features arise. Using the standard Noether coupling method (see, e.g. [10]), one derives the supercurrent

$$J_\mu = \tfrac{1}{4}\,\Gamma^{\rho\sigma}\Gamma_\mu\,\mathrm{Tr}(\chi F_{\rho\sigma}) \;, \qquad (4.1)$$

which describes the first-order coupling of the gravitino to the super-Yang-Mills spinor fields χ and the nonabelian field strength $F_{\mu\nu} = \partial_\mu A_\nu - \partial_\nu A_\mu - [A_\mu, A_\nu]$ (in our notation χ, $F_{\mu\nu}$ and A_μ are Lie-algebra valued and normalized such that the gauge field lagrangian equals $\frac{1}{4} g^{-2} \mathrm{Tr}(F_{\mu\nu}F^{\mu\nu})$; g is the Yang-Mills coupling constant). The supersymmetry variation of the supercurrent contains a term

$$\delta J_\mu = - \tfrac{1}{4} \Gamma_{\mu\nu\rho\sigma\lambda} \Omega^{\nu\rho\sigma\lambda} \varepsilon + \ldots\ldots , \tag{4.2}$$

where the dots denote other terms that do not concern us here, and

$$\Omega_{\mu\nu\rho\sigma} = \tfrac{1}{4} \mathrm{Tr}\left(F_{[\mu\nu} F_{\rho\sigma]}\right). \tag{4.3}$$

Note that Ω, which, in four dimensions, corresponds to the Pontryagin density, is closed (but not exact), i.e.

$$\partial_{[\mu}\Omega_{\nu\rho\sigma\lambda]} = 0 , \tag{4.4}$$

as a consequence of the Bianchi identity on F,

$$\partial_{[\mu}F_{\nu\rho]} - [A_{[\mu}, F_{\nu\rho]}] = 0 . \tag{4.5}$$

Locally it is possible to express Ω as the curl of a rank-3 antisymmetric tensor, known as the Chern-Simons term,

$$\Omega_{\mu\nu\rho\sigma} = \partial_{[\mu}\omega_{\nu\rho\sigma]} , \tag{4.6}$$

where

$$\omega_{\mu\nu\rho} = \mathrm{Tr}\left(A_{[\mu} \partial_\nu A_{\rho]} - \tfrac{2}{3} A_{[\mu} A_\nu A_{\rho]}\right) . \tag{4.7}$$

In order to have local supersymmetry the supercurrent coupling $\bar{\psi}_\mu J^\mu$ must be supplemented with similar couplings to the other fields of N=1 supergravity. To cancel the variation of $\bar{\psi}_\mu J^\mu$ induced by (4.2) one obviously needs the variation of some new coupling of the tensor gauge field $A_{\mu\nu}$ to $\Omega_{\rho\sigma\lambda\tau}$. However, for dimensional reasons no extra space-time derivatives are allowed in such a coupling; therefore no direct $A_{\mu\nu} - \Omega_{\rho\sigma\lambda\tau}$ coupling can be written down as the indices do not match. There are two ways to get around this problem. One is to exploit (4.6) and introduce the interaction

$$L' = -\frac{3}{2}\sqrt{2}\, g^{-2} e\, e^{2\phi}\, \omega^{\mu\nu\rho}\, \partial_\mu A_{\nu\rho} \tag{4.8}$$

(the factor $e^{2\phi}$ is only found after higher-order iterations in the Noether procedure, and will also multiply the $\bar{\psi}_\mu J^\mu$ interaction). This is the solution found in [9] for the abelian theory; its nonabelian extension was subsequently given in [11]. The supersymmetry variation of $A_{\mu\nu}$ induces a term in $\delta L'$ proportional to $\omega\, \partial(\bar{\varepsilon}\psi)$ which combines with variations of new interactions proportional to $\omega\, \psi^2$ and contributions from a new term in the supersymmetry transformation law of the gravitino field proportional to $\omega\varepsilon$. When added to the variation of the supercurrent coupling proportional to $\Omega\, \bar{\varepsilon}\psi$, all these terms combine into a total derivative, which may be dropped from the lagrangian. Continuing the Noether procedure reveals that (4.8) and similar higher-order interactions are precisely generated by replacing the field strength associated with $A_{\mu\nu}$ by a modified field strength

$$F'_{\mu\nu\rho} = \partial_{[\mu}A_{\nu\rho]} + \sqrt{2}\, g^{-2}\, \omega_{\mu\nu\rho}\ , \tag{4.9}$$

which satisfies the Bianchi identity

$$\partial_{[\mu}F'_{\nu\rho\sigma]} = \frac{\sqrt{2}}{4}\, g^{-2}\, \mathrm{Tr}\big(F_{[\mu\nu}F_{\rho\sigma]}\big)\ . \tag{4.10}$$

This modified field strength is invariant under both tensor and Yang-Mills gauge transformations, which read

$$\delta A_\mu = \partial_\mu \Lambda - [A_\mu, \Lambda]\ ,$$

$$\delta A_{\mu\nu} = \partial_{[\mu}\Lambda_{\nu]} - \sqrt{2}\, g^{-2}\, \mathrm{Tr}\big(\Lambda\, \partial_{[\mu}A_{\nu]}\big)\ . \tag{4.11}$$

Here Λ is the Lie-algebra valued Yang-Mills parameter, and Λ_μ characterizes the tensor gauge transformations. Observe that the tensor field is not invariant under Yang-Mills transformations.

Another way to circumvent the problem mentioned above, first presented in [12], is to change the field representation of supergravity by replacing the rank-2 tensor field by a rank-6 tensor. This formulation is dual to the previous one in the sense described in the previous section. An action, analogous to (3.25) which describes both formulations at once was already presented in [9], and worked out further in [13]. The terms pertaining to the antisymmetric tensor field read

$$L = \frac{3}{4} e\, e^{-2\phi} (t_{\mu\nu\rho})^2 - \frac{3}{2} e\, t^{\mu\nu\rho}\, \partial_\mu A_{\nu\rho}$$
$$- \frac{3}{2} \sqrt{2}\, g^{-2} e\, t^{\mu\nu\rho} \{\omega_{\mu\nu\rho} + \frac{1}{24} \mathrm{Tr}(\bar{\chi}\Gamma_{\mu\nu\rho}\chi) + \bar{\psi}\psi,\ \bar{\psi}\lambda\text{-terms}\} \qquad (4.12)$$

One may solve the field equation for $t_{\mu\nu\rho}$, which gives

$$t_{\mu\nu\rho} = e^{2\phi}\{\partial_{[\mu}A_{\nu\rho]} + \sqrt{2}\, g^{-2}\big(\omega_{\mu\nu\rho} + \frac{1}{24} \mathrm{Tr}(\bar{\chi}\Gamma_{\mu\nu\rho}\chi)\big) + \bar{\psi}\psi,\ \bar{\psi}\lambda\text{-terms}\}. \qquad (4.13)$$

This solution, which contains the modified tensor field strength (4.9), can be substituted into (4.12), thus leading to the previous formulation with a rank-2 tensor gauge field.

Alternatively, one may solve the field equation for $A_{\mu\nu}$ by making the substitution

$$t^{\mu\nu\rho} = \frac{\sqrt{2}}{3.6!} ie^{-1}\, e^{\mu\nu\rho\sigma_1\ldots\sigma_7}\, \partial_{\sigma_1} A_{\sigma_2\ldots\sigma_7}\ . \qquad (4.14)$$

Substituting this back into the lagrangian leads to the formulation with a 6-index tensor field. All terms depending on the Yang-Mills fields then take the form

$$L_{YM} = g^{-2}\, e\, e^{\phi} \{\mathrm{Tr}(\tfrac{1}{4}F_{\mu\nu}F^{\mu\nu} + \tfrac{1}{2}\bar{\chi}\Gamma^{\mu}D_{\mu}\chi) + \bar{J}_{\mu}(\phi_{\mu} + \tfrac{1}{6}\Gamma_{\mu}\lambda)\}$$

$$+ \frac{i}{6!} g^{-2} \varepsilon^{\mu_1 \cdots \mu_{10}} \Omega_{\mu_1 \cdots \mu_4} A_{\mu_5 \cdots \mu_{10}}$$

$$+ \frac{1}{4.6!} g^{-2} e\, \mathrm{Tr}(\bar{\chi}\, \Gamma^{\mu_1 \cdots \mu_7} \chi)\, \partial_{\mu_1} A_{\mu_2 \cdots \mu_7}$$

$$+ \text{quartic fermion terms} . \qquad (4.15)$$

There are obvious differences between the two formulations. Unlike the rank-2 tensor version the lagrangian (4.15) is only quadratic in the super-Yang-Mills fields. Furthermore, the tensor field remains inert under Yang-Mills transformations; invariance under tensor gauge transformations is simply ensured by (4.4). The cancellation of anomalies, which was examined originally in the version with rank-2 tensor field [15], has also been considered recently for the rank-6 tensor version, with identical results [16]. One of the interesting features of (4.15) concerns the presence of the factor $\exp(\phi)$ in the first line. As we shall discuss shortly, this factor is such that the lagrangian becomes conformally invariant (without this factor the Yang-Mills action is only conformally invariant in 4 dimensions). In addition the factor $\exp(\phi)$ appears in (4.12-15) in such a way that all expressions remain unchanged under a simultaneous rescaling of the Yang-Mills coupling constant, the tensor fields and $\exp(\phi)$:

$$g \to e^{\alpha} g \;,\quad e^{\phi} \to e^{2\alpha} e^{\phi} \;,$$

$$A_{\mu\nu} \to e^{-2\alpha} A_{\mu\nu} \;,\quad A_{\mu\nu\rho\sigma\lambda\tau} \to e^{2\alpha} A_{\mu\nu\rho\sigma\lambda\tau} \;,$$

$$t_{\mu\nu\rho} \to e^{2\alpha} t_{\mu\nu\rho} \,. \qquad (4.16)$$

By an appropriate choice for α the explicit dependence on the Yang-Mills coupling constant can thus be absorbed into ϕ, $A_{\mu\nu}$, $A_{\mu\nu\rho\sigma\lambda\tau}$ and $t_{\mu\nu\rho}$.

The above results have been taken from [13], where also the higher-order fermion terms are given, after making the following substitutions

$$A_{\mu_1\ldots\mu_6} \to \frac{1}{2.6!} A_{\mu_1\ldots\mu_6} ,$$

$$\lambda \to \frac{1}{6} \lambda ,$$

$$\phi \to \exp(\frac{w}{6} \phi) . \qquad (4.17)$$

For future use we also record the supersymmetry variations of the various fields, up to bilinear fermion terms. The transformations of the super-Yang-Mills fields are

$$\delta A_\mu = \tfrac{1}{2} \bar{\varepsilon}\Gamma_\mu \chi ,$$

$$\delta\chi = -\tfrac{1}{4} \Gamma^{\mu\nu} F_{\mu\nu}\varepsilon , \qquad (4.18)$$

while those of the supergravity fields read

$$\delta e_\mu{}^a = \tfrac{1}{2} \bar{\varepsilon}\Gamma^a \psi_\mu ,$$

$$\delta\psi_\mu = D_\mu\varepsilon + \frac{1}{8!} e^{-\phi}\left(\Gamma_\mu\Gamma^{(7)} - 2\Gamma^{(7)}\Gamma_\mu\right) R(A)_{(7)}\varepsilon ,$$

$$\delta A_{\mu_1\ldots\mu_6} = \frac{3}{2} e^{\phi} \bar{\varepsilon}\left(\Gamma_{[\mu_1\ldots\mu_5}\psi_{\mu_6]} + \frac{1}{6} \Gamma_{\mu_1\ldots\mu_6}\lambda\right) ,$$

$$\delta\lambda = \tfrac{1}{2} \not{\partial}\phi\, \varepsilon - \frac{1}{4.7!} e^{-\phi} \Gamma^{(7)} R(A)_{(7)} \varepsilon ,$$

$$\delta\phi = \tfrac{1}{2} \bar{\varepsilon}\lambda . \qquad (4.19)$$

In (4.19) we use the definition

$$R(A)_{\mu_1\ldots\mu_7} = 7\, \partial_{[\mu_1} A_{\mu_2\ldots\mu_7]} . \qquad (4.20)$$

From the above transformations it is easy to see how the supercurrent variation (4.2) is cancelled in (4.15) by the variation of the tensor field.

Another intriguing feature of the second version of d=10 supergravity becomes apparent when counting the off-shell degrees of freedom. With the rank-2 tensor field we have 45+1+36=82 bosonic and 144+16=160 fermionic field components (after subtracting gauge degrees of freedom), as is shown in table 4, while with a rank-6 tensor field we count 45+1+84=130 bosonic and, again, 160 fermionic field components. Let us compare these numbers to the number of physical states of the smallest massive N=1 supermultiplet. That multiplet consists of 128 bosonic and 128 fermionic states, decomposing into the **44**, **84** and **128** representations of SO(9). Therefore the theory with the rank-6 tensor field contains more field components than the number of states of a massive supermultiplet. It should therefore be possible to reduce this number somewhat and find an off-shell formulation of N=1 supergravity based on 128 + 128 degrees of freedom. Indeed, this turns out to be the case, and the resulting theory is conformal supergravity which was constructed in [13] (the first indication that this is possible comes from the analysis of the d=10 supercurrent multiplet [17]). We emphasize that, sofar, this situation where an on-shell formulation of Poincaré supergravity is based on more field components than the number of states of a massive supermultiplet, has only been encountered in 10 space-time dimensions. In all other cases one needs extra field components (auxiliary fields) to define an off-shell formulation of the theory. However, as we will discover shortly, the price that we have to pay is that the fields in d=10 conformal supergravity are subject to differential constraints.

In [13] it is extensively discussed how Poincaré supergravity follows from conformal supergravity in 10 dimensions. Here we will present things in opposite order and start from Poincaré supergravity. With a rank-6 tensor field the lagrangian of this theory takes the form (we set $\kappa=1$)

$$e^{-1}L_P = -\tfrac{1}{2}R - \tfrac{1}{2}\,\bar{\psi}_\mu\Gamma^{\mu\rho\sigma}D_\rho\psi_\sigma - \bar{\lambda}\not{D}\lambda - \frac{1}{7!}\,e^{-2\phi}\big(R(A)_{(7)}\big)^2 - (\partial_\mu\phi)^2$$

$$+\ \bar{\psi}_\mu\not{\partial}\phi\Gamma^\mu\lambda$$

$$+\ \frac{1}{4.7!}\,e^{-\phi}R(A)_{(7)}\big(\bar{\psi}_\mu\Gamma^{[\mu}\Gamma^{(7)}\Gamma^{\nu]}\psi_\nu + 2\bar{\psi}_\mu\Gamma^{(7)}\Gamma^\mu\lambda\big)$$

$$+\ \text{quartic fermion terms}, \qquad (4.21)$$

The action corresponding to (4.21) is invariant under supersymmetry transformations, which coincide with the transformations listed in (4.19) up to a fermionic bilinear in the variations of ψ_μ and λ. To be precise, the supersymmetry transformations of ψ_μ and λ corresponding to the combined Einstein-Yang-Mills lagrangian (defined as the sum of (4.15) and (4.21)) contain terms proportional to $Tr(\bar{\chi}\Gamma_{\mu\nu\rho}\chi)$, which are not present in the supersymmetry transformations relevant for pure Einstein supergravity. The latter, which henceforth will be called Q-supersymmetry transformations, can be derived entirely from off-shell arguments without the need to refer to a specific invariant action. This will be explained shortly. We should add that the Q-supersymmetry transformations can also be defined for the super-Yang-Mills fields and turn out to coincide entirely with the variations of χ and A_μ of Einstein-Yang-Mills supergravity, indicated in (4.18).

As explained above, the lagrangian (4.21) is based on 130 bosonic and 160 fermionic field components. To reduce these numbers we now introduce local dilatations (denoted by D) and S-supersymmetry (sometimes called conformal supersymmetry) transformations on the fields:

$$\delta e_\mu{}^a = -\Lambda_D e_\mu{}^a \quad , \qquad \delta\psi_\mu = -\tfrac{1}{2}\,\Lambda_D\psi_\mu - \Gamma_\mu\eta\ ,$$

$$\delta\lambda = \tfrac{1}{2}\,\Lambda_D\lambda + 6\eta \quad , \qquad \delta\phi = 6\Lambda_D\ ,$$

$$\delta\chi = \frac{3}{2}\,\Lambda_D\chi\ , \qquad (4.22)$$

where Λ_D and η are the infinitesimal parameters of these transformations. In (4.22) we have also included the variation of the Yang-Mills spinors χ. Note that the vector and tensor gauge fields A_μ and $A_{\mu\nu\rho\sigma\lambda\tau}$ are inert under D and S transformations.

The action of Poincaré supergravity is not invariant under dilatations and S supersymmetry; one finds

$$\delta S_P = \int d^{10}x\ e\{72\Lambda_D(C - \tfrac{1}{4}\ \bar{\phi}.\Gamma\Psi) + 36\bar{\eta}\Psi\} \quad , \tag{4.23}$$

where

$$\Psi = -\frac{1}{3}\ e^{-2\phi/3}\ \not{D}(e^{2\phi/3}\lambda) + \frac{1}{18.7!}\ \Gamma^{(7)}\lambda\ e^{-\phi}\ R(A)_{(7)} \quad , \tag{4.24}$$

and

$$C = e^{-2\phi/3}\{\frac{1}{4}\ D^2\ e^{2\phi/3} + \frac{1}{9}\ \bar{\lambda}\not{D}(e^{2\phi/3}\lambda)\} + \frac{1}{9.7!}\ e^{-2\phi}\left(R(A)_{(7)}\right)^2 \ , \tag{4.25}$$

The derivatives D_μ in (4.24) and (4.25) are fully covariant with respect to all superconformal symmetries. Observe that the covariant Dirac operator in (4.24) acts on $\exp(2\phi/3)\lambda$, which is a spinor with Weyl weight (i.e. scale dimension) $w=9/2$, whereas the D'Alembertian D^2 in (4.25) acts on $\exp(2\phi/3)$, which is a scalar with weight $w=4$. This is necessary for Ψ and C to transform covariantly under dilatations. It is important to realize that, because of S covariance, $\not{D}[\exp(2\phi/3)\lambda]$ contains a term $1/3 \exp(2\phi/3)\ \Gamma^{\mu\nu}\psi_{\mu\nu}$, where $\psi_{\mu\nu}$ is the Rarita-Schwinger field strength $\psi_{\mu\nu} = \partial_\mu\psi_\nu - \partial_\nu\psi_\mu$, while, because of D covariance, $D^2\exp(2\phi/3)$ contains a term $2/9\ R\exp(2\phi/3)$, where R is the Ricci scalar (the above results are all taken from [13] in a special gauge: $b_\mu = 0$).

An important observation is that the variations of Ψ and C under superconformal transformations remain proportional to Ψ and C, as is obvious from the transformation rules under dilatations, Q and S supersymmetry (cf. eq. (6.6) of [13]):

$$\delta\Psi = - C\varepsilon - \frac{17}{32}\,\varepsilon(\bar{\lambda}\Psi) - \frac{3}{64}\,\Gamma^{(2)}\varepsilon\,(\bar{\lambda}\Gamma_{(2)}\Psi)$$

$$+ \frac{1}{768}\,\Gamma^{(4)}\varepsilon\,(\bar{\lambda}\Gamma_{(4)}\Psi) + \frac{3}{2}\,\Lambda_D\Psi\,, \qquad (4.26a)$$

$$\delta C = - \frac{1}{4}\,\bar{\varepsilon}\not{D}\Psi - \frac{1}{2.8!}\,\bar{\varepsilon}\Gamma^{(7)}\Psi\, e^{-\phi}R(A)_{(7)}$$

$$+ \frac{1}{512}\,\bar{\varepsilon}\Gamma^{(3)}\Psi\,(\bar{\lambda}\Gamma_{(3)}\lambda) - \frac{3}{2}\,\bar{\eta}\Psi + 2\Lambda_D C\,, \qquad (4.26b)$$

where Λ_D, ε and η are the infinitesimal parameters of D, Q and S transformations, respectively.

According to (4.26) we can consistently put C and Ψ to zero, which reduces the original number of 130 + 160 degrees of freedom to 129 + 144. In addition, introducing the superconformal gauge symmetries (4.22) causes a further reduction by an equal amount. Therefore imposing $C = \Psi = 0$ and introducing local D and S symmetries leaves us with precisely 128 + 128 independent degrees of freedom. As these numbers coincide with the numbers of bosonic and fermionic states of a massive supermultiplet, it is plausible that the resulting field configuration defines an off-shell multiplet. It was shown in [13] that this is indeed the case: all superconformal transformations close if we impose the condition $\Psi = 0$. After appropriate gauge choices and linearizing the constraint equations $C = \Psi = 0$, one can show that their solutions are not restricted to be massless!

Conformal supergravity in 10 dimensions is thus defined in terms of the fields $e_\mu{}^a$, ψ_μ, $A_{\mu_1\ldots\mu_6}$, λ and ϕ, subject to the constraints $C = \Psi = 0$. Its gauge symmetries are general coordinate transformations, local Lorentz transformations, Q and S supersymmetry, dilations and tensor gauge transformations. Modulo field redefinitions, the full transformation rules as well as the explicit expressions for Ψ and C follow uniquely from requiring the closure of the superconformal transformations. To prove this, one starts from the linearized Q transformations, and proceeds by iteration. Therefore, conformal supergravity is defined in terms of a consistent field representation which is meaningful outside the context of a specific invariant action. The explicit form of the commutator algebra is rather complicated, and we refer to [13] for more details.

Because the super-Poincaré action is invariant under Q supersymmetry, the commutator of two Q transformations should decompose into the various symmetries of that theory, modulo terms that are proportional to the super-Poincaré field equations. On the other hand, from the above results we know that the $\{Q,Q\}$ commutator decomposes also into superconformal transformations, which do not all leave the super-Poincaré action invariant, up to terms proportional to Ψ (it turns out that C does not appear). This proves that Ψ and C must correspond to linear combinations of super-Poincaré field equations, as is indeed confirmed by (4.23). Furthermore the superconformal transformations in the $\{Q,Q\}$ commutator that do not leave the super-Poincaré action invariant, must have coefficients proportional to the super-Poincaré field equations. Explicit calculations have confirmed that this is the case [13].

Assuming the Q-supersymmetry transformation rule $\delta A_\mu = \frac{1}{2}\bar{\varepsilon}\Gamma_\mu\chi$, as given in (4.18), one can also implement the full superconformal algebra on A_μ, which requires the following Q variation of χ:

$$\delta_Q\chi = -\tfrac{1}{4}\Gamma^{\mu\nu}\varepsilon\left(\hat{F}_{\mu\nu} + \frac{7}{32}\,\bar{\lambda}\Gamma_{\mu\nu}\chi\right) + \frac{7}{64}\,\varepsilon\,(\bar{\lambda}\chi) - \frac{1}{1536}\,\Gamma^{\mu\nu\rho\sigma}\varepsilon\,\left(\bar{\lambda}\Gamma_{\mu\nu\rho\sigma}\chi\right), \tag{4.27}$$

where $\hat{F}_{\mu\nu}$ is the supercovariant field strength defined by

$$\hat{F}_{\mu\nu} = \partial_\mu A_\nu - \partial_\nu A_\mu - [A_\mu, A_\nu] - \bar{\phi}_{[\mu}\Gamma_{\nu]}\chi\ . \tag{4.28}$$

However, on χ the superconformal transformations close modulo an expression which turns out to correspond to the χ field equation that follows from the lagrangian (4.15). Therefore the super-Yang-Mills fields A_μ and χ define only an on-shell representation of the superconformal algebra. In view of this result it comes as no surprise that the action corresponding to (4.15) is invariant under <u>all</u> superconformal transformations, provided one imposes the constraint Ψ=0 on the supergravity fields. As we have already pointed out, the factor exp(ϕ) that multiplies the first line of (4.15) is crucial for D invariance. In the same fashion S supersymmetry becomes possible because the Noether current in (4.15) couples to an S invariant linear combination of ψ_μ and λ.

Hence, supersymmetric Yang-Mills theory couples only to the subset of supergravitational fields corresponding to conformal supergravity. If we refrain from imposing the constraint $\Psi=0$, then the action corresponding to (4.15) is no longer invariant under Q supersymmetry, and changes according to

$$\delta_Q S_{YM} = \tfrac{3}{64}\, g^{-2} \int d^{10}x\; e\; e^{\phi}\; \mathrm{Tr}(\bar{\chi}\Gamma^{\mu\nu\rho}\chi)\; \bar{\varepsilon}\Gamma_{\mu\nu\rho}\Psi \;, \tag{4.29}$$

Comparing (4.29) to (4.23) it is now easy to derive the supersymmetry transformations of Einstein-Yang-Mills supergravity, decomposed into superconformal transformations. Requiring that

$$\delta(\varepsilon)\left(S_P + S_{YM}\right) = 0 \quad , \tag{4.30}$$

one finds

$$\delta(\varepsilon) = \delta_Q(\varepsilon) + \delta_S(\eta) \quad , \tag{4.31}$$

where

$$\eta = -\frac{1}{768}\, g^{-2} e^{\phi}\; \bar{\chi}\Gamma^{\mu\nu\rho}\chi\; \Gamma_{\mu\nu\rho}\varepsilon \;. \tag{4.32}$$

For more general lagrangians than (4.15) and (4.21) it may be necessary to modify the transformation rules and the constraints $C = \Psi = 0$. Such lagrangians are useful for describing the interactions of the massless states of the superstring. The modifications to the Einstein-Yang-Mills lagrangian are then of higher orders in the string tension α' (this has been discussed extensively by many authors; some recent papers are given in [18]). It is not easy to construct complete and explicit expressions for the corresponding higher-order supersymmetric invariants (in fact, it is not always possible to do this for on-shell theories). A special class of interactions is characterized by the fact that the corresponding terms in the lagrangian are invariant under D and S. For instance, the Yang-Mills lagrangian (4.15) belongs to this class. In that case the constraints $C = \Psi = 0$ remain unmodified, simply because D and S invariance implies that the additional interactions depend only on the D and S invariant quantities $\exp(\phi/6)\, e_\mu{}^a$, $\exp(\phi/12)\,(\psi_\mu + \frac{1}{6}\Gamma_\mu\lambda)$, $\exp(-3\phi/2)\,\chi$, A_μ and $A_{\mu\nu\rho\sigma\lambda\tau}$. Therefore the additional terms do not affect the linear combinations of field equations that correspond to the constraints, so that we can employ standard off-shell arguments and conclude that the additional terms in the lagrangian are invariant under Q supersymmetry, order-by-order in α', up to terms proportional to Ψ and C (more detailed arguments indicate that this variation itself must be S invariant). The supersymmetry transformation corresponding to the full lagrangian is then again a linear combination of (field-dependent) superconformal transformations, in direct analogy with (4.31).

It is likely that the above remarks will apply to the supersymmetric completion of the term

$$L'' = \frac{i}{6!}\,\varepsilon^{\mu_1\ldots\mu_{10}}\, A_{\mu_1\ldots\mu_6}\, R^{ab}_{\mu_7\mu_8}(\omega)\, R^{ab}_{\mu_9\mu_{10}}(\omega) \quad , \tag{4.33}$$

where $R^{ab}_{\mu\nu}(\omega)$ is the curvature tensor corresponding to the spin connection field ω^{ab}_μ , in view of its close resemblance to the Ω-A coupling in the Yang-Mills lagrangian (observe that, after the duality transformation, (4.33) gives rise to a Chern-Simons term with respect to local Lorentz transformations, which is required for anomaly cancellation). Its invariance under tensor gauge transformations follows from the Bianchi identity for $R^{ab}_{\mu\nu}(\omega)$. To see whether (4.33) can also be D and S invariant, we write down the variation of (4.33)

$$\delta L'' = \frac{i}{7!}\,\varepsilon^{\mu_1\ldots\mu_{10}}\{7\,\delta A_{\mu_1\ldots\mu_6}\, R^{ab}_{\mu_7\mu_8}(\omega)\, R^{ab}_{\mu_9\mu_{10}}(\omega) + 4\,\delta\omega^{ab}_{\mu_1}\, R^{ab}_{\mu_2\mu_3}(\omega)\, R_{\mu_4\ldots\mu_7}(A)\} \tag{4.34}$$

As $A_{\mu\nu\rho\sigma\tau}$ is invariant under D and S, we concentrate on the second term. The local scale invariance of (4.33) follows if

$$R^{ab}_{[\mu\nu}(\omega)\, e_{\rho]a} = 0 \,, \tag{4.35}$$

which holds for a torsion-free connection ω_μ^{ab}. As this connection depends only on zehnbein derivatives, it is invariant under S supersymmetry. Hence we conclude that (4.33) is invariant under D, S and tensor gauge transformations, provided that ω_μ^{ab} contains no torsion. The next task is to prove that (4.33) can be completed to a Q-supersymmetric expression, once one imposes the constraints $C = \Psi = 0$. At that point one discovers the need for many new terms. Of particular interest are $R(\omega)\, R^2(A)$ terms, because they may form an obstacle in converting to the rank-2 tensor formulation by means of a duality transformation (the $R(\omega)\, R^2(A)$ terms emerge from the heterotic string as demonstrated in the last work of [18]). Results for the supersymmetric extension of (4.33) have been obtained both in the rank-6 [19] and the rank-2 formulation [20] (note that in the second work of [19] the $O(\alpha')$ modification of the transformation rules takes indeed the form of a field-dependent S transformation). We should add here that there are arguments indicating that superstring interactions in tree approximation are invariant under <u>constant</u> scale transformations [21].

In view of the differential nature of constraints $C = \Psi = 0$ the superconformal framework may be regarded as a partially off-shell formulation of Poincaré supergravity. As an intermediate step towards a fully off-shell formulation one may write the super-Poincaré lagrangian in a superconformally invariant form by introducing compensating fields A and ξ, which transform under Q, S and D as

$$\delta A = \bar{\varepsilon}\xi + 8\Lambda_D A$$

$$\delta\xi = \frac{1}{4}\,\not{\partial}A\varepsilon + \frac{5}{64}\,\Gamma^{(3)}\varepsilon\;\bar{\xi}\Gamma_{(3)}\lambda + \frac{21}{32}\,\Gamma^{a}\varepsilon\;\bar{\lambda}\Gamma_{a}\xi$$

$$- \frac{1}{2560}\,\Gamma^{(5)}\varepsilon\;\bar{\lambda}\Gamma_{(5)}\xi + 4A\eta + \frac{17}{2}\,\Lambda_D\xi \,. \tag{4.36}$$

The super-Poincaré lagrangian then takes the form

$$e^{-1}L_P = -9e\{A(C - \tfrac{1}{4}\bar{\phi}_\mu \Gamma^\mu \Psi) + \bar{\xi}\Psi\} , \tag{4.37}$$

which coincides with (4.21) after imposing the gauge conditions

$$A = 1 \quad , \xi = 0 . \tag{4.38}$$

The constraints $C = \Psi = 0$ can now be understood as the equations of motion corresponding to A and ξ, which act as Lagrange multipliers in (4.37). The form of (4.37) suggests an immediate generalization, in which one extends A and ξ to a full scalar multiplet S of Lagrange multipliers, whereas C and Ψ are regarded as the highest dimensional components of another scalar multiplet Φ. Together with the superconformal field representation Φ combines into an unconstrained off-shell multiplet. The lagrangian is then based on a product of the two corresponding scalar superfields,

$$L_P \propto \int d^{16}\theta \; S(x,\theta) \; \Phi(x,\theta) . \tag{4.39}$$

In the superconformal gauge (4.38) this lagrangian acquires the same structure as the linearized superspace lagrangian presented in [22]. From the extension (4.39) it is again possible to derive the decomposition rule (4.31), which is now induced by the gauge condition $\xi = 0$. In the presence of the auxiliary fields the gauge algebra will be modified [13]. As is well-known these modifications are directly related to the off-shell values of the superspace torsion tensor.

The total number of degrees of freedom has now increased substantially: in addition to the superconformal field configuration of 128 + 128 components, there are two scalar multiplets, each consisting of $2^{15} + 2^{15}$ components. To handle so many degrees of freedom obviously requires the use of superspace methods. Some recent papers on ten-dimensional superspace supergravity have been collected in [23].

4.3. Chiral N=2 supergravity

The supermultiplet that underlies chiral N=2 supergravity, shown in table 2, is described by complex chiral fermion fields ψ_μ and λ, satisfying

$$\Gamma^{(11)}\,\psi_\mu = \psi_\mu \quad , \quad \Gamma^{(11)}\,\lambda = -\lambda \ , \tag{4.40}$$

a zehnbein field $e_\mu{}^a$, a complex scalar field S, two real (or one complex) rank-2 gauge fields $A^\alpha_{\mu\nu}$ (α=1,2), and one rank-4 gauge field $A_{\mu\nu\rho\sigma}$, whose field strength is selfdual:

$$F_{\mu_1\ldots\mu_5} = \frac{ie}{5!}\,\varepsilon_{\mu_1\ldots\mu_5\nu_1\ldots\nu_5}\,F^{\nu_1\ldots\nu_5} \ , \tag{4.41}$$

where $e = \det e_\mu{}^a$. All these fields with the exception of $e_\mu{}^a$ and $A_{\mu\nu\rho\sigma}$ transform under rigid phase transformations, which coincide with the U(1) (or (SO(2)) automorphism of the supersymmetry algebra that rotates the two chiral supersymmetry charges (cf. section 2).

It is a common feature of many supergravity theories that the scalar fields are subject to a nonlinearly realized symmetry group. In the case at hand, one expects this group to be SU(1,1), because the complex plane is mathematically equivalent to the SU(1,1)/U(1) coset space. Indeed chiral N=2 supergravity turns out to exhibit this particular invariance group [5], and it is the aim of the subsequent discussion to elucidate some of its implications. We should mention that there are also other supergravity theories with SU(1,1) invariance, namely d=4, N=4 Poincaré and conformal supergravity [24,25].

Rather than with a complex scalar S, we start with a complex doublet field ϕ_α (α=1,2) satisfying the constraint

$$|\phi_1|^2 - |\phi_2|^2 = 1 \ . \tag{4.42}$$

This condition is invariant under phase transformations and SU(1,1) transformations. The latter are 2 × 2 matrices with unit determinant that satisfy

$$U^{-1} = \eta U^{\dagger} \eta \ , \tag{4.43}$$

where η is the diagonal matrix

$$\eta = \mathrm{diag}(1,-1) \ . \tag{4.44}$$

For further use it is convenient to introduce the notation

$$\phi^{\alpha} \equiv \eta^{\alpha\beta}(\phi_{\beta})^{*} = (\phi_1^{*}, -\phi_2^{*}) \ , \tag{4.45}$$

so that (4.42) takes the form $\phi_{\alpha}\phi^{\alpha} = 1$.

Variations of ϕ_{α} that commute with SU(1,1) can be written as

$$\delta\phi_{\alpha} = \Gamma \ \varepsilon_{\alpha\beta}\phi^{\beta} + i\Lambda \ \phi_{\alpha} \ , \tag{4.46}$$

or, alternatively,

$$\delta\phi^{\alpha} = - \Gamma^{*} \ \varepsilon^{\alpha\beta}\phi_{\beta} - i\Lambda \ \phi^{\alpha} \ , \tag{4.47}$$

where $\varepsilon_{\alpha\beta}$ is the rank-2 Levi-Civita tensor, and Λ is real. To clarify this result we combine ϕ_{α} and $\varepsilon_{\alpha\beta} \ \phi^{\beta}$ into an SU(1,1) matrix Φ,

$$\Phi \equiv \begin{pmatrix} \phi_1 & \phi_2^{*} \\ \phi_2 & \phi_1^{*} \end{pmatrix} \ , \tag{4.48}$$

and observe that (4.46) corresponds to an infinitesimal SU(1,1) transformation of Φ acting from the right. This in contradistinction with the SU(1,1) transformations on ϕ_{α}, which induce an SU(1,1) transformation on Φ from the left.

We now consider classes of ϕ_{α} that are equivalent up to an overall phase factor. This can be done by introducing a U(1) local gauge invariance

$$\phi_{\alpha} \rightarrow \phi'_{\alpha} = \exp(-i\Lambda) \ \phi_{\alpha} \tag{4.49}$$

so that ϕ_α and ϕ'_α are gauge equivalent. The SU(1,1) elements Φ, defined in (4.52), are then gauge equivalent if they differ by a (local) U(1) subgroup of SU(1,1) multiplied from the right, i.e.

$$\Phi' = \Phi \begin{pmatrix} e^{-i\Lambda} & 0 \\ 0 & e^{i\Lambda} \end{pmatrix} . \tag{4.50}$$

Such equivalence classes are called cosets, and a set of gauge inequivalent elements defines a parametrization of the coset space, denoted by SU(1,1)/U(1). Choosing in a parametrization amounts to imposing a gauge condition; a convenient condition is to choose ϕ_1 real, i.e.

$$\phi_1 = \phi_1^* . \tag{4.51a}$$

Denoting ϕ_2/ϕ_1 by S, we thus find

$$\phi_1 = \frac{1}{\sqrt{1-|S|^2}} , \quad \phi_2 = \frac{S}{\sqrt{1-|S|^2}} , \tag{4.52b}$$

and

$$\Phi = \frac{1}{\sqrt{1-|S|^2}} \begin{pmatrix} 1 & S^* \\ S & 1 \end{pmatrix} , \tag{4.51c}$$

where S is a complex field which takes its values in the unit disc ($|S| \leqslant 1$).

Let us now consider the form that supersymmetry and SU(1,1) transformations take in the gauge (4.51). Assuming that SU(1,1) commutes with supersymmetry transformations, the latter must take the form (4.46) with Γ independent of ϕ_α. Because we assume that ϕ_α is separately invariant under local phase transformations we discard the second term. Under supersymmetry ϕ_α must transform into the complex spinor λ: modulo redefinitions of ε and λ (not involving the scalar fields ϕ_α) the unique supersymmetry variation is therefore

$$\delta\phi_\alpha = - \bar{\varepsilon}\lambda \, \varepsilon_{\alpha\beta}\phi^\beta . \tag{4.52}$$

In the gauge (4.51) one substitutes (4.51b) and, to ensure that

$\delta\phi_1 = \delta\phi_1^*$, one includes a compensating field-dependent U(1) transformation. The parameter of this compensating transformation turns out to be equal to

$$\Lambda = \mathrm{Im}(\bar{\varepsilon}\lambda S^*) \ , \tag{4.53}$$

so that the supersymmetry transformation of S equals

$$\delta S = (1-|S|^2)\ \bar{\varepsilon}\lambda \ . \tag{4.54}$$

Infinitesimal SU(1,1) transformations on ϕ_α read

$$\delta\begin{pmatrix}\phi_1\\ \phi_2\end{pmatrix} = \begin{pmatrix} i\zeta & \gamma^* \\ \gamma & -i\zeta\end{pmatrix}\begin{pmatrix}\phi_1\\ \phi_2\end{pmatrix} \ , \tag{4.55}$$

where $\zeta = \zeta^*$. Again, to maintain the gauge (4.51), it is necessary to include a compensating U(1) transformation, now with parameter

$$\Lambda = \mathrm{Im}(i\zeta + \gamma^* S) \ . \tag{4.56}$$

Taking this into account the SU(1,1) variation of S becomes

$$\delta S = 2i\zeta S + \gamma(1-|S|^2) \ . \tag{4.57}$$

The rigid phase transformation depending on the parameter ζ is relevant for characterizing the physical states, and corresponds to the automorphism of the N=2 supersymmetry algebra. We shall return to this aspect shortly. The transformation depending on γ contains an inhomogeneous term, and the corresponding invariance will therefore not reflect itself in the spectrum.

The presence of a local U(1) symmetry (4.49) in the formulation based on ϕ_α requires a corresponding gauge field. There is a standard way of defining such fields for a coset space. Namely, one decomposes $\Phi^{-1}\partial_\mu\Phi$ into the generators of the SU(1,1) Lie algebra,

$$\Phi^{-1}\partial_\mu\Phi = \begin{pmatrix} -iQ_\mu & P^*_\mu \\ P_\mu & iQ_\mu \end{pmatrix} , \tag{4.58}$$

where

$$Q_\mu = -i\phi_\alpha\partial_\mu\phi^\alpha ,$$

$$P_\mu = \varepsilon^{\alpha\beta}\phi_\alpha\partial_\mu\phi_\beta ,$$

$$P^*_\mu = -\varepsilon_{\alpha\beta}\phi^\alpha\partial_\mu\phi^\beta . \tag{4.59}$$

Under local U(1) transformations (4.49) Q_μ transforms as a gauge field, so that we can define an U(1) covariant derivative

$$D_\mu\phi_\alpha = (\partial_\mu + iQ_\mu)\phi_\alpha . \tag{4.60}$$

With the help of this covariant derivative the definitions (4.59) can be written in covariant form

$$\phi_\alpha D_\mu\phi^\alpha = 0 ,$$

$$P_\mu = \varepsilon^{\alpha\beta}\phi_\alpha D_\mu\phi_\beta ,$$

$$P^*_\mu = - \varepsilon_{\alpha\beta}\phi^\alpha D_\mu\phi^\beta . \tag{4.61}$$

Applying a second derivative on the above relations, one can derive the Cartan-Maurer equations associated with the SU(1,1)/U(1) coset space

$$\partial_\mu Q_\nu - \partial_\nu Q_\mu = -i(P^*_\mu P_\nu - P^*_\nu P_\mu) ,$$

$$D_\mu P_\nu - D_\nu P_\mu = 0 . \tag{4.62}$$

It is now straightforward to write a U(1) × SU(1,1) invariant action for the scalar field, namely

$$L = \tfrac{1}{2}\mathrm{Tr}\left(D_\mu \Phi D^\mu \Phi^{-1}\right) , \tag{4.63}$$

which can be expressed as

$$L = - |P_\mu|^2 \tag{4.64}$$

by virtue of (4.58). In the gauge (4.51), we have

$$P_\mu = \frac{\partial_\mu S}{1-|S|^2} , \tag{4.65}$$

so that (4.64) reads

$$L = - \frac{|\partial_\mu S|^2}{\left(1-|S|^2\right)^2} . \tag{4.66}$$

This is a nonlinear sigma model lagrangian associated with a noncompact version of the one-dimensional complex projective space CP^1.

There are other fields in chiral N=2 supergravity that transform under SU(1,1), namely the tensor gauge fields $A^\alpha_{\mu\nu}$. These fields satisfy an SU(1,1) invariant reality condition (unlike for SU(2) it is possible to have a real doublet)

$$A^\alpha_{\mu\nu} = \varepsilon^{\alpha\beta} A_{\mu\nu\beta} , \tag{4.67}$$

where we use the same convention for raising and lowering of SU(1,1) indices as for the scalar doublet ϕ_α (cf. (4.45)). The reality condition (4.67) implies that

$$A^\alpha_{\mu\nu} A_{\mu\nu\alpha} = 0 , \tag{4.68}$$

so that the obvious SU(1,1) invariant action quadratic in $\partial_{[\mu} A^\alpha_{\nu\rho]}$ has no content. Another important aspect is that $A^\alpha_{\mu\nu}$ cannot transform under the local U(1) group for two reasons: for a real doublet such transformations cannot commute with SU(1,1), and local U(1) (or SO(2)) transformations will not commute with the tensor gauge transformations. In terms of separate components the SU(1,1) transformations of $A^\alpha_{\mu\nu}$ takes the form (cf. (4.55))

$$\delta A_{\mu\nu 1} = i\zeta\, A_{\mu\nu 1} + \gamma^*\, A_{\mu\nu 2}\ ,$$

$$\delta A_{\mu\nu 2} = -i\zeta\, A_{\mu\nu 2} + \gamma\, A_{\mu\nu 1}\ , \qquad (4.69)$$

which is indeed consistent with (4.67): if $A_{\mu\nu 2} = A^*_{\mu\nu 1}$ one has also $\delta A_{\mu\nu 2} = \delta A^*_{\mu\nu 1}$. As there are no compensating U(1) transformations acting on $A^\alpha_{\mu\nu}$, the result (4.69) remains unchanged in the gauge (4.51).

SU(1,1) invariance implies that the supersymmetry variation of $A^\alpha_{\mu\nu}$ must also transform as a real doublet. Therefore the variation must be proportional to ϕ^α or $\varepsilon^{\alpha\beta}\phi_\beta$, such that the resulting expression is real. Indeed one derives

$$\delta A^\alpha_{\mu\nu} = \phi^\alpha\{4\bar\varepsilon\Gamma_{[\mu}\psi^*_{\nu]} + \bar\varepsilon^*\Gamma_{\mu\nu}\lambda\} + \varepsilon^{\alpha\beta}\phi_\beta\{4\bar\varepsilon^*\Gamma_{[\mu}\psi_{\nu]} + \bar\varepsilon\Gamma_{\mu\nu}\lambda^*\}\ . \qquad (4.70)$$

This result was first derived in the second work of [5]; the conventions used here are as in [26].

To construct an SU(1,1) invariant action for $A^\alpha_{\mu\nu}$, we first contract the field strength $3\,\partial_{[\mu}A^\alpha_{\nu\rho]}$ with ϕ_α,

$$G_{\mu\nu\rho} = 3\,\phi_\alpha\,\partial_{[\mu}A^\alpha_{\nu\rho]}\ ,$$

$$G^*_{\mu\nu\rho} = -3\varepsilon_{\alpha\beta}\,\phi^\alpha\,\partial_{[\mu}A^\beta_{\nu\rho]}\ . \qquad (4.71)$$

The SU(1,1) invariant tensor $G_{\mu\nu\beta}$ and its complex conjugate satisfy the algebraic relation

$$\phi^\alpha G_{\mu\nu\rho} + \varepsilon^{\alpha\beta}\,\phi_\beta\, G^*_{\mu\nu\rho} = 3\,\partial_{[\mu}A^\alpha_{\nu\rho]}\ . \qquad (4.72)$$

In the gauge (4.51) one easily derives

$$G_{\mu\nu\rho} = \frac{3}{\sqrt{1-|S|^2}}\{\partial_{[\mu}A^*_{\nu\rho]1} - S\,\partial_{[\mu}A_{\nu\rho]1}\}\ . \qquad (4.73)$$

The SU(1,1) invariant action

$$L = -\frac{1}{3}|G_{\mu\nu\rho}|^2 \tag{4.74}$$

thus contains a kinetic term for the tensor field as well as nonlinear interactions with the scalar field.

Finally we draw attention to the way in which the U(1) subgroup of SU(1,1) acts in the gauge (4.51). Comparing (4.57) and (4.69) we see that, in obvious units, the scalar S has U(1) charge $|q| = 2$, while the tensor has $|q| = 1$. The two remaining boson fields, $e_\mu{}^a$ and $A_{\mu\nu\rho\sigma}$ are inert under both SU(1,1) and local U(1), so that they will have $q = 0$. The fermions are invariant under SU(1,1) but may transform under local U(1). Comparing both sides of the supersymmetry variations (4.54) and (4.70) shows that local U(1) must act according to

$$\psi_\mu \to \exp(\tfrac{1}{2}i\Lambda)\,\psi_\mu ,$$

$$\lambda \to \exp(-\tfrac{3}{2}i\Lambda)\,\lambda . \tag{4.75}$$

Therefore compensating U(1) transformations lead to $|q| = 1/2$ for ψ_μ and $|q| = 3/2$ for λ. One can show that this assignment pattern $q = -2$, $-3/2$, -1, $-1/2$, 0, 1/2, 1, 3/2, 2, coincides with the U(1) charges of the states belonging to the chiral N=2 supermultiplet shown in table 2.

We thank M. de Roo, O. Foda and D.J. Smit for useful discussions.

REFERENCES

1. P. van Nieuwenhuizen, in "Supergravity '81", p. 151, Eds. S. Ferrara and J.G. Taylor (Cambridge Univ. Press).

2. B. de Wit, in "Supersymmetry and Supergravity '84", p. 49, Eds. B. de Wit, P. Fayet and P. van Nieuwenhuizen (World Scientific).
3. R. Slansky, Phys. Rep. 79 (1981) 1.
4. L. Brink, J. Scherk and J.H. Schwarz, Nucl. Phys. B121 (1977) 77.
F. Gliozzi, J. Scherk and D. Olive, Nucl. Phys. B122 (1977) 253.
5. M.B. Green and J.H. Schwarz, Phys. Lett. 122B (1983) 143;
J.H. Schwarz and P.C. West, Phys. Lett. 126B (1983) 301;
J.H. Schwarz, Nucl. Phys. B226 (1983) 269;
P.S. Howe and P.C. West, Nucl. Phys. B238 (1984) 181.
6. E. Cremmer, B. Julia and J. Scherk, Phys. Lett. 76B (1978) 409.
7. S. Deser, R. Jackiw and G. 't Hooft, Ann. Phys. 152 (1984) 220.
8. A.H. Chamseddine, Nucl. Phys. B185 (1981) 403.
9. E. Bergshoeff, M. de Roo, B. de Wit and P. van Nieuwenhuizen, Nucl. Phys. B195 (1982) 97.
10. P. van Nieuwenhuizen, Phys. Rep. 68 (1981) 189.
11. G.F. Chapline and N.S. Manton, Phys. Lett. 120B (1983) 109.
12. A.H. Chamseddine, Phys. Rev. 24D (1981) 3065.
13. E. Bergshoeff, M. de Roo and B. de Wit, Nucl. Phys. B217 (1983) 489.
14. L.J. Romans, Phys. Lett. 169B (1986) 374.
15. M.B. Green and J.H. Schwarz, Phys. Lett. 149B (1984) 117.
16. S.J. Gates and H. Nishino, Phys. Lett. 157B (1985) 157.
17. E. Bergshoeff and M. de Roo, Phys. Lett. 112B (1982) 53.
18. P. Candelas, G.T. Horowitz, A. Strominger and E. Witten, Nucl. Phys. B258 (1985) 46.
B. Zwiebach, Phys. Lett. 156B (1985) 315.
E.S. Fradkin and A.A. Tseytlin, Phys. Lett. 158B (1985) 316; 160B (1985) 69; Nucl. Phys. B261 (1985) 1.
A.A. Tseytlin, Nucl. Phys. B276 (1986) 391.
D.J. Gross, J. Harvey, E. Martinec and R. Rohm, Nucl. Phys. B256 (1985) 253; B267 (1986) 75.
C.G. Callan, D. Friedan, E. Martinec and M.J. Perry, Nucl. Phys. B262 (1985) 593.
C.G. Callan, I.R. Klebanov and M.J. Perry, Nucl. Phys. B278 (1986) 78.

M.T. Grisaru, A. van de Ven and D. Zanon, Phys. Lett. 173B (1986) 423; Nucl. Phys. B277 (1986) 388, 409.

M.D. Freeman and C.N. Pope, Phys. Lett. 174B (1986) 48.

M.D. Freeman, C.N. Pope, M.F. Sohnius and K.S. Stelle, Phys. Lett. 178B (1986) 199.

D.J. Gross and E. Witten, Nucl. Phys. 277B (1986) 1.

S.J. Gates and H. Nishino, Phys. Lett. 173B (1986) 52.

P. Candelas, M.D. Freeman, C.N. Pope, M.J. Sohnius and K.S. Stelle, Phys. Lett. 177B (1986) 341.

D. Chang and H. Nishino, Maryland preprint PPN 86-178.

19. E. Bergshoeff, A. Salam and E. Sezgin, Nucl. Phys. B279 (1987) 659.

S.J. Gates and H. Nishino, Phys. Lett. 173B (1986) 52.

20. L.J. Romans and N.P. Warner, Nucl. Phys. B273 (1986) 320.

S.K. Han, J.K. Kim, I.G. Koh and Y. Tanii, preprint KAIST-85/22 (1985).

21. E. Witten, Phys. Lett. 155B (1985) 151.

R.E. Kallosh, Phys. Lett. 159B (1985) 111.

22. P. Howe, H. Nicolai and A. Van Proeyen, Phys. Lett. 112B (1982) 446.

23. R.E. Kallosh and B.E.W. Nilsson, Phys. Lett. 167B (1986) 47.

B.E.W. Nilsson and A.K. Tollstén, Phys. Lett. 169B (1986) 369, 171B (1986) 212.

B.E.W. Nilsson, Phys. Lett. 175B (1986) 319.

J.J. Atick, A. Dhar and B. Ratra, SLAC preprint-PUB 3839 (1985).

P.S. Howe and A. Umerski, Phys.Lett. 177B (1986) 163

24. E. Cremmer, J. Scherk and S. Ferrara, Phys. Lett. 74B (1978) 61.

25. E. Bergshoeff, M. de Roo and B. de Wit, Nucl. Phys. B182 (1981) 173.

26. B. de Wit, D.J. Smit, N.D. Hari Dass, Nucl. Phys. B, to appear.

SUPERSPACE AND SUPERSYMMETRY ON THE WORLD SHEET

M. T. Grisaru

Physics Department, Brandeis University
Waltham, MA 02254
USA

ABSTRACT

Supersymmetry and supergravity for two space-time dimensions are described in a superspace setting.

I. INTRODUCTION

A few years ago, when four-dimensional supersymmetry was beginning to gain wide acceptance and serious investigations of higher dimensional supersymmetric theories were beginning, S. J. Gates, M. Rocek, W. Siegel and myself wrote a book on superspace and superfield methods in four dimensions[1]. The book begins with what we considered to be a simple example, "A Toy Superspace", for describing supersymmetry in three-dimensional space-time. At the time, it was just a toy, and it did not even occur to us to descend to two dimensions. What would have been the point? Nowadays, superstrings have changed all that, and superspace provides a nice framework for dealing with some aspects of two-dimensional supersymmetry on the world sheet.

As in higher dimensions, one can do <u>component</u> supersymmetry and supergravity and many people are more comfortable with this formalism. However, superspace does provide a compact notation and has many formal advantages, and at the quantum level it is by far superior (viz. for example, recent four-loop calculations of the β-function in supersymmetric σ-models)[2].

The subject has exploded recently, and one cannot really give a complete account and at the same time attempt to be brief. Besides, a complete account would be pointless because much of the material is a mere replica of the four-dimensional theories. (In fact, it is almost identical to the three-dimensional theory mentioned above.) Therefore, in these notes, I shall concentrate on aspects of two-dimensional supersymmetry and superspace which are somewhat different from anything in higher dimensions. I have in mind (p,q) supersymmetry in particular. However, I will begin with a brief account of standard N=1 supersymmetry. For more details, the reader may refer to my lectures at the School in past years.[3] Although the treatment there was exclusively in four dimensions, the general philosophy and many of the technical details go through unchanged.

My superspace conventions are essentially identical to those of three-dimensions, as described in "Superspace" (see also [4]). In two dimensions the irreducible spinor representations of the Lorentz group are one-dimensional, carried by Weyl (left or right handed spinors χ^+, χ^-. A Dirac spinor is a two-component object

$$\psi^\alpha = \begin{pmatrix} \chi^+ \\ \chi^- \end{pmatrix} \tag{1.1}$$

in a basis where the γ-matrices are

$$(\gamma^0)_\alpha{}^\beta = \begin{pmatrix} 0 & i \\ i & 0 \end{pmatrix} \qquad \gamma^1 = \begin{pmatrix} 0 & -i \\ +i & 0 \end{pmatrix} \qquad \gamma^{"5"} = \begin{pmatrix} 1 & \\ & -1 \end{pmatrix} \tag{1.2}$$

A Majorana spinor satisfies $\psi^* = \psi$. In two dimensions a spinor can be both Majorana and Weyl (i.e., χ^+, χ^- can be real). Indices can be raised and lowered with the charge conjugation matrix $C_{\alpha\beta} = \sigma^2$:

$$\psi_\beta = \psi^\alpha C_{\alpha\beta} \tag{1.3}$$

We write

$$(\psi)^2 = \frac{1}{2}\, \psi^\alpha \psi_\alpha = \frac{1}{2}\, \psi^\alpha \psi^\beta C_{\beta\alpha} \; . \tag{1.4}$$

Vectors can be represented in spinor notation as

$$V_a \to V_{\alpha\beta} = (\gamma \cdot V)_{\alpha\beta} \tag{1.5}$$

and in the basis above this just gives light-cone components

$$(\gamma\cdot V)_{\alpha\beta} = \begin{pmatrix} V_0+V_1 & 0 \\ 0 & V_0-V_1 \end{pmatrix} = \begin{pmatrix} V_{++} & 0 \\ 0 & V_{--} \end{pmatrix} \tag{1.6}$$

II. POINCARE SUPERSYMMETRY

The super-Poincaré algebra is generated by P_a, J_{ab}, and spinorial charges Q^i_α, with

$$\{Q^i_\alpha, Q^j_\beta\} = \delta^{ij} P_{\alpha\beta} \qquad i,j = 1, \ldots N \tag{2.1}$$

for N-extended supersymmetry. However, if we denote by Q_+, Q_-, the supersymmetry charges of positive and negative chirality we can also define (p,q) supersymmetry[5]

$$\begin{aligned} \{Q^i_+, Q^j_+\} &= \delta^{ij} P_{++} \qquad & i,j &= 1 \ldots p \\ \{Q^{i'}_-, Q^{j'}_-\} &= \delta^{i'j'} P_{--} \qquad & i',j' &= 1 \ldots q \end{aligned} \tag{2.2}$$

In this notation ordinary supersymmetry is $p = q = N/2$.

The N=1 supersymmetry algebra can be represented on fields as follows:

a) Component representations, as transformations between bosons and fermions, e.g.,

$$\begin{aligned} \delta A(x) &= \varepsilon\psi(x) \\ \delta\psi &= \partial\!\!\!/ A\varepsilon + F\varepsilon \\ \delta F &= \varepsilon\partial\!\!\!/\psi \end{aligned} \tag{2.3}$$

In two dimensions we have similar representations for (p,q) supersymmetry, e.g. for (1,0)

$$\begin{aligned} \delta A &= \varepsilon_- \chi_+ \\ \partial\chi_+ &= \varepsilon_- \partial_{++} A \end{aligned} \tag{2.4}$$

or

$$\delta\psi_- = F\varepsilon_- \qquad \delta F = \varepsilon_-\partial_{++}\psi_- \tag{2.5}$$

for a "scalar" or "spinor" multiplet. In the second example F is an auxiliary field if ψ_- has canonical dimension; this is an example of a multiplet where "physical" bosonic and fermionic degrees of freedom do not balance.

b) Superspace representations on superfields, where the algebra acts as coordinate transformations in superspace, e.g., for N=1 supersymmetry

$$\Psi(x^a,\theta) \to \Psi(x^a-\varepsilon\gamma^a\theta,\ \theta+\varepsilon) \tag{2.6}$$

Here the spinorial coordinates θ are two-component Grassman variables. On N=1 superfields the supersymmetry generators are represented by

$$Q_\alpha = \frac{\partial}{\partial\theta^\alpha} - \theta^\beta i\partial_{\beta\alpha}\ , \qquad \{Q_\alpha,Q_\beta\} = 2i\partial_{\alpha\beta}\ . \tag{2.7}$$

Covariant derivatives (with respect to supersymmetry) are

$$D_\alpha = \frac{\partial}{\partial\theta^\alpha} + \theta^\beta i\partial_{\beta\alpha} \qquad \{D_\alpha,D_\beta\} = 2i\partial_{\alpha\beta} \qquad \{D_\alpha,Q_\beta\} = 0 \tag{2.8}$$

In superspace (Berezin) integration is equivalent to differentiation:

$$\int d^2x\ d^2\theta\ (\ ) = \int d^2x\ D^2(\ )|_{\theta=0}\ . \tag{2.9}$$

The component representations can be obtained by expanding the superfields in a (terminating) Taylor series with respect to the θ's. A more efficient way is to define components "by projection." For example, for a scalar superfield Ψ we define the components by

$$A(x) = \Psi(x,\theta)|_{\theta=0} \qquad \chi_\alpha(x) = D_\alpha\Psi(x,\theta)|_{\theta=0} \qquad F(x) = D^2\Psi(x,\theta)|_{\theta=0} \tag{2.10}$$

These definitions make it very easy to find component actions from their superspace counterparts.

For a scalar N=1 superfield the conventional kinetic action is given by a superspace integral

$$S = \frac{1}{4} \int d^2x\ d^2\theta\ D^\alpha \Psi D_\alpha \Psi \tag{2.11}$$

Using the definition of the Berezin integral we have

$$\begin{aligned} S &= \frac{1}{4} \int d^2x\ D^2 (D^\alpha \psi D_\alpha \Psi) \\ &= \frac{1}{4} \int d^2x\ \frac{1}{2}\ D^\beta D_\beta (D^\alpha \psi D_\alpha \Psi) \\ &= \frac{1}{4} \int d^2x \left[-D^\beta D^\alpha \Psi D_\beta D_\alpha \Psi + 2 D^\alpha \Psi D^2 D_\alpha \Psi \right]_{\theta=0} \end{aligned} \tag{2.12}$$

Using

$$\begin{aligned} D_\alpha D_\beta &= i\partial_{\alpha\beta} + C_{\beta\alpha} D^2 \\ (D^2)^2 &= \Box \\ D^2 D_\alpha &= -i\partial_{\alpha\beta} D^\beta \end{aligned} \tag{2.13}$$

and the definition of components, we get

$$S = \int d^2x \left[-\frac{1}{2} A \Box A + \frac{1}{2} F^2 - i\chi^\alpha \partial_{\alpha\beta} \chi^\beta \right] \tag{2.14}$$

More complicated actions with N=1 supersymmetry are non-linear σ-models:

$$S = \frac{1}{4} \int d^2x\ d^2\theta\ g_{ij}(\Psi)\ D^\alpha \Psi^i D_\alpha \Psi^j \tag{2.15}$$

where g_{ij} is an arbitrary function of the superfields Ψ^i. When worked down to components it gives

$$S \sim \int d^2x\ g_{ij}(A) [\frac{1}{2}\ D^a A^i D_a A^j + \chi^i D \chi^j] + R_{ijk\ell} \chi^i \chi^j \chi^k \chi^\ell \tag{2.16}$$

where R is the curvature tensor and D_a is a covariant derivative associated with the metric g_{ij} of the manifold with coordinates A^i.

We can also define nonlinear σ-models with extended supersymmetry: In N=1 superspace, with the above σ-model action, one has manifest supersymmetry under

$$\delta\Psi^i = [\varepsilon Q, \Psi^i]$$

$$= \delta^i{}_j \varepsilon\left(\frac{\partial}{\partial\theta} - i\theta\partial\right)\Psi^j , \qquad \{Q,Q\} = i\partial . \tag{2.17}$$

One can look for additional supersymmetries: With the Ansatz

$$\delta\Psi_i = f^i{}_j[\eta Q, \Psi^j] \tag{2.18}$$

one finds this works provided $f^i{}_j$ is a complex structure on the manifold, i.e., a covariantly constant tensor satisfying

$$f^i{}_j f^j{}_k = -\delta^i{}_k \tag{2.19}$$

If only one such tensor exists we have N=2 supersymmetry and the manifold is a Kähler manifold. If there are three (two always imply three), one gets N=4 supersymmetry on a hyperkähler manifold (see, e.g., J. Bagger's Bonn Lectures 1984,[6] as well as the corresponding discussion of four dimensions in "Superspace".) For example, in the case $g_{ij} = \delta_{ij}$ (flat space) one can take a pair of real superfields ψ^1, Ψ^2 with

$$\delta_\varepsilon\Psi^i = \varepsilon\left(\frac{\partial}{\partial\theta} - \theta i\partial\right)\Psi^i \tag{2.20}$$

and

$$\begin{aligned}\delta_\eta\Psi^1 &= i\eta\left(\frac{\partial}{\partial\theta} + \theta i\partial\right)\Psi^2 \\ \delta_\eta\Psi^2 &= -i\eta\left(\frac{\partial}{\partial\theta} + \theta i\partial\right)\Psi^1\end{aligned} \tag{2.21}$$

so here the complex structure is just $i\varepsilon_{ij}$.

The same result can be expressed in N=2 superspace by introducing a second Majorana spinor coordinate θ' and combining θ, θ' into a Dirac spinor. So we have now $\theta,\bar{\theta}$ and charges $Q,\bar{Q}$, and derivatives $D,\bar{D}$. Now a complex superfield

$$\Phi(x,\theta,\bar{\theta}) \tag{2.22}$$

can be made to satisfy a chirality constraint

$$\bar{D}_\alpha\Phi = 0 \tag{2.23}$$

which is just like the four-dimensional chirality constraint. In fact such a superfield can be obtained by trivial dimensional reduction from four dimensions. It allows writing

$$\Phi(x,\theta,\bar{\theta}) = e^{i\bar{\theta}\gamma\partial\theta}\ \phi(x,\theta) \tag{2.24}$$

with

$$\phi(x,\theta) = A + \theta\chi + \theta^2 F \tag{2.25}$$

where A, χ, F are essentially the components of the complex N=1 superfield $\Psi^1 + i\Psi^2$.

In terms of N=2 superfields the general action is

$$S = \int d^2x d^2\theta d^2\bar{\theta}\ K(\Phi,\bar{\Phi}) \tag{2.26}$$

where K is the Kähler potential. The reduction to components proceeds as follows: First

$$\begin{aligned} S &= \frac{1}{4}\int d^2x d\theta^\alpha d\bar{\theta}^\beta\ D_\alpha \bar{D}_\beta K \\ &= \frac{1}{4}\int d^2x d\theta^\alpha d\bar{\theta}^\beta \frac{\partial^2 K}{\partial\phi^i \partial\bar{\phi}^j}\ D_\alpha\phi^i \bar{D}_\beta \bar{\phi}^j \end{aligned} \tag{2.27}$$

reduces it to the N=1 form (2.15) after suitable relabeling of the θ's, and with the identification of g_{ij} as the Kähler metric:

$$g_{ij} = \frac{\partial^2 K}{\partial\phi^i \partial\bar{\phi}^j} \tag{2.28}$$

Reduction to components is then as in the N = 1 case.

Recently, Gates, Hull and Rocek[4] have introduced the concept of a "twisted" chiral multiplet, satisfying a chirality constraint with respect to one component of D_α and one component of $\bar{D}_\beta$. It has a number of interesting features and, in a nonlinear σ-model context, a rich mathematical structure.

With this I conclude this brief discussion of ordinary global supersymmetry in two dimensions. As mentioned above, most features are an exact replica of its four-dimensional counterpart, and the formalism is almost identical to that of the three-dimensional theory. Indeed the superspace approach can be studied by reading "A Toy

Superspace"[1] and simply replacing the words "three-dimensional" by "two-dimensional." In particular, the quantization procedure in superspace is discussed there in detail. The resulting supergraph rules can be used in two dimensions without any change.[2]

Before leaving the subject, there is one result worth mentioning: It is well known that in four dimensions, if supersymmetry is not spontaneously broken at the tree level, it is not broken by radiative corrections. This is not the case in two dimensions.[7] Quantum corrections can lead to an effective potential which breaks supersymmetry. The difference between the two cases is related to the manner in which the effective action depends on the auxiliary fields. I refer the interested reader to the reference.

III. LOCAL SUPERSYMMETRY

If the spinorial parameter ε is made a function of x, invariance of actions requires introducing a spin 3/2 gravitino field $\psi_{m\alpha}$ that is the gauge field of local supersymmetry and its spin 2 supersymmetry parameter, the graviton g_{mn} (or vielbein $e_m^{\ a}$. In two dimensions, these fields carry no dynamics normally.

The local supersymmetry transformations take a form analogous to four-dimensions:

$$\delta e_m^{\ a} = \varepsilon\gamma^a\psi_m$$
$$\delta\psi_m = D_m\varepsilon + \ldots \qquad (3.1)$$

where D_m is a suitable (gravitationally) covariant derivative.

N = 1 supergravity also contains a scalar auxiliary field S. On the other hand (1,0) supergravity does not.

The superspace description of supergravity is via geometry. Local supersymmetry transformations are viewed as general coordinate transformations in superspace

$$z^M \rightarrow z'^M = f^M(z)$$
$$z^M = (x^m, \theta^\mu) \qquad (3.2)$$

One has the notion of tangent space and supervielbein

$$E_A^{\ M}(x,\theta) \tag{3.3}$$

and covariant derivatives

$$\nabla_A = E_A^{\ M} D_M + \phi_A \tag{3.4}$$

where ϕ_A are Lorentz connections. The derivatives transform as

$$\nabla_A \to \nabla'_A = e^K \nabla_A e^{-K}$$
$$K = K^M D_M \qquad K^M = K^M(x,\theta) \tag{3.5}$$

This corresponds to general coordinate transformations on superfields

$$\psi(x,\theta) \to \psi(x',\theta') = \psi'(x,\theta) = e^K \psi(x,\theta) \tag{3.6}$$

Field strengths (torsions and curvatures) are defined via

$$\{\nabla_A, \nabla_B\} = T_{AB}^{\ \ C}\nabla_C + R_{AB} \tag{3.7}$$

One may impose constraints to reduce the representation content of $E_A^{\ M}$ (which contains much too many fields). These constraints take the form of setting some of the curvature components to zero and some torsion components to zero (or equal to some invariant tensor, e.g., $T_{\alpha\beta}^{\ \ c} = \delta_{\alpha\beta}^{c}$). Together with the Bianchi identities they imply further constraints on the fields.

The constraints are used for expressing the connections in terms of derivatives of the vielbein (conventional constraints) and also impose additional relations between components of the vielbein $E_A^{\ M}$. These can be solved in a supersymmetric fashion by expressing the vielbein in terms of unconstrained superfields, the prepotentials. For example, for N=1 supergravity, the prepotentials are a vector-spinor $H_a^{\ \alpha}(x,\theta)$ and a scalar $\psi(x,\theta)$.[8] Similar sets exist for (p,q) supergravity.[9,10] The local supersymmetry transformations are superspace gauge transformations, e.g.,

$$\delta H_a^{\ \alpha} = \partial_a K^\alpha + \ldots \tag{3.8}$$

Finally, the constraints, together with the Bianchi identities show that all the torsions and curvatures are expressible in terms of a single field strength, a covariant superfield that contains the gravitino field strength and the scalar curvature among its components.

The details of the construction will be illustrated in the next section when we discuss (1,0) supergravity. However, the general approach is similar (though simpler) to that of four dimensions, and the reader may refer to my lectures at the preceding school for a detailed discussion.[3]

IV. (1,0) SUPERSYMMETRY AND SUPERGRAVITY

The material in this section is based largely on a paper by Brooks, Gates and Muhammad.[9] We use the following conventions for (1,0) superspace: The single Majorana-Weyl fermionic coordinate is denoted by θ^+. The bosonic coordinates are light-cone coordinates

$$\sigma^{\ddagger} = \frac{1}{\sqrt{2}}(\tau+\sigma) \qquad \sigma^{=} = -\frac{1}{\sqrt{2}}(\tau-\sigma) \tag{4.1}$$

where $(\ddagger) \equiv (++)$, $(=) = (--)$ indicates the Lorentz charge. The superspace derivatives are

$$D_M = (D_+, \partial_{\ddagger}, \partial_=)$$
$$D_+ = \frac{\partial}{\partial\theta^+} + i\theta^+\partial_{\ddagger} \tag{4.2}$$

so that

$$\{D_+, D_+\} = 2i\partial_{\ddagger} \qquad \text{i.e.,}\ D_+D_+ = i\partial_{\ddagger} \tag{4.3}$$

Typical superfields are:

A scalar

$$\Phi = A + \theta^+\psi_+ \tag{4.4}$$

A spinor

$$\Psi_- = \eta_- + i\theta^+F \tag{4.5}$$

(Note that η_- has the same Lorentz properties as θ^+. Raising and lowering of indices is done with $\varepsilon_{\alpha\beta} \sim \varepsilon_{+-}$).

Components are best defined by projection:[1]

$$A = \Phi|_{\theta=0} \qquad \psi_+ = D_+\Phi|_{\theta=0}$$
$$\eta_- = \Psi_-| \qquad iF = D_+\Psi_-| \tag{4.6}$$

Matter Actions

The integration measure is $d^2\sigma d\theta^-$, where also $d\theta^- \equiv \frac{\partial}{\partial\theta^+}$ for the Berezin integration. Hence Lagrangians must have the Lorentz structure $L^+ \equiv L_-$. We write actions as

$$\int d^2\sigma\, d\theta^-\, L_- = \int d^2\sigma\, D_+ L_- \big|_{\theta=0} \tag{4.7}$$

Here are some examples, together with the corresponding component actions[9]:

Scalar Multiplet σ-model = Heterotic right movers

$$\begin{aligned} S_\sigma &= \frac{i}{2}\int d^2\sigma d\theta^-\, g_{ij}(\Phi)\, D_+\Phi^i \partial_= \Phi^j \\ &= \int d^2\sigma\, D_+ [\qquad\qquad]\big|_{\theta=0} \\ &= \int d^2\sigma\, \{g_{ij}\,[\, -\frac{1}{2}\,\partial_{\ddagger} A^i \partial_= A^j - \psi_+{}^i \partial_= \psi_+{}^j\,] \\ &\quad + g_{ij,k}\, \psi_+^k\, \psi_+^i\, \partial_= A^j\,\} \end{aligned} \tag{4.8}$$

where

$$g_{ij,k} = \frac{\partial g_{ij}}{\partial A^k} \tag{4.9}$$

For $g_{ij} = \delta_{ij}$ we have a free scalar multiplet.

Dirac Action = Heterotic Left Movers

$$\begin{aligned} S_D &= \frac{1}{2}\int d^2\sigma d\theta^-\, G_{mn}(\Phi)\, \Psi_-{}^m D_+ \Psi_-{}^n \\ &= \frac{1}{2}\int d^2\sigma\, D_+ [\qquad\qquad]\big|_{\theta=0} \\ &= \frac{1}{2}\int d^2\sigma\, G_{mn}[-F^m F^n - \eta_-{}^m i\partial_{\ddagger}\eta_={}^n] \\ &\quad - iG_{mn},_k\, \psi_+{}^k \eta_-{}^m F^n \end{aligned} \tag{4.10}$$

When $G_{mn} = \delta_{mn}$ we get the free Dirac action with F as auxiliary field. One can of course write other interaction terms.

Vector Multiplet

It is described by introducing a superfield connection to gauge covariantize the superspace derivative:

$$\nabla_+ = D_+ - i\Gamma_+ \qquad \nabla_{\ddagger} = \partial_{\ddagger} - i\Gamma_{\ddagger} \qquad \nabla_= = \partial_= - i\Gamma_= \tag{4.11}$$

As usual one can impose contraints so as to reduce the representation: We expect components $A_{\ddagger}$, $A_=$ and λ_-, and component field strength $f_{\ddagger=} \sim f$, a scalar. These should reside in superfield strengths W_- and F and, checking dimensions, one is led almost uniquely to the constraints

$$\begin{aligned} \{\nabla_+, \nabla_+\} &= 2i\nabla_{\ddagger} \\ \{\nabla_+, \nabla_{\ddagger}\} &= 0 \end{aligned} \tag{4.12}$$

which, together with the Bianchi identities give

$$\begin{aligned} \{\nabla_+, \nabla_=\} &= iW_- \\ \{\nabla_{\ddagger}, \nabla_=\} &= -iF \end{aligned} \tag{4.13}$$

as well as

$$\nabla_+ W_- = iF \qquad \nabla_+ F = \nabla_{\ddagger} W_- \ . \tag{4.14}$$

Therefore Γ_+, $\Gamma_=$ are independent superfields with the component expansions

$$\begin{aligned} \Gamma_+ &= \rho_+ + i\theta^+ A_{\ddagger} \\ \Gamma_= &= A_= + \theta^+(\lambda_- + \nabla_= \rho_+) \end{aligned} \tag{4.15}$$

while

$$\Gamma_{\ddagger} = -D_+\Gamma_+ - \frac{1}{2}\{\Gamma_+, \Gamma_+\} \tag{4.16}$$

As usual, the gauge transformations are ($\nabla_A \rightarrow \nabla'_A = e^{-iK}\nabla_A e^{iK}$)

$$\delta\Gamma_A = D_A K \qquad K = \Lambda + \theta^+ k_+ \ . \tag{4.17}$$

Therefore at the component level

$$\delta\rho_+ = D_+ K| = k_+ \tag{4.18}$$

can be used to gauge ρ_+ away (W-Z gauge) while

$$\delta A_{\ddagger} = -i\, D_+ D_+ K| = \partial_{\ddagger}\Lambda \tag{4.19}$$

the usual gauge transformation. The action is

$$\begin{aligned} S &= \frac{-i}{2}\int d^2\sigma d\theta^-\ FW_- = \frac{i}{2}\int d^2\sigma \nabla_+(FW_-)| \\ &= \frac{-1}{2}\int d^2\sigma\ \left[i(\nabla_{\ddagger}\lambda_-)\lambda_- + f^2\right] \end{aligned} \tag{4.20}$$

Supergravity

At the component level, in a Wess-Zumino gauge, we expect to describe the gravitino $\psi_{\ddagger}^{+}$, $\psi_{=}^{+}$, and the graviton, including its trace, $h_{\ddagger}^{=}$, $h_{=}^{\ddagger}$, h. Furthermore we expect two field strengths, the graviton curvature $r_{\ddagger=}$ and gravitino field strength $\psi_{\ddagger,=}^{+} - \psi^{+}{}_{=,\ddagger}$. We expect them to be $\theta=0$ components of superspace field strengths defined by commutators of covariant derivatives.

We start with

$$\nabla_A = E_A^{\ M} D_M + \phi_A\ , \qquad \phi_A = \omega_A M\ , \qquad M = \text{Lorentz operators} \tag{4.21}$$

and

$$[\nabla_A, \nabla_B] = T_{AB}{}^C \nabla_C + R_{AB}. \tag{4.22}$$

We impose constraints:[9]

$$\{\nabla_+, \nabla_+\} = 2i\nabla_{\ddagger} \tag{4.23}$$

generalizing flat superspace, and

$$[\nabla_+, \nabla_{\ddagger}] = 0 \tag{4.24}$$

because there is no component field strength that could be part of the right-hand side, and

$$[\nabla_+, \nabla_=] = -2i\Sigma^+ M \tag{4.25}$$

where Σ^+ could contain $\psi^+{}_{\ddagger=}$ as first component. Finally

$$[\nabla_{\ddagger}, \nabla_=] = -i(\Sigma^+\nabla_+ + RM) \tag{4.26}$$

R is where $r_{\ddagger=}$ could appear.

One checks that this is consistent with the Bianchi (Jacobi) identities provided also

$$\nabla_+\Sigma^+ = \frac{1}{2} R \quad , \quad \nabla_+ R = 2i\nabla_{\ddagger}\Sigma^+ \tag{4.27}$$

So there is just one independent field strength Σ^+. Its first component is the gravitino field strength, its θ component the curvature scalar.

One can now try to solve the constraints in terms of prepotentials, i.e., express $E_A{}^M$, ϕ_A, and Σ^+ in terms of unconstrained objects.

We look first at how they transform. General coordinate transformations on superfields can be represented as

$$\psi(\sigma,\theta) \to \psi'(\sigma,\theta) = e^K \psi(\sigma,\theta)$$
$$K = K^M D_M \tag{4.28}$$

and this implies that covariant derivatives should transform as

$$\nabla_A \to \nabla'_A = e^K \nabla_A e^{-K} \tag{4.29}$$

or

$$\delta\nabla_A = [K, \nabla_A] \tag{4.30}$$

For the vielbein, we get

$$\delta E_A{}^M = K^N(D_N E_A{}^M) + K^N E_A{}^L [D_N, D_L\}^M - E_A{}^N D_N K^M \tag{4.31}$$

or, at the linearized level

$$\delta E_A{}^M = - D_A K^M + K^N [D_N, D_A\}^M \tag{4.32}$$

The last term has a contribution only when N = A = + and M = $\ddagger$. Therefore, we obtain

$$\begin{aligned} \delta E_+{}^+ &= -D_+K^+ & \delta E_=^{\ +} &= -\partial_= K^+ \\ \delta E_+{}^{\ddagger} &= -D_+K^{\ddagger} + K^+ & \delta E_=^{\ \ddagger} &= -\partial_= K^{\ddagger} \\ \delta E_+{}^{=} &= -D_+K^{=} & \delta E_=^{\ =} &= -\partial_= K^{=} \end{aligned} \tag{4.33}$$

(<u>Note</u>: $E_{\ddagger}$ is fixed by the constraints.) From the second relation which contains a K^+ shift it is clear that one can make a supersymmetric gauge choice to set $E_+^{\ddagger}$ to zero. Henceforth we work in the Susy gauge

$$E_+^{\ddagger} = 0 \tag{4.34}$$

That means that K^+ is no longer available for independent gauging.

In this gauge we can write quite generally

$$E_+ = \psi^{-1/2}\,(D_+ + H_+^{=}\partial_{=})$$

where we have introduced a specific form for the coefficients. We can now start solving the constraints, and it is just a matter of straightforward algebra. We find, starting with the most general form for $E_=$,

$$E_= = \psi^{-1}(\partial_= + H_=^{\ddagger}\partial_{\ddagger}) + B_=^{+}E_+ \tag{4.36}$$

where

$$B_=^{+} = \frac{\Psi^{-\frac{1}{2}}}{2i}\;\frac{D_+H_=^{\ddagger} + H_+^{=}\partial_= H_=^{\ddagger} + H_=^{\ddagger}[\partial_= H_+^{=} + H_=^{\ddagger}\partial_{\ddagger}H_+^{=}]}{\left(1 + iH_=^{\ddagger}\,[D_+H_+^{=} + H_+^{\ddagger}\,\partial_{\ddagger}H_+^{=}]\right.} \tag{4.37}$$

We have introduced an additional arbitrary quantity $H_=^{\ddagger}$.

We also get expressions for the spin connections when solving the constraints. Finally we can compute

$$E^{-1} = \text{sdet}\ E_A^{M} = \psi^{3/2}\left[1 + iH_=^{\ddagger}\left(D_+H_+^{=} + H_+^{=}\partial_= H_+^{=}\right)\right]^{-1} \tag{4.38}$$

We have introduced 3 independent superfields. Here are their component expansions (up to gauge transformations).

$$\begin{aligned} H_+^{=} &= \rho_+^{=} + i\theta^+ h_{\ddagger}^{=} \\ \psi &= h + i\theta^+\psi_{\ddagger}^{+} \\ H_=^{\ddagger} &= h_=^{\ddagger} + i\theta^+\psi_=^{+} \end{aligned} \tag{4.39}$$

Let us examine their gauge transformations. Recall we are in a Susy gauge where we have gauged to zero $E_+^{\ddagger}$ so we have left (at the linearized level)

$$\delta E_+^+ = -D_+ K^+$$
$$\delta E_+^= = -D_+ K^= \qquad (4.40)$$

and

$$\delta E_=^+ = -\partial_= K^+ \qquad (4.41)$$

Note that having gauged to zero $E_+^{\ddagger}$, from now on, from

$$\delta E_+^{\ddagger} = -D_+ K^{\ddagger} + K^+ = 0 \qquad (4.42)$$

We have the constraint

$$K^+ = D_+ K^{\ddagger} \qquad (4.43)$$

so the independent gauge parameters are the two-vector $K^{\ddagger}$, $K^=$.

In addition to the general coordinate transformations induced by K, the vielbein also transforms under tangent space scale (Weyl) and Lorentz transformations. The former can be implemented by using Ψ as a scale compensator i.e. making it undergo an additional transformation

$$\delta\Psi = \sigma\Psi \qquad (4.44)$$

where the superfield σ parametrizes scale transformations. The Lorentz transformations can be implemented by introducing an additional pre-potential, the Lorentz compensator L, with transformation properties

$$\delta L = \Lambda \qquad (4.45)$$

and modifying the vielbein to

$$E_+ \to e^{L/2} E_+$$
$$E_= \to e^{-L} E_= \qquad (4.46)$$

We can now determine the transformation of the prepotentials. From the solution to the constraints, since

$$E_+^+ = \psi^{-1/2}, \qquad E_+^= = \psi^{-1/2} H_+^= \qquad (4.47)$$

we find first

$$\delta\psi^{-1/2} = -D_+K^+ = -D_+D_+K^{\ddagger} = -\frac{1}{2}\,\partial_{\ddagger}K^{\ddagger} \qquad (4.48)$$

and

$$\delta H_{+}^{=} = \delta\left(\psi^{1/2}\, E_{+}^{=}\right)$$
$$= -\,\psi^{1/2}\, D_{+}K^{=} + \psi^{-1/2}\, H_{+}^{=}\, D_{+}K^{+}\ . \qquad (4.49)$$

Similarly, from $E_{=} = \psi^{-1}\left[\ \ldots + H_{=}^{\ddagger}\,\partial_{\ddagger} + \ldots\ \right]$ we find

$$\delta H_{=}^{\ddagger} = \delta\left(\psi E_{=}^{\ddagger}\right)$$
$$= -\psi\partial_{=}K^{\ddagger} + (\partial_{\ddagger}K^{\ddagger})h_{=}^{\ddagger} \qquad (4.50)$$

At the linearized level these transformations become

$$\delta\psi = (\partial_{\ddagger}K^{\ddagger})$$
$$\delta H_{+}^{=} = -D_{+}K^{=} \qquad (4.51)$$
$$\delta H_{=}^{\ddagger} = -\partial_{=}K^{\ddagger}$$

Using (4.39)

$$\delta\rho_{+}^{=} = \delta H_{+}^{=}|_{\theta=0} = -D_{+}K^{=}\ | = k_{+}^{=} \qquad (4.52)$$

where k is the θ component of $K^{=}$. Therefore, one can gauge away $\rho_{+}^{=}$ by a non derivative transformation. In the resulting W-Z gauge we are left with the component fields of (1,0) supergravity

$$h_{\ddagger}^{=}\ ,\ h_{=}^{\ddagger}\ ,\ h\ ,\ \psi_{\ddagger}^{+}\ ,\ \psi_{=}^{+} \qquad (4.53)$$

whose transformation properties involve derivatives of gauge parameters, e.g.,

$$\delta h_{\ddagger}^{=} = \delta\left(-iD_{+}H_{+}^{=}\right)|_{\theta=0} = iD_{+}D_{+}K^{=}| \equiv -\partial_{\ddagger}k^{=}$$
$$\delta h_{=}^{\ddagger} = -\partial_{=}K^{\ddagger}|_{\theta=0} \equiv -\partial_{=}k^{\ddagger}$$
$$\delta\psi_{=}^{+} = \delta\left(-iD_{+}h_{=}^{\ddagger}\right)|_{\theta=0} = +i\partial_{=}D_{+}K^{\ddagger}| \equiv \partial_{=}\varepsilon^{+} \qquad (4.54)$$
$$\delta\psi_{\ddagger}^{+} = \delta\left(-iD_{+}\psi\right)|_{\theta=0} = iD_{+}\partial_{\ddagger}K^{\ddagger} \equiv \partial_{\ddagger}\varepsilon^{+}$$

It is now straightforward to write down covariant actions of the form

$$\int d^2\sigma d\theta^- \; E^{-1}L_- \tag{4.55}$$

and show that

$$\int d^2\sigma \; e^{-1}\left(\nabla_+ L_- - i\psi_{\ddagger}^{+} L_-\right) \tag{4.56}$$

gives the correct component action. In particular, the supergravity action is

$$\int d^2\sigma d\theta^- \; E^{-1}\Sigma^+ \tag{4.57}$$

This gives

$$\begin{aligned} &\int d^2\sigma e^{-1}\left[\nabla_+\Sigma^+ - i\psi_{\ddagger}^{+}\Sigma^+\right]_{\theta=0} \\ &\quad = \int d^2\sigma \; e^{-1}\;\left[r - i\psi_{\ddagger}^{+}\; D_{[=}\psi_{\ddagger]}^{+}\right] . \end{aligned} \tag{4.58}$$

A cosmological term is obtained by

$$\int d^2\sigma d\theta^- \; E^{-1}\lambda_- \tag{4.59}$$

where λ_- is an arbitrary spinor superfield. Finally, covariantized matter actions are, for example

$$\begin{aligned} &\int d^2\sigma d\theta^- E^{-1}\nabla_+\phi\nabla_=\phi \\ &\quad = \int d^2\sigma e^{-1}\left[\nabla_+(\nabla_+\phi\nabla_=\phi) - i\;\psi_{\ddagger}^{+}\;\nabla_+\phi\nabla_=\phi\right]\Big|_{\theta=0} \\ &\quad = \int d^2\sigma e^{-1}\left[\nabla_{\ddagger}A\nabla_=A + \ldots\right] \end{aligned} \tag{4.60}$$

V. (1,0) SUPERGRAPHS

The quantization of (1,0) theories and corresponding supergraph rules follow the standard pattern for four-dimensional theories.[1] For illustration we consider the scalar and spinor multiplets. Additional applications can be found in a forthcoming paper.[11] As usual, in a functional integral setting, we start with

$$Z(J) = \int [d\Psi]\; \exp[S(\Psi) + \int J\Psi] \tag{5.1}$$

where

$$S(\Psi) = \int d^2\sigma d\theta^- \, [L^o_-(\Psi) + L^{int}_-(\Psi)] \tag{5.2}$$

We rewrite Z(J) as

$$Z(J) = \exp \int d^2\sigma d\theta^- \, L^{int}_-(\delta/\delta J) \, Z_o(J) \tag{5.3}$$

where in Z_o we can perform the (Gaussian) integration over Ψ. The definition of functional differention is to some extent a matter of convenience. We choose

$$\frac{\delta J(z)}{\delta J(z')} = \delta(z-z') = \delta^{(2)}(\sigma-\sigma') \, \delta(\theta-\theta') \; . \tag{5.4}$$

Since $\delta(\theta-\theta') = \theta-\theta'$ the functional derivative is an odd Grassman element, even though the left hand side would a priori seem to be even.

For the scalar multiplet we have

$$\begin{aligned} Z_o(J) &= \int [d\Phi] \, \exp[\tfrac{i}{2} \int D_+\Phi\partial_=\Phi + \int J_-\Phi] \\ &= \exp \, [-\tfrac{i}{2} \int J_- \frac{D_+}{[\,]} J_-] \end{aligned} \tag{5.5}$$

Therefore the Φ propagator is

$$\langle\Phi(z) \; \Phi(z')\rangle = \frac{-D_+}{[\,]} \, \delta(z-z') \tag{5.6}$$

For the spinor multiplet we start with

$$\begin{aligned} Z_o(J) &= \int [d\psi] \, \exp[\tfrac{1}{2} \int \psi_- D_+ \psi_- + \int J\psi_-] \\ &= \exp \frac{1}{2} \int J \, \frac{iD_+\partial_=}{[\,]} \, J \end{aligned} \tag{5.7}$$

so that the propagator is

$$\langle\psi_-(z) \; \psi_-(z')\rangle = \frac{iD_+\partial_=}{[\,]} \, \delta(z-z') \tag{5.8}$$

We have used $D_+D_+ = i\partial_{\ddagger}$ and $\partial_{\ddagger}\partial_= = [\,]$.

There are two subtleties: The first one has to do with the Gaussian integral for spinor superfields ψ_- when it is not possible to put a measure on the space of such spinors. In the case of topologically trivial world sheets this is not a problem, but in general it is an issue that must be addressed. The second one is simpler: since the δ - functions that appear in the propagators are odd Grassman elements one encounters some minus signs which are unexpected. For example, one of the standard D-algebra manipulations takes the form

$$\delta(z-z')\overleftarrow{D}_+ = D_+\delta(z-z') \tag{5.9}$$

without a minus sign, in contrast to the situation in four dimensions or in p = q supersymmetry. For further discussion see Ref. [11].

REFERENCES

1. S. J. Gates, M. T. Grisaru, M. Rocek, W. Siegel, "Superspace", Benjamin-Cummings, Reading, MA (1983).

2. M. T. Grisaru and D. Zanon, Phys. Lett. 173B (1986) 423; Nucl. Phys. B (to be published).

3. M. T. Grisaru, in "Supergravity 1982, 1984" B. de Wit et al (eds). World Scientific, Singapore.

4. S. J. Gates, C. Hull and M. Rocek, Nucl. Phys. B248 (1984) 152.

5. C. Hull and E. Witten, Phys. Lett. 160B (1985) 398.

6. J. Bagger in "Supersymmetry", K. Dietz et al (eds), Plenum Press, N.Y. (1985).

7. M. T. Grisaru in "Unification of the Fundamental Particle Interaction II", J. Ellis and S. Ferrara (eds). Plenum Press, N.Y. (1981).

8. S. J. Gates and H. Nishino, Class. Quantum Gravity 3 (1986) 391; M. Rocek, P. van Nieuwenhuizen and S. C. Zhang, Stonybrook preprint ITP-SB-86-18 (1986).

9. R. Brooks, F. Muhammad and S. J. Gates, Nucl. Phys. B268 (1986) 599.

10. M. Evans and B. A. Ovrut, Rockefeller University preprints RU/86/149 and RU/86/BV153; G. Moore and P. Nelson, Harvard University preprint HUTP-86/A014.

11. M. T. Grisaru, L. Mezincescu and P. Townsend, "(1,0) Supergraphity", Brandeis University preprint.

String and Superstring Theory

Michael B. Green

Department of Physics, Queen Mary College,
Mile End Road, London E1 4NS

CONTENTS

1. Introduction; point particles.

1.1 Introductory Comments

These lectures are meant to provide an introduction to calculational techniques in superstring theory. Unfortunately, there is no time to include many topics of current interest such as string field theory, multi-loop diagrams and the beautiful techniques of conformal field theory. Some of these topics will be covered by other lecturers at this school. Before getting lost among the algebraic details I would like to emphasize some general points as well as summarize the origins of string theory.

At the present time the of understanding of string theory allows calculations at the level of string perturbation theory. There are good reasons for thinking that this is inadequate for calculating quantities of physical interest as well as being very far from providing a logical, compelling picture of the unity of the physical world. Nevertheless, the glimpses of consistency obtained from these primitive calculations have led to the prospect that a thorough understanding of the principles of the theory will result in a new formulation of the laws of physics on the smallest scales. As far as these lectures are concerned I shall talk about strings as if they are particles which exist independently of the background in which they are moving. This is like considering the graviton to be a spin 2 particle moving in a background space-time. We know that the graviton should really be considered as a wave in a fluctuating geometry – likewise, a string should presumably be something akin to a wave in some richer space. The obvious guess is that it is related to the geometry of the space of string configurations. Unlike with point particles there are already glimmerings of the connection between the propagating string and the background in which it is moving in the perturbative calculations that form the subject of these lectures.

The topics covered in these lectures are more fully described in the book 'Superstring Theory. vols. I and II'.* I shall not be giving references here in the hope that these lectures are self-contained, but a very extensive bibliography can be found in the same book.

1.2 The Origins of String Theory

The subject of string theory originates from the study of the strong interactions in the late 1960's. It had become clear that conventional field theory could not account for the plethora of high-spin hadronic particles discovered in accelerators. Any theory involving spins greater than one in four dimensions is non-renormalizable whether it is a theory of gravity with spin 2 or a hadronic theory in which there are higher spin particles. Alternatives to field theory were attempted based on general features of the S matrix and the insight gained by the parametrization of scattering amplitudes in terms of Regge poles. It was in this context that Veneziano (in 1968) suggested a form for the scattering amplitude that was later shown to be equivalent to a theory of interacting bosonic strings.

The subject of string theory developed rapidly in the early 1970's but certain problems seemed to be insurmountable. At the same time there was a rennaissance in conventional

* by M.B. Green, J.H. Schwarz and E. Witten (CUP, 1986).

field theory following the successes of the electroweak theory and of QCD. There no longer seemed to be a need for a string theory of hadrons and the idea that string theory might apply as a unified theory seemed far-fetched since there was no consistent theory of that type. The development of ideas concerning the toplogy of gauge fields and supergravity diverted work from string theory so that around the end of 1976 it had more or less died out. It is ironic that one of the last papers in 1976 suggested that, by a suitable truncation, the spinning string theory could be made supersymmetric in space-time.

The construction of string theories with space-time supersymmetry began around 1980 and has now mushroomed into a major enterprise.

1.3 Regge poles and duality

Several important issues in string theory are illustrated by reviewing the orignal hadronic motivation that led to Veneziano's dual resonance model.

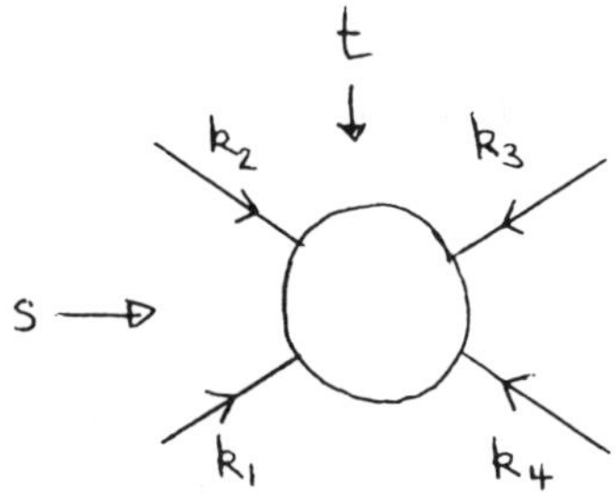

Figure 1.1. The kinematics for a four-particle amplitude.

Consider the scattering of four particles where the Mandelstam invariants are defined by

$$s = (k_1 + k_2)^2, \qquad t = (k_1 + k_4)^2, \qquad u = (k_2 + k_4)^2 \tag{1.3.1}$$

and

$$s + t + u = 4m^2, \tag{1.3.2}$$

where m is the mass of the external particles.

The scattering amplitude can be written as a sum of partial waves in the t channel as

$$A(s,t) = \sum (2J+1) P_J(\cos\theta_t) f_J(t) \tag{1.3.3}$$

where the t-channel center-of-mass scattering angle is given by $\cos\theta_t = (u - s)/(t - 4m^2)$. The essential ingredient in Regge theory (motivated originally by the work of Regge on potential scattering) is that the partial wave amplitude f_J can be analytically continued in the angular momentum, J. In particular, the Regge pole assumption is that the amplitude contains a pole at $J = \alpha(t)$

$$f_J(t) \sim \frac{1}{J - \alpha(t)}. \tag{1.3.4}$$

This corresponds to the contribution of spin J particle exchange to the scattering where the particle masses are given by

$$\alpha(M_J^2) = J. \tag{1.3.5}$$

The *trajectory function*, $\alpha(t)$, is therefore fixed at discrete positive values of t by the positions of the t-channel resonances.

Now consider the s-channel scattering process ($s > 0$, $t < 0$) at high energy ($s \to \infty$). The expression for the sum in (1.3.3) can be replaced by a contour integral using the Sommerfeld–Watson transform

$$A(s,t) \sim \int_C dJ\Gamma(-J)(2J+1)P_J(\cos\theta)f_J(t), \tag{1.3.6}$$

where the contour C encircles the poles of the Γ function at $J = 0, 1, \ldots$.

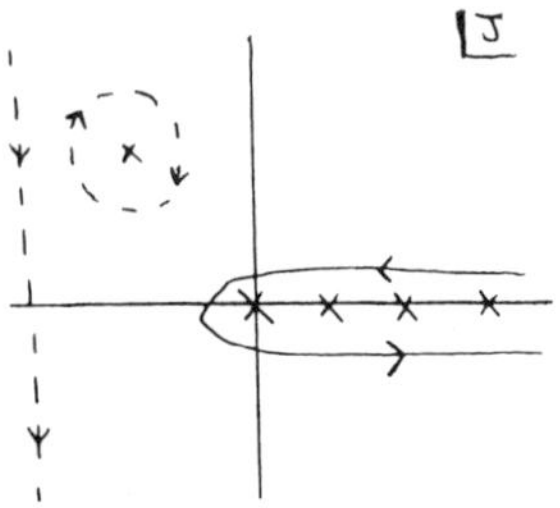

Figure 1.2. Contour of integration in the complex angular momentum plane appropriate to the Sommerfeld–Watson integral. The solid line denotes the original contour around the poles at the non-negative integers. The dashed lines denote the contour after distortion around the pole at $J = \alpha(t)$.

Substituting the Regge pole form for f_J into (1.3.6) and using the asymptotic form $\cos\theta_t \sim -s/m_0^2$ (where m_0 is a typical mass scale)

$$A(s,t) \sim \Gamma(-\alpha(t))s^{\alpha(t)}, \tag{1.3.7}$$

when $s \to \infty$ with t fixed. We see that a Regge pole not only determines the positions of the t-channel resonances but also determines the power behaviour of the high-energy s-channel scattering amplitude. Experimentally, the trajectory functions (determined both

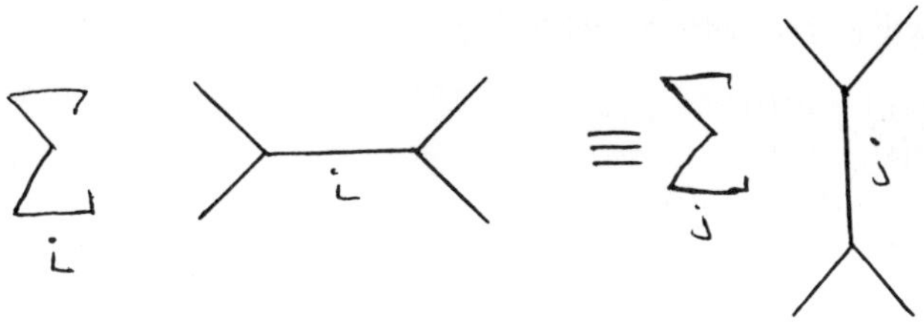

Figure 1.3. Equality of sum over s-channel poles and t-channel poles.

from high energy scattering and from the properties of resonances) are remarkably linear, *i.e.*

$$\alpha(t) = \alpha_0 + \alpha' t, \tag{1.3.8}$$

to a very good approximation.

By making use of the analyticity of scattering amplitudes Dolen, Horn and Schmid demonstrated that scattering data could be fitted by assuming that the amplitude can be expressed *either* as a sum over resonances in the s channel *or* as a sum over regge poles (corresponding to exchange of resonances in the t channel), as represented by figure 1.3.

This property is known as *duality*. Notice that, unlike Feynman rules in ordinary field theory, the contributions of s-channel and t-channel resonances are not added together - that would be double counting according to the rules of duality.

We now see how a theory containing this principle might avoid the traditional problem of high-spin field theory. A single spin-J particle contributes, as $s \to \infty$,

$$A \sim \frac{s^J}{t - M_J^2} \tag{1.3.9}$$

to a scattering amplitude at high energy. This is a power behaviour which leads to ultraviolet infinities in loop diagrams if $J > 1$. In a theory with duality there are an infinite number of high-spin particles and the sum of the s-channel resonances gives the leading regge pole behaviour of the form

$$A \sim \Gamma(-\alpha(t)) s^{\alpha(t)}. \tag{1.3.10}$$

Following these developments Veneziano made a very inspired guess at the form of the scattering amplitude describing the process $\pi\pi \to \pi\omega$. A slight modification of this is the four-particle amplitude with external scalar particles that, as we shall see, emerges from a theory of interacting open-ended bosonic strings

$$
\begin{aligned}
A(s,t) =& \frac{\Gamma(-\alpha(s))\Gamma(-\alpha(t))}{\Gamma(-\alpha(s)-\alpha(t))} \\
\simeq & -\sum_{n=0}^{\infty} \frac{(\alpha(t)+1)(\alpha(t)+2)\ldots(\alpha(t)+n)}{n!} \frac{1}{\alpha(s)-n}, \\
\simeq & \, \Gamma(-\alpha(t))(-\alpha(s))^{\alpha(t)}
\end{aligned}
\tag{1.3.11}
$$

where

$$
\Gamma(u) = \int_0^{\infty} dt t^{u-1} e^{-t}, \tag{1.3.12}
$$

which has poles at $u = 0, -1, \ldots$ and use has been made in the second step of Stirling's approximation

$$
\Gamma(u) \sim \sqrt{2\pi} u^{u-1/2} e^{-u}, \tag{1.3.13}
$$

valid for $|s| \to \infty$ with $\arg s \neq 0$.

These expressions embody the duality principle provided the trajectory function is of the form (1.3.8). Notice that the particle poles in the amplitude are on the real s axis and therefore have zero widths or infinite lifetimes. The amplitude is therefore like the Born approximation of a complete theory, to which must be added an infinite series of loop diagrams as in the Feynman diagram expansion of ordinary field theories. The weights of these diagrams are fixed by the requirements of unitarity. The amplitude (1.3.12) contains poles in both the s channel and the t channel. The complete, crossing symmetric, amplitude is obtained by adding two similar terms with s-channel and u-channel poles and with t-channel and u-channel poles, respectively.

Shortly after Veneziano's guess, Virasoro proposed an alternative dual amplitude describing four-particle scattering,

$$
A(s,t) = \frac{\Gamma(-\alpha(s)/2)\Gamma(-\alpha(t)/2)\Gamma(-\alpha(u)/2)}{\Gamma(-(\alpha(s)+\alpha(t))/2)\Gamma(-(\alpha(s)+\alpha(t))/2)\Gamma(-(\alpha(t)+\alpha(u))/2)}, \tag{1.3.14}
$$

which is fully crossing symmetric. This amplitude will emerge from the theory of interacting closed bosonic strings.

Historically, the path taken towards the string theory understanding of the dual resonance models followed after the four-particle amplitudes were generalized to an arbitrary number, M, of external particles. The form of these amplitudes is determined by tree-level unitarity which specifies that the residue of any intermediate pole in a tree amplitude must itself be the product of two tree amplitudes. In this way the full degeneracy of the states of the model was revealed and understood as the space spanned by the Fock spaces of an infinite number of harmonic oscillators with frequencies corresponding to the normal modes of a string-like extended particle. We will see in these lectures that the quantum treatment of this rather simple system only makes sense when very particular constraints are imposed: (i) The space-time dimension must take the value $D = 26$ for bosonic strings and $D = 10$ for superstrings. (ii) The massless states of open strings include a spin 1 particle while closed strings include a massless spin 2 state. (iii) The early string theories also have tachyon ground states, rendering them inconsistent as quantum field theories.

Even apart from the tachyon problem this spectrum is a disaster for a theory of hadronic physics since there are no massless hadrons with spin 1 or 2. However, it was realized that the spin 1 states behave like gauge particles while the spin 2 state behaves like a graviton. This led to the suggestion that string theory might describe gravity together with the non-gravitational gauge forces.

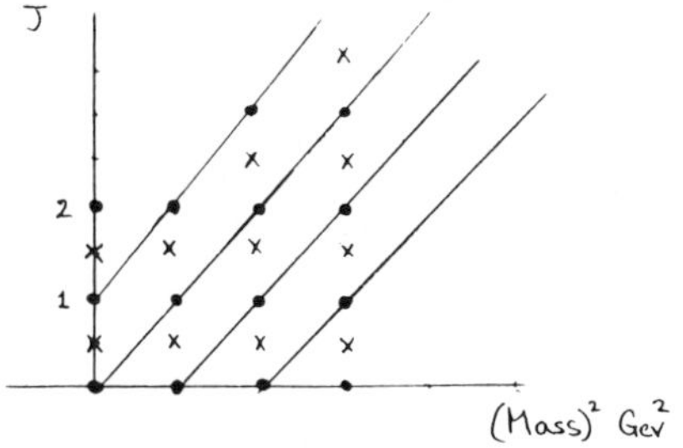

Figure 1.4. Generic spectrum of a superstring theory. Each dot represents many degenerate boson states while each cross represents degenerate fermion states.

Once the problem of tachyons is eliminated, as happens in superstring theories, the spectrum of states becomes something like figure 1.4. The spin 3/2 particle is the gravitino, the gauge particle of local supersymmetry.

In a theory of quantum gravity there is a natural scale of mass, namely the Planck scale which is about $10^{19} Gev$. Dual resonance models contain a natural scale set by α' which has dimensions $[length]^2$. In string theory this is related to the string tension, T, which has dimension $[length]^{-2} \equiv [mass]^2$ by

$$\alpha' = \frac{1}{2\pi T}. \tag{1.3.15}$$

In order for string theory to describe the force of gravity with approximately the correct strength this means that

$$T \sim (\triangle m)^2 \sim 10^{39} Gev^2$$

In many of the formulas in these lectures I will set $T = 1/\pi$ for convenience. This is merely a definition of units – the tension can always be reinstated into the formulas by elementary dimensional analysis.

In the course of these lectures I will be describing some methods used in the programme of constructing consistent superstring theories which will hopefully prove of relevance to observed physics. The objectives of superstring theory are to construct a consistent chiral quantum theory including gravity. The hope is that the consistency constraints on such a theory are so powerful that a more or less unique theory is picked out. The evidence so far is encouraging. The ten-dimensional theory can only have one of very symmetries ($E_8 \times E_8$, $SO(32)$ or $O(16) \times O(16)$ in a certain way of counting). The ten-dimensional space-time has not yet been proved to compactify but if six dimensions are assumed to curl up the resulting effective four-dimensional theory has some very realistic features.

Unfortunately I will not have time in my lectures to cover these topics that relate the theory to experimental observations.

1.4 The Relativistic Point Particle

I will begin with a brief review of the dynamics of a point particle of mass m as deduced from a covariant action principle. The simplest Poincaré-invariant action for a point particle is simply the length of the world-line traced out as the particle moves in space-time. This can be expressed as an integral over a single parameter, τ, as

$$S' = -m \int d\tau \sqrt{-\dot{X}(\sigma)^2}, \tag{1.4.1}$$

where $X^\mu(\tau)$ is the D-dimensional space-time coordinate of the particle ($\mu = 0, 1, \ldots, (D-1)$). This action is invariant under arbitrary reparametrizations, $\tau \rightarrow \tilde{\tau}(\tau)$, so that the theory can be thought of as one-dimensional general relativity in the τ space.

It proves useful to rewrite this action by introducing an auxiliary coordinate, $e(\tau)$, as

$$S = \frac{1}{2} \int d\tau \left[\frac{1}{e(\tau)} \eta_{\mu\nu} \dot{X}^\mu \dot{X}^\nu - m^2 e(\tau) \right], \tag{1.4.2}$$

which is just the action for one-dimensional general relativity, in which e plays the rôle of a metric density coupling to the D one-dimensional scalar 'matter' fields, $X^\mu(\tau)$. S is invariant under reparametrizations

$$\tau \rightarrow \tilde{\tau}(\tau) = \tau + \xi(\tau), \tag{1.4.3}$$

with (for infinitesimal $\xi(\tau)$)

$$\delta X = \xi \dot{X}, \qquad \delta e = \frac{d(\xi e)}{d\tau}. \tag{1.4.4}$$

The fact that S and S' define the same classical system is easy to see by eliminating e from the theory defined by S. The e equation of motion is the constraint $e = \sqrt{-\dot{X}^2/m^2}$. Substituting this expression back into S results in the action S'. Unlike S', the action S is not singular when $m \to 0$. S is also a useful form for considering particular parametrizations as choices of gauge for e. For example, a convenient choice of gauge is

$$e(\tau) = \frac{1}{m}. \tag{1.4.5}$$

The equations of motion, obtained by varying X^μ, are then

$$\ddot{X}^\mu = 0, \tag{1.4.6}$$

while the equation obtained by varying e and then setting $e = 1/m$, is the constraint equation

$$\dot{X}^2 + 1 = 0. \tag{1.4.7}$$

The momentum conjugate to X is $P \equiv \delta S'/\delta \dot{X} = m\dot{X}$. In the quantum theory this becomes $P = -i\partial/\partial X$. The particle is then described by a wave function Ψ and the constraint equation is simply the Klein–Gordon equation

$$\left(\frac{\partial}{\partial X^2} + m^2\right)\Psi = 0, \tag{1.4.8}$$

which is the free-field equation for a scalar field, Ψ.

2. The Nambu–Goto action; bosonic string theory; string functional integrals.

2.1 String Theory

A string sweeps out a world-sheet as it moves through space-time. The coordinates $X^\mu(\sigma,\tau)$ define a mapping of the sheet, (σ,τ), into D-dimensional space-time which, in general, will be curved. However, for most of this course I will assume that the string is moving in flat D-dimensional Minkowski space. The calculations will be analogous to perturbative calculations in general relativity obtained by expanding the Hilbert action around flat Minkowski space.

The simplest covariant action is the area of the world-sheet, as suggested by Nambu and Goto,

$$S' = T\int d\sigma d\tau \sqrt{\det \partial_\alpha X^\mu \partial_\beta X_\mu}, \tag{2.1.1}$$

where $\alpha, \beta = \tau, \sigma = 0, 1$. Since this is just the expression for the area of the world-sheet it is obviously independent of the parametrization of the sheet. This means that the action is unchanged by the infinitesimal transformations

$$\begin{aligned} \sigma &\to \sigma + \xi^1(\sigma, \tau) \\ \tau &\to \tau + \xi^0(\sigma, \tau) \end{aligned}, \tag{2.1.2}$$

provided X^μ transforms as a scalar,

$$\delta X^\mu(\sigma, \tau) = \xi^\alpha \partial_\alpha X^\mu, \tag{2.1.3}$$

where $\xi^\alpha(\sigma, \tau)$ is an arbitrary world-sheet vector. These are simply the transformations of two-dimensional general relativity.

The connection with general relativity in two dimensions can be made more explicit by introducing a two-dimensional metric, $h^{\alpha\beta}(\sigma, \tau)$, which is analogous to the introduction of $e(\tau)$ in the case of point particles. The new form of the action is

$$S = -\frac{T}{2} \int d\sigma d\tau \sqrt{h} h^{\alpha\beta} \partial_\alpha X^\mu \partial_\beta X^\nu \eta_{\mu\nu} \tag{2.1.4}$$

where $h = \det h_{\alpha\beta}$. The world indices are contracted with the Minkowski metric $\eta_{\mu\nu} = \mathrm{diag}(-1, 1, 1, \ldots, 1)$. The fact that the classical theories defined by S' and by S are equivalent can again be seen by eliminating $h^{\alpha\beta}$ from S by solving the $h^{\alpha\beta}$ equations of motion. This gives

$$h^{\alpha\beta} = f(\sigma, \tau) \partial_\alpha X^\mu \partial_\beta X_\mu, \tag{2.1.5}$$

where f is an arbitrary function. Since this is a constraint (*i.e.*, it commutes with the X equations of motion) it can be substituted back into the action, S, resulting in the action S'.

In addition to the action, S, we could consider adding other terms such as

$$S_1 = \lambda \int d\sigma d\tau \sqrt{h}, \tag{2.1.6}$$

$$S_2 = \frac{1}{2\pi} \int d\sigma d\tau \sqrt{h} R^{(2)}, \tag{2.1.7}$$

(where $R^{(2)}$ is the curvature scalar formed from $h_{\alpha\beta}$). The first of these terms is not allowed in the classical theory since it leads to inconsistent h equations of motion unless $\lambda = 0$. The term S_2 is the integral of a total derivative (the Gauss–Bonnet term). It is of great relevance when the world-sheet has handles as will arise in later developments but since it does not affect the equations of motion I will drop it for now.

2.2 World-sheet symmetries

The action is invariant under local reparametrizations, which are written in infinitesimal form as

$$\begin{aligned}\delta X^\mu &= \xi^\alpha \partial_\alpha X^\mu \\ \delta h^{\alpha\beta} &= \xi^\gamma \partial_\gamma h^{\alpha\beta} - \partial_\gamma \xi^\alpha h^{\gamma\beta} - \partial_\gamma \xi^\beta h^{\alpha\gamma}\end{aligned} \tag{2.2.1}$$

together with Weyl rescalings

$$\delta h^{\alpha\beta} = \Lambda(\sigma,\tau) h^{\alpha\beta}, \tag{2.2.2}$$

where $\Lambda(\sigma,\tau)$ is an arbitrary function.

The action is also invariant under the global symmetries

$$\delta X^\mu = a^\mu{}_\nu X^\nu + b^\mu, \tag{2.2.3}$$

where $a^{\mu\nu}$ is a constant antisymmetric matrix while b^μ is a constant. As far as the *two-dimensional* dynamics is concerned this is a global internal symmetry while, thinking of X^μ as D-dimensional coordinates, this is the statement of space-time Poincaré symmetry.

2.3 The conformal gauge

By a suitable choice of the arbitrary functions $\xi^0(\sigma,\tau)$ and $\xi^1(\sigma,\tau)$ the two-dimensional metric can always be *locally* transformed into the form

$$h_{\alpha\beta} = \eta_{\alpha\beta} e^{\phi(\sigma,\tau)} \equiv \begin{pmatrix} -1 & 0 \\ 0 & 1 \end{pmatrix} e^{\phi(\sigma,\tau)}. \tag{2.3.1}$$

In the classical theory the arbitrary function $\phi(\sigma,\tau)$ can also be eliminated by a Weyl transformation. As we shall see, in the quantum theory the Weyl symmetry is generally violated by a quantum anomaly except for special values of D (the bosonic theory is only invariant under Weyl transformations if $D = 26$).

It will often prove useful to use a two-dimensional light-cone notation by defining the combinations

$$\sigma^\pm = \tau \pm \sigma, \tag{2.3.2}$$

so that

$$\partial_\pm \equiv \frac{\partial}{\partial \sigma^\pm} = \frac{1}{2}\left(\frac{\partial}{\partial \tau} \pm \frac{\partial}{\partial \sigma}\right). \tag{2.3.3}$$

The components of an arbitrary two-vector, V^α, are written as

$$V^\pm = V^\tau \pm V^\sigma, \tag{2.3.4}$$

$$V_\pm = \frac{1}{2}\left(-V^\tau \pm V^\sigma\right). \tag{2.3.5}$$

In this notation the components of $h^{\alpha\beta}$ in the conformal gauge become

$$h_{++} = 0 = h_{--}, \qquad h_{+-} = \frac{1}{2}e^{\phi(\sigma,\tau)}. \tag{2.3.6}$$

In discussing the gauge-fixing involved in the choice of the conformal gauge in the quantum theory it is necessary to take account of factors in the functional measure for the integration over gauge-equivalent metrics. Equivalently, Fadeev–Popov ghost coordinates can be included in a manner that I will outline later.

In the conformal gauge there is still some residual gauge freedom. From (2.2.1) it is easy to see that the gauge conditions (2.3.6) are unaltered by reparametrizations and Weyl transformations restricted to satisfy

$$\partial^\alpha \xi^\beta + \partial^\beta \xi^\alpha = \Lambda \eta^{\alpha\beta}. \tag{2.3.7}$$

These conditions can be expressed as

$$\partial_- \xi^+ = 0 = \partial_+ \xi^-, \tag{2.3.8}$$

so that

$$\xi^+ = \xi^+(\sigma^+), \qquad \xi^- = \xi^-(\sigma^-) \tag{2.3.9}$$

as well as a condition that determines Λ in terms of $\xi^\pm$. These special coordinate transformations generate *(pseudo)-conformal* transformations $\xi^+(\sigma^+)\partial/\partial\sigma^+$ and $\xi^-(\sigma^-)\partial/\partial\sigma^-$. It is often useful to transform the τ variable $\tau \to i\tau$ and use the variables

$$z = \tau + i\sigma \qquad \bar{z} = \tau - i\sigma, \tag{2.3.10}$$

instead of $\sigma^\pm$. In this case the residual reparametrizations take the form of conformal transformations

$$z \to z + \xi(z), \qquad \bar{z} \to \bar{z} + \bar{\xi}(\bar{z}). \tag{2.3.11}$$

In these lectures I will often move between the euclidean and lorentzian conventions for the world-sheet signature with lack of concern.

The action, S, has a trivial form in a conformal gauge

$$S = -\frac{T}{2}\int d\sigma d\tau \partial_\alpha X^\mu \partial_\beta X_\mu \eta^{\alpha\beta}, \tag{2.3.12}$$

which must be supplemented by the constraints obtained from the $h_{\alpha\beta}$ equations of motion evaluated in the conformal gauge. The $h_{\alpha\beta}$ equations are just the two-dimensional

Einstein equations, which state that the energy-momentum tensor vanishes (since there is no dynamical part for the $h_{\alpha\beta}$ coordinates)

$$\begin{aligned} T_{\alpha\beta} &\equiv \frac{-2\pi}{\sqrt{h}} \frac{\delta S}{\delta h^{\alpha\beta}} \\ &= \partial_\alpha X^\mu \partial_\beta X_\mu - \frac{1}{2} h_{\alpha\beta} h^{\alpha'\beta'} \partial_{\alpha'} X^\mu \partial_{\beta'} X_\mu = 0 \end{aligned} \tag{2.3.13}$$

In terms of light-cone components this gives the two conditions

$$T_{++} = \partial_+ X^\mu \partial_+ X_\mu = 0, \tag{2.3.14}$$

$$T_{--} = \partial_- X^\mu \partial_- X_\mu. \tag{2.3.15}$$

The X^μ equations of motion that follow in the conformal gauge are

$$\nabla X^\mu \equiv \partial_\alpha \partial^\alpha X^\mu = 0. \tag{2.3.16}$$

The derivation of these Euler–Lagrange equations by varying the action requires the imposition of boundary conditions that eliminate surface terms as I will explain later.

The energy-momentum tensor is automatically traceless, *i.e.*,

$$h^{\alpha\beta} T_{\alpha\beta} = T_{+-} = 0. \tag{2.3.17}$$

Using the equations of motion it follows that $T_{\alpha\beta}$ is also conserved, *i.e.*,

$$\partial_\alpha T^\alpha_\beta = 0. \tag{2.3.18}$$

It is a very special feature of two dimensions that the two conditions, (2.3.17) and (2.3.18) imply the existence of an infinite set of conservation laws. From (2.3.17) and (2.3.18) it is easy to see that

$$\partial_+ \left(f^- T_{--}\right) = 0 = \partial_- \left(f^+ T_{++}\right), \tag{2.3.19}$$

where

$$f^+ \equiv f^+(\sigma^+), \qquad f^- \equiv f^-(\sigma^-). \tag{2.3.20}$$

Since $f^\pm$ are arbitrary functions (2.3.19) describes an infinite number of conserved quantities. The existence of this infinite number of conserved quantities is crucial for all subsequent discussions and is one of the reasons why the dynamics of strings is so special compared, say, with the behaviour of membranes.

2.4 Momentum Current

The global symmetries described by (2.2.3) are associated with conserved currents. The expressions for these currents can be obtained by the usual Noether procedure of treating the parameters of the transformations as functions of σ, τ instead of constants. For example, under a local translation

$$X^\mu \to X^\mu + b^\mu(\sigma, \tau), \tag{2.4.1}$$

the variation of the action has the form (in the conformal gauge)

$$\delta S = \int d\sigma d\tau P^\mu_\alpha \partial^\alpha b_\mu(\sigma, \tau), \tag{2.4.2}$$

where the momentum current is defined by

$$P^\mu_\alpha = T\partial_\alpha X^\mu. \tag{2.4.3}$$

The density of momentum on the string will be defined as the τ component of P_α, *i.e.*,

$$P^\mu(\sigma, \tau) \equiv P^\mu_\tau(\sigma, \tau) = T\dot{X}_\mu(\sigma, \tau). \tag{2.4.4}$$

The total momentum carried by the string, given by integrating over the string

$$p^\mu = \int_0^\pi d\sigma P^\mu(\sigma, \tau), \tag{2.4.5}$$

is conserved by virtue of the equations of motion.

2.5 Angular Momentum Current

In a similar manner the angular momentum current is obtained by considering infinitesimal Lorentz transformations which depend on σ and τ,

$$X^\mu \to X^\mu + a^\mu{}_\nu(\sigma, \tau) X^\nu. \tag{2.5.1}$$

The Noether current obtained in this manner is

$$J^{\mu\nu}_\alpha(\sigma, \tau) = T\left(X^\mu(\sigma, \tau)\partial_\alpha X^\nu(\sigma, \tau) - X^\nu(\sigma, \tau)\partial_\alpha X^\mu(\sigma, \tau)\right). \tag{2.5.2}$$

The density of angular momentum is the τ component of this current and the total angular momentum is given by the integral

$$J^{\mu\nu} \equiv T\int_0^\pi d\sigma \left(X^\mu \frac{\partial X^\nu}{\partial \tau} - X^\nu \frac{\partial X^\mu}{\partial \tau}\right). \tag{2.5.3}$$

Both the momentum and the angular momentum are conserved as a consequence of the X equations of motion.

2.6 Classical Poisson Brackets

From its definition the momentum density, $P^\mu(\sigma)$ is conjugate to $X^\mu(\sigma)$. The Poisson brackets

$$\left[P^\mu(\sigma,\tau), X^\nu(\sigma',\tau)\right] = -i\delta(\sigma-\sigma')\eta^{\mu\nu}, \tag{2.6.1}$$

follow from the action in the usual manner (where a factor of $-i$ has been inserted on the right-hand side in anticipation of the quantum theory). Making use of these brackets it follows that the energy-momentum tensor satisfies the classical commutation relation

$$[T_{++}(\sigma,\tau), T_{++}(\sigma',\tau)] = i(T_{++}(\sigma,\tau) + T_{++}(\sigma',\tau))\delta'(\sigma-\sigma') \tag{2.6.2}$$

with a similar relation involving T_{--}. This is the algebra of the two-dimensional conformal group – the algebra of diffeomorphisms of the circle. In the quantum theory two sources of additional (anomalous) terms will appear. One of these arises from operator ordering problems due to the fact that T_{++} is quadratic in the X^μ. This gives a c-number term proportional to D. In addition, the quantum theory involves additional coordinates, a Fadeev–Popov ghost and an anti-ghost coordinate. These also give rise to an anomaly, so that the total anomalous term will turn out to be

$$\frac{i\pi}{24}(26-D)\delta'''(\sigma-\sigma'). \tag{2.6.3}$$

This adds to the right-hand side of (2.6.2). Notice that this anomaly vanishes if $D = 26$, which is the special feature of the critical dimension.

2.7 Some Classical Motions

I will not have much to say about the classical behaviour of Nambu–Goto strings but it is worth studying a couple of specially simple configurations in order to illustrate why I have been calling T the tension and why $\alpha' = 1/2\pi T$ is the slope of the Regge trajectory.

First consider a closed string which is initially at rest in the shape of a circle of radius R in the (1,2) plane, $X^1 = R\cos\phi$, $X^2 = R\sin\phi$. The parametrizations,

$$\sigma = R\phi, \qquad \tau = \frac{X^0}{2\tau}, \tag{2.7.1}$$

satisfy the constraints (2.3.14) and (2.3.15) and the equation (2.3.16). The energy is given by

$$p^0 = \int d\sigma \dot{X}^0 T = 2\pi RT \tag{2.7.2}$$

so that T has the interpretation energy/length which is a tension.

The second example is that of an open string moving as a rod of length A at a given angular frequency in the $(1,2)$ plane, so that $X^1 = A\cos\tau\cos\sigma$, $X^2 = A\sin\tau\cos\sigma$ and

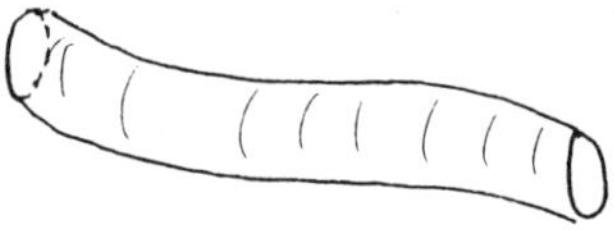

Figure 2.5. The propagation of a single closed string.

$X^0 = A\tau$. These expressions satisfy the constraints and the equations of motion. The energy is now given by

$$p^0 = \pi T A, \tag{2.7.3}$$

while the angular momentum is given by

$$J = \frac{1}{2}\pi A^2 T. \tag{2.7.4}$$

This is in fact the motion with the maximum angular momentum for a given mass. The ratio $J/M^2 = 1/2\pi T = \alpha'$.

2.8 Functional calculations

I will now describe how to calculate scattering amplitudes in string theory by a path integral method that generalizes the method of summing over histories in point particle physics. Following Feynman's prescription, the propagator for a free point particle is given by the functional integral,

$$\int_{\tau_i}^{\tau_f} De(\tau)DX(\tau)e^{-S[e,X]}, \tag{2.8.1}$$

where τ_i and τ_f denote the initial and final values of τ. I have used an imaginary τ variable and a euclidean space-time metric (*i.e.*, a Wick rotation) as is commonly done in order to avoid an oscillating phase in the exponent. This is just a trick which takes care of the causal boundary conditions automatically.

In a similar fashion the propagator describing a string in some initial configuration to propagate to some final configuration ought to be given by the two-dimensional functional integral,

$$\int_i^f Dh_{\alpha\beta}(\sigma,\tau)DX(\sigma,\tau)e^{-S[h,X]} \tag{2.8.2}$$

I have used a Euclidean signature world-sheet metric as well as space-time metric. It is an assumption that this is equivalent to the physically relevant lorentzian signatures as it is in the case of point particles.

2.9 Gauge Fixing and Ghost Coordinates

We would like to impose the conformal gauge conditions

$$h_{++} = 0 = h_{--} \qquad h_{+-} = \frac{1}{2}e^{\phi} \tag{2.9.1}$$

in the functional integral which can be written as

$$\int Dh_{++}Dh_{--}Dh_{+-}DXe^{-S[h,X]}. \tag{2.9.2}$$

Recall the infinitesimal coordinate transformations,

$$\delta h_{\alpha\beta} = \nabla_\alpha \xi_\beta + \nabla_\beta \xi_\alpha, \tag{2.9.3}$$

where the covariant derivatives are defined in terms of the Riemann–Christoffel connection by

$$\nabla_\alpha \xi_\beta = \partial_\alpha \xi_\beta - \Gamma^\gamma_{\alpha\beta}\xi_\gamma, \tag{2.9.4}$$

and, as usual, the Riemann–Christoffel connection is given by

$$\Gamma^\gamma_{\alpha\beta} = \frac{1}{2}h^{\gamma\delta}(\partial_\alpha h_{\beta\delta} + \partial_\beta h_{\alpha\delta} - \partial_\delta h_{\alpha\beta}. \tag{2.9.5}$$

In terms of light-cone coordinates

$$\delta h_{++} = 2\nabla_+ \xi_+ \qquad \delta h_{--} = 2\nabla_- \xi_-. \tag{2.9.6}$$

Following the Fadeev–Popov gauge-fixing procedure we would like to change variables from integrals over h_{++} and h_{--} to integrals over ξ_+ and ξ_- while the integral over h_{+-} is trivially equivalent to integrating over ϕ. Due to the gauge invariance of the theory the $\xi_\pm$ integrals are trivial, giving an (infinite) constant. More explicitly, let us insert unity into the functional integral in the form

$$1 = \int D\xi \delta(h^\xi_{++})\delta(h^\xi_{--}) \det\left(\frac{\delta h^\xi_{++}}{\delta\xi}\right) \det\left(\frac{\delta h^\xi_{--}}{\delta\xi}\right), \tag{2.9.7}$$

where h^ξ denotes the reparametrized metric where ξ^α is the parameter of the transformation and the integral is over the manifold of the group. Inserting this into the functional integral gives

$$\int D\xi \int Dh_{++}Dh_{--}D\phi Dx e^{-S[h,X]}\delta(h^{\xi}_{++})\delta(h^{\xi}_{--}) \det\left(\frac{\delta h^{\xi}_{++}}{\xi}\right) \det\left(\frac{\delta h^{\xi}_{--}}{\xi}\right). \quad (2.9.8)$$

Since $S[h, X] = S[h^{\xi}, X]$ we can change variables to integrate over ξ and h^{ξ}. The ξ dependence is now trivial, giving

$$\int DhDX\delta(h_{++})\delta(h_{--}) \det(\nabla_{+}) \det(\nabla_{-}) e^{-S[h,X]}. \quad (2.9.9)$$

The jacobian for the transformation, $\det\nabla_{+} \det\nabla_{-}$ can be expressed in terms of anti-commuting coordinates as the product of

$$\det\frac{\delta h_{++}}{2\xi_{+}} = \det\nabla_{+} = \int Dc^{-}Db_{--}e^{\frac{-1}{\pi}\int c^{-}\nabla_{+}b_{--}d\sigma d\tau}, \quad (2.9.10)$$

and

$$\det\frac{\delta h_{--}}{2\xi_{-}} = \det\nabla_{-} = \int Dc^{+}Db_{++}e^{\frac{-1}{\pi}\int c^{+}\nabla_{-}b_{++}d\sigma d\tau}. \quad (2.9.11)$$

In these equations the determinants have been represented as functional integrals over *Grassmann* variables, $c^{\pm}$ and $b_{\pm\pm}$. Grassmann variables are defined by the condition that $\theta_1\theta_2 = -\theta_2\theta_1$ where θ_1 and θ_2 are two arbitrary Grassmann variables. Integration over Grassmann variables is defined by the Berezin conditions

$$\int d\theta = 0, \qquad \int \theta d\theta = 1. \quad (2.9.12)$$

In a covariant notation the *ghost* coordinates $c^{\pm}$ form a two-vector $c^{\alpha}(\sigma,\tau)$. The *anti-ghost* coordinates $b_{\pm\pm}$ form a traceless symmetric tensor, $b_{\alpha\beta}(\sigma,\tau)$, (where $b_{+-} = 0$).

The determinants (2.9.10) and (2.9.11) are expressed as contributions of the ghost and antighost coordinates to the action. This is reflected in ghost contributions to the energy-momentum tensor which turn out to be

$$\begin{aligned} T^{gh}_{++} &= \frac{1}{2}c^{+}\partial_{+}b_{++} + \left(\partial_{+}c^{+}\right)b_{++} \\ T^{gh}_{--} &= \frac{1}{2}c^{-}\partial_{-}b_{--} + \left(\partial_{-}c^{-}\right)b_{--}. \end{aligned} \quad (2.9.13)$$

As is well-known from work on other gauge field theories, the gauge-fixed action, including the ghost terms, contains all the information concerning gauge invariance. This is encoded by the BRST (Becchi, Rouet, Stora and Tyupkin) symmetry. I will not pursue the approach to strings based on this symmetry since that is the subject of other lecturers at this school. It is a very efficient method for displaying the gauge symmetries of the theory. Furthermore, it seems to be an essential feature for understanding the second-quantized field theory of strings.

For many simple calculations the effect of the ghost modes is very trivial and they can effectively be ignored.

2.10 Interactions

Interactions between closed strings are assumed to take place locally by the strings splitting or joining when two points touch.

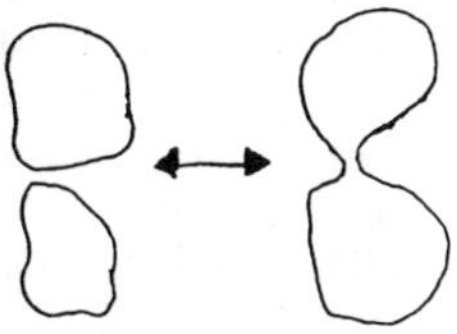

Figure 2.6. The interaction between a pair of closed strings, forming a third closed string.

In terms of the world-sheet swept out as the string moves the interaction looks like

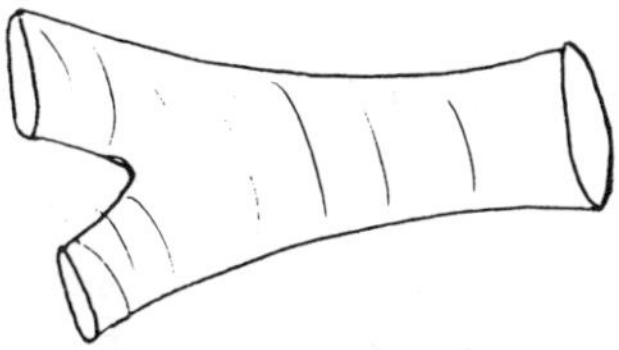

Figure 2.7. The world-sheet for three interacting closed strings.

An important feature of this diagram is that there is no special point on the surface at which the interaction can be said to take place in an objective sense. The interaction point depends on the way in which the time coordinate is defined which depends on the frame of the observer. This is in sharp contrast with point particle interactions and is one of the most basic distinguishing features of string physics.

2.11 Tree scattering amplitudes by functional integration

I will now describe the calculation of closed-string scattering amplitudes according to the prescription of Polyakov. This method is an *ansatz* which is not derived from a fundamental priciple. It certainly gives correct answers in the cases that can be compared with other calculational techniques (such as the light-cone methods of Mandelstam). This method seems to have great potential for describing world-sheets of arbitrary genus. As we will see the prescription does not obviously guarantee unitarity of scattering amplitudes but nevertheless it leads to unitary expressions once the overall coefficients of the expressions are normalized appropriately. Polyakov's method makes extensive use of the geometric properties of the string world-sheet with conformal invariance being of particular significance.

Figure 2.8. The mapping of a tree diagram to a a spherical sheet with singular points.

A tree diagram can be represented by the world-sheet on the left in the above figure. This figure might represent either the configuration of the sheet in space-time or in parameter space, in which case the parametrization has been conveniently chosen so that τ is roughly the same as the time coordinate in the laboratory frame. If the incoming and outgoing strings at $\tau = \pm\infty$ are considered to be points the world-sheet is topologically a sphere. It follows from a theorem of Riemann that an arbitrary sheet of this type can be conformally mapped to a standard sphere, *i.e.*, a general metric can be written in terms of the standard metric on a sphere, $\hat{h}$, as

$$h^{\alpha\beta} = e^{\phi}\hat{h}^{\alpha\beta}, \tag{2.11.1}$$

(where I am assuming in this discussion that the world-sheet has euclidean signature).

The strings at infinity are mapped into singular points on the surface of the sphere by this conformal transformation. The Polyakov method associates a *vertex operator*, $V_\Lambda(k)$, with the emission of the particle with quantum numbers denoted by Λ and momentum k at a singular point. A scattering amplitude is then assumed to be given by the correlation function of the vertex operators for the scattering particles. This *two-dimensional* correlation function is supposed to describe the scattering of particles in *26 space-time dimensions*! The vertex operator is assumed to be an integral of a local operator that is consistent with Poincaré invariance as well as being a scalar with respect to the world-sheet reparametrization symmetries. It has the general form

$$V_\Lambda(k) = \int d^2\sigma\sqrt{h}W_\Lambda(\sigma,\tau)e^{ik\cdot X(\sigma,\tau)}, \tag{2.11.2}$$

where the factor of $\exp(ik\cdot X)$ is needed if the vertex is to describe the emission of momentum k. This factor gives the usual change of phase, $e^{ik\cdot a}$, under translations $X^\mu \to X^\mu + a^\mu$. The prefactor W_Λ depends on the species of particle being emitted. In the simplest case $W = 1$, which describes the emission of a scalar state. If we take $W = \varsigma_{\mu\nu}\partial_\alpha X^\mu\partial^\alpha X^\nu$ the vertex describes the emission of a tensor state with polarization tensor $\varsigma^{\mu\nu}$. This will turn out to be the massless graviton. Higher-rank tensor states are obtained in a similar fashion. In general we might expect that the ghost modes enter into the vertex. This is true, for example, in the fermion emission vertices in superstring theories. However, for the simple vertices that we are considering the ghost modes can be neglected.

The M closed-string scattering amplitude is *assumed* to be given by

$$\begin{aligned} A(1,2,\ldots M) &= \kappa^{M-2} < \prod_{r=1}^{M} V_{\Lambda_i}(k_i) > \\ &= \kappa^{M-2}\int DX(\sigma,\tau)Dh_{\alpha\beta}(\sigma,\tau) \\ &\quad \exp\left(\frac{-1}{2\pi}\int d^2\sigma\sqrt{h}h^{\alpha\beta}\partial_\alpha X^\mu\partial_\beta X_\mu\right)\prod_{i=1}^{M}V_{\Lambda_i}(k_i), \end{aligned} \tag{2.11.3}$$

where the string tension has been chosen so that $T = 1/\pi$, or $\alpha' = 1/2$ as I will often do in subsequent formulas.

2.12 Evaluation of tree diagrams

The fact that the theory is invariant under conformal transformations is crucial in evaluating the functional integral (2.11.3). We have seen (but not yet proved) that the quantum theory is conformally invariant only if $D = 26$. In terms of the functional integral this is seen from the fact that the dependence on ϕ cancels out of (2.11.3) only in 26 dimensions. I will not show this in detail. Assuming that the theory is conformally invariant allows the gauge choice in which the metric, h, is chosen to be the standard metric on a sphere. Even simpler is the choice of metric on a plane, which is simply the stereographic projection of a sphere. In this case we have

$$h^{\alpha\beta} = e^\phi\begin{pmatrix}1 & 0\\ 0 & 1\end{pmatrix}, \tag{2.12.1}$$

i.e., the element of length is given by

$$ds^2 = e^\phi(dx^2+dy^2) = e^\phi dzd\bar{z}. \tag{2.12.2}$$

Since the ghost modes do not enter the expression for the vertices they can be trivially integrated out of the expression for the amplitude and just contribute to the overall normalization. Since this is in any case arbitrary we shall ignore the ghosts at this point. The ghost modes give non-trivial factors in the calculation of loop amplitudes.

I will consider the simplest case in which the external states are scalar states with vertex operators containing $W_\Lambda = 1$. After gauge fixing the tree amplitude becomes

$$\begin{aligned} A(1,2,..M) &= \kappa^{M-2} \int DX e^{\frac{-1}{2\pi}\int dz^2 \partial_\alpha X_\mu \partial^\alpha X^\mu} \prod_{i=1}^{M} \int e^{ik_i\cdot X(z_i,\bar{z}_i)} d^2 z_i \\ &\equiv \kappa^{M-2} < \prod_i V(k_i) > . \end{aligned} \tag{2.12.3}$$

Even after the choice of parametrization and Weyl gauge in which the metric is that of a flat plane there is a residual invariance. This arises from the fact that there are nontrivial nonsingular mappings of the complex plane onto itself. These are the transformations

$$z \to z' = \frac{az+b}{cz+d}, \tag{2.12.4}$$

where $ad - bc = 1$. These transformations form a group with three independent complex parameters. This is the group, $SL(2,C)$, of complex 2×2 matrices of unit determinant, often called the Möbius group. It is a subgroup of the full conformal group (which is the set of arbitrary conformal transformations, including those which generate singularities). The infinitesimal version of the transformation is given by

$$z \to z' = z + \lambda_0 + \lambda_1 z + \lambda_2 z^2. \tag{2.12.5}$$

In the gauge (2.12.1) the functional integral can be evaluated by completing the square in the exponent using

$$\begin{aligned} &\int DX(z,\bar{z}) \exp\left(i \int d^2 z J_\mu(z) X^\mu(z) + \frac{1}{2\pi} \int d^2 z X_\mu \partial^2 X^\mu \right) \\ &\qquad = \exp\left(\frac{1}{4} \int d^2 z d^2 z' J_\mu(z) G(z;z') J^\mu(z') \right), \end{aligned} \tag{2.12.6}$$

where

$$J^\mu(z) = \sum k_i^\mu \delta(z - z_i) \tag{2.12.7}$$

and

$$\partial^2 G(z;z') = 2\pi \delta^2(z - z'), \tag{2.12.8}$$

so that

$$G = \ln(|z - z'|\mu), \tag{2.12.9}$$

where μ is an infrared cutoff. Substituting this expression and (2.12.6) into (2.12.3) gives

$$A = K^{M-2} \int \prod dz_i \prod_{i<j} |z_i - z_j|^{k_i \cdot k_j/2}. \tag{2.12.10}$$

In evaluating this expression the infrared cutoff has cancelled with singular factors arising from the terms with $r = s$. This only happens if the emitted states have momenta satisfying the condition

$$k_i^2 = 8 = -(\text{Mass})^2. \tag{2.12.11}$$

This constraint on the mass of the external scalar states is also required for another reason. I argued earlier that the expression for the amplitude ought to be invariant under $SL(2,C)$ transformations. It is easy to check that under these transformations

$$\begin{aligned} dz \to d^2z' &= \frac{d^2z}{|cz+d|^4} \\ |z_i - z_j| \to |z_i' - z_j'| &= \frac{|z_i - z_j|}{|cz_i + d|^2|cz_j + d|^2} \end{aligned} \tag{2.12.12}$$

Substituting these transformations into the expression for the amplitude we see that the condition, (2.12.11) is essential if the amplitude is to be invariant.

Although the residual $SL(2,C)$ symmetry is not really a local symmetry, it is nevertheless associated with an infinite factor due to the infinite volume of the group. Since the group has three complex parameters the overcounting can be eliminated by arbitrarily fixing three of the complex integration variables at arbitrary complex points. This means changing variables from any three of the z variables, z_A, z_B and z_C to the parameters λ_0, λ_1 and λ_2 in (2.12.5). This gives rise to the jacobian

$$\frac{\partial(z_A, z_B, z_C)}{\partial(\lambda_0, \lambda_1, \lambda_3)} = |z_A - z_B|^2 |z_B - z_C|^2 |z_C - z_A|^2. \tag{2.12.13}$$

Substituting this expression into (2.12.10) together with $\delta(z_A - a)\delta^2(z_B - b)\delta^2(z_C - c)$ gives a finite expression for the amplitude. In particular, choosing $z_1 = 0, z_2 = 1$ and $z_3 = \infty$ gives the Koba-Nielsen formula

$$A = K^{M-2} \int \prod_{i=4}^{M} d^2 z_i \prod_{j=4}^{M} |z_j|^{k_1 \cdot k_j/2} |1 - z_j|^{k_2 \cdot k_j/2} \prod_{4 \le i < j \le M} |z_i - z_j|^{k_i \cdot k_j/2} \tag{2.12.14}$$

Notice that the infinite factors arising from $z_3 = \infty$ cancel when the mass-shell condition $k_i^2 = 8$ is imposed. For M = 4 this is Virasoro's amplitude quoted earlier as an expression in terms of ratios of products of Γ functions.

2.13 Conformal weights of vertices

The fact that the amplitude is $SL(2,C)$ invariant provided the external particles satisfy a particular mass-shell condition can be seen directly from the expressions for the vertices,

$$V(k_i) = \int d^2 z e^{ik_i \cdot X(z,\bar{z})}, \tag{2.13.1}$$

which must be invariant under $SL(2,C)$ transformations. Consider the special $SL(2,C)$ transformation $z \to \lambda z$ *i.e.*$d^2 z \to \lambda^2 d^2 z$ so that invariance of the integrated vertex requires

$$e^{ik\cdot X(z,\bar{z})} \to \lambda^{-2} e^{ik\cdot X(\lambda z, \lambda \bar{z})}. \tag{2.13.2}$$

Although X is a conformal scalar the exponential transforms non-trivially in the quantum theory due to operator ordering effects. This means that the integrand of the vertex can have an anomalous *conformal weight* (*i.e.*, anomalous dimension) due to quantum effects. The easiest way to calculate the anomalous dimension is to calculate the correlation function

$$\begin{aligned} < e^{ik\cdot X(z,\bar{z})} e^{-ik\cdot X(y,\bar{y})} > &= |z-y|^{k^2/2} \\ &= \lambda^{-k^2/2} |\lambda z - \lambda y|^{k^2/2}. \end{aligned} \tag{2.13.3}$$

If $k^2 = 8 = -(\text{Mass})^2$ each of the exponential factors scales like λ^{-2}, as required.

In considering the emission of a second-rank tensor state (a 'graviton') the vertex must include the prefactor describing the polarization states of the emitted particle, *i.e.*,

$$V(\varsigma, k) = \int d^2 z \eta^{\alpha\beta} \varsigma_{\mu\nu} \partial_\alpha X^\mu \partial_\beta X^\nu e^{ik\cdot X}. \tag{2.13.4}$$

The scale dimension of the measure, $d^2 z$, is now compensated by the dimensions of the two factors of ∂_α. This means that $e^{ik.X}$ must have dimension 0. By the same argument as above, (2.13.3) requires $k^2 = -(\text{Mass})^2 = 0$. This is the reason for calling the tensor state a graviton.

2.14 Open strings

In the case of the tree level scattering of open strings the world-sheet is topologically equivalent to a disk. By means of reparametrizations and conformal transformations it can be mapped onto a unit disk with the external particles attached to the boundary. For a given diagram these points on the boundary are integrated keeping a given sequence. Equivalently, the world-sheet can be mapped to the upper-half plane with particles attached to the real axis in a given order.

The upper-half plane is mapped into itself by $SL(2,R)$ transformations which have the form

$$z \to z' = \frac{az+b}{cz+d}, \tag{2.14.1}$$

with real parameters, a, b, c and d satisfying $ad - bc = 1$. The vertex operator for emitting a scalar open-string state is given by

$$V(k) = \int dx e^{ik\cdot X(x)}. \tag{2.14.2}$$

In order for this to have zero conformal weight the exponential factor must have conformal weight 1. This requires $k^2 = -(\text{Mass})^2 = 2$. This corresponds to the mass of the ground-state tachyon of the Veneziano model. The scattering amplitude deduced in this manner is the Veneziano model in the case that $M = 4$.

The vertex for emitting a vector state with polarization vector ς^μ is given by

$$V(\varsigma, k) = \int dx \varsigma_\mu \frac{dX^\mu(x)}{dx} e^{ik\cdot X(x)}. \tag{2.14.3}$$

In this case the requirement that the vertex be a conformal scalar again results in the condition $k^2 = -(\text{Mass})^2 = 0$. The particle is a massless gauge boson.

Open strings can carry charges at their endpoints so that amplitudes are associated with group theory factors. An amplitude comes in various pieces corresponding to the different possible cyclic orderings of the external particles along the boundary of the world-sheet. When the symmetry group is $SO(n), U(n)$ or $Sp(n)$ each external particle, r, is associated with a matrix, $\lambda^r_{\alpha\beta}$, where the indices take the values $1, \ldots, n$. The λ^r's describe the quantum numbers of the external states. They are contracted into each other around the boundary of the diagram so that there is a trace, such as $\text{tr}(\lambda^1\lambda^2 \ldots \lambda^M)$, associated with a diagram. This is the Chan–Paton procedure. The constraint that the intermediate particles propagated in tree diagrams should be consistent turns out to restrict the possible groups to the classical ones listed above.

The basic reason why these groups are picked out can be seen from the requirement that the massless vector particles are gauge particles with quantum numbers in the adjoint representation of the symmetry group. The string carries a pair of indices in the n-dimensional defining representation. For the group $U(n)$, $n \otimes \bar{n}$ is the adjoint representation. For the other classical groups ($SO(n)$ and $Sp(n)$) there is no $\bar{n}$ representation and the string quantum numbers are given by $n \otimes n = adjoint + \ldots$. Ignoring the group theory factors the open-string states have a definite symmetry under the interchange of the string endpoints. This means that the symmetry content is either $[n \otimes n]_{\text{symm}}$ or $[n \otimes n]_{\text{antisymm}}$. It is only for the classical groups that the adjoint representation is uniquely projected out by either antisymmetrizing (for $SO(n)$) or symmetrizing (for $Sp(n)$).

In constructing the tree scattering amplitudes for four-particle scattering there are three types of terms distinguished by their group theory content, corresponding to the three inequivalent cyclic orderings of four particles around the boundary of a disk. Each of these terms has poles in the channels defined by the ways in which pairs of particles can approach each other on the boundary. One term has a group theory factor $G = \text{tr}\lambda_1\lambda_2\lambda_3\lambda_4$ and has poles in s and t as in (1.3.11). The other pieces of the full amplitude consist of the term with group theory factor $\text{tr}\lambda_1\lambda_2\lambda_4\lambda_3$ which has poles in s and u and the term with group theory factor $\text{tr}\lambda_1\lambda_3\lambda_2\lambda_4$ which has poles in the t and u channel.

In superstring theories the consistency of string loops imposes stronger restrictions on the possible symmetry groups. In particular, the group $U(n)$ is not permitted, even before the question of chiral anomalies is addressed.

2.15 One-loop diagrams

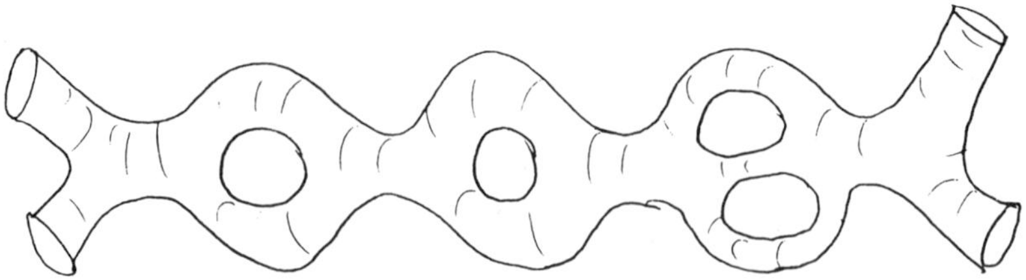

Figure 2.9. A closed-string multi-loop diagram.

The loop corrections to the tree diagrams correspond to world-sheets which are topologically non-trivial. In a theory with only closed strings the L-loop diagram is a sheet with L handles.

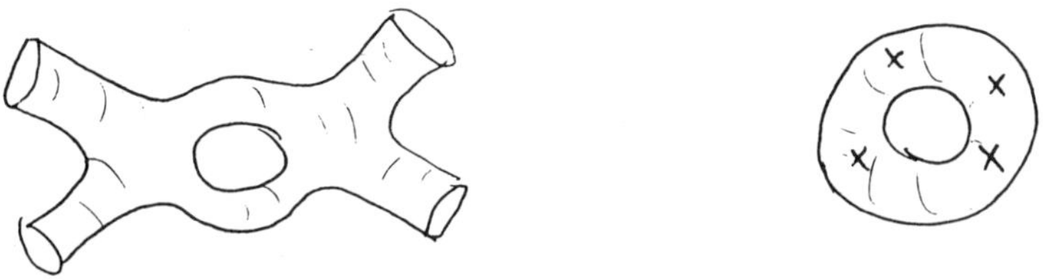

Figure 2.10. A one-loop closed-string diagram can be transformed into a torus.

Consider, for example, the one-loop case. The world-sheet can be transformed to that of a flat torus by making appropriate reparametrizations and conformal transformations. The external particles are mapped to singular points on the surface of the torus which must be integrated over. A flat torus can be described by identifying points in the complex z

plane,

$$z \simeq z + n\lambda_1 + m\lambda_2. \tag{2.15.1}$$

Unlike the case of a sphere, not all tori are related by reparametrizations and scale transformations. One of the complex parameters can be eliminated by a complex scaling $z \to kz$, leaving a single complex parameter $\tau = \lambda_2/\lambda_1$. Tori with different values of τ are inequivalent up to a discrete remnant of the original reparametrization and Weyl symmetry. This disrete symmetry that relates different τ's can be seen as follows. Consider new values of the parameters of the torus, λ_1' and λ_2', related to the old ones by

$$\begin{pmatrix} a & b \\ c & d \end{pmatrix} \begin{pmatrix} \lambda_1 \\ \lambda_2 \end{pmatrix} = \begin{pmatrix} \lambda_1' \\ \lambda_2' \end{pmatrix}, \tag{2.15.2}$$

where a, b, c and d are *integers* satisfying $ad - bc = 1$ so that the (2.15.2) is a $SL(2, Z)$ transformation. We may write the torus defined by (2.15.1) as

$$\begin{aligned} z &\simeq z + (n, m)\begin{pmatrix} \lambda_1 \\ \lambda_2 \end{pmatrix} \\ &= z + (n, m)\begin{pmatrix} d & -b \\ -c & a \end{pmatrix}\begin{pmatrix} a & b \\ c & d \end{pmatrix}\begin{pmatrix} \lambda_1 \\ \lambda_2 \end{pmatrix}, \\ &= z + n'\lambda_1' + m'\lambda_2' \end{aligned} \tag{2.15.3}$$

where $n' = nd - mc$ and $m' = -nb + ma$. We see that the original torus with $\tau = \lambda_1/\lambda_2$ is equivalent to a new torus with

$$\tau' = \frac{a\tau + b}{c\tau + d}. \tag{2.15.4}$$

$SL(2, Z)$ transformations of this type are known as *modular transformations.*

A general modular transformation can be built up by iterating two special transformations. These can be chosen to be

$$\tau \to \tau + 1, \tag{2.15.5}$$

and

$$\tau \to -1/\tau. \tag{2.15.6}$$

These transformations correspond to reparametrizations of the world-sheet which cannot be continuously deformed into the identity. They are often referred to as *large* diffeomorphisms.

For example, (2.15.5), corresponds to the diffeomorphism which is described by slicing through the torus as shown in figure 2.11, twisting through an angle of 2π and gluing the cut portions together.

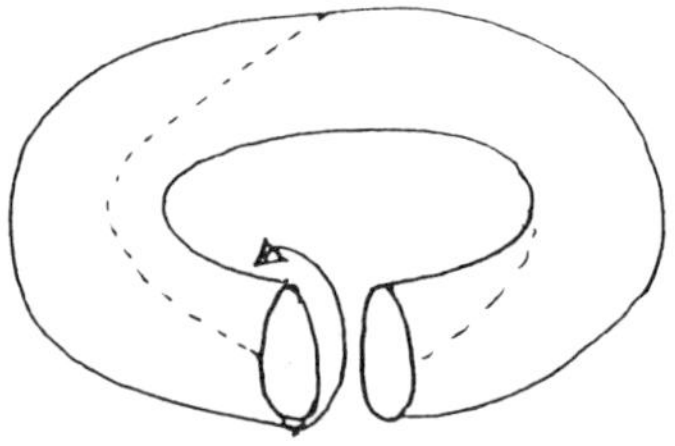

Figure 2.11. An example of a diffeomorphism of the torus which is not connected continuously to the identity.

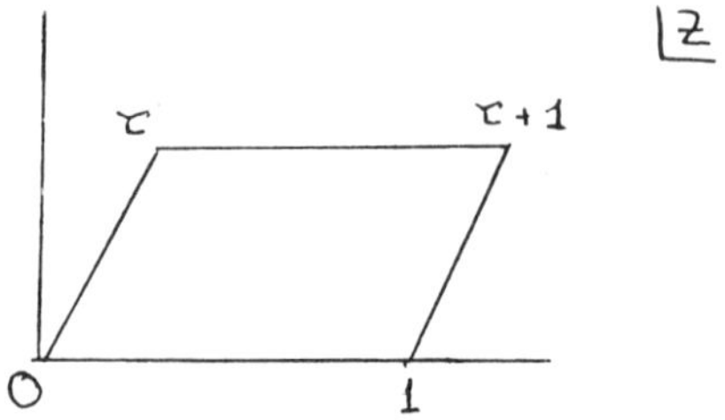

Figure 2.12. Representation of a torus in the complex z plane.

This is represented in terms of the flat z plane by cutting the parallelogram with a horizontal cut and displacing the upper part of the parallelogram by one unit relative to the bottom part as shown in figure 2.12. The transformation (2.15.6) corresponds to the transformation that interchanges the two axes in figure 2.12 (up to a sign and a complex rescaling).

Because of this discrete modular symmetry it is important not to integrate τ over all possible values since this would be overcounting.

Figure 2.13 shows the regions of the τ plane which are related by modular transformations. Any one region is equivalent to any other and it is often convenient to restrict the τ integral to the fundamental region marked F in the figure.

The parameter τ is known as a Teichmüller parameter and the upper-half τ plane is Teichmüller space. The fundamental region, which is (Teichmüller space/modular transformations) is known as moduli space.

For any fixed value of τ there is a $U(1) \times U(1)$ invariance of the loop integrand corresponding to rigid shifts of the origin of the z_i parameters which represent the points at which the external states are attached. This remnant of the $SL(2, C)$ invariance that arose in the case of the spherical world-sheet. This means that one complex z_i parameter can

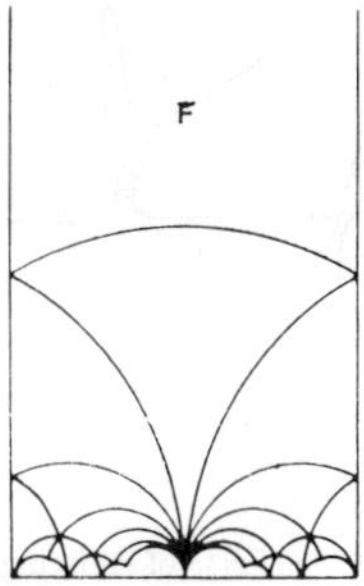

Figure 2.13. The region of τ integration, showing regions related by modular transformations.

be fixed, leaving $M-1$ z_i parameters to be integrated in addition to τ.

2.16 Multi-closed-string loops

More generally, for an L-loop amplitude (with $L>1$) there are $3L-3$ Teichmüller parameters. There are also M complex integration parameters corresponding to the positions of the external particles. When $L>1$ there are no residual invariances analogous to the $SL(2,C)$ symmetry of the $(L=0)$ tree diagrams or the $U(1)\times U(1)$ symmetry of the $(L=1)$ one-loop diagrams. In fact, the number of residual symmetries of the world-sheet is given by the number of conformal Killing vectors, which can be seen to be related to the number of zero modes of the ghost fields $c^{\pm}$. The number of Teichmüller parameters can likewise be shown to be equal to the number of zero modes of the anti-ghost field $b_{\alpha\beta}$. These quantities are given by

$$\begin{array}{rll} \text{when } L=0, & 1, & L>1 \\ \text{no.ofzero modes of } c^{\alpha}=6, & 2, & 0 \\ \text{no. of zero modes of } b_{\alpha\beta}=0, & 2, & 6L-6. \end{array} \quad (2.16.1)$$

The total number of *complex* integration variables is $M-$ # zero modes of c^{α} + # zero modes of $b_{\alpha\beta}=M+3L-3$, for any value of L. This number of parameters is in line with the number expected on the basis of the Riemann–Roch theorem.

For open string loops the corresponding number of parameters is $M+3L-3$ real variables.

Other lecturers at this school will explain this numerology in more detail so I will not pursue it here. Suffice to say that this illustrates the power of the ghost formalism in describing the geometrical and topological properties of world-sheets.

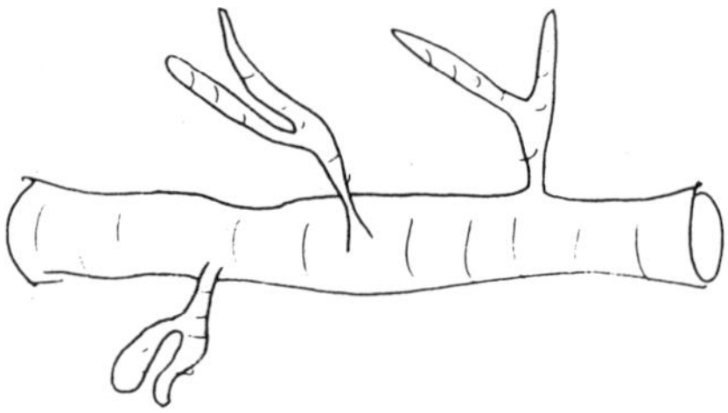

Figure 2.14. A closed-string propagator includes the effects of emitting closed-string tadpoles into the vacuum.

2.17 Simple consequences of conformal invariance

In calculating the tree amplitude for closed strings conformal invariance restricted the integrations to $M-3$ complex variables. As a result it is clear that the 3-particle on-shell closed-string amplitude is just a constant, *i.e.*,

$$A(1,2,3)=\kappa, \tag{2.17.1}$$

while

$$A(1,2)=0, \qquad A(1)=0. \tag{2.17.2}$$

The vanishing of the 1 and 2-particle amplitudes arises in Polyakov's approach from dividing by the infinite volume of the $SL(2,C)$ group. More explicitly, consider the special $SL(2,C)$ transformations, $z\rightarrow\lambda z$, from which it follows that

$$<V(z,\bar{z})>=\lambda^{-2}<V(\lambda z,\lambda\bar{z})>, \tag{2.17.3}$$

for closed strings (where V is the vertex for emitting a massless state with $k=0$). Choosing $z=0$ leads to an inconsistency unless

$$V=0. \tag{2.17.4}$$

For open strings coupling to open strings the analogous statement is

$$<V>=\lambda^{-1}<V>=0, \tag{2.17.5}$$

(which follows by again considering the vertex at the point $z=0$).

Other one-particle amplitudes do not vanish. For example, there is no scaling argument which would cause the vanishing of the vertex coupling the dilaton to the upper-half plane (or the interior of a disk, which is conformally equivalent to the upper-half plane).

The fact that the expectation value of the closed-string vertex operator vanishes on a sphere (or on the whole complex plane) due to the scale symmetry of the string world-sheet implies that string tadpoles vanish automatically at the tree level in any consistent closed-string theory. This means that the string theory is automatically defined with respect to a stable perturbative vacuum state.

Notice that, even for a single string, the string functional integral includes configurations of the world-sheet which correspond to summing over all tadpoles which describe the emission of strings into the vacuum as shown in figure 2.14. In ordinary field theories such tadpole diagrams are a feature of the interacting theory and the condition that the tadpoles vanish is equivalent to the statement that the equations of motion are satisfied by the classical background. Therefore, at the level of tree amplitudes, the string conformal invariance ensures that the backgound space-time satisfies the appropriate equations of motion. This applies equally well when the component fields have non-zero expectation values so that the string is moving in a curved background. I will return to this point when I describe strings propagating in curved backgrounds.

3. Operator methods for free bosonic strings; the Virasoro algebra.

3.1 Operator methods

Recall that the string action in the conformal gauge is given by

$$S = \frac{-T}{2} \int d\sigma d\tau \partial_\alpha X \cdot \partial^\alpha X. \tag{3.1.1}$$

The equations of motion, $\partial^2 X^\mu = 0$ that follow by varying this action must be supplemented by boundary conditions that make the surface terms $\partial_\sigma X \cdot \delta X \mid_{\sigma=0}^{\sigma=\pi}$ vanish. The appropriate boundary conditions for closed strings are simply periodicity of the coordinates, *i.e.*, $X^\mu(\sigma,\tau) = X^\mu(\sigma+\pi,\tau)$. For open strings the boundary conditions are

$$X'^\mu(\sigma,\tau) = 0 = X'^\mu(\sigma+\pi,\tau). \tag{3.1.2}$$

The constraint equations can be written as

$$T_{++} = \frac{1}{4}(\dot{X} + X')^2 = 0, \qquad (3.1.3)1$$

$$T_{--} = \frac{1}{4}(\dot{X} - X')^2 = 0, \qquad (3.1.4)2$$

where $\dot{X} = \partial_\tau X$ and $X' = \partial_\sigma X$.

The general solution of the closed-string equations and boundary conditions can be written as

$$X^\mu(\sigma,\tau) = X_R^\mu(\sigma^-) + X_L^\mu(\sigma^+). \tag{3.1.5}$$

The arbitrary periodic functions, X_R^μ and X_L^μ can be written as normal mode expansions in the form

$$X_R = \frac{1}{2}x^\mu + \frac{1}{2}p^\mu(\tau - \sigma) + \frac{i}{2}\sum_{n\neq 0}\frac{1}{n}\alpha_n^\mu e^{-2in(\tau-\sigma)}, \tag{3.1.6}$$

$$X_L = \frac{1}{2}x^\mu + \frac{1}{2}p^\mu(\tau + \sigma) + \frac{i}{n}\sum_{n\neq 0}\frac{1}{n}\tilde{\alpha}_n^\mu e^{-2in(\tau+\sigma)}. \tag{3.1.7}$$

The normal modes satisfy the conditions $\alpha_{-n}^\mu = \alpha_n^{*\mu}$ and $\tilde{\alpha}_{-n}^\mu = \tilde{\alpha}_n^{*\mu}$ which follow from the reality of X^μ. These conditions become $\alpha_{-n}^\mu = \alpha_n^{\dagger\mu}$ and $\tilde{\alpha}_{-n}^\mu = \tilde{\alpha}_n^{\dagger\mu}$ in the quantum theory (where the modes will become operators).

Recalling that $P^\mu = T\dot{X}^\mu = T(\dot{X}_R^\mu + \dot{X}_L^\mu)$ and substituting the mode expansions into the Poisson (or commutator) brackets

$$[P^\mu(\tau,\sigma), X^\nu(\tau,\sigma')] = -i\delta(\sigma - \sigma')\eta^{\mu\nu}, \tag{3.1.8}$$

gives (for $m, n \neq 0$)

$$[\alpha_m^\mu, \alpha_n^\nu] = m\eta^{\mu\nu}\delta_{m+n}, \tag{3.1.9}$$

$$[\tilde{\alpha}_m^\mu, \tilde{\alpha}_n^\nu] = m\eta^{\mu\nu}\delta_{m+n}, \tag{3.1.10}$$

and $[\alpha_m^\mu, \tilde{\alpha}_n^\nu] = 0$. These are the commutation relations of an infinite set of harmonic oscillators, with frequencies that depend on m. The zero modes are the overall momentum and position which satisfy

$$[p^\mu, x^\nu] = -i\eta^{\mu\nu}. \tag{3.1.11}$$

It will often prove convenient to define $\alpha_0 = \tilde{\alpha}_0 = p/2$ for closed strings.

For open strings the mode expansion satisfies the boundary conditions (3.1.2) and has the form

$$X^\mu(\sigma,\tau) = x^\mu + p^\mu\tau + i\sum_{n\neq 0}\frac{1}{n}\alpha_n^\mu e^{-in\tau}\cos n\sigma, \tag{3.1.12}$$

so that

$$\partial_\pm X^\mu \equiv \frac{1}{2}(\partial_\tau X^\mu \pm \partial_\sigma X^\mu) = \frac{1}{2}\sum_{n=-\infty}^{\infty}\alpha_n^\mu e^{-in(\tau\pm\sigma)}. \tag{3.1.13}$$

From this it is clear that

$$\partial_- X^\mu(\sigma,\tau) \equiv \partial_+ X(-\sigma,\tau), \tag{3.1.14}$$

and

$$T_{--}(\sigma,\tau) \equiv T_{++}(-\sigma,\tau). \tag{3.1.15}$$

Both $\partial_\pm X^\mu$ and $T_{\pm\pm}$ are functions which are periodic functions in the range $-\pi < \sigma < \pi$.

3.2 States of the quantised string

I will now describe the string states, beginning with the closed-string sector. The ground state, denoted $|0;k\rangle$, is an eigenstate of p^μ with eigenvalue k^μ and satisfies

$$\alpha_n^\mu|0;k>=0, \qquad n>0. \tag{3.2.1}$$

Excited states are obtained by applying chains of creation operators to the ground state

$$\alpha_{-n_1}^{\mu_1}\alpha_{-n_2}^{\mu_2}\cdots|0;k\rangle. \tag{3.2.2}$$

One obvious problem with these states is that they include states of negative norm created by the timelike oscillator modes, α_{-n}^0. For example, the state $\alpha_{-n}^0|0;k\rangle$ has a norm

$$\langle k';o|\alpha_n^0\alpha_{-n}^0|0;k\rangle=\eta^{00}\delta^{26}(k-k')=-\delta^{26}(k-k'). \tag{3.2.3}$$

Such states are called $\mathcal{GHOSTS}$.* They spell disaster for any theory since a negative-normed state has a negative probability of being produced. The problem of negative metric states ($\mathcal{GHOSTS}$) also arises in Yang–Mills theories where the component A^0 of the vector potential creates such states. In that case the consistency of the theory is assured by the gauge symmetry which guarantees that the $\mathcal{GHOSTS}$ decouple from the physical spectrum.

In the case of strings there are an infinite number of $\mathcal{GHOST}$ states in the full Fock space of the infinite number of oscillator modes. Correspondingly, there has to be an infinitely larger gauge symmetry to ensure that they decouple from the physical space of states. This is the reason why the existence of the infinite number of conservation laws noted earlier is so important.

From the earlier discussion it is evident that the conserved charges are associated with the moments of the energy-momentum tensor. These moments are known as the *Virasoro* generators. For closed strings we define (for $m\neq 0$)

$$L_m=\frac{1}{2\pi}\int_0^\pi e^{-2im\sigma}T_{--}d\sigma=\frac{1}{2}\sum_{n=-\infty}^{\infty}\alpha_{m-n}^\mu\alpha_{n\mu}, \tag{3.2.4}$$

$$\tilde{L}_m=\frac{1}{2\pi}\int_0^\pi e^{2im\sigma}T_{++}d\sigma=\frac{1}{2}\sum_{n=-\infty}^{\infty}\tilde{\alpha}_{m-n}^\mu\tilde{\alpha}_{n\mu}. \tag{3.2.5}$$

There is an ambiguity in the definition of L_0 and $\tilde{L}_0$ in the quantum theory due to the fact that the expressions contain factors which do not commute. The zero modes are *defined*

* I am using the ghost-like script to avoid confusing these negative-metric ghosts with those of Fadeev and Popov.

to be normal ordered

$$L_0 = \frac{1}{2}\alpha_0^2 + \sum_{n=1}^{\infty} \alpha_{-n}^{\mu}\alpha_{n\mu}, \tag{3.2.6}$$

$$\tilde{L}_0 = \frac{1}{2}\tilde{\alpha}_0^2 + \sum_{n=1}^{\infty} \tilde{\alpha}_{-n}^{\mu}\tilde{\alpha}_{n\mu}. \tag{3.2.7}$$

Reality of the energy-momentum tensor requires

$$L_n^{\dagger} = L_{-n}, \qquad \tilde{L}_n^{\dagger} = \tilde{L}_n. \tag{3.2.8}$$

The zero mode of the energy-momentum tensor, $L_0 + \tilde{L}_0$ is the 'Hamiltonian' of the two-dimensional theory. In other words it is the operator which generates translations in τ so that $L_0 + \tilde{L}_0 \sim i\partial_\tau$.

For open strings the discussion is similar. In this case we saw from (3.1.15) that $T_{++}(\sigma,\tau) = T_{--}(-\sigma,\tau)$. The modes are now defined (for $m \neq 0$) by

$$\begin{aligned} L_m &= \frac{1}{\pi}\int_0^{\pi} \left(e^{-im\sigma}T_{--} + e^{im\sigma}T_{++}\right) d\sigma = \frac{1}{\pi}\int_{-\pi}^{\pi} e^{im\sigma}T_{++}d\sigma \\ &= \frac{1}{2}\sum_{n=-\infty}^{\infty} \alpha_{m-n}^{\mu}\alpha_{n\mu}, \end{aligned} \tag{3.2.9}$$

and

$$L_0 = \frac{1}{2}\alpha_0^2 + \sum_{n=1}^{\infty} \alpha_{-n}^{\mu}\alpha_{n\mu}, \tag{3.2.10}$$

where, for open strings

$$\alpha_0^{\mu} = p^{\mu}. \tag{3.2.11}$$

The algebra of the energy-momentum can now be deduced from the commutation relations of the L_m and $\tilde{L}_m$,

$$[L_m, L_n] = (m-n)L_{m+n} + \frac{D}{12}(m^3 - m)\delta_{m+n,0}, \tag{3.2.12}$$

with an identical algebra for the $\tilde{L}_m$. Furthermore, $[L_m, \tilde{L}_n] = 0$.

Equation (3.2.12) expresses the modes of $[T_{++}(\tau,\sigma), T_{++}(\tau,\sigma')]$ which was quoted in the last lecture. The algebra (3.2.12) is obtained by substituting the expressions for L_m in terms of the oscillators. Only the anomaly (proportional to D) is tricky to calculate. It arises in the calculation from the difference of two infinite sums that arise from reordering the oscillators. The easiest way to calculate it is to use the following trick. The general form of the algebra is

$$[L_m, L_n] = (m-n)L_{m+n} + A(m)\delta_{m+n,0}. \tag{3.2.13}$$

The form for the anomaly is constrained by the Jacobi identity which requires

$$[[L_m, L_n], L_k] + [[L_n, L_k], L_m] + [[L_k, L_m], L_n] = 0. \tag{3.2.14}$$

Substituting for the commutators gives

$$(m-n)A(k) + (n-k)A(m) + (k-m)A(n) = 0. \tag{3.2.15}$$

Now consider the special case $k = 1$, $m = -n - 1$ so that

$$A(n+1) = \frac{(n+2)A(n) - (2n+1)A(1)}{(n-1)}. \tag{3.2.16}$$

This equation can be solved in terms of two arbitrary constants,

$$A(n) = an^3 + bn. \tag{3.2.17}$$

The constants can be determined by considering special matrix elements between zero-momentum ground states, $\langle 0;0|[L_{-1}, L_1]|0;0\rangle$ and $\langle 0;0|[L_{-2}, L_2]|0;0\rangle$. By substituting the explicit expressions for the L_m and the commutation relations (3.2.13) we find

$$A(1) = 0, \qquad A(2) = \frac{D}{2}. \tag{3.2.18}$$

This determines $a + b = 0$ and $6a = D/2$ so that

$$A(n) = \frac{D}{12}(n^3 - n). \tag{3.2.19}$$

This analysis ignores the contribution of the ghost modes to the Virasoro generators. In the formalism which includes the ghosts we must add the terms to L_m and $\tilde{L}_m$ which come from the modes of T^{gh}_{++} and T^{gh}_{--}, defined in (2.9.13). This gives

$$L^{gh} = \sum_{n=-\infty}^{\infty} (m-n) b_{m+n} c_{-n}, \tag{3.2.20}$$

where c_m and b_m are the modes of the ghost coordinate, $c^+(\sigma,\tau)$, and the anti-ghost coordinate, $b_{++}(\sigma,\tau)$, respectively. These modes satisfy

$$\{c_m, b_n\} = \delta_{m+n,0}, \tag{3.2.21}$$

from which it follows that

$$\left[L^{gh}_m, L^{gh}_n\right] = (m-n)L^{gh}_{m+n} + \frac{1}{6}(m - 13m^3). \tag{3.2.22}$$

Combining the anomaly terms on the right-hand sides of (3.2.12) and (3.2.22) gives a total anomaly of $\frac{1}{12}(D-26)(m^3 - m)$, in agreement with the formula quoted in the last lecture.

3.3 Physical states

The constraint equations of the classical theory, $T_{++} = 0 = T_{--}$ translate into conditions on the expectation values of these constraints between states of the quantum theory. I will describe the states in the original formalism in which ghost modes are ignored. In the formalism that includes ghosts the states I shall describe are those in the subspace of the states that have zero ghost number (more accurately, those with ghost-number -1/2). In order to impose the conditions on the expectation values it is sufficient to require the positive m conditions

$$L_m|\text{phys}\rangle = 0, \tag{3.3.1}$$

$$\tilde{L}_m|\text{phys}\rangle = 0, \tag{3.3.2}$$

together with the zero-mode conditions

$$(L_0 - a)|\text{phys}\rangle = 0 = (\tilde{L}_0 - \tilde{a})|\text{phys}\rangle. \tag{3.3.3}$$

The constants, a and $\tilde{a}$, have been inserted because of the normal ordering effects associated with the definitions of L_0 and $\tilde{L}_0$. The consistency of the theory will determine the values of these constants to be 1. Notice that it would not be consistent to impose the negative m conditions on the states as well as the positive m conditions. The fact that only half the conditions (for example, those with positive m) should be imposed arises quite naturally in the BRST method of treating the theory with ghost modes.

3.4 Spectrum of states

The zero-mode conditions, (3.3.3) determine the masses of the states since. For example, for open strings the condition

$$(L_0 - a)|\text{phys}\rangle = 0, \tag{3.4.1}$$

can be written as

$$M^2 \equiv -p^2 = 2N - 2a, \tag{3.4.2}$$

where the number operator is defined by

$$N = \sum_1^\infty \alpha_{-n} \cdot \alpha_n. \tag{3.4.3}$$

The ground state of the open-string sector, $|0;k\rangle$, has $M^2 = -2a$ and $L_n|0;k\rangle = 0$ for all $n > 0$.

The first excited state has a vector index which is associated with a polarization vector $\varsigma^\mu(k)$. The state $\varsigma_\mu(k)\alpha_{-1}^\mu|0;k\rangle$ has $M^2 = 2 - 2a$. For $n > 0$ the L_n conditions are automatically satisfied, but the condition $L_1|\ \rangle = 0$ requires $\varsigma_\mu k^\mu = 0$. This means that there are only $D-1$ independent polarizations. The normalization of these states is given by $\varsigma \cdot \varsigma$. To determine the sign of this normalization consider a special choice of the particle momentum $k^\mu = (k_0, 0, \ldots, 0, k_{D-1})$. The $D-2$ states with polarization vectors, ς^i which lie in the transverse directions $i = 1, 2, \ldots, D-2$ manifestly have positive norm. The sign of the norm of the last state depends on the value of k^2. If $k^2 > 0$ (which means that $a < 1$) this last state is a tachyon. This follows from the fact that k can be chosen to be $(0, 0, \ldots, k_{D-1})$ so that $\varsigma \cdot k = 0$ implies that ς is time-like ($\varsigma = (\varsigma_0, 0, \ldots, 0)$ and $\varsigma^2 < 0$). If $k^2 < 0$ (*i.e.*, $a > 1$) the last ς is space-like and $\varsigma^2 > 0$. Finally, if $k^2 = 0$ (*i.e.*, $a = 1$) the last ς^μ is proportional to k^μ so that $\varsigma^2 = 0$. In this case the last state is null so that there are only $D-2$ physically relevant states. This is, of course, the case that corresponds to Yang–Mills field theory where the massless vector gauge particle corresponds to transversely-polarized states.

This analysis can be extended to all the states in the string theory. The proofs (by Brower and by Goddard and Thorn) that the gauge conditions can decouple all the $\mathcal{GHOST}$ states was one of the milestones of the early development of string theories. I will not repeat the proof here since it will arise in another guise in my discussion of the light-cone gauge. The result of the analysis is that there are no $\mathcal{GHOSTS}$ in the physical spectrum provided $D = 26$ and $a = 1$ or $D < 26$ and $a \geq 1$.

I will delay the discussion of the closed-string spectrum until I have introduced the light-cone gauge. The advantage of the light-cone gauge treatment is that the spectrum is explicit without the need to account for the rather complicated subsidiary constraints.

4. The light-cone gauge; density of states; four-particle amplitudes; background curvature.

4.1 Light-cone gauge operators

The conformal gauge choice still permits conformal reparametrizations which satisfy $\partial_- \xi^+ = 0 = \partial_+ \xi^-$. This means that the residual reparametrization invariance is given for closed strings by the transformations

$$\begin{aligned} \tau \to \hat{\tau} &= \tau + \frac{1}{2}\xi^+(\sigma^+) + \frac{1}{2}\xi^-(\sigma^-) \\ \sigma \to \hat{\sigma} &= \sigma + \frac{1}{2}\xi^+(\sigma^+) - \frac{1}{2}\xi^-(\sigma^-) \end{aligned}, \tag{4.1.1}$$

i.e.,

$$\partial_+\partial_-\hat{\tau} = 0, \qquad \hat{\sigma} = \int_0^\sigma d\sigma' \partial_\tau \hat{\tau}(\sigma', \tau), \tag{4.1.2}$$

(in the case of open strings the transformations must respect the boundary conditions). Since $X^+(\sigma, \tau)$ satisfies the two-dimensional wave equation $\partial_\alpha \partial^\alpha X^+ = 0$ we can choose the new parameters so that*

$$X^+(\sigma,\tau) = x^+ + p^+\hat{\tau}. \tag{4.1.3}$$

This choice of parametrization has the property that X^+ does not depend on $\hat{\sigma}$. Each point on the string has a common value of the light-cone 'time' variable, X^+, in this parametrization. The constant, p^+, is the + component of the total momentum. This follows from the fact that $P^\mu(\hat{\sigma},\hat{\tau}) = T\dot{X}^\mu(\hat{\sigma},\hat{\tau})$ and setting $T = 1/\pi$ which has been my convention. Since all subsequent formulas will only involve hatted variables it is notationally convenient to drop all the hats.

An obvious disadvantage of this choice of parametrization is that a particular spatial direction is picked out and treated asymmetrically. As a result, the formalism will only be manifestly invariant under the $SO(D-2)$ Lorentz transformations transverse to the + and − directions. At every stage the full Poincaré symmetry of the theory is obscured and it is important to verify that it is maintained since it could be destroyed by quantum anomalies.

The advantage of using the light-cone gauge is that the constraint equations, $(\dot{X}\pm X')^2 = 0$, can be solved explicitly. Noting that $(\dot{X}\pm X')^+ = p^+$ it is easily seen that

$$(\dot{X}\pm X')^- = \frac{1}{2p^+}(\dot{X}^i \pm X'^i)^2. \tag{4.1.4}$$

Subtracting these two equations gives

$$X^-(\sigma,\tau) = \frac{1}{2p^+}\int_0^\sigma d\sigma' \underline{\dot{X}}\cdot\underline{X'} + \text{const.} \tag{4.1.5}$$

where $\underline{X} \equiv X^i$. In terms of the open-string mode expansion, (3.1.12),

$$\alpha_n^+ = 0, \qquad n > 0, \tag{4.1.6}$$

$$\alpha_n^- = \frac{1}{p^+}\left(\frac{1}{2}\sum_{m=-\infty}^{\infty} :\alpha_{n-m}^i\alpha_m^i: + \delta_{n,0}\frac{D-2}{2}\sum_{m=1}^{\infty} m\right). \tag{4.1.7}$$

In particular,

$$\alpha_0^- = \frac{1}{p^+}(\underline{k}^2 + N - a), \tag{4.1.8}$$

where N is defined by (3.4.3) and

$$a = -\frac{1}{2}(D-2)\sum_1^\infty m. \tag{4.1.9}$$

In the light-cone gauge approach the time variable is chosen to be $X^+ = -i\partial/\partial p^-$. This

---

* The conventions used for space-time light-cone components are that for an arbitrary vector V^μ, $V^\pm = \frac{1}{\sqrt{2}}(V^0 \pm V^{D-1})$ so that $V^+ = -V_-$, $V^- = -V_+$. The scalar product of two vectors is given by $W\cdot V \equiv W^\mu V_\mu = -W^+V^- - W^-V^+ + W^iV^i$, where $i = 1, 2, \ldots, D-2$.

means that the hamiltonian, the operator that generates time translations, is

$$h \equiv p^- = \int_0^\pi P^- d\sigma = \int_0^\pi T\dot{X}^- d\sigma = \alpha_0^- . \tag{4.1.10}$$

The mass-shell condition follows from this equation by substituting for α_0^- from (4.1.8), giving

$$(\text{Mass})^2 \equiv -p^2 = 2p^+p^- - \underline{p} = 2(N + \frac{D-2}{2}\sum_{m=1}^{\infty} m). \tag{4.1.11}$$

The normal ordering infinity, which arises from the ultraviolet frequencies, needs careful consideration. One way of dealing with it is to introduce a cutoff by considering a lattice approximation to the string parameter, σ (following the method of Giles and Thorn). Suppose the string is approximated by M points spaced by l so that $Ml = \pi$. In the end we are interested in taking the limit $M \to \infty$ with $l = \pi/M \to 0$. The wave equation reduces to a finite set of equations for the normal modes which can be solved exactly. I will not show the details but the result is that (as $M \to \infty$)

$$(\text{Mass})^2 = 2p^+p^- - \underline{p}^2 = 2\left(N - \frac{D-2}{24} + \frac{4M(D-2)p^+}{\pi l} - \frac{p^+}{l}\right). \tag{4.1.12}$$

In this expression there are two terms which are not Lorentz invariant (since they contain p^+) and which are infinite as $M \to \infty$. The first of these terms is proportional to the number of sites on the lattice and could be eliminated by redefining $P^-(\sigma, \tau)$ by subtracting $2(D-2)/\pi l$ at each of the M sites. Actually, it isnt necessary to eliminate this term from P^- since it does not lead to any measurable effects. It merely contributes an infinite phase to any scattering amplitude. The second infinite term in (4.1.12) is $-2p^+/l$. It can be eliminated by subtracting $-p^+/2$ from P^- at each string endpoint. After subtracting these terms the mass spectrum is Lorentz invariant. Redefining P^- in this manner is equivalent to using the formula

$$\sum_{m=1}^{\infty} m = -\frac{1}{12} \tag{4.1.13}$$

in (4.1.11). This bizarre-looking formula also results if it is assumed that it is consistent to use ζ function regularization. In that case one considers the Riemann ζ function, $\sum_{m=1}^{\infty} m^{-s} = \zeta(s)$, in the case $s = -1$. From (4.1.9) it follows that the parameter $a = (D-2)/24$.

4.2 Spectrum of open strings - The critical dimension

It is now easy to ennumerate the states of the open-string theory which are made by acting with the transverse creation operators, α^i_{-m} acting on the ground state. There are no longer any constraints to worry about. The ground state, $|0;k\rangle$, for which $N = 0$, has a mass given by

$$(\text{Mass})^2 = -\frac{D-2}{12}. \tag{4.2.1}$$

The first excited state, $\alpha^i_{-1}|0;k\rangle$, has $N = 1$ so that its mass is given by

$$(\text{Mass})^2 = 2 - \frac{D-2}{12}. \tag{4.2.2}$$

This state only has $D - 2$ (transverse) polarizations. We know from general properties of the Lorentz group that the only Lorentz covariant states which only have transverse indices are massless gauge particles. Therefore, the mass of the state must vanish, so that $D = 26$! Notice that this condition is stronger than the condition that arises in the covariant argument given earlier. The light-cone method, as described here, is predicated on the transversality of the spectrum. In the case that $D < 26$ there would also be longitudinal states and the method would have to be modified accordingly.

Continuing the ennumeration of states, one possible type of state at the second excited level is $\alpha^i_{-1}\alpha^j_{-1}|0;k\rangle$, which has $N = 2$ and has $(\text{Mass})^2 = 4 - (D-2)/12 = 2$ (if $D = 26$). There are $\frac{1}{2}(D-1)(D-2)$ states of this type (a symmetric $(D-2) \times (D-2)$ matrix). In addition there are $D - 2$ states of the form $\alpha^i_{-2}|0;k\rangle$ which also have $(\text{Mass})^2 = 4 - (D-2)/12$. The combination of these $SO(D-2)$ tensor and vector states form the $(D-2)(D+1)/2$ components of a single massive state transforming as a tensor of $SO(D-1)$, which is the appropriate little group for classifying massive states in D dimensions. In fact, all of the infinite number of massive states created by applying the transverse oscillators, combine in this way into $SO(D-1)$ representations. The fact that this is true is part of the great beauty of string theory. It is a consequence of the fact that although the theory does not look Lorentz invariant in the light-cone gauge it really is, at least when $D = 26$.

4.3 Spectrum of closed strings

In considering closed-string states there is one important new feature that must be discussed. The boundary conditions now require periodicity of the coordinates in σ. Clearly the origin of the σ coordinate must not affect the physics since no point on the string is special. This is a remnant of reparametrization invariance which is still of significance in the light-cone gauge. It implies that states must be unaltered by a rigid shift

$$\sigma \to \sigma + \sigma_0. \tag{4.3.1}$$

It is easy to construct the operator which translates σ in this way since

$$e^{i\sigma_0(N-\tilde{N})} X^i(\sigma) e^{-i\sigma_0(N-\tilde{N})} = X^i(\sigma + \sigma_0), \tag{4.3.2}$$

as can be seen by substituting the expression for $X^i(\sigma)$ in terms of the transverse oscillator modes. The number operators, N and $\tilde{N}$ are the same operators, made out of α_n and $\tilde{\alpha}_n$, as the open-string operator, (3.4.3). The statement that physics is independent of such shifts in σ is therefore the condition on any physical closed-string state

$$(N - \tilde{N})|\ \rangle = 0. \tag{4.3.3}$$

In terms of the coordinate representation the condition means that the string wave functional, $\Psi[X^i(\sigma)]$ must satisfy

$$\Psi[X^i(\sigma)] = \Psi[X^i(\sigma + \sigma_0)]. \tag{4.3.4}$$

If σ_0 is infinitesimal this can be expanded in a Taylor series to give

$$\int_0^\pi d\sigma' X^{i\prime}(\sigma') \frac{\delta}{\delta X^i(\sigma')} \Psi[\underline{X}] = 0. \tag{4.3.5}$$

The operator acting on $\Psi[\underline{X}]$ can be rewritten using $P^i = -i\delta/\delta X^i = \dot{X}^i/\pi$. Using the mode expansions this operator reduces to $N - \tilde{N}$. The normal ordering constants cancel mode by mode. This is the only constraint that connects the 'left-moving' modes (those with behaviour $\exp(in(\tau+\sigma))$) and the 'right- moving' modes (which have the behaviour $\exp(in(\tau-\sigma))$). In no other part of the analysis of the simplest closed-string theories is there any link between these two sectors. This statement is not true, however, of some of the recently discovered theories.

In the case of closed strings the mass-shell condition similar to (4.1.11) is (again using $\sum_1^\infty m = -1/12$)

$$(\text{mass})^2 \equiv -p^2 = 2p^+p^- - \underline{p}^2 = 4\left(N + \tilde{N} - \frac{D-2}{12}\right). \tag{4.3.6}$$

The ground state, $|0;k\rangle$, has $N = \tilde{N} = 0$ so that its mass satisfies $(\text{Mass})^2 = -(D-2)/3$.

The first excited state is $\alpha^i_{-1}\tilde{\alpha}^j_{-1}|0;k>$ which is the state with $N = \tilde{N} = 1$. (The state with $N = 0$ and $\tilde{N} = 1$ is not an allowed state since it violates the subsidiary condition, $N = \tilde{N}$.) The mass of this state is given by $(\text{Mass})^2 = 8 - (D - 2/3)$. As in the case of the vector open-string state, this transverse tensor can only make sense in a Lorentz-covariant theory if it is a massless state so that once more it is necessary that $D = 26$. These tensor states include the state created by the graviton field, g^{ij}, as the symmetric, traceless piece. In addition, the trace is a scalar state corresponding to a 'dilaton' field, ϕ and the antisymmetric piece corresponds to an antisymmetric tensor field, A^{ij}.

The higher-mass states are created by applying strings of α_{-n} and $\tilde{\alpha}_{-n}$ to the ground state,

$$\alpha^{i_1}_{-m1}\cdots\tilde{\alpha}^{j_1}_{-n_1}\cdots|0;k\rangle, \tag{4.3.7}$$

subject to the subsidiary constraint $N = \tilde{N}$ which implies

$$\sum_{r=1}^{\infty} m_r = \sum_{s=1}^{\infty} n_s. \tag{4.3.8}$$

The mass of these states is given by

$$(\text{Mass})^2 = 4\left(\sum m_r + \sum n_s - 2\right) \tag{4.3.9}$$

4.4 Lorentz invariance in the light-cone gauge.

A Lorentz transformation of the coordinates is defined, up to an arbitrary reparametrization by

$$\delta X^\mu = a^\mu{}_\nu X^\nu(\sigma,\tau) + \xi^\alpha \partial_\alpha X^\mu(\sigma,\tau). \tag{4.4.1}$$

In any conformal gauge the ξ^α are restricted to satisfy $\partial_-\xi^+ = \partial_+\xi^- = 0$. In the light-cone gauge a Lorentz transformation must be accompanied by a particular ξ^α term so that the Lorentz-transformed gauge condition is restored to the original gauge condition. This is called a *compensating* gauge transformation. This means that we require that $X^+ = x^+ + p^+\tau$ be restored in the transformed frame, ie

$$\delta X^+ = a^+{}_\nu x^\nu + a^+{}_\nu p^\nu \tau, \tag{4.4.2}$$

so that, using (4.4.1),

$$a^+{}_\nu X^\nu + \xi^0 p^+ = a^+{}_\nu (x^\nu + p^\nu \tau), \tag{4.4.3}$$

therefore,

$$\xi^0 = \frac{a^+{}_\nu}{p^+}\left(x^\nu + p^\nu\tau - X^\nu(\sigma,\tau)\right), \tag{4.4.4}$$

and

$$\xi^1 = \int_0^\sigma d\sigma' \frac{\partial \xi^0}{\partial \tau}. \tag{4.4.5}$$

Having determined ξ^α in this way we see from (4.4.1) that $\delta X^i(\sigma,\tau)$ involves quadratic terms for those transformations associated with $a^+_{\ i}$ which are generated by J^{i-}. We saw earlier that the Lorentz generators, which satisfy the algebra

$$[J^{\mu\nu}, J^{\rho\omega}] = -i\eta^{\nu\rho}J^{\mu\lambda} + i\eta^{\mu\rho}J^{\nu\lambda} + i\eta^{\nu\lambda}J^{\mu\rho} - i\eta^{mu\lambda}J^{\nu\rho}, \tag{4.4.6}$$

are given by $J^{\mu\nu} = \frac{1}{\pi}\int_0^\pi d\sigma[X^\mu\dot{X}^\nu - X^\nu\dot{X}^\mu]$. Clearly J^{i-} is indeed complicated since $X^- \sim \int \dot{X}^i X'^i d\sigma$ so that J^{i-} contains terms cubic in oscillators. In checking that the light-cone gauge expressions for the Lorentz generators satisfy (4.4.6) the most difficult one to check is $[J^{i-}, J^{j-}] = 0$. The reader should check that the commutator actually gives

$$[J^{i-}, J^{j-}] = -\frac{1}{(p^+)^2}\sum_{m=1}^{\infty}\left(\alpha^i_{-m}\alpha^j_m - \alpha^j_{-m}\alpha^i_m\right)\left(m\frac{(26-D)}{12} + \frac{1}{m}\left(\frac{D-26}{12} + 2(1-a)\right)\right), \tag{4.4.7}$$

which only vanishes if $D = 26$ and $a = 1$.

4.5 Density of states

Clearly the number of states of a given mass increases rapidly as the mass increases. Let d_n be the total number of states with mass, M, given by $M^2 = 2\pi T(n-1)$. A generating function for the d_n is easily written since (using the fact that the trace is the sum over all states)

$$\mathrm{tr}\omega^N = \sum_{n=0}^{\infty} d_n \omega^n. \tag{4.5.1}$$

Substituting the expression for the number operator gives (for $D = 26$)

$$\begin{aligned}\mathrm{tr}\omega^N \equiv \prod_{i=1}^{D-2}\prod_{n=1}^{\infty}\mathrm{tr}\omega^{n\alpha^i_{-n}\alpha^i_n} &= \prod_{i=1}^{24}\prod_{n=1}^{\infty}\sum_{r=0}^{\infty}\omega^{rn} \\ &= \prod_{n=1}^{\infty}(1-\omega^n)^{-24} \equiv [f(\omega)]^{-24}.\end{aligned} \tag{4.5.2}$$

The function $f[\omega]$ is the partition function, defined by

$$f(\omega) = \prod_{n=1}^{\infty}(1-\omega^n). \tag{4.5.3}$$

Setting $\omega = e^{2i\pi\tau}$ (or $\tau = \ln\omega/2i\pi$) we see that this is closely related to Dedekind's function

$$\eta(\tau) = e^{i\pi\tau/12}\prod_{n=1}^{\infty}\left(1 - e^{2\pi i n\tau}\right). \tag{4.5.4}$$

From (4.5.1) the number of states at level n is given by

$$d_n = \frac{1}{2i\pi}\oint\frac{\mathrm{tr}\omega^N}{\omega^{n+1}}d\omega, \tag{4.5.5}$$

where the integral encircles the origin in the ω plane. As $n \to \infty$ the integral is dominated by the region near $\omega = 1$. This is easiest to analyse by changing variables by performing the modular transformation $\tau \to \tau' = -1/\tau$, *i.e.*, $\omega \to \omega' = \exp(4\pi^2/\ln\omega)$. The η function transforms simply under modular transformations (as can be seen by expressing it as an infinite series and evaluating its fourier transform)

$$\eta(-1/\tau) = (-i\tau)^{1/2}\eta(\tau), \tag{4.5.6}$$

so that

$$f(\omega) = \left(\frac{-2\pi}{\ln\omega}\right)^{1/2} \omega^{-1/24} q^{1/12} f(q^2), \tag{4.5.7}$$

where

$$q = e^{2\pi^2/\ln\omega}. \tag{4.5.8}$$

As $\omega \to 1$ $(q \to 0)$

$$f^{-24}(\omega) \sim (\ln\omega)^{12} e^{-4\pi^2/\ln\omega}, \tag{4.5.9}$$

so that

$$\begin{aligned} d_n &\sim \frac{1}{2i\pi}\oint (\ln\omega)^{12} e^{-\frac{4\pi^2}{\ln\omega}-(n+1)\ln\omega} d\omega \\ &\sim \text{const } n^{-27/4} e^{4\pi\sqrt{n}}. \end{aligned} \tag{4.5.10}$$

The density of states of mass $M = \sqrt{2n}$ is therefore given by

$$\rho(M)dM \sim \text{const } M^{-13} e^{4\pi M/\sqrt{2}} dM. \tag{4.5.11}$$

This result shows an exponentially increasing number of states which is similar to the behaviour of the density of states postulated by Hagedorn in his 'statistical bootstrap' model of hadronic physics. Hagedorn noted that a system of this type has very unusual thermodynamics. In particular there is a maximum possible temperature, an 'ultimate' temperature, determined by the string tension. This can be seen by constructing the canonical partition function. Roughly speaking the number of massive states is so huge that as energy is pumped into the system it is entropically favourable to create a massive, static state rather than cause the light states to have greater kinetic energy. Furthermore, the thermodynamic behaviour is very sensitive to the power of M in the prefactor in (4.5.10). When this power is large enough there is a transition to a wildly fluctuating phase at finite energy.

These considerations are all in the context of the free string theory. Whether the notion of an ultimate temperature survives when string interactions are included is not known.

4.6 Operator calculations

Earlier, I described Polyakov's functional method for calculating scattering amplitudes. I would now like to describe the operator method for amplitude calculations. I will illustrate the method by calculating open-string tree diagrams with external ground-state tachyons. As with Polyakov's method the rules for these calculations are not derived from a fundamental principle but are arranged to give S-matrix elements which satisfy crossing symmetry and unitarity. The rules resemble ordinary Feynman rules which join vertices together with propagators.

The propagator, Δ, is an operator in the string Fock space defined by

$$\Delta = \frac{1}{2L_0 - 2} = \frac{1}{2}\int_0^\infty d\tau e^{-\tau(L_0 - 1)}, \tag{4.6.1}$$

(recalling that $L_0 = p^2/2 + \sum_{n=1}^{\infty} \alpha_{-n} \cdot \alpha_n$). This obviously has poles at the positions of the string states.

The vertex operators are of the same form as the vertices introduced earlier in the context of the functional approach. Open strings are emitted from the endpoint $\sigma = 0$ or $\sigma = \pi$ at an arbitrary proper time, τ. For tachyon emission we have

$$\begin{aligned} V(k,\tau) &=: e^{ik.X(\sigma=0,\tau)} : \\ &= e^{k_\mu \sum_1^\infty \frac{1}{n}\alpha^\mu_{-n} e^{n\tau}} e^{ik\cdot x + k\cdot p\tau} e^{-k_\mu \sum_1^\infty \frac{1}{n}\alpha^\mu_n e^{-n\tau}} . \end{aligned} \tag{4.6.2}$$

The exponential involving the zero modes can be written as $e^{ik\cdot x} e^{-k\cdot p\tau} e^{-k^2\tau/2}$ by using the relation

$$e^A e^B = e^{A+B+\frac{1}{2}[A,B]}, \tag{4.6.3}$$

which is valid if $[A, B]$ commutes with A and B.

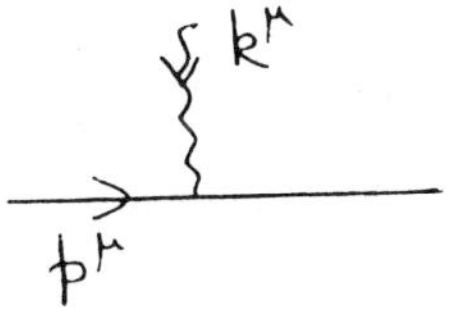

Figure 4.15. The kinematics associated with the emission vertex.

The value of τ is shifted by the operator $e^{\tau L_0}$ so that

$$V(k,\tau)=e^{\tau L_0}V(k,0)e^{-\tau L_0}, \tag{4.6.4}$$

(which follows from $[L_0,\alpha_m]=-m\alpha_m$). Another important relation that is needed if the tree diagrams are to transmit the gauge conditions that decouple the $\mathcal{GHOSTS}$ is

$$[L_m,V(k,\tau)]=\frac{d}{d\tau}\left(\tau^m V(k,\tau)\right). \tag{4.6.5}$$

Consider a scattering amplitude in which the external strings, $1,2,\ldots,M$, are ordered around the boundary of the world-sheet so that there is an overall group theory factor of $G=\mathrm{tr}\left(\lambda^1\lambda^2\ldots\lambda^M\right)$. The expression for this piece of the amplitude is

$$A(1,2,\ldots M)=g^{M-2}G<0;k_1|V(k_2;0)\Delta V(k_3;0)\ldots|0;k_M>. \tag{4.6.6}$$

For example, consider the case $M=4$,

$$A(1\ldots 4)=g^2\int_0^\infty d\tau<0;k_1|V(k_2;0)V(k_3;\tau)|0;k_4\rangle. \tag{4.6.7}$$

The zero-mode factor in the integrand is $x^{-1+(k_3+k_4)^2/2}=x^{-1-s/2}$ where $x=e^{-\tau}$ (so that $d\tau=dx/x$). The non-zero modes contribute the factor

$$\begin{aligned}\prod_{n=1}^{\infty}\langle 0;k_1|e^{-\frac{1}{n}k_{2\mu}\alpha_n^\mu}e^{\frac{1}{n}k_{3\mu}\alpha_n^{\dagger\mu}x^n}|0;k_4\rangle&=\prod_{n=1}^{\infty}\exp\left(-k_2\cdot k_3\frac{x^n}{n}\right)\\ =(1-x)^{k_2\cdot k_3}=(1-x)^{-2-t/2}.\end{aligned} \tag{4.6.8}$$

The result is

$$\begin{aligned}A(1,2,3,4)&=g^2G\int_0^1 dx x^{-2-s/2}(1-x)^{-2-t/2}\\ &=g^2G\frac{\Gamma(-s/2-1)\,\Gamma(-t/2-1)}{\Gamma(-2-s/2-t/2)}.\end{aligned} \tag{4.6.9}$$

This is the Veneziano amplitude which has poles at $s=-2,0,2,\ldots$ and $t=-2,0,2,\ldots$.

4.7 Background fields

I will now turn to a brief discussion of a string moving in a curved background with metric $G_{\mu\nu}(X)$. The obvious generalization of the Nambu–Goto action is

$$\begin{aligned} S_1 &= -\frac{T}{2}\int d\sigma d\tau \sqrt{h} h^{\alpha\beta}\partial_\alpha X^\mu \partial_\beta X^\nu G_{\mu\nu}(X) \\ &= -\frac{T}{2}\int d\sigma d\tau \partial_\alpha X^\mu \partial^\alpha X^\nu G_{\mu\nu}(X) \end{aligned}, \tag{4.7.1}$$

where the second expression is in the conformal gauge $g_{\alpha\beta} = \eta_{\alpha\beta}$. This is the action for a non-linear sigma model which maps the two-dimensional space of the world-sheet into the D-dimensional space spanned by the sigma-model fields, the coordinates $X^\mu(\sigma, \tau)$.

For this to be a sensible string theory there must be no quantum violation of the classical conformal symmetry of the theory, in particular the symmetry of the theory under scaling of the world-sheet $z \rightarrow \lambda z$. This condition is related to the vanishing of the renormalization group β function. In a two-dimensional sigma model there are effectively an infinite number of interaction terms expressed by expanding the function $G_{\mu\nu}(X)$ in power series in $X(\sigma, \tau)$. The renormalization of these couplings is compactly expressed in terms of a functional $\beta_{\mu\nu}[G]$ as originally discussed by Friedan in his thesis. The Callan–Symanzyk equation for the partition function demonstrates the relation between the β functional and the scale-dependence of the theory

$$\left(\lambda\frac{\partial}{\partial\lambda} + \int d\sigma d\tau \beta_{\mu\nu}\frac{\delta}{\delta G^{\mu\nu}}\right) Z = 0. \tag{4.7.2}$$

If Z does not depend on λ, β must vanish.

The function β can be calculated in perturbation theory in the sigma-model coupling, $\alpha' = 1/2\pi T$. The result is

$$\beta_{\mu\nu} = -\frac{1}{4\pi^2 T}(R_{\mu\nu} - \frac{1}{4\pi T}R_{\mu\kappa\lambda\tau}R_\nu{}^{\lambda\kappa\tau}) + 0(\frac{1}{T^2}). \tag{4.7.3}$$

The condition $\beta = 0$ therefore resembles Einstein's equation for $G_{\mu\nu}$, generalized to include higher derivative terms! It is at this point that the connection between string theory and general relativity becomes apparent.

More generally, there could be further terms in the sigma model, such as

$$S_2 = -\frac{T}{2}\int d\sigma d\tau \epsilon^{\alpha\beta}\partial_\alpha X^\mu \partial_\beta X^\nu B_{\mu\nu}(X), \tag{4.7.4}$$

(where $B_{\mu\nu}$ is an antisymmetric field) and

$$S_3 = \frac{1}{2\pi}\int d\sigma d\tau \sqrt{g} R^{(2)}\Phi(X), \tag{4.7.5}$$

(where Φ is a scalar function and $R^{(2)}$ is the curvature of the world-sheet). Unlike S_1 and S_2, S_3 is not Weyl invariant in the classical theory. However, its lack of Weyl symmetry is compensated for in the quantum theory if (to lowest order in $1/T$)

$$\begin{aligned}\beta_{\mu\nu}(G) &= R_{\mu\nu} - \frac{1}{4}H_\mu^{\lambda\rho}H_{\nu\lambda\rho} + 2D_\mu D_\nu\Phi = 0\\ \beta_{\mu\nu}(B) &= D_\lambda H^\lambda_{\mu\nu} - 2(D_\lambda\Phi)H^\lambda_{\mu\nu} = 0\\ \beta(\Phi) &= 4(D_\mu\Phi)^2 - 4D_\mu D^\mu\Phi - R + \frac{1}{12}H_{\mu\nu\rho}H^{\mu\nu\rho} = 0,\end{aligned} \tag{4.7.6}$$

where $H_{\mu\nu\rho} = \partial_\mu B_{\nu\rho} + \partial_\rho B_{\mu\nu} + \partial_\nu B_{\rho\mu}$. These are just the usual field equations for massless G, B, Φ fields.

More generally still, one could define sigma models involving non-zero values for backgrounds corresponding to the massive string fields. These would not be classically conformally invariant. However, there could be conformally-invariant theories defined at isolated points in the infinite-dimensional space of couplings of such a theory.

At this stage it is instructive to return to the calculation of the Virasoro anomaly and study its connection to the breakdown of scale invariance. I will first consider the case of a flat background space-time, *i.e.*, $G_{\mu\nu} = \eta^{\mu\nu}$. The easiest way to calculate the anomaly in the Virasoro algebra is to make use of the Ward identity that relates the commutator of two energy-momentum tensors to a derivative of a correlation function,

$$\partial_- < T(T_{++}(\sigma,\tau)T_{++}(\sigma',\tau')) >= \frac{1}{2}\delta(\tau-\tau') < [T_{++}(\sigma,\tau), T_{++}(\sigma',\tau)] > . \tag{4.7.7}$$

The correlation function is easy to calculate since, in the conformal gauge, $T_{++} \sim \partial_+ X \cdot \partial_+ X$ and X^μ is a free two-dimensional Bose field. The resulting value of the right-hand side of (4.7.7) is the same as that quoted for the anomalous term earlier.

The Virasoro anomaly can be directly related to the presence of a trace in the energy-momentum tensor when the world-sheet has non-zero curvature. Suppose $R^{(2)} \neq 0$ and $h^{\alpha\beta} = \eta^{\alpha\beta} + f^{\alpha\beta}$, where $f^{\alpha\beta}$ is small. In that case the action for a flat world-sheet metric gets an additional contribution so that the total action is $S' = S + \int d^2z f^{\alpha\beta}T_{\alpha\beta}$. The expresstion for the expectation value of a single energy-momentum tensor gets a contribution from the $f^{\alpha\beta}$ source, so that

$$< \partial_- T_{++} > \sim c\partial_+ R^{(2)}, \tag{4.7.8}$$

$$< \partial_+ T_{--} > \sim c'\partial_- R^{(2)}, \tag{4.7.9}$$

where c and c' are the coefficients of the Virasoro anomalies in the right-moving and left-moving sectors. Notice that if $c \neq c'$ (*i.e.*, the theory is chiral on the world-sheet) these equations imply that there is a breakdown in the conservation of the energy-momentum tensor, *i.e.*, $D^\alpha T_{\alpha\beta} \neq 0$. This is a chiral two-dimensional gravitational anomaly. When $c = c'$ the non-conservation of $T^{\alpha\beta}$ can easily be compensated by adding local counterterms to the action which generate appropriate values for T_{+-} and T_{-+}. However, these give rise to a non-zero trace,

$$< h^{\alpha\beta} T_{\alpha\beta} > = < T_{+-} > \sim cR^{(2)}. \tag{4.7.10}$$

Since this would represent a breakdown in scale invariance it must not be present if the string theory is to make sense, which means that c must vanish.

If the background space is curved with a metric $G_{\mu\nu}(X)$ the above arguments generalize so that the trace of the energy-momentum tensor is given by

$$< T_{+-} > \sim c(G) R^{(2)} + \beta_{+-}(G). \tag{4.7.11}$$

The condition for scale invariance now requires both $\beta = 0$ and $c = 0$.

5. Fermions; super-Virasoro algebra; the GSO projection

5.1 Theories with fermions

There are two distinct procedures for introducing fermions into string theories. The first method, originating with the work of Ramond and of Neveu and Schwarz (RNS), introduces two-dimensional world-sheet Majorana spinors, $\psi^\mu(\sigma,\tau)$. The index μ is a space-time *vector* index which is inert under world-sheet transformations. After a suitable truncation, introduced by Gliozzi, Scherk and Olive (GSO) this theory is supersymmetric in space-time. The other method, introduced by Green and Schwarz introduces superspace Grassmann coordinates which are *space-time* chiral spinors, $\theta^a(\sigma,\tau)$. The index a is a spinor index which, in the critical dimension $D = 10$ has 16 values. These two procedures give equivalent theories in all the cases that have been considered so far – that is, in string perturbation theory around either flat ten-dimensional space-time or around certain fixed backgrounds. I will first discuss the RNS procedure.

The spinor variables have components

$$\psi^\mu(\sigma,\tau) \equiv \begin{pmatrix} \psi_-(\sigma,\tau) \\ \psi_+(\sigma,\tau) \end{pmatrix}. \tag{5.1.1}$$

A Majorana spinor is one which has real components in a particular basis of Dirac matrices – the Marjorana basis. I will denote two-dimensional Dirac matrices by ρ^α to distinguish them from the ten-dimensional ones, Γ^μ. The ρ^α satisfy

$$\{\rho^\alpha, \rho^\beta\} = -2\eta^{\alpha\beta}. \tag{5.1.2}$$

In the Majorana basis the ρ^α's are given by

$$\rho^0 = \begin{pmatrix} 0 & -i \\ i & 0 \end{pmatrix}, \qquad \rho^1 = \begin{pmatrix} 0 & i \\ i & 0 \end{pmatrix}. \tag{5.1.3}$$

The action for the fermionic string theory is given by

$$S = -\frac{T}{2}\int d\sigma d\tau\left(\partial_\alpha X^\mu \partial^\alpha X_\mu - i\bar{\psi}^\mu \rho^\alpha \partial_\alpha \psi_\mu\right), \tag{5.1.4}$$

where $\bar{\psi} = \psi\rho_0$. This action is just the action for the simplest kind of supersymmetric theory in two dimensions. There are D copies of a simple supersymmetry multiplet. The expressions are more compact when expressed in terms of a superspace formulation but I will not take that path.

The constraints analogous to the Virasoro constraints are not apparent in (5.1.4) because the action has been written directly in the conformal gauge (later I will give the formulation with the local invariances that give rise to these constraints).

The momentum conjugate to ψ is defined by $p_\psi = \delta S/\delta\psi$. Since ψ is a Majorana spinor there is a phase-space constraint,

$$p_\psi = \psi, \tag{5.1.5}$$

(whereas for a Dirac spinor $p_\psi = \psi^\dagger$, which is an independent phase-space variable). The fermionic anticommutation relations that follow from the action taking this constraint into account differ from the unconstrained case by a factor of two,

$$\{\psi^\mu_A(\sigma,\tau), \psi^\nu_B(\sigma,\tau)\} = \pi\eta^{\mu\nu}\delta_{AB}\delta(\sigma-\sigma'), \tag{5.1.6}$$

where $A, B = 1, 2$ are the two-dimensional spinor indices. The anticommutation relations appropriate to an unconstrained Dirac spinor (where $\psi^\dagger$ is independent of ψ) have an extra factor of two on the right-hand side.

5.2 World-sheet supersymmetry

I mentioned above that the theory is supersymmetric. This means that it is invariant under the transformations

$$\begin{aligned} \delta X^\mu &= \bar{\epsilon}\psi^\mu \\ \delta\psi^\mu &= -i\rho^\alpha\partial_\alpha X^\mu \epsilon, \end{aligned} \tag{5.2.1}$$

where the anti-commuting parameter is a two-component spinor

$$\epsilon = \begin{pmatrix} \epsilon_- \\ \epsilon_+ \end{pmatrix}. \tag{5.2.2}$$

The superalgebra can be checked explicitly by commuting two transformations on the coordinates,

$$\begin{aligned} [\delta_1, \delta_2]X^\mu &= \delta_1\left(\bar{\epsilon}_2\psi^\mu\right) - (1 \leftrightarrow 2) \\ &= \left(2i\bar{\epsilon}_1\rho^\alpha\epsilon_2\right)\partial_\alpha X^\mu \\ [\delta_1, \delta_2]\psi^\mu &= \left(2i\bar{\epsilon}_1\rho^\alpha\epsilon_2\right)\partial_\alpha\psi^\mu \end{aligned} \quad . \tag{5.2.3}$$

The derivation of the last relation requires the use of the ψ equation of motion, $\rho\cdot\partial\psi^\mu = 0$, as is familiar in other supersymmetric contexts when auxiliary fields are ignored.

Just as with the Poincaré symmetry we can use the Noether method to derive a conserved supersymmetry current. Taking the parameter ϵ to be a function of σ, τ the variation of the action becomes

$$\delta S = \frac{2}{\pi} \int d\sigma d\tau (\partial_\alpha \bar{\epsilon}) J^\alpha. \tag{5.2.4}$$

The current J^α is a spinor, *i.e.*,

$$J^\alpha_- = \begin{pmatrix} J^\alpha_+ \\ J^\alpha_- \end{pmatrix} = \frac{1}{2} \rho^\beta \rho_\alpha \psi^\mu \partial_\beta X_\mu. \tag{5.2.5}$$

The current is conserved, so that $\partial_\alpha J^\alpha = 0$, and furthermore $\rho^\alpha J_\alpha = 0$, as can be seen by noting that $\rho^\alpha \rho^\beta \rho_\alpha = 0$.

5.3 Energy-momentum tensor

The energy-momentum tensor is the Noether current associated with translations, $\tau \to \tau + \delta\tau$, $\sigma \to \sigma + \delta\sigma$. It is given by

$$T_{\alpha\beta} = \partial_\alpha X^\mu \partial_\beta X_\mu + \frac{i}{4} \bar{\psi}^\mu \rho_\alpha \partial_\beta \psi_\mu + \frac{i}{4} \bar{\psi}^\mu \rho_\beta \partial_\alpha \psi^\mu - \text{trace}. \tag{5.3.1}$$

As before, $\frac{1}{2} h^{\alpha\beta} T_{\alpha\beta} = T_{+-} = T_{-+} = 0$. In two-dimensional light-cone notation (5.3.1) becomes

$$\begin{aligned} T_{++} &= \partial_+ X^\mu \partial_+ X_\mu + \frac{i}{2} \psi^\mu_+ \partial_+ \psi_{+\mu} \\ T_{--} &= \partial_- X^\mu \partial_- X_\mu + \frac{i}{2} \psi^\mu_- \partial_- \psi_{-\mu}. \end{aligned} \tag{5.3.2}$$

5.4 Equations of motion

The X equations of motion are the same as for the bosonic theory, with closed-string solutions of the form

$$X^\mu(\sigma, \tau) = X^\mu_R(\sigma^-) + X^\mu_L(\sigma^-). \tag{5.4.1}$$

The fermions satisfy the two-dimensional Dirac equation

$$\begin{aligned} (\partial_\sigma + \partial_\tau)\psi_- &= 0 \\ (\partial_\tau - \partial_\sigma)\psi_+ &= 0. \end{aligned} \tag{5.4.2}$$

Some of the most interesting theories are chiral in the two-dimensional sense which means that ψ is an eigenstate of $\rho^3 \equiv \rho^1 \rho^2$,

$$\rho^3 \psi_\pm = \mp \psi_\pm. \tag{5.4.3}$$

In this case either ψ_- or ψ_+ vanishes. We shall see that this is a property of the heterotic string.

The equations of motion for the bosonic and fermionic closed-string coordinates can be written in the suggestive manner

$$\begin{aligned} \partial_+\psi^\mu_- &= \partial_+(\partial_- X^\mu_R) = 0 \\ \partial_-\psi^\mu_+ &= \partial_-(\partial_+ X^\mu_L) = 0, \end{aligned} \tag{5.4.4}$$

which illustrates the fact that the left-moving and right-moving modes are separately supersymmetric. This can be defined to be $(1,1)$ supersymmetry, where the first entry indicates the number of left-moving supercharges and the second entry the number of right-moving supercharges. More generally, we might contemplate models with (m,n) supersymmetry. In the case of the heterotic string, to be considered later, the theory has $(0,1)$ supersymmetry.

5.5 Constraints

In the fermionic theory there are extra negative-normed states created by the modes of $\psi^0_\pm$. In order for the theory to make sense there must be extra sets of gauge constraints that eliminate these states from the space of physical states. These extra gauge constraints must be fermionic. It is reasonable to expect them to be given by requiring the vanishing of the supercurrent, since J^α is in the same supermultiplet as the energy-momentum tensor whose vanishing gives the Virasoro constraints.

In the two-dimensional light-cone notation the supercurrent has components

$$J_\pm = \psi^\mu_\pm \partial_\pm X_\mu, \tag{5.5.1}$$

and its conservation is expressed by $\partial_+ J_- = \partial_- J_+ = 0$. Together with T_{++}, J_+ generates a supersymmetric extension of the Virasoro algebra known as the *superconformal* algebra (with an independent algebra generated by T_{--} and J_-),

$$\begin{aligned} \{J_+(\sigma), J_+(\sigma\prime)\} &= \pi\delta(\sigma-\sigma\prime)T_{++}(\sigma) \\ \{J_-(\sigma), J_-(\sigma\prime)\} &= \pi\delta(\sigma-\sigma\prime)T_{--}(\sigma). \end{aligned} \tag{5.5.2}$$

5.6 Boundary conditions

The question of boundary conditions for the bosonic space-time coordinates is the same in the fermionic theory as in the bosonic one. However, the fermionic conditions require special consideration. The boundary terms encountered upon varying the action to deduce the Euler–Lagrange equations have the form

$$(\psi_+\delta\psi_+ - \psi_-\delta\psi_-)\,|^{\sigma=\pi}_{\sigma=0} = 0. \tag{5.6.1}$$

There are different kinds of possible solutions of these conditions. This should not be surprising since a spinor field need not be single-valued. In general, upon transporting the field around a closed curve there can be a sign change.

I will discuss the case of open strings here in some detail. All the considerations generalize to the case of closed strings very simply. The condition (5.6.1) allows $\psi_+ = \pm\psi_-$ at each end of an open string. This leads to two independent sectors.

(i) Periodic (Ramond) sector.

If the Grassmann coordinates are chosen so that

$$\psi_+^\mu(0,\tau) = \psi_-^\mu(0,\tau), \qquad \psi_+^\mu(\pi,\tau) = \psi_-^\mu(\pi,\tau), \tag{5.6.2}$$

then the solutions of the Dirac equation can be expanded in a series of integer modes,

$$\begin{aligned} \psi_-^\mu &= \frac{1}{\sqrt{2}}\sum_{-\infty}^{\infty} d_n^\mu e^{-in(\tau-\sigma)} \\ \psi_+^\mu &= \frac{1}{\sqrt{2}}\sum_{-\infty}^{\infty} d_n^\mu e^{-in(\tau+\sigma)} \end{aligned} \tag{5.6.3}$$

The Grassmann normal mode coefficients satisfy $d_m = d_{-m}^*$ (or $d_{-n} = d_n^\dagger$ in the quantum theory) which ensures the reality of ψ. In order for the anticommutation relations (5.1.6) to be satisfied the modes must satisfy

$$\{d_m^\mu, d_n^\nu\} = \eta^{\mu\nu}\delta_{m+n,0}. \tag{5.6.4}$$

The zero mode relation,

$$\{d_0^\mu, d_0^\nu\} = \eta^{\mu\nu}, \tag{5.6.5}$$

is particularly noteworthy since it is of the same form as the Clifford algebra satisfied by D-dimensional Dirac matrices*. This means that

$$d_0^\mu = -\frac{i}{\sqrt{2}}\Gamma^\mu. \tag{5.6.6}$$

The states on which d_0 acts are therefore $SO(D-1,1)$ (space-time) spinors. The inclusion of world-sheet fermions has led to space-time fermion states!

The ground state in this sector is denoted by $|0;k\rangle_{\alpha d} u^a(k) \equiv |0;k\rangle_\alpha \otimes |0\rangle_d u^a(k)$. This is the ground state of all the α_n and d_n for $n \neq 0$ together with a spinor, $u^a(k)$, which accounts for the fermionic zero modes (and a momentum dependence which accounts for the bosonic zero modes as usual). The excited states are obtained by operating with the creation modes on the ground states

$$\alpha_{-n}^{\mu_1} \ldots d_{-n_1}^{\mu} \ldots |0;k>_{\alpha d} u^a(k), \tag{5.6.7}$$

subject to the gauge constraint conditions.

---

* The D-dimensional Dirac matrices, Γ^μ, are defined to satisfy $\{\Gamma^\mu, \Gamma^\nu\} = -2\eta^{\mu\nu}$. They have dimension $2^{D/2} \times 2^{D/2}$.

(ii) Anti-periodic (Neveu-Schwarz) sector
The other possible choice for the fermion boundary conditions is

$$\psi_+^\mu(0,\tau) = \psi_-^\mu(0,\tau), \qquad \psi_+^\mu(\pi,\tau) = -\psi_-^\mu(\sigma,\tau). \tag{5.6.8}$$

The choice which interchanges the $\sigma = 0$ and $\sigma = \pi$ conditions is physically equivalent. In this case the mode expansions are in half-integer modes

$$\begin{aligned} \psi_-^\mu(\sigma,\tau) &= \frac{1}{\sqrt{2}} \sum_{r\in\frac{1}{2}+Z} b_r^\mu e^{-ir(\tau-\sigma)} \\ \psi_+^\mu(\sigma,\tau) &= \frac{1}{\sqrt{2}} \sum_{r\in\frac{1}{2}+Z} b_r^\mu e^{-ir(\tau+\sigma)}, \end{aligned} \tag{5.6.9}$$

(where $b_{-r} = b_r^*$, or $b_{-r} = b_r^\dagger$ in the quantum theory). Once again the ψ anti-commutation relations demand that

$$\{b_r^\mu, b_s^\nu\} = \delta_{rs}\eta^{\mu\nu}. \tag{5.6.10}$$

There is now no zero mode and the ground state, $|0;k\rangle_{\alpha,b}$ is unique. The absence of a zero mode in this sector of states is a signal that these boundary conditions break two-dimensional supersymmetry.

5.7 Super-Virasoro operators

The generators of the super-conformal algebra can be defined in terms of the modes of the energy-momentum tensor and the supercurrent,

$$L_m = \frac{1}{\pi}\int_0^\pi d\sigma \left(e^{im\sigma}T_{++} + e^{-im\sigma}T_{--}\right) = \int_{-\pi}^{\pi} d\sigma e^{im\sigma}T_{++}, \tag{5.7.1}$$

$$F_m = \frac{\sqrt{2}}{\pi}\int_0^\pi d\sigma \left(e^{im\sigma}J_+ + e^{-im\sigma}J_-\right) = \int_{-\pi}^{\pi} d\sigma e^{im\sigma}J_+, \tag{5.7.2}$$

(for the Ramond sector) and

$$G_r = \frac{\sqrt{2}}{\pi}\int_0^\pi d\sigma \left(e^{ir\sigma}J_+ + e^{-ir\sigma}J_-\right) = \int_{-\pi}^{\pi} d\sigma e^{ir\sigma}J_+, \tag{5.7.3}$$

in the Neveu–Schwarz sector. The components of the energy-momentum tensor are given by (5.3.2) and those of the supercurrent by (5.5.1). Substituting the mode expansions of these operators gives, for the Ramond sector,

$$F_m = \sum_{n=-\infty}^{\infty} \alpha_{-n} \cdot d_{m+n} \tag{5.7.4}$$

(so that $F_0 = \Gamma \cdot p + \ldots$ is the string generalization of the Dirac operator) and

$$L_m^R = \frac{1}{2}\sum_{-\infty}^{\infty} : \alpha_{-n} \cdot \alpha_{m+n} : + \frac{1}{2} \sum_{n=-\infty}^{\infty} (n + \frac{1}{2}m) : d_{-n} \cdot d_{m+n} : . \tag{5.7.5}$$

In the Neveu–Schwarz sector the mode expansions are

$$G_r = \sum_{n=-\infty}^{\infty} \alpha_{-n} \cdot b_{r+n} \tag{5.7.6}$$

(where r is a half-integer) and

$$L_m^{NS} = \frac{1}{2}\sum_{-\infty}^{\infty} : \alpha_{-n} \cdot \alpha_{m+n} : + \frac{1}{2} \sum_{r=-\infty}^{\infty} (r + \frac{1}{2}m) : b_{-r} \cdot b_{m+r} : . \tag{5.7.7}$$

The algebra of the operators in the Ramond sector is defined by the (anti)commutation relations

$$\begin{aligned} \left[L_m^R, L_n^R\right] &= (m-n)L_{m+n}^R + \frac{D}{8}m^3\delta_{m+n,0}, \\ \left[L_m^R, F_n\right] &= \left(\frac{1}{2} - n\right) F_{m+n}, \\ \{F_m, F_n\} &= 2L_{m+n}^R + \frac{D}{2}m^2\delta_{m+n,0}. \end{aligned} \tag{5.7.8}$$

The anomaly term in the commutation of the L_n's now has a coefficient of $D/8$, which is the sum of $1/12$ for each boson mode and $1/24$ for each fermion mode. The fact that each boson contributes twice the anomaly of a fermion will prove to be of great importance. The operators of the Neveu–Schwarz sector, L_m^{NS} and G_r, satisfy a similar set of (anti)commutation relations. Note that the zero-mode relation, $F_0^2 = L_0^R$ generalizes the relation between the usual Dirac and the Klein–Gordon operators, $(\Gamma \cdot p)^2 = -p^2$.

The physical states, $|\phi\rangle_{R,NS}$ (where the subscripts R or NS indicate which sector the state belongs to), of the fermionic theory are restricted by constraints that generalize the Virasoro gauge constraints, namely

$$\begin{aligned} L_m^{R,NS}|\phi\rangle_{R,NS} &= 0, \qquad m > 0 \\ F_m|\phi\rangle_R &= 0, \qquad m \geq 0 \\ G_r|\phi\rangle_{NS} &= 0, \qquad r > 0, \\ (L_0^{R,NS} - a_{R,NS})|\phi\rangle_{R,NS} &= 0. \end{aligned} \tag{5.7.9}$$

The constants a_R and a_{NS} are the normal-ordering constants for each sector. The Ramond sector condition,

$$F_0|\phi\rangle_R = 0, \tag{5.7.10}$$

does not have a normal ordering ambiguity, from which we see that

$$F_0^2|\phi\rangle_R = L_0|\phi\rangle_R = 0 \tag{5.7.11}$$

so that $a_R = 0$, a condition that can also be derived from requiring the decoupling of the negative-metric $\mathcal{GHOST}$ states. In this case the proof of absence of such states turns out to require *either* $D = 10$ and $a_R = 0$ (ground-state fermion $(\text{mass})^2 = 0$), $a_{NS} = 1/2$ (ground-state boson $(\text{mass})^2 = -\pi T$) *or* $D < 10$ and $a_{NS} \leq 1/2$. The former conditions are the ones that will be encountered in the light-cone gauge treatment to be described later.

5.8 Reparametrization-invariant action

I will now discuss the form of the action from which all the constraints can be derived as gauge constraints. In this case the vanishing of the supercurrent will arise as the equation of motion of a gravitino in the same way as the vanishing of the energy-momentum tensor comes from the equation of motion of the graviton. This means that we are interested in extending the fermionic action to one which has a *local* supersymmetry, which means that it is the action of two-dimensional supergravity. To do this recall that under a local supersymmetry transformation with Grassmann parameter ϵ the variation of the action is given by

$$\delta S = \frac{2}{\pi}\int d\sigma d\tau (\partial_\alpha \bar{\epsilon}) J^\alpha. \tag{5.8.1}$$

As is familiar from supergravity field theories this variation can be cancelled by adding a term to the action which couples the supercurrent to a vector-spinor field, χ_A^α,

$$S' = -\frac{2}{\pi}\int d\sigma d\tau \sqrt{h}\bar{\chi}_\alpha J^\alpha. \tag{5.8.2}$$

A small digression is needed to recall how spinors are described in general relativity. In general relativity with spinor fields it is necessary to make the theory invariant under local Lorentz transformations since spinors have no nice transformation properties under general linear transformations. In order to acheive this it is necessary to introduce *vielbein* fields, e_α^a (where 'viel' is 'vier' in familiar four dimensional theories but is 'zwei' in the two-dimensional context of relevance here). These convert world indices, labelled by greek letters, to flat indices, labelled by roman letters, so that it is possible to define Dirac matrices with world indices by

$$\rho_\alpha = e_\alpha^a \rho_a. \tag{5.8.3}$$

The vielbein is related to the metric tensor by

$$h_{\alpha\beta} = \eta_{ab} e_\alpha^a e_\beta^b. \tag{5.8.4}$$

The gauge potential associated with local Lorentz symmetry is the spin connection, ω_μ^{ab}.

Using this we can define derivatives which are covariant under local Lorentz transformations

$$\nabla_\alpha\psi = \left(\partial_\alpha + \frac{1}{4}\omega_\alpha^{ab}\rho_{ab}\right)\psi, \tag{5.8.5}$$

where $\rho_{ab} = \frac{1}{2}(\rho_a\rho_b - \rho_b\rho_a)$.

(i) Local World-sheet supersymmetry: The variation (5.8.1) is cancelled by the variation of χ_α in (5.8.2) if χ_α is chosen to transform under supersymmetry as

$$\delta\chi_\alpha = \nabla_\alpha\epsilon. \tag{5.8.6}$$

That is not the whole story since we must now also consider the variation of J^α in (5.8.2). Another term, quadratic in χ and in ψ must still be added to get a supercovariant action. The complete action is given by

$$\begin{aligned} S = -\frac{1}{2\pi}\int d\sigma d\tau e \Big\{ & h^{\alpha\beta}\partial_\alpha X^\mu \partial_\beta X_\mu - i\bar{\psi}^\mu\rho^\alpha\partial_\alpha\psi_\mu \\ & +2\bar{\chi}_\alpha\rho^\beta\rho^\alpha\psi^\mu\partial_\beta X_\mu + \frac{1}{2}\left(\bar{\psi}_\mu\psi^\mu\right)\left(\bar{\chi}_\alpha\rho^\beta\rho^\alpha\chi_\beta\right)\Big\}, \end{aligned} \tag{5.8.7}$$

where e is the determinant of the *zweibein*, e_α^a, so that $e = \sqrt{h}$.

The local supersymmetry transformations are given by

$$\begin{aligned} &\delta\chi_\alpha = \nabla_\alpha\epsilon, \qquad \delta e_\alpha^a = -2i\bar{\epsilon}\rho^a\chi_\alpha \\ &\delta X^\mu = \bar{\epsilon}\psi^\mu\delta\psi^\mu = -i\rho^\alpha\epsilon\left(\partial_\alpha X^\mu - \bar{\psi}^\mu\chi_\alpha\right). \end{aligned} \tag{5.8.8}$$

(ii)Local world-sheet Lorentz symmetry: The local Lorentz transformations are given by

$$\delta\psi^\mu = \frac{1}{4}\theta^{ab}\rho_{ab}\psi^\mu = \frac{1}{2}\theta\rho_3\psi, \tag{5.8.9}$$

$$\delta X^\mu = 0, \qquad \delta e_\alpha^a = \theta^a{}_b e_\mu^b = \epsilon^{ab}e_{\mu b}\theta, \tag{5.8.10}$$

where θ is the single parameter of two-dimensional Lorentz transformations.

(iii) Weyl symmetry: The action is invariant under Weyl transformations. Unlike the X^μ coordinates, the fermionic coordinates transform with non-zero weight. The infinitesimal transformations of the fields are

$$\begin{aligned} &\delta X^\mu = 0, \qquad \delta\psi^\mu = -\frac{1}{2}\Lambda\psi^\mu \\ &\delta e_\alpha^a = \Lambda e_\mu^a, \qquad \delta\chi_\alpha = \frac{1}{2}\Lambda\chi_\alpha. \end{aligned} \tag{5.8.11}$$

(iv) Super-Weyl symmetry: The action is also invariant under the supersymmetric analogue of Weyl symmetry,

$$\begin{aligned} \delta\chi_\alpha &= i\rho_\alpha\eta \\ \delta e_\alpha^\mu &= \delta\psi^\mu = \delta X^\mu = 0, \end{aligned} \tag{5.8.12}$$

where the parameter $\eta(\sigma,\tau)$ is a two-component spinor.

5.9 Gauge fixing – the superconformal gauge

Both the zweibein, e^a_α, and the gravitino, χ_α, have four independent components. There are four bosonic gauge parameters, ξ^α, θ and Λ as well as four fermionic gauge parameters, ϵ_A and η_A. This means that e^a_α can be gauged to δ^a_α (at least locally) and χ_α can be gauged away. More generally, by means of reparametrizations and local supersymmetry transformations, it is possible to choose a *super*conformal gauge,

$$e^a_\alpha = e^\phi \delta^a{}_\alpha, \qquad \chi_\alpha = \rho_\alpha \eta. \tag{5.9.1}$$

This is analogous to the conformal gauge in the bosonic theory. The field $\eta(\sigma, \tau)$ is an arbitrary two-component spinor. The (super)-Weyl invariance ensures that ϕ and η cancel out of the action in the classical theory.

In this gauge the equations of motion for χ_α and e^a_α are the constraint equations

$$\begin{aligned} \frac{\delta S}{\delta \chi^\alpha} &= -\frac{2e}{\pi} J_\alpha = 0 \\ e^a_\beta \frac{\delta S}{\delta e^\alpha_a} &= -\frac{1}{\pi} T_{\alpha\beta} = 0. \end{aligned} \tag{5.9.2}$$

These are just the conditions that were imposed earlier but which now emerge as gauge constraints on the states of the theory.

The discussion of gauge-fixing in the quantum theory again requires the introduction of Fadeev–Popov ghost coordinates. These come in two types. The first are the anticommuting ghosts, c^α and antighosts, $b_{\alpha\beta}$, for the reparametrizations introduced earlier. Recall that the tensor structure of these coordinates arose from the fact that the ghosts were introduced in order to represent the Jacobian for the transformation from a symmetric, traceless tensor $(\sqrt{h} h^{\alpha\beta})$ to a vector (ξ^α).

The ghosts for local supersymmetry are commuting quantities which describe the Jacobian for the transformation from the integration over two components of $\chi_{\alpha A}$ which satisfy $\rho^\alpha_{AB}\chi_{\alpha B} = 0$ to the components of ϵ_A. These commuting ghosts are denoted by a spinor, γ_A, and a spinor-vector, $\beta_{\alpha A}$ (satisfying $\rho^\alpha \beta_{\alpha A} = 0$).

The Polyakov method of calculating amplitudes by integrating over world-sheets can be generalized to integration over super world-sheets. This is probably the simplest way of calculating amplitudes. For a world-sheet of arbitrary genus g there are integrations over $2g - 2$ *supermoduli* in addition to the $3g - 3$ moduli described earlier. I will not describe this approach which will be the subject of other lectures at this school. In order to discuss the spectrum of the theory it is more convenient to use the light-cone method in which all the constraints can be eliminated explicitly.

5.10 Light-Cone gauge

As with the bosonic theory the light-cone gauge uses the residual conformal invariance of the conformal gauge to choose the parametrization in which

$$X^+ = x^+ + p^+\tau. \tag{5.10.1}$$

In addition the supersymmetric part of the conformal symmetry can be chosen so that

$$\psi_A^+ = 0. \tag{5.10.2}$$

The constraint equations can then be solved to determine X^- and ψ^- in terms of the physical transverse degrees of freedom, X^i and ψ^i. The constraints

$$J_{\pm A} \equiv \psi_A^\mu \partial_\pm X_\mu = 0, \tag{5.10.3}$$

give

$$\psi_A^- = \frac{1}{p^+}\psi_A^i \partial_\pm X^i, \tag{5.10.4}$$

The constraints

$$T_{\pm\pm} = 0 = \partial_\pm X^\mu \partial_\pm X_\mu + \frac{i}{2}\psi_A^\mu \partial_\pm \psi_{\mu A}, \tag{5.10.5}$$

lead to

$$\partial_\pm X^- = \frac{1}{p^+}(\partial_\pm X^i \partial_\pm X^i + \frac{i}{2}\psi_A^i \partial_\pm \psi_A^i). \tag{5.10.6}$$

By subtracting the minus components of these equations from the plus components the coordinates X^- and ψ_A^- are expressed in terms of the transverse components, X^i and ψ_A^i, and their σ (*i.e.*, spatial) derivatives.

The generators of the global Poincaré symmetry, P^μ and $J^{\mu\nu}$, can be calculated using the expressions obtained by the Noether method and substituting the constrained expressions for X^- and ψ^-. For example,

$$P^- = \frac{1}{\pi}\dot{X}^- = \frac{1}{\pi}(\partial_+ X^- + \partial_- X^-), \tag{5.10.7}$$

which has a mode expansion (for open strings with NS boundary conditions)

$$\alpha_m^- = \frac{1}{2p^+}\left(\sum_{n=-\infty}^{\infty} : \alpha_{m-n}^i \alpha_n^i : + \sum_{r\in Z} + \frac{1}{2}\left(r - \frac{m}{2}\right) : b_{m-r}^i b_r^i : -2a^{NS}\delta_{n,0}\right). \tag{5.10.8}$$

The constant, a^{NS}, arises from the ambiguity in normal ordering both the bosonic and the fermionic operators in passing from the expression for P^- in terms of X^i, ψ^i and their derivatives to the oscillator expression.

5.11 Open-String spectrum

I will now describe the spectrum of states in the open-string sector of the fermionic theory, leaving the closed-string sector until I have discussed supersymmetry in some detail.

(i) NS bosons.

For open strings with NS boundary conditions (5.10.8) leads to the hamiltonian

$$h \equiv p^- = \alpha_0^- = \frac{1}{p^+}\left(\frac{\underline{p}^2}{2} + N^\alpha + N^b - a^{NS}\right), \tag{5.11.1}$$

where the number operators are defined by

$$N^\alpha = \sum_{n=1}^{\infty} \alpha_{-n}^i \alpha_n^i, \qquad N^b = \sum_{r=\frac{1}{2}}^{\infty} r b_{-r}^i b_r^i. \tag{5.11.2}$$

The normal ordering constant is written formally as the sum of two infinite terms

$$\frac{D-2}{2}\sum_{m=1}^{\infty} m - \frac{D-2}{2}\sum_{r=\frac{1}{2}}^{\infty} r, \tag{5.11.3}$$

where the first term comes from the bosonic oscillators and the second from the fermionic ones. As with the bosonic theory these infinite quantities can be consistently regulated to give a finite expression for physically relevant quantities. For the first sum in (5.11.3) we can use the mnemonic, $\sum_1^\infty m = -1/12$, discussed earlier. The second sum then follows from $\sum_1^\infty 2n = -\frac{1}{6}$, so that $\sum_{\text{odd}n} n = -\frac{1}{12} + \frac{1}{6} = \frac{1}{12}$ from which it follows that

$$\sum_{r=\frac{1}{2}}^{\infty} r = \frac{1}{24}. \tag{5.11.4}$$

As a result the total normal ordering term is given by

$$(D-2)\left(-\frac{1}{24} - \frac{1}{48}\right) = -\frac{D-2}{16}, \tag{5.11.5}$$

so that $a^{NS} = (D-2)/16$.

The masses of the open strings in the NS sector are therefore given by

$$(\text{Mass})^2 \equiv 2p^+p^- - \underline{p}^2 = 2N - \frac{D-2}{8}, \tag{5.11.6}$$

where

$$N = N^\alpha + N^b. \tag{5.11.7}$$

The ground state, $|0;k\rangle_{NS}$, has $(\text{Mass})^2 = -(D-2)/8$ (in units in which $T = 1/\pi$, as usual) and is a tachyon. The first excited state is $b_{-1/2}^i|0;k\rangle_{NS}$ which has $N = 1/2$

and hence has $(\text{Mass})^2 = 1 - (D-2)/8$. Since the state is a transverse vector it must be massless if the theory is to be Lorentz invariant. This means that the dimension must be constrained to $D = 10$. As in the case of the bosonic string the light-cone method only gives the theory in the critical dimension. In sub-critical dimensions the theory is not transverse and if it is possible to make sense of it there must be extra, longitudinal, modes.

The higher excited states are obtained by applying powers of creation modes α_m^i and b_r^i to the ground state. The states fall into two classes. States with an even power of b modes which have masses which are odd integers and those with an odd power of b modes which have masses which are even integers. These two kinds of states are distinguished by the sign of the 'G-parity' operator (the nomenclature dates to the days when this theory was thought to describe mesons),

$$G = -(-1)^{\sum b_{-r}^i b_r^i}. \tag{5.11.8}$$

States of $G = -1$ have $(\text{Mass})^2 = -1, 1, 3, \ldots$ while those of $G = +1$ have $(\text{Mass})^2 = 0, 2, 4, \ldots$.

In order to eliminate the tachyon state it is possible to consistently throw away half the states in the spectrum by imposing the extra constraint

$$\left(\frac{1-G}{2}\right) |\,\rangle_{NS} = 0, \tag{5.11.9}$$

on any physical state. This keeps states with $G = +1$ so it throws away the tachyon. This truncation is the bosonic sector piece of the truncation of Gliozzi, Scherk and Olive.

(ii) R fermions

When the ψ fields satisfy the R boundary conditions that lead to space-time fermions the hamiltonian is given by

$$h \equiv p^- = \alpha_0^- = \frac{1}{p^+}\left(\frac{p^2}{2} + N^\alpha + N^d - a^R\right). \tag{5.11.10}$$

The d-moded number operator is defined by $N^d = \sum_1^\infty n \underline{d}_{-n} \cdot \underline{d}_n$. The normal ordering term coming from the d modes is the same magnitude as that from the α modes but opposite in sign (since d's anticommute). As a result $a^R = 0$, as anticipated earlier.

The ground state, $|0; k>_R u^a(k)$ has $(\text{Mass})^2 = 0$. In the light-cone gauge half of the components of a spinor satisfying the Dirac equation can be expressed in terms of the other half. I will return to this point in more detail later. This means that solutions of the zero-mode constraint $F_0|0; k>_R u^a(k) = 0$ (which implies $\Gamma \cdot u^a(k) = 0$), are associated with $SO(D-2)$ spinors of dimension $2^{(D-2)/2}$ (or $2^{(D-4)/2}$ if the spinor is chiral).

Excited states are obtained by acting on the ground state with the creation modes, d_{-n}^i

and α^i_{-n}. There is an analogue of the G-parity operator in the R sector

$$\bar{\Gamma} = \Gamma_{11}(-1)^{-\sum_{n=1}^{\infty} d^i_{-n} d^i_n}, \tag{5.11.11}$$

where

$$\Gamma_{11} = \Gamma_0 \Gamma_1 \ldots \Gamma_9. \tag{5.11.12}$$

The operator $\bar{\Gamma}$ is the string generalization of Γ_{11} and is defined so that

$$\{\bar{\Gamma}, \bar{d}^\mu_n\} = 0. \tag{5.11.13}$$

Using this operator it is possible to define a string generalization of chirality by truncating the theory so that all the states satisfy

$$\frac{1}{2}(1 \pm \bar{\Gamma}) \mid \rangle_R u = 0. \tag{5.11.14}$$

Notice that the sign in (5.11.14) is ambiguous (in contrast to the G-parity projection of the Neveu–Schwarz sector where the sign was determined by the elimination of the tachyon state).

I would like to enlarge slightly on the use of ten-dimensional spinors. The anticommutation relations determine that the Dirac matrices are imaginary in the Majorana basis where the spinors are real and that Γ^0 is antisymmetric while Γ^μ ($\mu \neq 0$) is symmetric. A Majorana representation can be found in dimensions $2, 3, 4$ mod 8. In any even dimension chiral spinors can be defined which are eigenstates of $\Gamma_{D+1} = \Gamma_0 \Gamma_1 \ldots \Gamma_{D-1}$. It is easy to see, however, that $\Gamma^2_{D+1} = (-1)^{(D-2)/2}$. This means that the eigenvalues of Γ_{D+1} are ± 1 if $D = 2$ mod 4 and are $\pm i$ if $D = 4$ mod 4. This means that it is not consistent to require a spinor to be chiral and real in 4 mod 4 dimensions. Hence, the only dimensions in which the Majorana condition can be imposed on a chiral spinor are $D = 2$ mod 8 as stated earlier. A ten-dimensional Majorana–Weyl spinor has $2^{(D-4)/4} = 8$ independent components.

The space-time spinor, u^a, in (5.11.14) is taken to be both a chiral spinor satisfying $(1 \pm \Gamma_{11})u = 0$ and a Majorana spinor (so that its components are real in the Majorana basis). We have seen that it is consistent to do this in ten dimensions. One intriguing feature of string theory is that Majorana–Weyl spinors seem to be of fundamental importance in both the two and the ten-dimensional sense.

The massless states satisfying the condition (5.11.14) are chiral. The massive excitations cannot be chiral so there must be an equal number of left-handed and right-handed excitations at each mass level. However, in the string generalization of chirality the states of one handedness have an odd number of d excitations while those of the opposite handedness have an even number. For example, the states at the first excited level are

$$\alpha^i_{-1}|0; k\rangle u^a, \qquad d^i_{-1}|0; k\rangle u^a. \tag{5.11.15}$$

These two states are eigenstates of $\bar{\Gamma}$ with eigenvalues $+1$ and -1.

5.12 Number of states of a given mass.

The number of states at any mass level is determined in the same way as for the bosonic theory.

(i) NS bosons

The partition function which describes the number of bosonic states at any mass level, d^n_{NS} is given by (including the GSO projection)

$$\begin{aligned} f_{NS} &= \sum d^n_{NS}\omega^n = \mathrm{tr}\frac{1+G}{2}\omega^{N-1/2} \\ &= \frac{1}{2\sqrt{\omega}}\left[\prod_{m=1}^{\infty}\left(\frac{1+\omega^{m-1/2}}{1-\omega^m}\right)^8 - \prod_{m=1}^{\infty}\left(\frac{1-\omega^{m-1/2}}{1-\omega^m}\right)^8\right]. \end{aligned} \tag{5.12.1}$$

(ii) R fermions

The number of states at any mass in the fermion sector, d^n_R, is given by

$$\begin{aligned} f_R &= \sum_{n=1}^{\infty} d^n_R\omega^n = 16\mathrm{tr}\left(\frac{1+\bar{\Gamma}}{2}\right)\omega^N \\ &= 8\prod_{m=1}^{\infty}\left(\frac{1+\omega^m}{1-\omega^m}\right)^8. \end{aligned} \tag{5.12.2}$$

The factor of 16 comes from the trace over the chiral fermion ground states (and the trace involving Γ_{11} is zero).

The partition functions can be expressed in terms of Jacobi Θ functions. It can then be seen that

$$f_R = f_{NS}, \tag{5.12.3}$$

as a result of a famous identity due to Jacobi. (This identity can be derived from the bozonization of the two-dimensional fermion fields.) The relation (5.12.3) shows that the truncation of GSO results in a spectrum with an equal number of fermion states and boson states at each mass level. Although this does not prove that the theory is supersymmetric in ten-dimensional space-time it certainly suggests it.

In recent times it has become apparent that the GSO projection is required in order to obtain a consistent theory. One argument for this is based on considering the closed-string loop amplitudes. As I mentioned earlier these have to be invariant under large diffeomorphisms (modular transformations) in order to give unitary amplitudes. Modular invariance requires the theory to be projected in the GSO manner.

6. Space-time supersymmetry; superstrings.

6.1 $D = 10$ Supersymmetric Yang–Mills theory.

Before considering string theories with space-time supersymmetry I would like to describe a simpler ten-dimensional supersymmetric theory, namely, D = 10 supersymmetric Yang–Mills theory. The action is given by

$$S = \int d^{10}x \left(-\frac{1}{4}F^2 + \frac{i}{2}\bar{\psi}\Gamma \cdot D\psi \right), \tag{6.1.1}$$

where ψ^a is a Majorana–Weyl spinor field (which has 16 independent real components) and $\bar{\psi} = \psi\Gamma^0$. The action is invariant under the global supersymmetry transformations

$$\begin{aligned} \delta A^a_\mu &= i\bar{\epsilon}\Gamma_\mu\psi \\ \delta\psi^a &= \frac{1}{4}F^a_{\mu\nu}\Gamma_{\mu\nu}\psi, \end{aligned} \tag{6.1.2}$$

where $\Gamma_{\mu\nu} = \frac{1}{2}[\Gamma_\mu, \Gamma_\nu]$.

The light-cone gauge is defined by the choice

$$A^+ = 0. \tag{6.1.3}$$

In this case the equations of motion of the free theory determine that

$$\partial_\mu A^\mu = 0, \tag{6.1.4}$$

so that

$$A^- = \frac{\partial^i A^i}{\partial^+} = \frac{p^i A^i}{p^+}. \tag{6.1.5}$$

This means that there are $D-2$ independent transverse components, $A^i (i = 1, 2 \ldots D-2)$, in the vector potential.

The light-cone components of the Γ matrices, defined by

$$\Gamma^\pm = \frac{\Gamma^0 \pm \Gamma^{D-1}}{\sqrt{2}}, \tag{6.1.6}$$

satisfy

$$(\Gamma^+)^2 = (\Gamma^-)^2 = 0, \tag{6.1.7}$$

so that

$$\frac{\Gamma^+\Gamma^-}{2} + \frac{\Gamma^-\Gamma^+}{2} = 1, \tag{6.1.8}$$

and a spinor can be decomposed into two projected pieces

$$\psi = \frac{\Gamma^+\Gamma^-}{2}\psi + \frac{\Gamma^-\Gamma^+}{2}\psi. \tag{6.1.9}$$

The Dirac equation for the free fermion fields,

$$\Gamma \cdot p\psi = 0, \tag{6.1.10}$$

can usefully be written as the sum of two terms

$$\Gamma \cdot p\frac{\Gamma^+\Gamma^-}{2}\psi + \Gamma \cdot p\frac{\Gamma^-\Gamma^+}{2}\psi = 0. \tag{6.1.11}$$

By multiplying (6.1.11) by Γ^+ it is possible to express one of the projected components of ψ in terms of the other

$$\frac{\Gamma^+\Gamma^-}{2}\psi = \frac{1}{2p^+}\Gamma^+\Gamma^i p^i \frac{\Gamma^-\Gamma^+}{2}\psi. \tag{6.1.12}$$

Thus, only half the components of the spinor, $\Gamma^-\Gamma^+\psi$ are independent dynamical variables. Multiplying (6.1.11) by Γ^- then gives

$$p^2\frac{\Gamma^-\Gamma^+}{2}\psi = 0. \tag{6.1.13}$$

The fact that (6.1.12) is a constraint equation depends on the choice of X^+ as the time variable, *i.e.*, the choice of the light-cone coordinate system. There are no X^+ derivatives (and therefore no "time" derivatives on the right-hand side of the equation (recall that $p^+ = i\partial/\partial X^-$).

Since I did not give the proof that it is possible to find a Majorana representation for the Γ^μ matrices I will show the representation explicitly. These can be expressed as the tensor product of the 2×2 Pauli matrices, σ_i. Firstly, the $32 \otimes 32$ Γ matrices can be written in the form $\Gamma = \sigma \otimes \gamma$ where σ is a 2×2 Pauli matrix or the 2×2 unit matrix while γ is a 16×16 $SO(8)$ matrix. Explicitly,

$$\begin{aligned}\Gamma^0 &= \sigma_2 \otimes I_{16}\\ \Gamma^i &= i\sigma_1 \otimes \gamma^i\\ \Gamma^9 &= i\sigma_3 \otimes I_{16},\end{aligned} \tag{6.1.14}$$

where I_{16} denotes the 16×16 unit matrix and $i = 1, 2, \ldots, 8$. The $SO(8)$ matrices γ^i satisfy

$$\left\{\gamma^i, \gamma^j\right\} = 2\delta^{ij}. \tag{6.1.15}$$

In turn, these matrices can be written in terms of two sets of 8×8 matrices, $\gamma^i_{a\dot{a}}$ and $\gamma^i_{\dot{a}a}$

as

$$\gamma^i = \begin{pmatrix} 0 & \gamma^i_{a\dot{a}} \\ \gamma^i_{\dot{a}a} & 0 \end{pmatrix}. \tag{6.1.16}$$

The matrices $\gamma^i_{a\dot{a}}$ and $\gamma^i_{\dot{a}a}$ are Clebsch–Gordon coefficients which couple the $SO(8)$ vector (with index i) to the two inequivalent spinor representations (with indices a and $\dot{a}$). These matrices are given explicitly by

$$\begin{aligned} \gamma^1 &= -i\sigma_2 \otimes \sigma_2 \otimes \sigma_2, & \gamma^2 &= iI_2 \otimes \sigma_1 \otimes \sigma_2 \\ \gamma^3 &= iI_2 \otimes \sigma_3 \otimes \sigma_2, & \gamma^4 &= i\sigma_1 \otimes \sigma_2 \otimes I_2 \\ \gamma^5 &= i\sigma_3 \otimes \sigma_2 \otimes I_2, & \gamma^6 &= i\sigma_2 \otimes I_2 \sigma_1 \\ \gamma^7 &= i\sigma_2 \otimes I_2 \otimes \sigma_3, & \gamma^8 &= I_2 \otimes I_2 \otimes I_2. \end{aligned} \tag{6.1.17}$$

These 8×8 matrices will be of relevance in the description of superstrings in the light-cone gauge in a later lecture. The spinors on which the $\gamma_{a\dot{b}}$ and $\gamma_{\dot{a}b}$ matrices act are the two halves of the full Dirac spinor defined by (6.1.9). I will later use a notation in which $\frac{1}{2}\Gamma^-\Gamma^+\psi \equiv \psi^a$ and $\frac{1}{2}\Gamma^+\Gamma^-\psi \equiv \psi^{\dot{a}}$, where the indices a and $\dot{a}$ are the two inequivalent 8-component spinor indices associated with the $SO(8)$ subgroup of $SO(9,1)$.

6.2 Ten-dimensional space-time supersymmetry

Ten-dimensional supersymmetry can be represented in superspace by appending one or two Grassmann-valued Majorana–Weyl spinor coordinates θ^{Aa} to the space-time coordinates X^μ. The index $A = 1,2$ labels the two possible spinors. These may either have the same chirality or opposite chirality, *i.e.*,

$$\frac{1}{2}(1 + \eta^A\Gamma_{11})\theta^A = 0, \tag{6.2.1}$$

where $\eta^A = \pm 1$. The variables θ^{1a} and θ^{2a} each have sixteen components (which are real in the Majorana representation). This leads to several different kinds of field theories which can be classified in a way that also applies to the corresponding superstring theories (in which the coordinates will be functions of the position along the string). The type IIA theory has $N = 2$ supersymmetry (there are two conserved ten-dimensional supercharges) and it is automatically a supergravity theory. It has $\eta^1 = -\eta^2$ so there is no net chirality which makes it of little obvious interest for describing observed physics. The corresponding string theory only describes closed strings. The type IIB theory also has $N = 2$ supersymmetry but it has $\eta^1 = \eta^2$ and is chiral. There is no possibility of incorporating internal Yang–Mills symmetry so its possible relevance to observed physics is also unclear. The corresponding string theory also only describes closed strings. It would be relevant if the internal symmetry is generated in the process of curling up the six extra dimensions (although this is likely to destroy the chirality of the theory). By a suitable truncation, the type IIB theory can be reduced to the supergravity sector (or closed-string sector) of type I theories. The supersymmetric Yang–Mills sector which couples to $N = 1$ supergravity corresponds to the open-string sector of type I superstring theories where the internal symmetry is incorporated by attaching charges to the string endpoints. There is only one conserved supercharge. Finally, in the heterotic string theory, one of the θ's is absent so that there is again only one supersymmetry. In this case the theory only describes closed strings. These can carry quantum numbers which arise as charge densities smeared out on the world-sheet. As I will describe in detail, in the heterotic theory the internal symmetry is incorporated by means of other coordinates, either fermionic or bosonic, in addition to X^μ and $\theta^1 \equiv \theta$.

6.3 Super-Poincaré transformations

The supersymmetric extension of the Poincaré transformations is given by

$$\begin{aligned}\delta X^\mu &= a^\mu{}_\nu X^\nu + b^\nu + i\bar{\epsilon}^A\gamma^\mu\theta^A \\ \delta\theta^A &= -\frac{1}{2}a_{\mu\nu}\gamma^{\mu\nu}\theta^A + \epsilon^A,\end{aligned} \tag{6.3.1}$$

where the parameters $\epsilon^A \equiv \epsilon^{Aa}$ $(A = 1,2)$ are spinor parameters. This same global algebra will be a symmetry of superstring theories, in which the coordinates are functions of the string parameters.

As an explicit example I will remind you of supersymmetric Yang–Mills theory. The action,

$$S = \int d^{10}x\left(-\frac{1}{4}F^2 + \frac{i}{2}\bar{\psi}\Gamma\cdot D\psi\right), \tag{6.3.2}$$

is invariant under the supersymmetry transformations (6.1.2). In order to verify that these are symmetries of the action it is necessary to prove that, if ψ_1, ψ_2 and ψ_3 are Majorana–Weyl spinors then

$$\bar{\epsilon}\Gamma_\mu\psi_{[1}\bar{\psi}_2\Gamma^\mu\psi_{3]} = 0. \tag{6.3.3}$$

The proof of this relation is of importance in several different areas of the subject. It is an example of a Fierz transformation. It may be proved as follows. First, consider the expression on the left-hand side of (6.3.3) with the factors of $\bar{\epsilon}^a$ and ψ_3^b deleted. This leaves a matrix

$$M_{ab} = \Gamma^\mu_{ab}\bar{\psi}_1\Gamma_\mu\psi_2 + (\Gamma^\mu\psi_1)_a(\bar{\psi}_2\Gamma_\mu)_b - (\Gamma^\mu\psi_2)_a(\bar{\psi}\Gamma_\mu)_b. \tag{6.3.4}$$

M_{ab} can be expanded in complete basis products of antisymmetrized products of Γ matrices

$$(\Gamma_{\mu_1\mu_2\ldots\mu_k})_{ab} = \frac{1}{k!}\Gamma_{[\mu_1}\Gamma_{\mu_2}\ldots\Gamma_{\mu_k]}. \tag{6.3.5}$$

However,

$$\Gamma_{\mu_1\ldots\mu_k} = \epsilon_{\mu_1\ldots\mu_{10}}\Gamma^{\mu_{k+1}\ldots\mu_{10}}\Gamma_{11}. \tag{6.3.6}$$

This means that when the indices are contracted with chiral spinors (so $\Gamma_{11} = \pm 1$) the only independent elements are those with $k \le 5$. Furthermore, terms involving an even number of Γ matrices, of the form $\bar{\psi}_1\Gamma^\mu\Gamma^\nu\psi_2$ vanish since there is a mismatch in chirality. This means that k must be odd. The remaining possibilities for $\Gamma^0 M_{ab}$ are combinations of

$$\left(\Gamma^0\Gamma^\mu\right)_{ab}, \qquad \left(\Gamma^0\Gamma^{\mu_1\ldots\mu_5}\right)_{ab}, \tag{6.3.7}$$

which are symmetric under $a \leftrightarrow b$ and

$$\left(\Gamma^0\Gamma^{\mu_1\mu_2\mu_3}\right)_{ab}, \tag{6.3.8}$$

which is antisymmetric under $a \leftrightarrow b$. But $(\Gamma^0 M)_{ab}$ is symmetric under $a \leftrightarrow b$. It has $16 \times 17/2 = 136$ components. The term with $k = 1$ has 10 components and the term with $k = 5$ has $10!/2 \cdot 5! = 126$ components (where the factor of $1/2$ comes from the self-duality relation that follows from (6.3.6)). Now consider each of the two terms separately.

The $k = 1$ term is isolated by multiplying M_{ab} by $(\Gamma_\rho)_{ab}$, giving

$$\begin{aligned} &\operatorname{tr}(\Gamma^\mu\Gamma_\rho)\bar{\psi}_1\Gamma_\mu\psi_2 - \bar{\psi}_2\Gamma_\mu\Gamma_\rho\Gamma^\mu\psi_1 + \bar{\psi}_1\Gamma_\mu\Gamma_\rho\Gamma^\mu\psi_2 \\ &= 16\bar{\psi}_1\Gamma_\mu\psi_2 + 8\bar{\psi}_2\Gamma_\rho\psi_1 - 8\bar{\psi}_1\Gamma_\rho\psi_2 = 0, \end{aligned} \tag{6.3.9}$$

(using $\bar{\psi}_2\Gamma_\rho\psi_1 = -\bar{\psi}_1\Gamma_\rho\psi_2$).

The $k = 5$ term is isolated by multiplying M_{ab} by $(\gamma_{\rho_1\ldots\rho_5})_{ab}$ giving,

$$-\bar{\psi}_2\gamma_\mu\gamma_{\rho_1\ldots\rho_5}\gamma^\mu\psi_1 + \bar{\psi}_1\gamma_\mu\gamma_{\rho_1\ldots\rho_5}\gamma^\mu\psi_2 = 2\bar{\psi}_1\gamma_\mu\gamma_{\rho_1\ldots\rho_5}\gamma^\mu\psi_1. \tag{6.3.10}$$

This vanishes when $D = 2k$ by using

$$\begin{aligned} \gamma_\mu\gamma_{\rho_1\ldots\rho_k}\gamma^\mu &= (-1)^k(D - 2k)\gamma_{\rho_1\ldots\rho_k} \\ &= 0. \end{aligned} \tag{6.3.11}$$

Hence the required identity works when $D = 10$.

6.4 Superstrings in superspace

I will now describe the approach to superstring theory which starts from a manifestly supersymmetric set of superspace coordinates. Unfortunately, so far this formulation has defied attempts at quantization in a covariant gauge. This is one of the most profound unsolved aspects of the theory. It is, however, possible to tackle the quantum theory in the light-cone gauge. Since space-time supersymmetry is manifest in this formulation many of the calculations are much more compact than in the covariant formulation based on the RNS coordinates together with the GSO truncation.

The starting point is to guess at a generalization of the Nambu–Goto action which possesses the global super-Poincaré symmetry of (6.3.1). The obvious guess at a sensible action is

$$S_1 = -\frac{1}{2\pi}\int \sqrt{h}h^{\alpha\beta}\Pi^\mu_\alpha \Pi_{\beta\mu} d\sigma d\tau, \tag{6.4.1}$$

where

$$\Pi^\mu_\alpha = \partial_\alpha X^\mu - i\bar{\theta}^A \Gamma^\mu \partial_\alpha \theta^A. \tag{6.4.2}$$

The Grassmann coordinates $\theta^{1a}(\sigma,\tau)$ and $\theta^{2a}(\sigma,\tau)$ are world-sheet scalars (in contrast to the fermionic variables, ψ^μ, of the RNS theory). The quantity Π^μ_α is the supersymmetric element of length so that the fact that S_1 is invariant under the transformations of (6.3.1) is easy to show, in particular

$$\delta_\epsilon S_1 = 0, \tag{6.4.3}$$

so that the action is invariant under global N=2 supersymmetry transformations. It is also invariant under reparametrizations of the world-sheet and Weyl scaling. However the theory defined by S_1 is not a sensible string theory. The subsequent analysis will show that its quantum behaviour is not conformally invariant, which should not be surprising since it looks like a complicated interacting two-dimensional theory.

There is one other term that can be added to the action which is also supersymmetric and classically has the required local symmetries. This term is

$$S_2 = \frac{1}{\pi}\int d\sigma d\tau \left(-i\epsilon^{\alpha\beta}\partial_\alpha X^\mu \left(\bar{\theta}^1\Gamma_\mu\partial_\beta\theta^1 - \bar{\theta}^2\Gamma_\mu\partial_\beta\theta^2\right) + \epsilon^{\alpha\beta}\bar{\theta}^1\Gamma^\mu\partial_\alpha\theta^1\bar{\theta}^2\Gamma_\mu\partial_\beta\theta^2\right). \tag{6.4.4}$$

The proof that S_2 is invariant under supersymmetry transformations is more complicated than it was for S_1. In checking the ϵ variation it is necessary to prove the vanishing of

$$\int d\sigma d\tau \epsilon^{\alpha\beta}\bar{\epsilon}^1\Gamma^\mu\partial_\alpha\theta^1\bar{\theta}^1\Gamma_\mu\partial_\beta\theta^1 \equiv \int d\sigma d\tau \bar{\epsilon}^1\Gamma_\mu\psi_{[1}\bar{\psi}_2\Gamma^\mu\psi_{3]}. \tag{6.4.5}$$

The right-hand side is exactly the combination of spinors that was shown to vanish in considering the supersymmetry of supersymmetric Yang–Mills theory earlier. The vanishing can be proved to occur in special dimensions, namely:- $D = 3$ when the spinors are Majorana, $D = 4$ when the spinors are either chiral or Majorana, $D = 6$ when the spinors are chiral, and $D = 10$ with Majorana–Weyl spinors. In the cases that involve chiral (but not Majorana) spinors $\bar{\theta}\Gamma^\mu\partial_\alpha\theta$ is replaced in all formulas by $\frac{1}{2}\bar{\theta}\Gamma^\mu\partial_\alpha\theta - \frac{1}{2}\partial\bar{\theta}\Gamma^\mu\theta$.

The appropriate superstring action is the sum

$$S = S_1 + S_2. \tag{6.4.6}$$

Precisely this combination of terms, with exactly the coefficients given in (6.4.1) and (6.4.4), has a large amount of extra local symmetry.

(i) The reparametrization symmetry (with parameters ξ^α) is manifest (with the transformations described earlier for the bosonic string supplimented by $\delta_\xi\theta^A = \xi\cdot\theta^A$).

(ii) In addition there is a local fermionic symmetry associated with Grassmann parameters $\kappa^{A\alpha a}$. These parameters satisfy

$$\kappa^{1\alpha} = P_-^{\alpha\beta}\kappa_\beta^1, \qquad \kappa^{2\alpha} = P_+^{\alpha\beta}\kappa_\beta^2, \tag{6.4.7}$$

where the projectors $P_\pm^{\alpha\beta}$ are defined by

$$P_\pm^{\alpha\beta} = \frac{1}{2}\left(h^{\alpha\beta} \pm \frac{\epsilon^{\alpha\beta}}{\sqrt{h}}\right). \tag{6.4.8}$$

Multiplying an arbitrary two-vector, v_α by $P_+^{\alpha\beta}$ ($P_-^{\alpha\beta}$) projects it into a self-dual (anti self-dual) vector, *i.e.*,

$$v_+^\alpha = P_+^{\alpha\beta}v_\beta = \frac{1}{2}\begin{pmatrix} v_0 + v_1 \\ -v_0 - v_1 \end{pmatrix}, \qquad v_-^\alpha = P_-^{\alpha\beta}v_\beta = \frac{1}{2}\begin{pmatrix} v_0 - v_1 \\ -v_0 + v_1 \end{pmatrix}. \tag{6.4.9}$$

The transformations of the fields in the action have the form

$$\begin{aligned} \delta_\kappa\theta^A &= 2i\Gamma \cdot \Pi_\alpha\kappa^{A\alpha} \\ \delta_\kappa X^\mu &= i\bar{\theta}^A\Gamma^\mu\delta\theta^A \\ \delta_\kappa\left(\sqrt{h}h^{\alpha\beta}\right) &= 16\sqrt{h}\left(P_-^{\alpha\beta}\bar{\kappa}^{1\beta}\partial_\gamma\theta^1 + P_+^{\alpha\beta}\bar{\kappa}^{2\beta}\partial_\gamma\theta^2\right). \end{aligned} \tag{6.4.10}$$

In proving the invariance of S under these transformations the same Fierz transformation must be used as in the proof of the global supersymmetry of S_2. The proof requires the combination of S_1 and S_2 with precisely determined coefficients.

(iii) The local symmetries described so far do not form a closed algebra. In particular the commutator of two κ transformations does not give one of the listed transformations. There is actually another set of bosonic symmetries with parameters λ^α. I will not describe them here since they can be shown to be unimportant in the subsequent discussion. After fixing the covariant gauge this symmetry becomes inactive– it does not eliminate any extra components of the fields. The full significance of this symmetry has yet to be elucidated.

The status of the action $S_1 + S_2$ is reminiscent of the non-linear sigma model (S_1) on a group manifold in the presence of a Wess–Zumino term (S_2). Such a model has an action of the form

$$\begin{aligned} S_1 &= \frac{1}{g}\int G_{ab}\partial_\alpha\phi^a\partial^\alpha\phi^b d_\sigma^2, \\ S_2 &= K\int d^3\sigma \mathrm{tr}\left(U^{-1}\partial_\alpha UU^{-1}\partial_\beta UU^{-1}\partial_\gamma U\right)\epsilon^{\alpha\beta\gamma}. \end{aligned} \tag{6.4.11}$$

The β function for such a theory is roughly as in figure 6.16. The β function is negative for small g (asymptotic freedom) but it turns up and goes through zero at $g = 1$. When $g = 1$ there is an infinite dimensional local symmetry (associated with a Kac-Moody

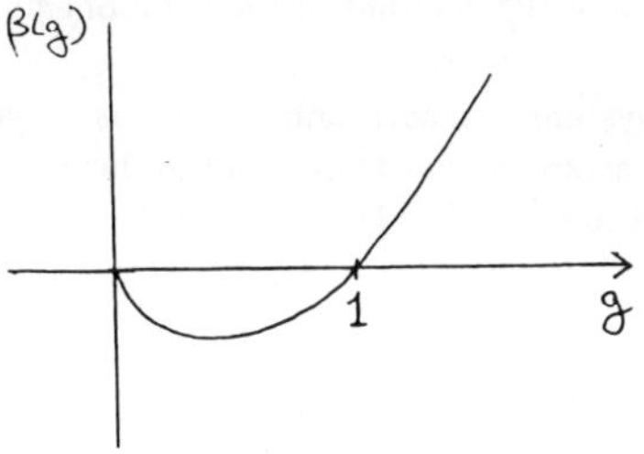

Figure 6.16. Illustration of the β function for a sigma model with a Wess–Zumino term.

algebra). As stressed by Witten this is related to the fact that the theory can be expressed in terms of free fermion fields at this point. The Wess–Zumino term (S_2) can be described in terms of a torsion on the field manifold. In that language the point $g = 1$ is the point at which the Riemann curvature of the field manifold vanishes when it is calculated from the connection which contains the torsion. The torsion is said to 'parallelize' the manifold.

In superstring theory the action, $S_1 + S_2$, is a kind of sigma model which maps the world-sheet into $N = 1$ or $N = 2$ superspace. The Wess–Zumino term, S_2, is interpretable in terms of the torsion on the supermanifold. When g=1 we have seen that there is an enlargement of the local symmetry. I will show that this symmetry allows the choice of the light-cone gauge in which the theory is also free field theory. At least in the case of the heterotic theory (where there is only one θ) this can again be interpreted by thinking of the term S_2 as a Wess–Zumino term, corresponding to a torsion which parallelizes the supermanifold. As yet the equivalence of the theory to a free theory has not been demonstrated in a manifestly lorentz-invariant manner.

6.5 Equations of motion

The equations of motion that follow from varying the θ variables in the action are

$$\Gamma \cdot \Pi_\alpha P_-^{\alpha\beta} \partial_\beta \theta^1 = 0 = \Gamma \cdot \Pi_\alpha P_+^{\alpha\beta} \partial_\beta \theta^2. \tag{6.5.1}$$

The X^μ equations of motion are

$$\partial_\alpha \left[\sqrt{h} (h^{\alpha\beta} \partial_\beta X^\mu - 2i P_-^{\alpha\beta} \bar{\theta}^1 \Gamma^\mu \partial_\beta \theta^1 - 2i P_+^{\alpha\beta} \bar{\theta}^2 \Gamma^\mu \partial_\beta \theta^2 \right] = 0. \tag{6.5.2}$$

The $h^{\alpha\beta}$ equations of motion are constraint equations. Since $h^{\alpha\beta}$ only occurs in S_1 this is the only term that contributes to these equations

$$\Pi_\alpha^\mu \cdot \Pi_{\beta\mu} - \frac{1}{2} h_{\alpha\beta} h^{\gamma\delta} \Pi_\gamma^\mu \Pi_{\delta\mu} = 0. \tag{6.5.3}$$

This expression can be written as the equations

$$\Pi_+^\mu \Pi_{+\mu} = 0 = \Pi_-^\mu \Pi_{-\mu}, \tag{6.5.4}$$

where $\Pi_{\pm}^{\mu\alpha} \equiv P_{\pm}^{\alpha\beta}\Pi_{\beta}^{\mu}$ and where Π_{+}^{μ} is either of the components of $\Pi_{+}^{\mu\alpha}$ (and similarly for Π_{-}^{μ}).

The quantization of this system in a covariant gauge has proved elusive. The reason for this is the existence of some awkward constraints in the theory. This is seen from the form of the momentum conjugate to θ^A, defined by

$$P_{\theta^A} = \frac{\delta S}{\delta \partial_\tau \bar{\theta}^A}. \tag{6.5.5}$$

This is a constraint since the right-hand side is a function of the other phase-space variables and their σ derivatives, θ^A, $\theta^{A\prime}$, X^μ, P^μ and $X^{\mu\prime}$ (*i.e.*, it does not depend on $\partial_\tau\theta^A$). According to Dirac's procedure for dealing with constrained systems, those constraints which do not commute with all the other ones (the 'second-class constraints') must be solved explicitly or their Poisson brackets must be replaced by Dirac brackets. The problem here is that the first and second-class constraints are mixed up. No convincing way has been found to disentangle them in a covariant manner although suggestions have been made. The choice of light-cone gauge is a non-covariant way of converting the constraints to second-class one which can be solved explicitly.

6.6 Light-cone gauge

The fermionic local symmetry (the κ symmetry) can be used to choose

$$\theta^a \equiv \frac{1}{2}\Gamma^-\Gamma^+\theta^A = 0, \tag{6.6.1}$$

where the undotted spinor index $a = 1, \ldots, 8$, as discussed earlier. Superficially the θ transformation laws in (6.4.10) eliminate too many components of θ^{Aa} since $\kappa^{A\alpha a}$ has the same number of components as θ^{Aa} (when the self-duality of κ is taken into account). However, the form of the θ^{1a} transformations in (6.4.10) are proportional to $\Gamma \cdot \Pi_-$. This means that when multiplied by $\Gamma \cdot \Pi_-$ there is a factor of $\Pi_- \cdot \Pi_-$ which vanishes since it is a constraint. This means that the transformations only act on half the θ components. The + components have similar form.

With the choice of gauge in (6.6.1) the X^+ equation of motion is simply the wave equation

$$\partial_\alpha\partial^\alpha X^+(\sigma,\tau) = 0. \tag{6.6.2}$$

As before, this means that the light-cone parametrization can be chosen, *i.e.*,

$$X^+(\tau,\sigma) = x^+ + p^+\tau. \tag{6.6.3}$$

It is convenient to choose a new normalization for the Grassmann coordinates by defining S^{Aa} (for the type IIB theory) by

$$\begin{aligned} \sqrt{p^+}\theta^1 &\to S^{1a} \\ \sqrt{p^+}\theta^2 &\to S^{2a}. \end{aligned} \tag{6.6.4}$$

For the typeIIA theory one of the spinors is taken to be dotted, *e.g.* $S^{2\dot{a}}$. From (6.6.1) the

S variables satisfy

$$\Gamma^+ S^A = 0. \tag{6.6.5}$$

It will be useful to note the following properties of products of spinors in the light-cone gauge.

$$\begin{aligned} \bar\theta\Gamma^\mu\partial_\alpha\theta &= \theta\Gamma^+\Gamma^-\partial_\alpha\theta, \qquad && \mu = - \\ &= 0, && \text{otherwise.} \end{aligned} \tag{6.6.6}$$

All the components of $\bar\theta\Gamma^{\mu\nu\rho}\theta$ vanish except $\bar\theta\Gamma^{-ij}\theta \sim S\Gamma^{ij}S$.

Using these properties it is easy to see that the equations of motion assume a very simple form in the light-cone gauge

$$\partial_+ S^{1a} = 0 = \partial_- S^-, \tag{6.6.7}$$

$$\partial^\alpha\partial_\alpha X^i = 0. \tag{6.6.8}$$

The constraint equations (6.5.4) can again be solved explicitly for X^- and hence for the hamiltonian, $h \equiv p^- = \int d\sigma T\dot{X}^-$.

6.7 The light-cone gauge action

The light-cone gauge equations of motion follow from a simple

$$S = -\frac{1}{2}\int d\sigma d\tau \left(T\partial_\alpha X^i \partial^\alpha X^i - \frac{1}{\pi}\bar{S}^a \rho^\alpha \partial_\alpha S^a \right). \tag{6.7.1}$$

In this expression the two S^A variables have been combined into a two-component spinor

$$S^a \equiv \begin{pmatrix} S^{1a} \\ S^{2a} \end{pmatrix}, \tag{6.7.2}$$

and

$$\bar{S}^a = (S\rho^0)^a. \tag{6.7.3}$$

Note that S^{Aa} is a world-sheet spinor whereas its components originated from two independent world-sheet scalars, θ^A. The factor of $\sqrt{p^+}$ in (6.6.4) is essential for S^{Aa} to transform properly as a two-dimensional spinor on the index A. The anticommutation relaztions that follow from the action are

$$\left\{ S^{Aa}(\sigma,\tau), S^{Bb}(\sigma',\tau) \right\} = \pi\delta^{ab}\delta^{AB}\delta(\sigma - \sigma'). \tag{6.7.4}$$

6.8 Canonical quantization

I will begin by describing the open-string sector. The boundary conditions that follow from the action are like the conditions in the Ramond sector of the fermionic string theory

$$\begin{aligned} S^{1a}(0,\tau) &= S^{2a}(0,\tau) \\ S^{1a}(\pi,\tau) &= S^{2a}(\pi,\tau). \end{aligned} \tag{6.8.1}$$

In this case the NS boundary conditions cannot be used if space-time supersymmetry is to be preserved since there would be no zero mode, S_0^{Aa}. Although there are interesting heterotic theories which break supersymmetry in this way (in particular, a theory with $O(16) \times O(16)$ symmetry) they involve more structure that I have not yet mentioned. I will assume that supersymmetry is desirable, which requires the above conditions.

The mode expansions for the solutions of the S equations of motion are

$$\begin{aligned} S^{1a}(\sigma,\tau) &= \frac{1}{\sqrt{2}} \sum S_n^a e^{-in(\tau-\sigma)} \\ S^{2a}(\sigma,\tau) &= \frac{1}{\sqrt{2}} \sum S_n^a e^{-in(\tau+\sigma)}. \end{aligned} \tag{6.8.2}$$

The reality of S requires $S_{-m}^a = (S_m^a)^\dagger$. The modes of the anticommutation relations (6.7.4) give

$$\left\{S_m^a, S_n^b\right\} = \delta^{ab}\delta_{m+n,0}. \tag{6.8.3}$$

For closed strings the boundary conditions are

$$S^{Aa}(\sigma,\tau) = S^{Aa}(\sigma+\pi,\tau). \tag{6.8.4}$$

There are now two independent sets of modes, S_n^a and $\tilde{S}_n^a$. The subsidiary constraint that ensures that the theory is independent of the origin of the σ coordinate is

$$N^\alpha + N^S = \tilde{N}^\alpha + \tilde{N}^S, \tag{6.8.5}$$

where $N^S = \sum_1^\infty n S_{-n}^a S_n^a$ with a similar expression for $\tilde{N}$ in terms of $\tilde{S}$.

In comparing the present formulation with that of the RNS theory recall that the light-cone degrees of freedom in that case were X^i and ψ^i which are in the $\mathbf{8}_v$ representation of SO(8) while S^a is a $SO(8)$ spinor in the $\mathbf{8}_s$ representation. The other spinor $(S^{\dot{a}})$ is the $\mathbf{8}_c$ representation. It is a special feature of $SO(8)$, known as *triality*, that the vector representation has the same dimension as the two spinors.

6.9 Super-Poincaré algebra

A single 16-component supercharge decomposes into eight-component $SO(8)$ spinors $\epsilon \to (\eta^a, \epsilon^{\dot{a}})$. The light-cone description of the $N = 2$ supersymmetry transformations decomposes into eight-component pieces in a corresponding fashion

$$\begin{aligned} \delta S^a &= \sqrt{2p^+}\eta^a \\ \delta X^i &= 0 \end{aligned} \tag{6.9.1}$$

and

$$\begin{aligned} \delta S^a &= \frac{-i}{\sqrt{p^+}}\rho \cdot \partial X^i \gamma^i_{a\dot{a}} \epsilon^{\dot{a}} \\ \delta X^i &= \frac{1}{\sqrt{p^+}}\gamma^i_{a\dot{a}}\bar{\epsilon}^{\dot{a}} S^a. \end{aligned} \tag{6.9.2}$$

These transformations include the effects of compensating local supersymmetry transformations which are needed to restore the gauge condition $\gamma^+\theta = 0$, *i.e.*, the global supersymmetry transformation of θ is of the form

$$\delta\theta = \epsilon + 2i\gamma \cdot \pi_\alpha \kappa^\alpha, \tag{6.9.3}$$

with a particular choice for the parameter κ that ensures that the transformed θ satisfies the gauge condition, (6.6.1).

The superalgebra can be seen by investigating the effect of commuting two transformations with parameters $(\eta^{(1)a}, \epsilon^{(1)\dot{a}})$ and $(\eta^{(2)a}, \epsilon^{(2)\dot{a}})$

$$\begin{aligned} [\delta_1, \delta_2]X^i &= \xi^\alpha \partial_\alpha X^i + a^i \\ [\delta_1, \delta_2]S^a &= \xi^\alpha \partial_\alpha S^a, \end{aligned} \tag{6.9.4}$$

where

$$\xi^\alpha = -2i\epsilon^{(1)}\rho^\alpha \epsilon^{(2)}, \tag{6.9.5}$$

which is a translation on the world-sheet, while

$$a^i = \sqrt{2}\eta^{(2)}\gamma^i\epsilon^{(1)} - \sqrt{2}\epsilon^{(2)}\gamma^i\eta^{(1)}, \tag{6.9.6}$$

which is a translation in transverse space-time.

The Noether current associated with these supersymmetry transformations can be deduced in the manner used earlier in deriving the two-dimensional supercurrent in the RNS theory. The integral of the supercharge densities then gives the generators of the supersymmetry transformations. In a covariant notation the algebra of these charges has the form $\{Q^a, \bar{Q}^b\} \sim \frac{1}{2}\left((1 \pm \Gamma_{11})\Gamma \cdot p\right)^{ab}$. In the light-cone gauge a single supercharge decomposes into $SO(8)$ spinors, $Q \to (Q^a, Q^{\dot{a}})$. The expressions for these operators in terms of modes in open superstring theories are

$$Q^a = (2p^+)^{1/2} S_0^a, \qquad Q^{\dot{a}} = \frac{1}{(p^+)^{1/2}} \gamma^i_{\dot{a}a} \sum_{n=-\infty}^{\infty} S^a_{-n} \alpha^i_n. \tag{6.9.7}$$

The algebra formed by these operators can be deduced from these expansions, giving

$$\begin{aligned} \left\{Q^a, Q^b\right\} &= 2p^+ \delta^{ab} \\ \left\{Q^a, Q^{\dot{b}}\right\} &= \sqrt{2}\gamma^i_{a\dot{b}} P^i \\ \left\{Q^{\dot{a}}, Q^{\dot{b}}\right\} &= 2p^- \delta^{ab}. \end{aligned} \tag{6.9.8}$$

The hamiltonian in this expression is the same as that obtained by integrating $\dot{X}^-$

$$h \equiv p^- = \alpha_0^- = \frac{1}{2p^+}(\underline{p}^2 + 2N^\alpha + 2N^S). \tag{6.9.9}$$

There are no normal ordering ambiguities in this expression since they cancel mode by mode (just as in the Ramond sector of the NRS theory).

The generators of Lorentz transformations, J^{ij}, J^{+i}, J^{+-} and J^{i-} can be constructed by the Noether method as in the earlier cases. In the light-cone gauge those transformations which transform the gauge conditions are more complicated than the others for reasons that I described earlier. In particular, the J^{i-} components need careful study. In verifying the closure of the algebra the fact that the model is defined in ten dimensions with massless ground states is crucial. Recall that at the classical level the covariant theory could be defined in $D = 3, 4, 6$ and 10. The light-cone gauge quantum Lorentz algebra only closes properly for the ten-dimensional case. This indicates that the other cases have conformal anomalies. This leaves the possibility that they can be sensibly defined as quantum theories by the procedure suggested by Polyakov which involves extra dynamical, 'longitudinal' modes.

The $SO(8)$ subgroup of the Lorentz group is generated by J^{ij}. The piece of this which acts on the Grassmann coordinates is iR^{ij}, where

$$R^{ij} \equiv \frac{1}{4\pi} \int d\sigma S^A(\sigma) \gamma^{ij} S^A(\sigma), \tag{6.9.10}$$

and $\gamma^{ij} = \frac{1}{2}\left(\gamma^i_{a\dot{b}} \gamma^j_{\dot{b}b} - i \leftrightarrow j\right)$. The zero-mode piece of this is given by

$$R_0^{ij} = \frac{1}{4} S_0^a \left(\gamma_{ij}\right)_{ab} S_0^b, \tag{6.9.11}$$

which satisfies the Lorentz algebra

$$\left[R_0^{ij}, R_0^{kl}\right] = \delta^{il} R_0^{jk} - \delta^{ik} R_0^{jl} + \delta^{jk} R_0^{il} - \delta^{jl} R_0^{ik}. \tag{6.9.12}$$

The operator R_0^{ij} rotates the helicities of the states without changing the mass level.

6.10 Spectrum

The ground state spectrum of states is determined by properties of the zero modes of the S_0^a operators which satisfy

$$\left\{S_0^a, S_0^b\right\} = 2\delta^{ab}, \tag{6.10.1}$$

which is a Clifford algebra, differing from that of the γ matrices in the interchange of a spinor index for a vector index. This means that these operators may be represented by matrices in a 16×16-dimensional space spanned by the $\mathbf{8}_V + \mathbf{8}_C$ representations of $SO(8)$,

$$S_0^a \sim \begin{pmatrix} 0 & \gamma_{i\dot{a}}^a \\ \gamma_{\dot{a}i}^a & 0 \end{pmatrix}. \tag{6.10.2}$$

This converts bosons in the $\mathbf{8}_V$ representation into fermions in the $\mathbf{8}_C$ representation and vice versa.

(i) Open Strings

For open strings the mass-shell condition obtained from the hamiltonian is

$$(\text{Mass})^2 = 2N^\alpha + 2N^S. \tag{6.10.3}$$

The ground states are the $\mathbf{8}_V$ representation, $|i\rangle$ and the $\mathbf{8}_C$ representation, $|\dot{a}\rangle$, which are normalized so that

$$\langle i|j\rangle = \delta_{ij}, \qquad \langle \dot{a}|\dot{b}\rangle = \delta_{\dot{a}\dot{b}}. \tag{6.10.4}$$

Since R^{ij} rotates the helicity of the states they are defined to obey

$$\begin{aligned} R_0^{ij}|k> &= \delta^{jk}|i> - \delta^{ik}|j> \\ R_0^{ij}|\dot{a}> &= -\frac{1}{2}\gamma_{\dot{a}\dot{b}}^{ij}|\dot{b}>. \end{aligned} \tag{6.10.5}$$

Furthermore, the undotted piece of the supercharge ($Q^a = \sqrt{2p^+ S_0^a}$) rotates the states into each other

$$\begin{aligned} S_0^a|\dot{a}> &= \frac{1}{\sqrt{2}}\gamma_{a\dot{a}}^i|i> \\ S_0^a|i> &= \frac{1}{\sqrt{2}}\gamma_{a\dot{a}}^i|\dot{a}>. \end{aligned} \tag{6.10.6}$$

The normalizations of these last two equations can be determined by requiring consistency with (6.10.5) by applying a second factor of S_0^a and using

$$\begin{aligned} S_0^a S_0^b &= \frac{1}{2}\left\{S_0^a, S_0^b\right\} + \frac{1}{2}\left[S_0^a, S_0^b\right] \\ &= \frac{1}{2}\delta^{ab} + \frac{1}{16} S_0^c \gamma_{cd}^{ij} S_0^d \gamma_{ab}^{ij}. \end{aligned} \tag{6.10.7}$$

The excited states are obtained by applying the higher moded operators, α_{-n}^i and S_{-n}^a, to the ground states. The spectrum is guaranteed to be supersymmetric by the closure of the super-Poincaré algebra. For example, the states at the first excited level consist of 128 bosons

$$\alpha_{-1}^i|j>, \qquad S_{-1}^a|\dot{a}>, \tag{6.10.8}$$

and 128 fermions

$$\alpha_{-1}^i|\dot{a}>, \qquad S_{-1}^a|i>. \tag{6.10.9}$$

These 256 states fall into representations of $SO(9)$, the appropriate little group for classifying massive particles. Notice that this is the same classification of states as for the massless states of 11-dimensional supergravity for which the little group is also $SO(9)$.

For closed strings the mass formula turns out to be given by

$$(\text{Mass})^2 = 4(N^\alpha + N^S + \tilde{N}^\alpha + \tilde{N}^S), \tag{6.10.10}$$

which is suplemented with the subsidiary condition $N^\alpha + N^S = \tilde{N}^\alpha + \tilde{N}^S$. The states of this theory are naturally expressed as tensor products of the left-moving and right-moving states.

This means that the massless closed-string ground states can be written as tensor products of $|i\rangle$ and $|\dot{a}\rangle$. The type II theories have two supercharges and therefore the ground states are the same as those in the type II supergravity theories. There are 64 massless bosons given by

$$|i\rangle \otimes |\tilde{j}\rangle, \tag{6.10.11}$$

which describe the massless fields, g^{ij}, A^{ij} (an antisymmetric tensor potential) and ϕ (a scalar dilaton) as in the bosonic theory.

There are 64 more boson states made out of spinors,

$$|\dot{a}\rangle \otimes |\tilde{\dot{a}}\rangle, \tag{6.10.12}$$

for the type IIB theory (for the type IIA theory one of the spaces has an undotted spinor). These spinor bilinears can be decomposed into fields with vector indices. The decomposition depends on whether the two spinors have the same chirality (type IIB) or opposite chirality (type IIB). For the type IIB theory the terms that contribute are $\gamma_{ijkl}^{\dot{a}\dot{b}} A^{ijkl}$ (where the fourth-rank antisymmetric tensor potential satisfies the self-duality condition, $A^{ijkl} = \epsilon^{ijklmnop} A_{mnop}$), $\gamma_{ij}^{\dot{a}\dot{a}} B^{ij}$ (where B^{ij} is another antisymmetric tensor potential) and $\delta^{\dot{a}\dot{b}}\phi'$ (another dilaton).

For the type IIA theory the terms that contribute are $\gamma^{a\dot{b}}_{ijk}A^{ijk}$ and $\gamma^{a\dot{b}}_{i}A^{i}$.
There are also 64 fermion states

$$|\dot{a}\rangle \otimes |\tilde{i}\rangle, \tag{6.10.13}$$

which describe a gravitino field $\psi^{\dot{a}}_i$ (satisfying the 'spin 3/2' condition, $\gamma^i\psi_i = 0$) and a spinor, $\lambda^{\dot{a}}$. Likewise the states

$$|i\rangle \otimes |\tilde{\dot{a}}\rangle, \tag{6.10.14}$$

describe another gravitino, $\psi'^{\dot{a}}_i$ and spinor field, $\lambda^{\dot{a}}$. For the type IIA theory one set of fermions are described by dotted spinors and the other by undotted ones.

The excited states are obtained by applying α^i_{-m}, $\tilde{\alpha}^i_{-m}$, S^a_{-m} and $\tilde{S}^a_{-m}$ to the ground states (subject to $N^\alpha + N^S = \tilde{N}^\alpha + \tilde{N}^S$).

The closed-string sector of the type I closed string theory is obtained by symmetrizing the states of the type IIB theory between the two Fock spaces. This truncation is equivalent to the statement that the string is unoriented. This means that the string states satisfy $X^i(\sigma)|\ \rangle = X^i(-\sigma)|\ \rangle$ as can be seen explicitly from the mode expansions. In this case the ground states consist of 36 bosons made from

$$|i\rangle \otimes |\tilde{j}\rangle + |j\rangle \otimes |\tilde{i}\rangle. \tag{6.10.15}$$

These states include the graviton, g^{ij} and the dilaton, ϕ. There are 28 more boson states made from

$$|\dot{a}\rangle \otimes |\tilde{\dot{b}}\rangle + |\dot{b}\rangle \otimes |\tilde{\dot{a}}\rangle, \tag{6.10.16}$$

which describe the antisymmetric tensor, $\gamma^{\dot{a}\dot{b}}_{ij}B^{ij}$. There are 64 fermion states

$$|i\rangle \otimes |\dot{a}\rangle + |\dot{a}\rangle \otimes |i\rangle, \tag{6.10.17}$$

which describe a single gravitino field, $\psi^{\dot{a}}_i$ and a spinor, $\lambda^{\dot{a}}$. This list of states is the content of N=1 supergravity, as expected. Together with the Yang–Mills supermultiplet it completes the list of ground-states in the type I theories. It also coincides with the massless states of the heterotic string which also has $N = 1$ supersymmetry.

7. Superstring tree amplitudes; heterotic strings in the fermion representation.

7.1 Propagators and vertex operators for superstring theories

In order to calculate scattering amplitudes in either the functional approach or in the operator formalism it is necessary to construct the vertices describing the emission of the relevant states. I have described the form of these operators in the bosonic open and closed-string theory in earlier lectures and I will now turn to the fermionic and superstring theories. I will present the form of the vertices in the operator formalism but the translation of these expressions into the functional formalism is straightforward.

Recall the general form for the M-particle tree amplitude in the operator approach is

$$A(1,2,\cdots M) = g^{M-2}\langle 0;k_1|V(k_2,0)\Delta\ldots\Delta V(k_{M-1},0)|0;k_M\rangle, \tag{7.1.1}$$

where Δ is the propagator and $V(k,\tau)$ is the vertex for emitting an external state of momentum k from the endpoint $\sigma = 0$.

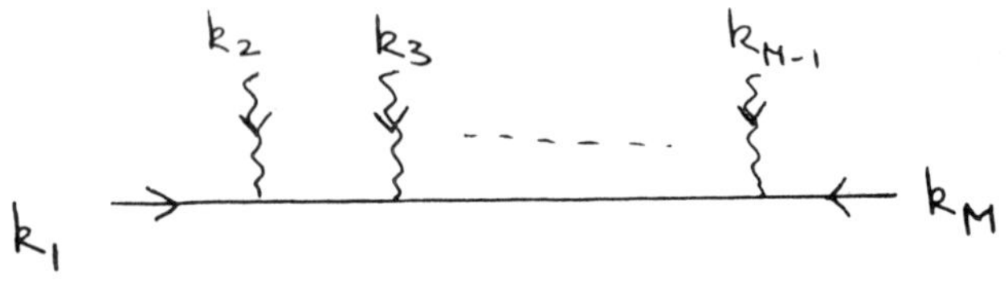

Figure 7.17. An M-particle tree amplitude.

(i) Fermionic strings (The RNS theory)

The open-string propagators may be obtained by guessing operators which have poles at the appropriate positions. These are given by

$$\Delta^{NS} = \frac{1}{L_0^{NS} - \frac{1}{2}} = \int_0^1 dx x^{L_0^{NS} - 3/2} \tag{7.1.2}$$

when the string propagates NS bosons and

$$\Delta^R = \frac{1}{F_0} = \frac{F_0}{L_0^R} = F_0 \int_0^1 dx x^{L_0^R - 1} \tag{7.1.3}$$

when the string propagates R fermions.

These propagators must be supplemented by the GSO projection operators $(1+G)/2$ or $(1 \pm \bar{\Gamma})/2$ if the amplitude is to be supersymmetric in space-time.

The vertices are determined by the same kinds of considerations that determined the vertices in the bosonic theory. This means that the vertices ought to respect Lorentz invariance and transform with weight 1 under conformal transformations generated by the super-Virasoro operators. (The easiest way to describe these vertices is to use the two-dimensional superspace formalism in which the world-sheet is extended to a superspace world-sheet with parameters σ, τ and θ^A, where θ^A is a two-component Majorana spinor with components θ and $\tilde{\theta}$. I do not have space to describe this approach here.)

For open strings, particles are emitted from the boundary of the world-sheet, $\sigma = 0$ or $\sigma = \pi$. The vertex for emitting a ground-state tachyon from a NS string is simply given by

$$V^{NS}(k,\tau) = k \cdot \psi(\sigma=0,\tau) \colon e^{ik\cdot X(\sigma=0,\tau)} \colon = \{G_r, V_0\}\,, \tag{7.1.4}$$

where V_0 is the bosonic string vertex $: e^{ik\cdot X(\sigma=0,\tau)} :$ and $k^2 = 1$. Clearly, this vertex is not relevant if the GSO projection is included in the propagator since the tachyon does not couple in the projected theory.

The condition that the gauges pass through the vertices in an appropriate manner so that no negative-normed states couple to a tree diagram is that they transform appropriately under G_r and L_n^{NS} (in the case of the NS sector). The proof of these gauge properties relies on the facts that

$$\left\{G_r, V^{NS}\right\} = [L_{2r}, V_0]\,, \tag{7.1.5}$$

and

$$\left[L_m, V^{NS}(k,\tau)\right] = -i\frac{d}{d\tau}\left(e^{im\tau}V^{NS}(k,\tau)\right). \tag{7.1.6}$$

The vertex for emitting a tachyon from a Ramond fermion is given by

$$V^R(k,\tau) =: e^{ik\cdot X(\sigma=0,\tau)} :, \tag{7.1.7}$$

where $k^2 = 1$.

Similarly vertices for the emission of excited states can easily be obtained. For example, the vertex for emitting a massless vector state with polarization ς^μ (satisfying $\varsigma \cdot k = 0 = k^2$)is

$$V^{NS}(k,\eta,\tau) = \left(\varsigma \cdot \dot{X} - \varsigma \cdot \psi k \cdot \psi\right) e^{ik\cdot X}, \tag{7.1.8}$$

when the state is emitted from a Neveu–Schwarz boson and

$$V^R(k,\eta,\tau) = \varsigma \cdot \psi e^{ik\cdot X}, \tag{7.1.9}$$

when the state is emitted from a Ramond fermion.

The closed-string vertices are obtained as products of open-string ones. There are four types of vertices of this type since the string from which the particle is emitted is described by the tensor product of NS and R sectors. These four terms may be denoted (NS, NS), (NS, R), (R, NS) and (R, R). For example the vertex for emitting a massless spin 2 graviton from a (NS, NS) string is

$$W^{NS,NS} = \int_0^\pi d\sigma \varsigma_{\mu\nu} \left(\partial_+ X^\mu - \psi_+^\mu k \cdot \psi_+\right) \left(\partial_- X^\nu - \psi_-^\nu k \cdot \psi_-\right) e^{ik \cdot X}, \tag{7.1.10}$$

where $\varsigma_{\mu\nu}$ is the polarization tensor of the graviton. The other vertices are obtained in an obvious way as integrals over products of open-string vertices.

The rules I have just described constitute what is known as the 'F_2 picture'. There is a physically equivalent formalism, known as the 'F_1 picture', in which the rules for Δ^{NS} and V^R are different. The F_1 NS propagator is given by $\Delta^{NS} = (L_0^{NS} - 1)^{-1}$. This appears to include a state with $(\text{mass})^2 = -2$, but this turns out to decouple. The F_1 vertex for the emission of a tachyon from a Ramond string is given by $V^R(k, \tau) = k \cdot \psi(\tau) : \exp(k \cdot X(\tau)) :$ which has the same form as the NS tachyon emission vertex. I will not pursue the relation between these two pictures but it is an important feature that arises in the description of the vertex for emitting a fermion state. This vertex poses special problems and it is only with the recent work of Friedan, Martinec and Shenker and of Knizhnik that the construction of the vertex was completed. I do not have time to describe their work, but it is described by other lecturers at this school. In the light-cone formalism the emission of fermions is on an equal footing with the emission of bosons.

7.2 Supersymmetric light-cone formalism

Although the superspace formalism cannot yet be covariantly quantized we have seen that quantization poses no problems in the light-cone gauge. The light-cone gauge description of the bosonic and the RNS theories was pioneered by Mandelstam. Unlike the covariant fuctional approach or the operator approach, this light-cone formalism is a way of deducing scattering amplitudes from the ordinary principles of quantum mechanics. The amplitudes emerge as quantum-mechanical transitions between the incoming string states at $\tau = -\infty$ and the outgoing states at $\tau = \infty$. The description of the interaction between three arbitrary open strings is conveniently described by rescaling the parameters σ and τ so that the length of a string in σ is proportional to the p^+ carried by the string, *i.e.*, $0 \leq \sigma \leq 2\pi |p^+|$.

The interaction in which two strings join at their endpoints to form a third string can then be represented in parameter space by figure 7.18. Notice that the total width of the string in σ is preserved in an interaction due to the conservation of p^+. A scattering amplitude is composed of vertices like this joined together by propagators which are strips. Loops are represented by slits in the world-sheet in this parametrization

In this approach the world-sheet has been parametrized in such a way that the curvature vanishes everywhere except at the interaction vertices where it is infinite.

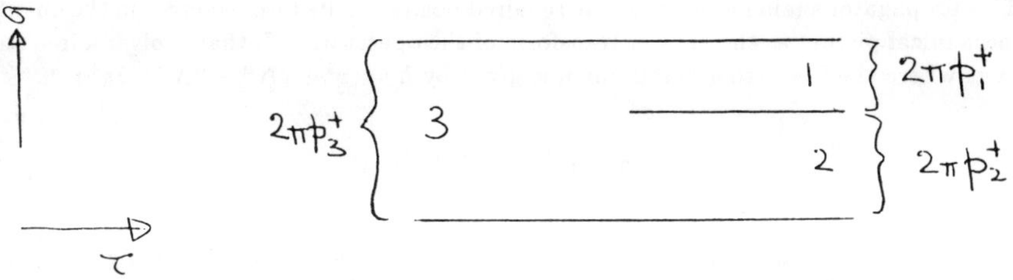

Figure 7.18. Open-string interaction in the light-cone parametrization.

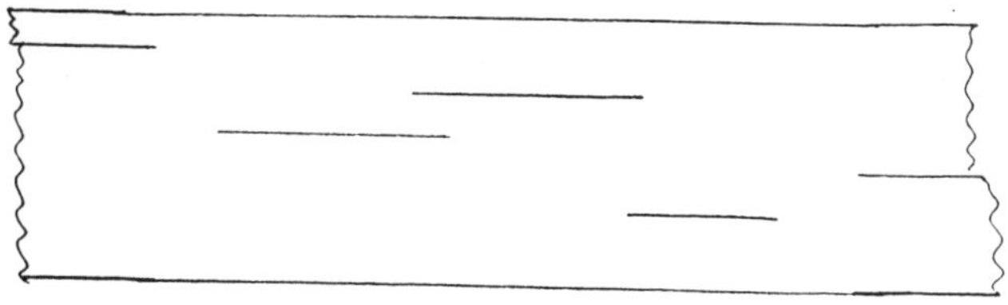

Figure 7.19. An open-string multi-loop amplitude in the light-cone parametrization.

If the particle labelled 2 in the vertex has infinitesimal k^+ then the vertex looks like the covariant vertex in which particle 2 is emitted from the end of the wide string. In the bosonic theory the tachyon-emission vertex is given by

$$V_0 =: e^{ik\cdot X(\sigma=0,\tau)} : . \tag{7.2.1}$$

This looks very much like the vertex in the covariant formalism. However, in the light-cone gauge the expression for X^- is a bilinear in the transverse coordinates. This means that the expression (7.2.1) is only well-defined if $k^+ = 0$. The more general vertex with arbitrary k^+ requires the more general formalism of Mandelstam. So long as there are not too many external particles, the restriction on the values of the k^+ for the all but two of the external particles in a scattering process amounts only to a choice of Lorentz frame. If the theory is assumed to be Lorentz invariant then the use of this frame is legitimate.

In this restricted light-cone formalism the open-superstring propagator is given by

$$\begin{aligned}\Delta = \frac{1}{L_0} &= \frac{2}{p^2 - (\text{Mass})^2} \\ &= \frac{1}{\underline{p}^2/2 + N^\alpha + N^S - p^+p^-}.\end{aligned} \tag{7.2.2}$$

This propagator again has poles at the required positions. Its form emerges in the quantum mechanical picture as the Fourier transform of the operator, $e^{ih\tau}$, that evolves a free string in τ (where the free-string hamiltonian is given by $h = (\underline{p}^2 + 2N^\alpha + 2N^S)/2p^+$)

$$\Delta = i \int_0^\infty e^{ih\tau - ip^-\tau} d\tau / 2p^+. \tag{7.2.3}$$

The vertices for emitting a ground-state massless particle from a superstring are uniquely determined by the global super-Poincaré invariance. I will denote the vertex for emission of a vector boson with polarization ς^μ by $V_B(\varsigma, k)$ and that for emitting a fermion described by a spinor u^a by $V_F(u, k)$.

The kinematics in the frame $k^+ = 0$ has special features. Firstly, note that since $k^2 = 0$ the momentum must satisfy

$$(k^i)^2 = 0, \tag{7.2.4}$$

so that the momentum will have to be taken to be complex in order for it to be non-zero. This appears to pose no problems. The polarization vector satisfies the gauge condition $\varsigma^+ = 0$ and the on-shell condition, $\varsigma \cdot k = 0$, so that ς^- is determined by

$$\varsigma^- = \frac{1}{k^+} \varsigma^i k^i. \tag{7.2.5}$$

When $k^+ = 0$ we will take $\varsigma^i k^i = 0$ in which case only seven of the ς^i are independent but ς^- is now the eighth independent polarization vector. Likewise the spinor u^a satisfies the Dirac equation, $\Gamma \cdot ku = 0$, so that only half of its components are independent. Recall that we can write (with $SO(8)$ spinor notation)

$$u^a = \frac{1}{k^+} \gamma^i k^i u^{\dot{a}}. \tag{7.2.6}$$

When $k^+ = 0$ we can take $\gamma^i k^i u^{\dot{a}} = 0$ so that only four of the components of $u^{\dot{a}}$ are independent. Multiplying (7.2.6) by $\gamma^i k^i$ and using (7.2.4) we see that half the components of u^a are also independent.

The conditions that follow from imposing space-time supersymmetry are

$$\begin{aligned} \delta_\epsilon V_F(u, k) &= V_B(\tilde{\varsigma}, k) + \frac{\partial}{\partial \tau} W_B \\ \delta_\epsilon V_B(\varsigma, k) &= V_F(\tilde{u}, k) + \frac{\partial}{\partial \tau} W_F, \end{aligned} \tag{7.2.7}$$

where $\tilde{\varsigma} = \delta_\epsilon u$ and $\tilde{u} = \delta_\epsilon \varsigma$ are the transformed wave-functions for the on-shell emitted states (which are just the supersymmetry transformations of supersymmetric Yang–Mills theory, discussed earlier). The τ derivatives are permitted since they do not contribute to the supersymmetry of tree diagrams which are integrals over τ. (The fact that the boundary terms that arise from integrating these terms vanish is a feature of string theories not shared by point-particle theories.)

The unique solution of (7.2.7) turns out to be

$$V_B(\varsigma, k) = \varsigma^\mu B_\mu e^{ik\cdot X}, \tag{7.2.8}$$

$$V_F(u, k) = (u^a F^a + u^{\dot a} F^{\dot a}) e^{ik\cdot X}. \tag{7.2.9}$$

The functions B_μ, F^a and $F^{\dot a}$ are given by

$$\begin{aligned} \varsigma \cdot B &= \varsigma \cdot \dot X - \varsigma^i R^{ij} k^j) \\ F^{\dot a} &= \frac{1}{\sqrt{2p^+}} \left\{ (\gamma \cdot \dot X S)^{\dot a} + \frac{1}{3} : (\gamma^i S)^{\dot a} R^{ij} k^j : \right\} \\ F^a &= (p^+/2)^{1/2} S^a. \end{aligned} \tag{7.2.10}$$

The irrelevant functions W_B and W_F are also determined in this manner.

The tree amplitudes are calculated in the operator formalism in the same manner as I earlier described for the bosonic theory. The extra complications arise from the fermionic modes. I will not present the details here.

7.3 The heterotic string

The heterotic string theory has $(0, 1)$ supersymmetry. This means that the right-moving modes are supersymmetric so that the solutions of the equations of motion are described in the light-cone gauge by coordinates $X^i_R(\tau+\sigma)$ and $S^a(\tau+\sigma)$. The left-moving modes are not supersymmetric. Obviously there must be left-moving space-time coordinates, $X^i_L(\tau - \sigma)$. There are also other left-moving coordinates which may be described in various ways. The clearest way of describing them is in terms of a set of 32 fermion two-dimensional Majorana spinors, $\lambda^A_-(\tau - \sigma)$ $(A = 1, \ldots, 32)$. In that case the action can be written in the light-cone gauge as

$$S = -\frac{1}{2\pi} \int d\sigma d\tau \left(\partial_\alpha X^i \partial^\alpha X^i - S^a \partial_- S^a - 2 \sum_{A=1}^{32} \lambda^A_- \partial_+ \lambda^A_- \right). \tag{7.3.1}$$

This action could equally well be written in covariant form. If the RNS formalism is used the right-moving S^a variables are replaced by terms involving ψ^μ variables as usual. In this case each component of ψ^μ is a two-dimensional chiral spinor, *i.e.*, $\rho^3 \psi^\mu = \psi^\mu$. In the conformal gauge they are described by the chiral Dirac action $\bar\psi^\mu \rho \cdot \partial_\alpha \psi^\mu \sim \psi^\mu \partial_- \psi^\mu$, just like the action for the spinors λ^A (which satisfy $\rho^3 \lambda^A = \lambda^A$). If the covariant superspace formalism is used the part of the action involving the superspace coordinates X^μ and θ^a is the same as in (6.4.1) + (6.4.4) with θ^{2a} set to zero. The λ^A variables are treated by a chiral Dirac action once again.

I showed in an earlier lecture that the conformal symmetry of string theory requires the separate cancellation of the Virasoro anomalies in the left-moving and right-moving sectors when the anomalies arising from the gauge-fixing ghosts are taken into account. The right-moving sector of the heterotic string is free of such anomalies if $D = 10$ since it is identical to the right-moving sector of type II superstring theories. The anomalies cancel between the coordinates X^μ, ψ^μ, the boson ghosts c^α and $b_{\alpha\beta}$ and the fermion ghosts γ and β_α. In the left-moving sector the only ghosts are c^α and $b_{\alpha\beta}$, associated with the reparametrization symmetry. This means that there must be 26 left-moving boson coordinates (as in the bosonic theory). These divide into ten space-time dimensions and 16 extra variables. We have already seen that the Virasoro anomaly arising from a fermion is 1/2 that of a boson. This means that instead of 16 extra bosonic variables it is possible to use 32 internal fermionic coordinates, λ^A. This is the case that I will consider first, returning later to the alternative way of including the internal symmetry which uses 16 extra left-moving bosonic variables instead of the 32 fermion spinors.

7.4 $SO(32)$ spectrum

If all the 32 components of λ^A are treated on an equal footing the internal symmetry described by the action (7.3.1) is $SO(32)$. In this case there are two sectors just as for the RNS fermion variables. These are a periodic sector, labelled P (analogous to the Ramond sector), and the anti-periodic sector, labelled A (analogous to the Neveu–Schwarz sector).

The spectrum is determined by similar considerations to those which determine the spectrum of type II theories. The terms involving the eight physical left-moving Grassmann variables are replaced by similar terms made out of the 32 λ^A variables. The mass-shell condition is now

$$N + \tilde{N} + \frac{p^2}{4} - a - \tilde{a} = 0, \tag{7.4.1}$$

together with the subsidiary constraint

$$N - a = \tilde{N} - \tilde{a}, \tag{7.4.2}$$

where the number operators are defined by

$$N = \sum_1^\infty \left(\alpha^i_{-n} \cdot \alpha^i_n + nS^a_{-n}S^a_n\right), \tag{7.4.3}$$

and

$$\tilde{N} = \sum_1^\infty \left(\tilde{\alpha}^i_{-n} \cdot \tilde{\alpha}^i_n + n\lambda^A_{-n}\lambda^A_n\right). \tag{7.4.4}$$

In the type II theories the normal ordering constants a and $\tilde{a}$ in the right-moving and left-moving sectors were equal (and cancelled out of the subsidiary condition of the form (7.4.2)). The heterotic string involves left-moving internal fermion coordinates, λ^A.

These normal ordering constants are determined by the same algebra as in the other theories. The right-moving constant is obviously

$$a = 0, \tag{7.4.5}$$

since the right-moving sector has an equal number of boson and integer-moded fermion modes. The constant arising from left-moving modes comes from eight boson coordinates, each contributing 1/24 and 32 fermions, each contributing 1/48 in the A sector or $-1/24$ in the P sector. These values for the normal ordering constants of the fermion variables are the same as those found previously, when the fermions were the ψ variables of the RNS model. The result is

$$\tilde{a}_A = \frac{8}{24} + \frac{32}{48} \qquad \text{in A sector} \tag{7.4.6}$$

$$\tilde{a}_P = \frac{8}{24} - \frac{32}{24} \qquad \text{in P sector.} \tag{7.4.7}$$

As a result the subsidiary constraint requires

$$N = \tilde{N} - 1 \qquad \text{in A sector} \tag{7.4.8}$$

$$N = \tilde{N} + 1 \qquad \text{in P sector.} \tag{7.4.9}$$

The mass-shell condition then becomes

$$(\text{Mass})^2 = -p^2 = 8N. \tag{7.4.10}$$

Notice that there are no tachyons in the spectrum.

The ground states are the states with $N = 0$. From (7.4.8) it is clear that there are no states with $N = 0$ in the P sector since $\tilde{N}$ must be positive. We see that the A sector ground states have $\tilde{N} = 1$. These states, which are tensor products of states for right-movers and left-movers, fall into several types. The right-moving sector consists of the usual superstring states, $|i\rangle$ and $|\dot{a}\rangle$. The left-moving states with mode number $\tilde{N} = 1$ are $\tilde{\alpha}^i_{-1}|0\rangle$ and $\lambda^A_{-\frac{1}{2}}\lambda^B_{-\frac{1}{2}}|0\rangle$. As a result there are 64 massless boson states

$$|i\rangle \otimes \tilde{\alpha}^j_{-1}|0\rangle, \tag{7.4.11}$$

and 64 massless fermion states

$$|\dot{a}\rangle \otimes \tilde{\alpha}^i_{-1}|0\rangle, \tag{7.4.12}$$

which form the $N = 1$ supergravity multiplet. In addition there are 8×496 boson states,

$$|i\rangle \otimes \lambda^A_{-\frac{1}{2}}\lambda^B_{-\frac{1}{2}}|0\rangle \tag{7.4.13}$$

(noting that $\lambda^A\lambda^B$ has $32 \times 31/2 = 496$ components) and 8×496 fermion states

$$|\dot{a}\rangle \otimes \lambda^A_{-\frac{1}{2}}\lambda^B_{-\frac{1}{2}}|0\rangle, \tag{7.4.14}$$

which form the states in the Yang–Mills supermultiplet in the adjoint representation of the group $SO(32)$.

Just as in the fermionic string theory there is an operator

$$\bar{\Gamma}_A = (-1)^{\sum_{r=1/2}^{\infty} \lambda_{-r}^A \lambda_r^A}, \tag{7.4.15}$$

which is ± 1 depending on whether the states in the A sector have an odd or an even number of λ excitations. Notice that the states in the A sector which satisfy $N = \tilde{N} - 1$ cannot have half-integer values of $\tilde{N}$ since N is an integer. This means that the physical states automatically satisfy the analogue of the GSO projection condition $(1 + \bar{\Gamma}_A)|\ \rangle = 0$.

In the P sector the operator

$$\bar{\Gamma}_P = \bar{\lambda}_0 (-1)^{\sum_1^{\infty} \lambda_{-n}^A \lambda_n^B} \tag{7.4.16}$$

(where $\bar{\lambda}_0 = \lambda_0^1 \lambda_0^2 \dots \lambda_0^{32}$) is analogous to the operator $\bar{\Gamma}$ introduced in the study of Ramond fermions. It turns out that it is necessary to use the analogue of the GSO projection in this case also. This requirement only becomes apparent from the requirement that the loop amplitude be invariant under modular transformations. In this case the sign in the projection is unambiguously determined (in contrast to the Ramond sector of the fermionic string). Furthermore, modular invariance requires that the states of this sector obey the same statistics as those of the A sector, a crucial requirement if the two sectors relate to internal symmetries. In contrast, the states of the Ramond sector of the fermionic string theory are space-time fermions whereas those of the Neveu–Schwarz sector are bosons. These issues are discussed by Seiberg and Witten.

The states at the first excited level include both the A and the P sectors. The states in the A sector have $N = 1$ and $\tilde{N} = 2$. The right-moving states are the 256 states at the first massive level of the open superstring discussed earlier. These are tensored with the left-moving states with $\tilde{N} = 2$ are

$$\begin{aligned} &\tilde{\alpha}_{-2}^i|0\rangle, \qquad \tilde{\alpha}_{-1}^i \tilde{\alpha}_{-1}^j|0\rangle, \qquad \tilde{\alpha}_{-1}^i \lambda_{-\frac{1}{2}}^A \lambda_{-\frac{1}{2}}^B|0\rangle, \\ &\lambda_{-\frac{1}{2}}^A \lambda_{-\frac{3}{2}}^B|0\rangle, \qquad \lambda_{-\frac{1}{2}}^A \lambda_{-\frac{1}{2}}^B \lambda_{-\frac{1}{2}}^C \lambda_{-\frac{1}{2}}^D|0\rangle. \end{aligned} \tag{7.4.17}$$

The states in the P sector at this mass level are those with $N = 1$ and $\tilde{N} = 0$. The right-moving sector is just the same as those in the A sector while the $\tilde{N} = 0$ states form the ground states of the left-moving sector. As with the R sector in the fermionic string the ground states form a spinor, which in this case is the spinor of $SO(32)$, denoted $|A\rangle$ which has 2^{15} components.

The total number of components at the first massive level of the heterotic string adds up to 18883554!.

7.5 The E_8 algebra

So far I have assumed that all 32 components of λ^A are treated symmetrically, leading to $SO(32)$ symmetry. I will now describe another possibility (among very many) in which the symmetry group is $E_8 \otimes E_8$. The reason that these two examples are particularly important is not apparent at the level of the free theory or at the level of the tree diagrams. We shall see that at the level of the loop diagrams, where modular invariance provides strong constraints, these particular groups are picked out in the heterotic theory. They are the self-same groups that are also selected by very general arguments about the absence of chiral anomalies.

To begin with I will give one description of the algebra of E_8. Later, I will give an alternative description of the algebra. The generators are easily thought of by starting with the 120 generators, J_{ij}, of $SO(16)$ which is a subgroup of E_8 satisfying

$$[J_{ij}, J_{kl}] = J_{il}\delta_{jk} - J_{jk}\delta_{il} + J_{jk}\delta_{il} - J_{ik}\delta_{jl}. \tag{7.5.1}$$

To these we now append 128 operators, Q_α, which form a $SO(16)$ spinor. This means that they transform as

$$[J_{ij}, Q_\alpha] = (\sigma_{ij})_{\alpha\beta} Q_\beta, \tag{7.5.2}$$

where σ_{ij} is the antisymmetrized product of the $SO(16)$ γ matrices. The 248 generators J_{ij} and Q_α are the generators of E_8. The commutation relation that completes the specification of the algebra is

$$[Q_\alpha, Q_\beta] = (\sigma_{ij})_{\alpha\beta} J_{ij}. \tag{7.5.3}$$

In order for this to be consistent it is necessary to check the Jacobi identities. The only non-trivial one is the relation

$$[[Q_\alpha, Q_\beta], Q_\gamma] + [[Q_\beta, Q_\gamma], Q_\alpha] + [[Q_\gamma, Q_\alpha], Q_\beta] = 0. \tag{7.5.4}$$

In order to check this using the commutation relations of the algebra it is necessary to use a Fierz transformation. The structure of this relation is the S0(16) version of the relation that I proved earlier in the context of discussing the supersymmetry of supersymmetric Yang–Mills theory.

7.6 $E_8 \times E_8$ symmetry in the heterotic string

I will now consider what happens if the 32 λ^A variables are split into groups which are treated independently as far as their boundary conditions are concerned. The example I shall consider is the decomposition

$$SO(32) \supset SO(n) \times SO(32-n), \tag{7.6.1}$$

obtained by taking one group with n components, λ^A $(A = 1, \ldots, n)$, and the other group with $(32-n)$ components, λ^A $(A = n+1, \ldots, 32)$. The components in each of these groups all satisfy either A or P boundary components so there are more sectors than in the earlier example which was equivalent to the case $n = 0$. In all cases the normal-ordering constant in the right-moving sector is still $a = 0$. In the left-moving sector the possibilities are

$$\tilde{a}_{AA} = \frac{8}{24} + \frac{n}{48} + \frac{32-n}{48} = 1, \tag{7.6.2}$$

$$\tilde{a}_{AP} = \frac{8}{24} + \frac{n}{48} - \frac{32-n}{24} = \frac{n}{16} - 1, \tag{7.6.3}$$

$$\tilde{a}_{PA} = \frac{8}{24} + \frac{32-n}{48} - \frac{n}{24} = 1 - \frac{n}{16}, \tag{7.6.4}$$

$$\tilde{a}_{PP} = \frac{8}{24} + \frac{32-n}{48} + \frac{n}{48} = -1, \tag{7.6.5}$$

where the subscripts indicate the boundary conditions on the two groups of components.

Now consider the special case $n = 16$. Recalling the subsidiary condition $N = \tilde{N} - \tilde{a}$ and $(\text{Mass})^2 = 0$ we see that there are massless states ($\tilde{N} = 0$) with several types of structure in the left-moving sector. (i) In the AA sector there are states made from $\lambda^i_{-\frac{1}{2}}\lambda^j_{-\frac{1}{2}}|o\rangle$. These fall into the representations

$$(\mathbf{120}, \mathbf{1}) \qquad \text{if } i = 1, \ldots, 16 \text{ and } j = 1, \ldots, 16, \tag{7.6.6}$$

$$(\mathbf{1}, \mathbf{120}) \qquad \text{if } i = 17, \ldots, 32 \text{ and } j = 17, \ldots, 32, \tag{7.6.7}$$

$$(\mathbf{16}, \mathbf{16}) \qquad \text{if } i = 1, \ldots, 16 \text{ and } j = 17, \ldots, 32. \tag{7.6.8}$$

These states have the same content as $SO(32)$ decomposed into its $SO(16) \times SO(16)$ subgroups. (ii) There are more massless states which come from the AP and PA sectors. These states have $N = \tilde{N} = 0$. The P piece of the left-moving sector is therefore described by either of the spinors of $SO(16)$, which have 128 components. These spinors have opposite 'chiralities', *i.e.*, they are eigenstates of $\bar{\lambda}_0 = \lambda^1_0 \ldots \lambda^{16}_0$ or $\bar{\lambda}'_0 = \lambda^{17}_0 \ldots \lambda^{32}_0$, with eigenvalues ± 1. The content of the massless states is therefore

$$(\mathbf{128}, \mathbf{1}) \oplus (\mathbf{128}', \mathbf{1}), \tag{7.6.9}$$

and

$$(\mathbf{1}, \mathbf{128}) \oplus (\mathbf{1}, \mathbf{128}'). \tag{7.6.10}$$

There is again the possibility of projecting with a generalization of the GSO operator. As in the previous examples, this is seen to be necessary in the study of the modular invariance of loop amplitudes. This time there is an independent GSO operator in each sector of the form $\frac{1}{2}(1 + (-1)^{F_1})$ and $\frac{1}{2}(1 + (-1)^{F_2})$, where F_1 and F_2 are defined in terms of the λ modes in the usual fashion. It is understood that in the P sector the factor $(-1)^F$

includes the prefactor consisting of the product of the 16 zero modes. The only states which survive in the A sector are those with an even number of λ excitations. This means that the survivng AA states are (7.6.6) and (7.6.7),

$$(\mathbf{120},\mathbf{1}) \oplus (\mathbf{1},\mathbf{120}). \tag{7.6.11}$$

In the P sector the GSO projection picks out one particular chirality so that the surviving states are

$$(\mathbf{128},\mathbf{1}) \oplus (\mathbf{1},\mathbf{128}). \tag{7.6.12}$$

The states (7.6.11) and (7.6.12) constitute the content of the 496 states in the adjoint of $E_8 \times E_8$. Each E_8 is built from 120 states in the adjoint of $SO(16)$ and 128 states in a spinor of $SO(16)$, as described earlier. In the heterotic string these states are tensored with the right-moving supersymmetry multiplet, giving the 496-dimensional supermultiplet of Yang–Mills gauge bosons. The supergravity sector is not affected by the λ degrees of freedom so it is described in the same way as in the $SO(32)$ heterotic theory.

In the examples considered so far the left-moving and right-moving sectors have been treated independently. There are obviously many possibilities for imposing boundary conditions on the left-moving and right-moving fermions which are correlated. Of all the possible ways of doing this there is only one which leads to modular-invariant loop amplitudes and which has no tachyons. That is a theory in which the right-moving supersymmetry is broken by choosing antisymmetric boundary conditions for the S^a variables. The internal symmetry group in that case is $O(16) \times O(16)$. Although the one-loop amplitude is modular invariant it is divergent. I will not pursue it here.

8. E_8 and spin$(32)/Z_2$ lattices; the heterotic string in the bosonic formulation; vertex operators for the heterotic string; the constraint of modular invariance.

8.1 Compactification on a torus

In order to understand the formulation of the heterotic string in terms of bosonic internal symmetry coordinates it is useful to describe the way in which extra dimensions can be curled-up onto a toroidal space-time. In a different context, this is obviously important as one way of understanding the compactification of the extra dimensions in string theory. Even though compactification on a torus is somewhat trivial, there are interesting aspects of string theory which do not arise in the analogous compactification in a theory with point particles.

For simplicity, consider the case in which only one spatial dimension is curled up. A torus can be considered to be a flat plane subject to the boundary conditions

$$x \approx x + 2\pi R m, \tag{8.1.1}$$

where R is the radius of the torus and m is an integer. In this case a closed string can wind around the toroidal dimension. This means that the mode expansion for $X(\sigma,\tau)$ must be modified since it no longer has to be periodic in σ. When σ changes by π, $X(\sigma,\tau)$ can wind around the string n times, *i.e.*,

$$X(\sigma+\pi) \equiv X(\sigma,\tau) + 2\pi Rn, \tag{8.1.2}$$

where n is the winding number. This leads to the mode expansion in the extra dimension

$$X(\sigma,\tau) = x + p\tau + 2L\sigma + \frac{i}{2}\sum_{n\neq 0}\frac{1}{n}\left(\alpha_n e^{-2in(\tau-\sigma)} + \tilde{\alpha}_n e^{-2in(\tau+\sigma)}\right). \tag{8.1.3}$$

In this expression the total momentum is quantized in the usual Kaluza–Klein manner,

$$p = \frac{m}{R}, \tag{8.1.4}$$

since the string wave function has a factor $e^{ip\cdot x}$ which must be single-valued on the circle of circumference $2\pi R$, so that $e^{ip\cdot x} = e^{ip\cdot(x+2\pi Rn)}$. The constant L in (8.1.3) is determined by (8.1.2) to be

$$L = nR. \tag{8.1.5}$$

The left-moving and right-moving components of the coordinates, which are functions of $(\tau-\sigma)$ and $(\tau+\sigma)$ are given by

$$X_R(\tau-\sigma) = x_R + \left(\frac{p}{2} - L\right)(\tau-\sigma) + \frac{i}{2}\sum_{n\neq 0}\frac{1}{n}\alpha_n e^{-2in(\tau-\sigma)}, \tag{8.1.6}$$

$$X_L(\tau+\sigma) = x_L + \left(\frac{p}{2} + L\right)(\tau+\sigma) + \frac{i}{2}\sum_{n\neq 0}\frac{1}{n}\tilde{\alpha}_n e^{-2in(\tau+\sigma)}. \tag{8.1.7}$$

The states of the theory depend on the Kaluza–Klein charges and the winding numbers. This means that compactification gives rise to many more extra states than in the compactification of a point field theory. This can result in an enlargement of the symmetry group in a manner that is at the heart of the heterotic string.

8.2 Bosonic formulation of heterotic string

One way of thinking of the bosonic formulation of the heterotic string is to think of the extra 16 left-moving coordinates as if they were extra bosonic coordinates compactified on a torus. Of course, it is a little wierd to have 26 left-moving dimensions and only 10 right-moving ones.

The theory has the content of the $D=10$ superstring for the right-moving modes which can be described in terms of $X^i_R(\sigma,\tau)$ $(i=1,\ldots,8)$ and $S^a(\sigma,\tau)$ $(a=1,\ldots,8)$. The left-moving coordinates are $X^i_L(\sigma,\tau)$ and $X^I_L(\sigma,\tau)$ $(L = 1,\ldots,16)$. The extra bosonic coordinates have the mode expansion

$$X^I_L = x^I_L + p^I_L(\tau+\sigma) + \frac{1}{2}\sum_{n\neq 0}\frac{1}{n}\tilde{\alpha}^I_n e^{-2in(\tau+\sigma)}, \tag{8.2.1}$$

where, from (8.1.7),

$$p^I_L = \frac{p^I}{2} + L^I. \tag{8.2.2}$$

Since there are no right-moving modes in the extra dimensions, it follows that $p^I_R = \frac{1}{2}p^I - L^I = 0$, so that

$$p^I_L = 2L = 2nR \tag{8.2.3}$$

The field X^I_L is a 'chiral' Boson field satisfying the self-duality constraint,

$$P^{\alpha\beta}_+\partial_\beta X^I_L \equiv (\partial^\alpha - \epsilon^{\alpha\beta}\partial_\beta)X^I_L \tag{8.2.4}$$

There are problems in writing an action for such a field which is why I have phrased this section in terms of the solutions of the equations of motion. The commutation relations of the $\tilde{\alpha}_n$ are the usual oscillator relations while the zero modes satisfy

$$[x^I_L, p^J_L] = i\frac{1}{2}\delta^{IJ}. \tag{8.2.5}$$

The factor of 1/2 in this equation can be seen in various ways. For example, by combining (8.1.6) and (8.1.7) we must reconstruct the usual relation $[x_R + x_L, p_R + p_L] = i$. Alternatively, one can set up the quantization in the Dirac manner to take account of the constraint (8.2.4).

There are obvious generalizations of the vertex operators in the extra dimensions. These play an important rôle in understanding the symmetry in the theory as well as in the construction of amplitudes. The simplest vertex is

$$V_0(K,\tau+\sigma) =: e^{2iK\cdot X_L(\tau+\sigma)} :, \tag{8.2.6}$$

where K^I is an eigenvalue of p^I_L which means that

$$K^I = 2L^I. \tag{8.2.7}$$

The factor of 2 in the exponent arises because $p_L \sim -\frac{i}{2}\partial/\partial x_L$ due to (8.2.5), so that

$$\left[p^I_L, V_0(\tau+\sigma)\right] = K^I V_0(\tau+\sigma). \tag{8.2.8}$$

8.3 Lattices

A lattice defines a torus, R^{16}/Γ, which can be described by 16 basis vectors, $\bar{e}_i \equiv e_i^I$, $(i = 1 \ldots 16)$. An arbitrary lattice point is described by

$$q^I = \pi \sum_i n_i e_i^I, \tag{8.3.1}$$

so that

$$x^I \approx x^I + q^I \approx x^I + 2\pi L^I. \tag{8.3.2}$$

The *metric* on the lattice is defined by

$$g_{ij} = \sum_{I=1}^{16} e_i^I e_j^I \equiv \bar{e}_i \cdot \bar{e}_j. \tag{8.3.3}$$

The *dual* lattice is defined by the basis vectors e_i^* which satisfy

$$\sum_{I=1}^{16} e_i^{*I} e_j^I \equiv \bar{e}_i^* \cdot \bar{e}_j = \delta_{ij}. \tag{8.3.4}$$

The metric, g_{ij}^*, on the dual lattice, $\tilde{\Gamma}$, satisfies

$$g_{ij}^* = g_{ij}^{-1}. \tag{8.3.5}$$

The wave function contains the factor $e^{2iK^I x^I}$ which must be single valued, which requires $K^I e_i^I =$ integer for all i. This means that K^I can be written in the form

$$K^I = \sum_{i=1}^{16} m_i e_i^{*I}. \tag{8.3.6}$$

Together with (8.2.7) this means that K^I is both a point on the lattice Γ and on the dual lattice $\tilde{\Gamma}$, *i.e.*,

$$K^I = \sum_i m_i e_i^{*I} = \sum_i n_i e_i^I. \tag{8.3.7}$$

In general there are no non-trivial values for K^I other than the origin of Γ and $\tilde{\Gamma}$ (which coincide by convention). However, if g_{ij} is an integer for all i and j (when the lattice is said to be an *integral* lattice) so that

$$\bar{e}_i \cdot \bar{e}_j = \text{integer}, \tag{8.3.8}$$

then

$$\bar{e}_i = \sum m_{ij} \bar{e}_j^*. \tag{8.3.9}$$

In this case $\Gamma \subset \tilde{\Gamma}$ (the points in Γ are all in $\tilde{\Gamma}$).

The mass-shell condition and the subsidiary constraint that define the spectrum of states in the heterotic theory can be written as

$$N^{\alpha} + N^{S} + \frac{\sum_i (p^i)^2}{8} = \tilde{N}^{\alpha} + \frac{\sum_i (p^i)^2}{8} + \frac{\sum_I (p^I)^2}{2} - 1 = 0. \tag{8.3.10}$$

The constant -1 is the normal ordering constant that comes from the 26 left-moving coordinates (the right-moving constant is zero). From (8.3.10) $\vec{p}_L^2 = 2 \times$ integer, so that the lattice Γ is an *even* lattice (a lattice with points which are separated by a distance which is the square root of an even integer). The mass spectrum is defined by

$$(\text{Mass})^2 = -p^2 = 8N. \tag{8.3.11}$$

The ground states have $N = 0$. These fall into two classes. The first class consists of those states with left-moving excitations, so that $\tilde{N} = 1$ and (from (8.3.10)) $p_L^I = 0$. These are the states

$$|i\rangle \text{ or } |\dot{a}\rangle \ \otimes \tilde{\alpha}_{-1}^{i}|0\rangle, \tag{8.3.12}$$

which form the 128 states in the supergravity multiplet and

$$|i\rangle \text{ or } |\dot{a}\rangle \ \otimes \tilde{\alpha}_{-1}^{I}|0\rangle, \tag{8.3.13}$$

which form 16 of the massless vector states which are gauge particles (together with their fermionic super-partners). These 16 gauge particles are just the ones which arise in ordinary Kaluza–Klein compactification on a torus. They are associated with the rather boring isometry group of the torus, $[U(1)]^{16}$. These vector states are described by a piece of the 26-dimensional metric tensor, $A_I^{\mu} \sim g^{\mu I}$. The second class of massless states are those with $N = 0$ and $\tilde{N} = 0$ which, from (8.3.10), requires $\left(p_L^I\right)^2 = 2$. These states are

$$|i\rangle \text{ or } |\dot{a}\rangle \ \otimes |\underline{m}\rangle. \tag{8.3.14}$$

There are obviously many choices for the 16-component vector, $\underline{m}$, depending on the lattice. We see that the number of massless vector states is larger than that usually obtained in the Kaluza–Klein method. This allows interesting, non-Abelian symmetries to be described.

8.4 Root lattices

There is a convenient basis for describing any Lie algebra in which the generators are divided into two sets, H_I and $E_{\vec{K}}$. The first set are a maximal set of d commuting generators

$$[H_I, H_J] = 0, \tag{8.4.1}$$

which is the *Cartan subalgebra* with dimension equal to the rank of the group ($I = 1, \ldots, d$). The Cartan subgroup is $[U(1)]^d$. The eigenvalues of the operators H_I form points in

d-dimensional *weight* space. A state is specified by its coordinates in this space. The generators $E_{\bar{K}}$ are *step* operators, satisfying

$$[H_I, E_{\bar{K}}] = K^I E_{\bar{K}}, \tag{8.4.2}$$

where the eigenvalues K^I are points in weight space which form the *root vector* $\bar{K}$. In the cases that I will deal with the roots are even, *i.e.*, $(\bar{K})^2 = 2$, for reasons discussed above. The algebra is completed by the relations

$$\begin{aligned}
[E_{\bar{K}}, E_{\bar{K}'}] &= 0 && \text{if } (\bar{K}+\bar{K}')^2 \geq 4 \ (i.e.,\ \bar{K}\cdot\bar{K}' \geq 0), \\
&= \epsilon(\bar{K},\bar{K}') E_{\bar{K}+\bar{K}'} && \text{if } (\bar{K}+\bar{K}')^2 = 2 \ (i.e.,\ \bar{K}\cdot\bar{K}' = -1), \\
&= \sum_I K^I H_I && \text{if } (\bar{K}+\bar{K}') = 0 \ (i.e.,\ \bar{K}\cdot\bar{K}' = -2).
\end{aligned} \tag{8.4.3}$$

8.5 Representation by vertex operators

I will now describe how the above decomposition of the algebra can be represented by the use of the vertex operator (8.2.6). This is based on developments by physicists and mathematicians (Halpern; Lepowski and Wilson; Frenkel and Kac; Segal; Goddard and Olive). Firstly we can identify the generators of the Cartan subalgebra with the momentum operator

$$H^I = p_L^I = \int_0^\pi \frac{\dot{X}_L}{\pi}(\sigma). \tag{8.5.1}$$

The step operators can be written as

$$E_{\bar{K}} = A_{\bar{K}} c_{\bar{K}}, \tag{8.5.2}$$

where the $c_{\bar{K}} \equiv c_{\bar{K}}(p_L^I)$ are sets of signs determined by the closure of the algebra. The $A_{\bar{K}}$ are defined by

$$A_K = \frac{1}{\pi} \oint V_0(\bar{K}, z) \frac{dz}{2\pi i z}, \tag{8.5.3}$$

which is an integral of a density defined by the vertex operator

$$V_0(\bar{K}, z) = e^{2i\bar{K}\cdot\bar{x}_L + \bar{K}\cdot\bar{p}_L \ln z} e^{\sum_{n\neq 0} \frac{1}{n}\bar{K}\cdot\bar{\alpha}_{-n} z^n} e^{-\sum_{n\neq 0} \frac{1}{n}\bar{K}\cdot\bar{\alpha}_n z^{-n}}. \tag{8.5.4}$$

I do not have space to do more than indicate why the H_I and E_K defined in this way have the correct commutation relations. The commutators involving the generators H_I is trivially satisfied by using (8.2.8). The heart of the proof of the closure of the algebra depends on understanding the algebra of the densities, (8.5.1) and (8.5.4). In fact these form a current algebra (a special example of a Kac–Moody algebra). Consider the product

$$\begin{aligned} A_{\bar{K}}A_{\bar{K}'} &= \frac{1}{\pi^2}\oint \frac{dz}{2\pi i z}\frac{dw}{2i\pi w} V_0(\bar{K},z)V_0(\bar{K}',w) \\ &= \frac{1}{\pi^2}\oint (wz)^{-\bar{K}\cdot\bar{K}'/2}(z-w)^{\bar{K}\cdot\bar{K}'} : e^{2i\left(\bar{K}\cdot\bar{X}(z)+\bar{K}'\cdot\bar{X}(w)\right)} : dzdw, \end{aligned} \tag{8.5.5}$$

for $|z| > |w|$. The product in the reverse order is given by the same expression except that now the contours are defined with $|z| < |w|$ and there is an extra factor of $(-1)^{\bar{K}\cdot\bar{K}'}$. As a result

$$A_{\bar{K}}A_{\bar{K}'} - (-1)^{\bar{K}\cdot\bar{K}'}A_{\bar{K}'}A_{\bar{K}} = \oint_{z=w} (wz)^{-\bar{K}\cdot\bar{K}'/2}(z-w)^{\bar{K}\cdot\bar{K}'} : e^{2i\left(\bar{K}\cdot\bar{X}(z)+\bar{K}'\cdot\bar{X}(w)\right)} : dzdw. \tag{8.5.6}$$

When $\bar{K}\cdot\bar{K}' = -1$ the integrand has a pole at $z = w$ and the result is just its residue, which is simply $A_{\bar{K}+\bar{K}'}$. When $\bar{K}\cdot\bar{K}' = -2$ the integrand has a double pole at $z = w$. A careful analysis shows that in this case the right-hand side of (8.5.6) is proportional to $\bar{K}\cdot\bar{H}$. Finally, if $\bar{K}\cdot\bar{K}' \geq 0$ (it must be an integer) there is no singularity at $z = w$ and the integral vanishes. These are the three possibilities listed in (8.4.3) so that apart from the signs contained in $c_{\bar{K}}$ the closure of the algebra is verified.

The $c_{\bar{K}}$ have to satisfy

$$c_{\bar{K}}(\bar{p}-\bar{K}')c_{\bar{K}'}(\bar{p}) = \epsilon(\bar{K},\bar{K}')c_{\bar{K}+\bar{K}'}(\bar{p}), \tag{8.5.7}$$

and

$$c_{\bar{K}}(\bar{p}-\bar{K}')c_{\bar{K}'}(\bar{p}) = (-1)^{\bar{K}\cdot\bar{K}'}c_{\bar{K}'}(\bar{p}-\bar{K})c_{\bar{K}}(\bar{p}). \tag{8.5.8}$$

These equations can easily be solved for $c_{\bar{K}}(\bar{p})$ and $\epsilon(\bar{K},\bar{K}')$.

8.6 The O(2n) root lattice

I will now describe the root vectors of the orthogonal groups of integral rank n, namely $SO(2n)$ (or D_n) in terms of a complete basis of orthonormal vectors, u_i^I $(i = 1,\ldots,n)$, satisfying

$$u_i^I u_j^I = \delta_{ij}. \tag{8.6.1}$$

The root vectors are given by

$$R = \pm\bar{u}_i \pm \bar{u}_j, \qquad i \neq j, \tag{8.6.2}$$

so that

$$R^2 = 2. \tag{8.6.3}$$

There are $n \times (2n-2)$ roots. The remaining generators are the n generators of the Cartan subalgebra, making a total of $n(2n-1)$ generators in the algebra, as expected. The root vectors coincide with the adjoint weights in the weight lattice.

The spinor representations are associated with the weights

$$S = \frac{1}{2}\left(u_1 \pm u_2 \pm u_3 \ldots \pm u_n\right). \tag{8.6.4}$$

There are 2^{n-1} states with an even number of + signs and 2^{n-1} states with odd no of + signs. These two sets of states define the two spinor weights of $O(2n)$. The length of these spinor weights is given by

$$S^2 = \frac{n}{4}. \tag{8.6.5}$$

The weights of the vector representation are the vectors $\pm u_i$ with $2n$ components and length 1. This completes the ennumeration of the four *conjugacy* classes of $O(2n)$. A conjugacy class is the set of weight vectors which can be connected to each other by a combination of root vectors, or adjoint weights. The group $O(2n)$ has four conjugacy classes, namely, those defined by the adjoint, the two independent spinors and the vector.

Notice that when $n = 4$ the spinor weights and the vector weight have the same length which is a signal of the property of *triality* of $O(8)$. This property is crucial for the fact that the eight transverse degrees of freedom, ψ^i, of the RNS model describe the same supersymmetric theories (after the GSO projection) as the eight spinor components of S^a in the superspace formulation.

When $n = 8$ (*i.e.*, for O(16)) the spinor weights have length 2. In this case one of the spinor weights together with the root vectors form a lattice which is the root lattice of E_8, called Γ_8. The lattice Γ_8 consists of 112 vectors $\pm u_i \pm u_j$ $(i \neq j)$ which form the root lattice of $O(16)$ and the 128 vectors $\frac{1}{2}(u_1 \pm u_2 \ldots \pm u_8)$ with an even number of + signs, which form one of the spinor weights of $O(16)$. Together with the 8 generators in the Cartan subalgebra these account for the 248 generators of E_8. Notice the correspondence between this description of the E_8 algebra and the earlier one in terms of the $O(16)$ algebra.

8.7 Self-duality

From the definition of the dual lattice vectors, $\bar{e}^*_i \cdot \bar{e}_j = \delta_{ij}$, the determinant of the metric is seen to satisfy

$$\det g_{ij} = \left(\det e^I_i\right)^2 = \left(\det e^{*I}_i\right)^{-2} = (\det g^*_{ij})^{-1}. \tag{8.7.1}$$

We see that if $\det g_{ij} = 1$ then $\det g^*_{ij} = 1$. In this case the lattice is said to be self-dual, so that $\Gamma = \tilde{\Gamma}$. For E_8 the *Cartan* matrix $g_{ij} = \bar{e}_i \cdot \bar{e}_j$ is given by

$$g_{ij} = \begin{pmatrix} 2 & -1 & 0 & 0 & 0 & 0 & 0 & 0 \\ -1 & 2 & -1 & 0 & 0 & 0 & 0 & 0 \\ 0 & -1 & 2 & -1 & 0 & 0 & 0 & 0 \\ 0 & 0 & -1 & 2 & -1 & 0 & 0 & 0 \\ 0 & 0 & 0 & -1 & 2 & -1 & 0 & -1 \\ 0 & 0 & 0 & 0 & -1 & 2 & -1 & 0 \\ 0 & 0 & 0 & 0 & 0 & -1 & 2 & 0 \\ 0 & 0 & 0 & 0 & -1 & 0 & 0 & 2 \end{pmatrix}, \qquad (8.7.2)$$

which satisfies $\det g = 1$. The E_8 lattice, Γ_8, is the (unique) rank-eight, even, self-dual lattice. The group $E_8 \times E_8$ has the root lattice $\Gamma_8 \oplus \Gamma_8$. This is obviously self-dual since each Γ_8 is. It has 480 root vectors which, together with the 16 generators in the Cartan subalgebra form the 496 generators of the group. Since the lattice Γ_8 can also be thought of as the root lattice of $O(16)$ with a spinor weight added, the group $O(16) \times O(16)$ may be expected to have special features if $E_8 \times E_8$ does.

There is one other rank-16, even, self-dual lattice, known as Γ_{16}, which is the root lattice of $SO(32)$ supplemented by one set of spinor weights. This lattice consists of the root vectors $\pm u_i \pm u_j$ $(i, j = 1, \ldots, 16\ i \neq j)$ and the spinor weights $\frac{1}{2}(u_1 \pm u_2 \ldots \pm u_{16})$ (with an even number of + signs). These are two of the four conjugacy classes of $SO(32)$. This is summarized by saying that Γ_{16} is the root lattice of the group spin$32/Z_2$. The term spin$(2n)$ refers to the group $SO(2n)$ when it acts on representations in any of the four conjugacy classes. Dividing by Z_2 accounts for the fact that in the case of interest only two of the four classes are present. (Note that if we had included three conjugacy classes then the fourth would have automatically been required.)

8.8 Why have self-duality?

The reason that self-duality is an important consideration is not apparent until loop amplitudes are considered. I stated in an early lecture that a closed-string theory must be invariant under modular transformations in order to satisfy perturbative unitarity. Since the heterotic string theory is a closed-string theory it is important to verify that the loop amplitudes are modular invariant.

As I described in the second lecture, loop diagrams can be constructed in the functional method by considering world-sheets which are topologically non-trivial. In the case of an oriented closed-superstring theory (types IIA, IIB or heterotic) there is only one topologically distinct diagram at each order in perturbation theory. At L loops the world-sheet is topologically a sphere with L handles attached.

The operator method is very cumbersome for considering multi-loop amplitudes. This is particularly true in the light-cone formalism since beyond one loop it is no longer possible to restrict the momenta of the external states in order for one particle at each vertex to have zero k^+. However, the one-loop amplitudes are easy to construct in the operator light-cone formalism since they have no internal vertices. The advantage of the operator construction is that the integration measure is automatically specified and it is easy to normalize the diagram to be in accord with unitarity.

The expression for a loop diagram is given by sewing together a tree diagram. This is just like making a loop diagram in ordinary field theories. Starting with a tree diagram with $P+M+Q$ particles attached it is easy to factorize it on arbitrary internal states shown by wiggly lines in the figure.

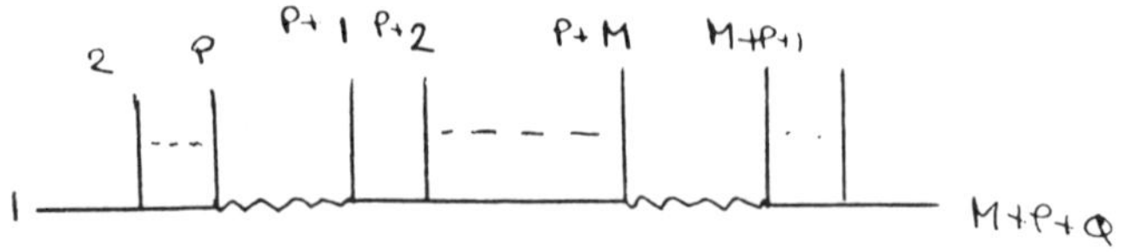

Figure 8.20. Factorization of a general tree diagram in order to make a loop.

The loop is obtained by taking a trace over the states marked by the wiggly lines after inserting a propagator in the link joining the wiggly lines. The trace includes an integration over the zero-mode momentum. The result is the diagram

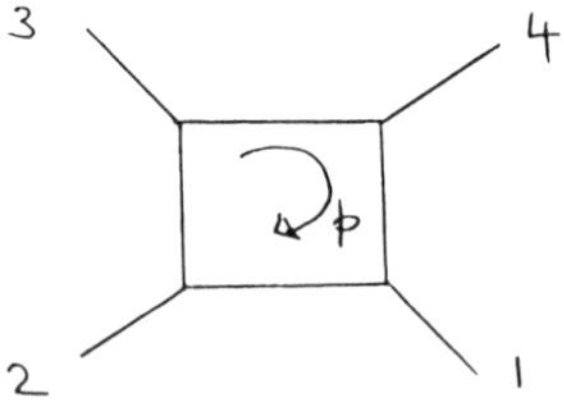

Figure 8.21. A one-loop amplitude with $M = 4$.

The expression for this amplitude is

$$\int d^{10}p \sum_{p^I \in \Gamma} \mathrm{Tr}\left(\Delta V_1 \Delta V_2 \ldots \Delta V_M \Delta\right). \tag{8.8.1}$$

In addition it is necessary to add the terms related by non-cyclic permutations . An

expression of the form of (8.8.1) applies in either the covariant formalism or the light-cone formalism. In the covariant treatment it is essential to include the ghost coordinates which serve to cancel the unphysical states circulating around the loop. In the light-cone formalism the external states are taken to have $k^+ = 0$. I shall only consider the light-cone formalism here.

In (8.8.1) Δ is the string propagator and V_r is the vertex for the emission of a closed string evaluated at $\sigma = 0$ and $\tau = 0$. In the case of the heterotic string the propagator is given by

$$\begin{aligned}\Delta = \frac{1}{L_0}\delta_{N-\tilde{N}-P'^2/2} &= \frac{1}{4\pi}\int_{|z|\le 1} d^2z\, z^{L_0-1}\bar{z}^{\tilde{L}_0-2} \\ &= \int_0^\infty d\tau e^{-(\frac{p^2}{4}+N+\tilde{N}+P'^2/2-1)\tau}\int_0^{2\pi} d\sigma e^{i(N-\tilde{N}-P'^2/2+1)\sigma}\end{aligned}, \tag{8.8.2}$$

where

$$z = e^{\tau - i\sigma}. \tag{8.8.3}$$

As in the case of tree diagrams the propagators can be eliminated by circulating them through the vertices so that the expression for the loop is

$$\int d^{10}p \int \frac{d^2w}{w\bar{w}^2}\int\left(\prod_{r=2}^{M} d^2z_r\right)\mathrm{Tr}\left(|w|^{N+\tilde{N}}V(k_1;1,1)V(k_2;z_2,\bar{z}_2)\ldots V(k_M;z_M,\bar{z}_M)\right). \tag{8.8.4}$$

The vertices have the structure of an integral of the product of left-moving and right-moving pieces

$$V(k;z,\bar{z}) = \int_0^{2\pi} W_L(\frac{k}{2};e^{\tau+i\sigma})W_R(\frac{k}{2};e^{\tau-i\sigma})d\sigma. \tag{8.8.5}$$

When the external states are massless ground states the right-moving piece, $W_R(k;e^{\tau-i\sigma})$, is given by the usual open superstring vertex describing the emission of either a vector or spinor factor. The left-moving piece has the form

$$W_L(k;\bar{z}) = \varsigma^i\dot{X}^i(\bar{z})e^{k^iX^i(\bar{z})} \tag{8.8.6}$$

if the emitted state is one in the supergravity multiplet (so that its left-moving part is $\alpha^i_{-1}|K^I = 0\rangle$). If the left-moving factor is $\alpha^I_{-1}|K^I = 0\rangle$ W_L has the form

$$W_L(k;\bar{z}) = \varsigma^I\dot{X}^I(\bar{z})e^{ik^iX^i(\bar{z})}, \tag{8.8.7}$$

while if the emitted state is one of the 480 states described by a length 2 vector in the weight lattice the vertex contains

$$W_L(K;k;\bar{z}) = c(K) : e^{2iK^IX^I(\bar{z})}e^{ik^iX^i(\bar{z})} :, \tag{8.8.8}$$

where $(K^I)^2 = 2$.

In any loop amplitude the external ground states can be any combination of these three types. For simplicity, I shall take them to be states in the supergravity multiplet with $\varsigma^I = K^I = 0$ (but the following statements generalize in a straightforward manner). In this case the dependence on the loop lattice momentum, P^I, is all contained in the factors of

$$\bar{z}^{(P^I)^2/2} \tag{8.8.9}$$

in the propagator.

The oscillator algebra involved in the trace in (8.8.1) is of a type common to many loop calculations in string theory. The main distinctive feature of the heterotic string loop amplitude compared to the other string theories is the sum over the discrete values of the momentum P^I. This arises in the form

$$\begin{aligned}\mathcal{L} &= \sum_{\underline{P}\in\Gamma} \prod_{r=1}^{M} \bar{z}_r^{(P^I)^2} \\ &= \sum_{\underline{P}\in\Gamma} e^{-i\pi\tau \underline{P}^2},\end{aligned} \tag{8.8.10}$$

(where $i\pi\bar{\tau} = \ln w \equiv \ln \sum z_r$). It is a straightforward exercise in Fourier transforms to prove that this expression can be rewritten as a sum over the dual lattice, $\tilde{\Gamma}$, as

$$\mathcal{L} = \frac{1}{\sqrt{\det g}} \left(-\frac{i}{\tau}\right)^8 \sum_{\underline{P}\in\tilde{\Gamma}} e^{i\pi \underline{P}^2/\tau}, \tag{8.8.11}$$

where $\det g \equiv \det e_i^I e_j^I$ is the volume per lattice site. We therefore see that only if the lattice is self-dual, *i.e.*, $\Gamma = \tilde{\Gamma}$ (and $\det g = 1$), then $\mathcal{L}$ transforms very simply under the modular transformation $\tau \to -1/\tau$,

$$\mathcal{L}(\tau) = \left(-\frac{i}{\tau}\right)^8 \mathcal{L}(-1/\tau). \tag{8.8.12}$$

From its definition in (8.8.10), $\mathcal{L}$ clearly transforms trivially under the other independent modular transformation, $\tau \to \tau + 1$ (which leaves $w = \prod z_r$ unchanged). Combining these transformations of $\mathcal{L}$ with the transformations of the rest of the integrand in the expression for the loop amplitude leads to the conclusion that the integrand is invariant under modular transformations only for self-dual lattices (which have to also be even lattices from the earlier arguments). It is at this stage that the restriction to the groups $E_8 \times E_8$ or spin$(32)/Z_2$ enters since the lattices $\Gamma_8 + \Gamma_8$ and Γ_{16} are the unique rank 16, even, self-dual lattices.

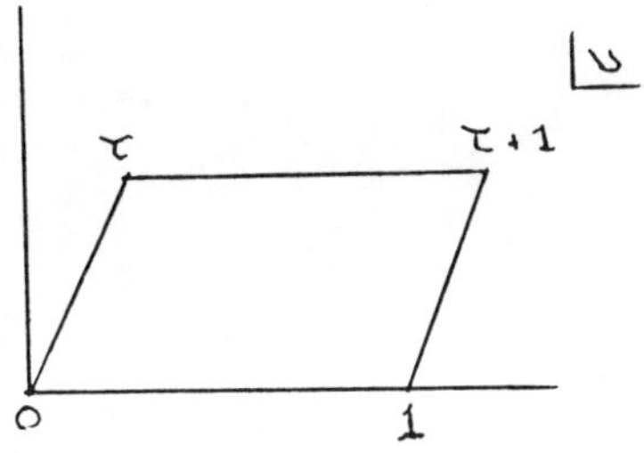

Figure 8.22. Integration region corresponding to the surface of a torus.

At this stage the expression for the loop is an integral over the positions of the particles on the surface of the torus. Defining ν_r by

$$z_r = e^{2i\pi\nu_r}, \tag{8.8.13}$$

the integration region can be represented by the parallelogram in figure 8.22. The complex parameter τ must still be integrated, but it would clearly be wrong to integrate it over the whole complex strip $-1/2 \leq \mathrm{Re}\tau \leq 1/2$, $0 \leq \mathrm{Im}\tau \leq \infty$. This overcounts by an infinite factor since the regions indicated in figure 2.13 are mapped into each other by modular transformations, under which the amplitude is invariant. In order to get the finite answer prescribed by unitarity it is necessary to truncate the region of the τ integral so that it covers just one fundamental region. It is often taken to be the region marked F in figure 2.13. Notice that from this point of view the restriction on the τ integration is a consequence of unitarity whereas we saw earlier that it is necessary for the geometrical interpretation of the theory in the functional method described earlier. Obviously, this requirement of modular invariance should emerge naturally in any unitary field theory of strings that generates the torus diagram in perturbation theory. At the moment this is one of the challenges facing the string field theorists. In the light-cone gauge approach to string field theory, pioneered by Mandelstam, the amplitudes automatically emerge in a form consistent with unitarity and are therefore automatically modular invariant.

In general there can still be divergences associated with the integrals over the ν_r variables. In particular, the modular-invariant torus amplitude of the bosonic theory, studied originally by Shapiro, has a divergence originating from an endpoint at which all the ν_r's approach each other on the torus. By a conformal transformation the toroidal world-sheet can be mapped into a sheet in which the loop is separated from the external particles by a long neck as shown below.

The divergence can be thought of as arising from the propagation of an on-shell massless dilaton state in the neck. (Actually, there is another component to the divergence in the bosonic theory arising from the tachyon in the neck. This does not arise in theories with no tachyons.) This kind of divergence is not of a kind that is familiar in ordinary field theories. In a sense it is an infrared divergence since it involves a low energy dilaton

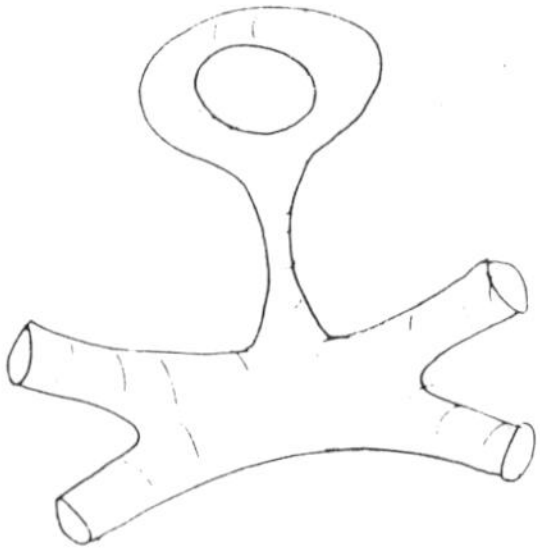

Figure 1. The closed-string divergence seen as a dilaton emission into the vacuum.

coupling to the vacuum. Theories with such divergences are defined around the wrong vacuum state. Just as in ordinary field theories the correct perturbative vacuum is the one in which such dilaton tadpoles do not arise. In superstring theories the tadpole is absent which signals the absence of a one-loop divergence. The statement is equivalent to a statement that the theory is supersymmetric since supersymmetry forbids an expectation value for the scalar dilaton field. In fact, at an arbitrary order in perturbation theory the condition is equivalent to a statement that supersymmetry is preserved and probably also to the statement that the theory is finite. The expectation is that the breaking of supersymmetry, as well as the compactification of space is an effect that is not seen in perturbation theory.

Figure 2. Finiteness of a closed superstring theory probably requires the vanishing of the dilaton tadpole to all orders.

The cosmological constant is related to the torus diagram with no external particles. It is proportional to the dilaton tadpole and also vanishes in any supersymmetric theory.

The $O(16) \times O(16)$ theory does have a modular-invariant one-loop amplitude but it does not have space-time supersymmetry. As a result it does have a divergence of the type described. It also has a non-zero, but finite, cosmological constant.

LECTURES ON ALGEBRAS, LATTICES AND STRINGS

David I. Olive
Blackett Laboratory, Imperial College,
London SW7 2BZ.

INTRODUCTION

Originally the dual resonance theory (or string theory) was constructed to describe the observed spectrum of hadronic physics but it soon developed into a theory that was quite revolutionary, both in terms of the mathematical structure it embodied and in terms of the physics it described. Throughout its history it has undergone a series of metamorphoses according to which its symmetry has been enlarged in unexpected ways, thereby rendering this physics more interesting and special.

At first the theory possessed an O(2,1) symmetry but it was Virasoro[1)] who realised that if the string intercept parameter equalled 1 then this algebra could be enlarged to become infinite dimensional. This changed the complexion of the theory since it now had massless spin 1 particles due to open strings, and further, massless spin 2 particles due to closed strings if the space-time dimension was chosen equal to 26[2)]. Thus it became clear to all workers in the field that the theory promised to provide a unified theory of all the fundamental forces if indeed these were due to the exchange of gauge particles and gravitons[3)].

Furthermore, with intercept 1 and space-time dimension 26, the spectrum was ghost-free (so that probabilities were positive as consistency demanded) and "transverse" in a precise and neat sense, due to Ward identities associated with Virasoro's algebra[4].

At the same time a new version of the theory embodying fermions had been constructed[5] and seen to be ghost-free and generally self-consistent if the space-time dimension was 10 and the ground-state fermions massless[6]. This theory exhibited supersymmetry on the two-dimensional space-time world sheet of the string[7]. A full 10-dimensional space-time supersymmetry could also hold if the fermions were real and chiral (a possibility special to 2, 10, 18, 26... dimensions) and the bosonic sector was similarly projected thereby eliminating the tachyon, the last remaining awkward feature at that time[8,9].

Each increase in symmetry had made the theory more consistent mathematically and/or more realistic physically, yet the theory then foundered, principally I think, because it still appeared so revolutionary. Nevertheless the ideas it had embodied, supersymmetry, string structure in hadrons, critical dimensions, neo-Kaluza-Klein interpretations, etc., lived on in less revolutionary theories such as supergravity, lattice gauge theories and the general theory of topological objects in quantum field theory.

Meanwhile, other aspects of theoretical physics were developed, the theory of anomalies, the BRST theory of quantising gauge theories and the grand unification theory of particle interactions. At the same time, in pure mathematics, the theory of Kac-Moody algebras and their representations was created[10] and provided, through the theory of vertex operators[11,12] for example, a mathematical framework related to string theory.

About five years ago these separate strands began to

entwine and string theory as a result was reborn and underwent yet one more metamorphosis whereby anomalies cancelled provided the gauge groups were chosen to be spin(32)/Z_2 or $E_8 \times E_8$[13]. Remarkably these groups contained in a natural way the fashionable grand unified groups. This cancellation was related to the fact that the weight lattices of these groups were even and possessed the extra symmetry of being self-dual[14].

Historically the theory was developed in terms of its algebraic structure since this provided the most precise and reliable formulation, despite the existence of the more appealing and geometrical string picture at the back of everyone's mind. This picture has become better understood because of the interest of workers in supergravity but here I want to talk about this algebraic structure which once seemed so strange and revolutionary, but which has now become "respectable" because of its relationship to the representation theory of infinite dimensional algebras.

More precisely, I shall talk about the vertex operator construction and its many variations which link the representation theory of affine Kac-Moody algebras, string theory and, more recently, other branches of theoretical physics. This is now a nice subject on its own and indeed already too large for a detailed treatment in the four lectures available to me.

Peter Goddard and I have presented many aspects of the theory in conference talks and review lectures and during the last year we have put all these together in an extended review article about to be published[15]. Sections 6 and 7 deal with vertex operators and were written with my prospective Trieste lectures in mind and provide all the details.

Because of the availability of this I shall deem it unnecessary to repeat myself in print and merely summarize what I said in Trieste referring to this review article for the detailed treatment.

Lecture 1 : Introduction, definitions, and framework. This follows sections 1.1 and 1.4 with a smattering from sections 2 and 3.

Lecture 2 : The vertex operator construction of level 1 representations of simply-laced affine untwisted Kac-Moody algebras i.e. the current algebras of su(n), so(2n), E_6, E_7, and E_8. This follows section 6.

Lecture 3 : The vertex operator construction of fermi fields with applications in the theory of solitons and of superstrings. This follows section 7.

Lecture 4 : More advanced topics including applications to non-simply-laced algebras and Lorentzian algebras, based on a recent paper[16)] and on the 1983 Berkeley vertex operator conference notes[14)]. Because the new treatment is somewhat different and hopefully improved I shall present this here.

MORE ADVANCED ASPECTS OF VERTEX OPERATORS (LECTURE 4)

We have seen that the vertex operator

$$U^{\alpha}(z) = z^{\alpha^2/2} :e^{i\alpha \cdot Q(z)}: \qquad (1)$$

can be consistently defined when α lies on an integral lattice. When $\alpha^2 = 1$, it describes fermion fields or the superstring fermion emission vertex in the transverse formalism; when $\alpha^2 = 2$, Kac-Moody fields for level 1 representations of $\hat{su}(n)$, $\hat{so}(2n)$, $\hat{E}_6$, $\hat{E}_7$, or $\hat{E}_8$; and when $\alpha^2 = 3$ it can describe a complex, N=2 supercharge[17].

This sort of analysis has revealed a remarkable wedding between the theories of Lie algebras and of quantum fields. I now wish to discuss further elaborations of the construction which are particularly relevant to string theory.

The first concerns the construction of level 1 representations of $\hat{g}$ when g is no longer simply-laced i.e. is of so(2n+1), sp(n), F_4 or G_2 type. The analysis is due to Goddard, Olive and Schwimmer[16] and to Bernard and Thierry-Mieg[18] independently. In brief, the first step is to note that g possesses a subalgebra g_L of the same rank whose roots are the long roots of g. This means that the previous construction is immediately applicable to g_L as it is simply-laced, leaving only the step fields for the short roots of g to be found by the ansatz

$$E^{\alpha} \sim U^{\alpha}(z)\Psi_{\Omega}(z) \qquad (2)$$

For so(2n+1), sp(n) and F_4 the short roots satisfy $\alpha^2=1$ (if the long roots have length $\surd 2$) so that Ψ_{Ω} must have conformal weight 1-1/2 = 1/2 and hence be a fermi field in order to ensure an overall conformal weight of 1. We find that the same fermi field Ψ_{Ω} is assigned to all the short roots belonging to a single orbit Ω of the short roots of g under the action of the Weyl group of g_L, that is reflections in the long roots of g. Thus there are as many fermions as these orbits but the fermions have the unusual properties of not being independent of each other thereby explaining why the Sugawara c-number for this representation,

$$c = \text{rank } g + \frac{n(n+1)}{n+3} , \qquad (3)$$

where n is the number of short simple roots of g, is not an integer or half integer in general. The problem of correcting signs by Klein transformations is more acute than in the simply-laced case involving complex phase factors or cocycles. The interest of this construction is two-fold: (1) it reveals an unsuspected connection with division algebras, the complex numbers, quaternions and octonions and (2) it sheds light on the structure of the fermionic string theory, suggesting how it may be obtained from a bosonic string theory in higher dimensions.

The second extension of the vertex operator construction concerns a covariant reformulation of the construction of $\hat{g}$ (with g simply-laced) which can be extended to Lorentzian rather than Euclidean lattices[14,19].

Recall the Virasoro transformation properties of the tachyon vertex operators associated with points α of length $\sqrt{2}$.

$$[L_m, U_n^\alpha] = -nU_{m+n}^\alpha \quad \text{if} \quad U_m^\alpha = \oint \frac{dz}{2\pi iz} z^m U^\alpha(z)$$

putting m=0, $U^{\alpha}=U^{\alpha}_{o}$, we have

$$[L_m, U^{\alpha}] = 0 \tag{4}$$

These U^{α} generate (modulo signs) the finite dimensional Lie algebra g with roots α, rather than $\hat{g}$ i.e. they directly construct g from $\Lambda_R(g)$, the root lattice of g.

The particular interest of this to Peter Goddard and I was that it made explicit an idea we had had with Jean Nuyts concerning the relation between Lie groups and lattices[20].

The context of this idea was the occurrence of magnetic monopoles as solitons in a spontaneously broken gauge theory with weight lattice Λ. If the soliton could be truly a quantum mechanical particle it would have to possess a quantum field operator satisfying renormalisable equations of motion which, we conjectured[20], corresponded to the gauge theory whose weight lattice was Λ^*, the dual of Λ, because of the Dirac quantisation condition. A particularly interesting symmetry between the two formulations of the same theory arose if Λ was self-dual, and further seemed to require extended supersymmetry. We shall return to this.

The algebra generated by the U^{α} in (4) closes and is finite dimensional essentially because an integral lattice in Euclidean space has only a finite number of points of given length, here $\sqrt{2}$.

We shall now consider integral lattices Λ in spaces which are no longer Euclidean but have metrics which are

(a) singular i.e. $\eta = \mathrm{diag}\,(0,1,1,1,\ldots)$ (5a)

(b) Lorentzian i.e. $\eta = \mathrm{diag}\,(-1,1,1,1,1,\ldots)$ (5b)

Let us further suppose δ is a point of such a lattice which is light like and primitive i.e. with $\delta^2=0$ and such that if $\lambda\delta \in \Lambda$, $|\lambda|=1,2,3,\ldots$ In case (5a) δ is unique up to a sign. Let us construct the singular lattice

$$\Lambda = \Lambda_R(g) \oplus \mathbb{Z}\,\delta \tag{6}$$

embedded in a Lorentzian lattice of type (5b). The points of (6) with length $\sqrt{2}$ are

$$a = \alpha + n\delta\ ,\ \alpha \text{ a root of } g,\ n \in Z \tag{7}$$

As in the case that Λ is Euclidean.

$$U^a\,U^b - (-1)^{a.b}U^bU^a = 0 \qquad a.b = 0,1,2,\ldots \tag{8a}$$

$$= U^{a+b} \qquad a.b=-1 \tag{8b}$$

corresponding to $(a+b)^2 \geqslant 4$ in (8a) and $(a+b)^2=2$ in (8b). $a.b = -2$ means $(a+b)^2=0$ which in Euclidean space implies $a+b=0$, whereas in a singular space (5a) it implies

$$a + b = m\delta \tag{9}$$

for some integer m. Using Cauchy's theorem whereby the contour around a double pole is the derivative of the residue we find

$$[U^a, U^b] = \oint \frac{dz}{2\pi iz} :a.P(\zeta)\ e^{im\delta .Q(\zeta)}:$$

Note that

$$\oint \frac{d\zeta}{2\pi i\zeta} :\delta .P(\zeta)\ e^{im\delta .Q(\zeta)}: = 0 \qquad (10)$$

as the integrand is proportional to a total derivative. Hence the commutator can be written

$$[U^a, U^b] = (\frac{a-b}{2}).V^{m\delta} \qquad (11)$$

where

$$\varepsilon .V^{m\delta} = \oint \frac{d\zeta}{2\pi i\zeta}\ \varepsilon .P(\zeta)e^{im\delta .Q(\zeta)} \qquad (12)$$

is the contour integral of the emission vertex for a "photon" with momentum $m\delta$ and polarisation ε, rather than the tachyon associated with U[21]. In (11) the polarisation is $(a-b)/2$ and automatically transverse to the momentum $m\delta$

$$\frac{a-b}{2}.m\delta = \frac{(a-b).(a+b)}{2} = \frac{a^2-b^2}{2} = 0 \quad . \qquad (13)$$

Notice that the longitudinal polarisation $\varepsilon = \delta$ in (12) vanishes by (10) and that normal ordering in (12) is indeed unnecessary for transverse polarisations. When m=0, (12) reduces to

$$\varepsilon .V^0 = a.p = \alpha .p + n\delta .p \qquad (14)$$

if $a=\alpha+n\delta$. This shows that the central term is

$$k = \delta.p \quad . \tag{15}$$

The algebra of the U^a and the $\varepsilon.V^{m\delta}$ close on the $\hat{g}$ Kac-Moody algebra (the affinisation of the g whose root lattice occurred in (6)) with the correspondences

$$U^{\alpha}_{m} \rightarrow U^{\alpha+m\delta} \quad , \; \varepsilon.H_m \rightarrow \varepsilon.V^{m\delta} \quad . \tag{16}$$

What has been achieved here is a "covariant" version of the Frenkel-Kac-Segal construction[11,12] which can be regarded as a transverse or light cone gauge treatment involving oscillators in two less dimensions than the natural number. The advantage is the clear correspondence between the lattice Λ and the algebra $g(\Lambda)$ obtained by assigning roots with $a^2 = 2$ or 0 (called respectively real or imaginary roots by mathematicians) to contour integrals of vertex operators for the emission of a "tachyon" or "photon" respectively. The disadvantage is that a larger Fock space is used, one which, in fact carries not only a representation of $g(\Lambda)$ but $g(\Lambda)\oplus V$ where V is the "covariant" Virasoro algebra commuting with $g(\Lambda)$ by (4). Since, apart from the scalar representation, V has only infinite dimensional representations, each $g(\Lambda)$ representation occurs with infinite multiplicity. This is equally true if Λ is Euclidean or singular, of type (6). In the latter case the quantities $\delta.\alpha_n$ also commute with $g = g(\Lambda)$.

Now let us proceed one step further and consider even

integral Lorentzian lattices, possibility (5b). There may now be many primitive light-like vectors on the lattice instead of just $\pm\delta$ in case (5a), but the previous calculation of the corresponding vertex operator (12) holds good for each of these if we start with roots of length $\sqrt{2}$ represented by contour integrals of tachyon vertices. Now the product of two tachyon vertex operators exhibits a pole factor

$$(z_1-z_2)^{a_1.a_2} = (z_1-z_2)^{[(a_1+a_2)^2-a_1^2-a_2^2]/2}$$

Now $(a_1+a_2)^2$ can be a negative even integer and so the calculation will now lead to step operators for roots a with $a^2 = -2, -4,$. etc. expressed as contour integrals of expia.Q(z) times a polynomial in $P^i(z)$ and its derivatives, all normal ordered. Because of the free index on P^i there will also be increasing degeneracy as a^2 decreases through negative even integers. The contour integrals of these vertices will also commute with the covariant Virasoro generators since they are commutators or multiple commutators of the U^a, $a^2 =2$, which do , (4). Hence these vertices also have conformal weight one and are therefore vertex operators for the emission of physical states of the string theory.

As the Virasoro algebra again commutes with the generators of the algebra $g(\Lambda)$ which we now call "Lorentzian" because the lattice is, the Fock space supports an infinite number of representations of $g(\Lambda)$. Unfortunately there are negative norm or ghost states in the Fock space created by the time component oscillators and we would like to have a representation in a positive definite space. If we consider the subspace defined by the physical state conditions of string theory

$$L_n|\psi\rangle = 0, \quad n \geq 1; \quad L_0|\psi\rangle = |\psi\rangle \quad , \tag{17}$$

we can deduce from the no-ghost theorem of string theory that this space is positive definite providing the dimension of the lattice is 26 or less. Thus if dim $\Lambda \leq 26$, (17) furnishes a space with positive norm carrying a unitary representation of the Lorentzian algebra $g(\Lambda)$. This representation is actually the adjoint representation since $L_0=1$, the same as the conformal weight of the vertex operators, and saying that it is positive definite is akin to saying that a compact finite-dimensional Lie algebra is compact if its adjoint representation is unitary and positive definite[14]. Thus we see that the critical dimension, 26, of string theory also plays an important role in the theory of infinite dimensional Lie algebras, something first anticipated by Winnberg[22], the teacher of Kac, in his analysis of Weyl groups.

We can summarise the situation for the simply-laced algebras of Lie, affine untwisted and Lorentzian type by depicting them in what physicists have called a Chew-Frautschi plot which relates the spin of an elementary particle to its mass squared. This is achieved by thinking of a root a as corresponding to the momentum of an elementary particle, a^2 to minus its mass squared and its degeneracy to its spin. The plot is actually that appropriate to string theory.

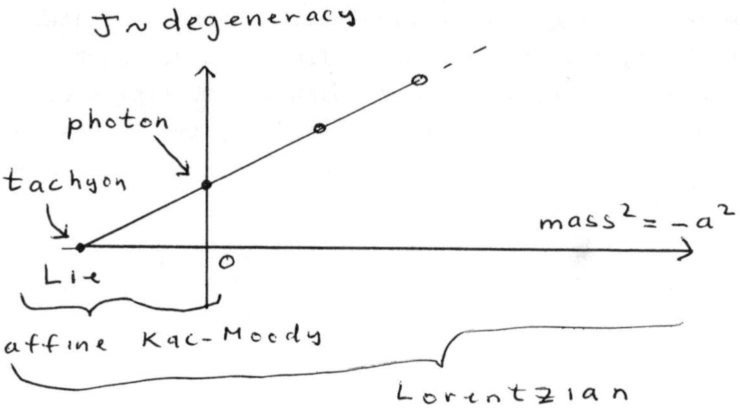

An alternative approach, developed by mathematicians, is to note that affine Lie and affine Kac-Moody algebras can be defined in terms of a Cartan matrix or Dynkin diagram, and to define generalised algebras in terms of generalised Cartan matrices or Dynkin diagrams. Some of these matrices may have one negative eigenvalue, symptomatic of a Lorentzian algebra, and some may have more. On the other hand a Lorentian algebra as we have defined it in terms of a lattice need not have a Cartan matrix as we shall see. Hence we have different sorts of generalisation of affine Kac-Moody algebra.

It was mentioned that self-dual Euclidean lattices are of interest because of a conjectured electromagnetic-duality symmetry that could arise in connection with the magnetic monopole solitons of a spontaneously broken gauge theory. Extended supersymmetry seems to be another requirement for the realisation of this symmetry[23]. For this reason an extended supersymmetric E_8 gauge theory was proposed as a grand unified theory[24]. The E_8 root lattice, $\Lambda_R(E_8)$, is even as well as self-dual. The combination of these two properties relates to modular invariance and it seems desirable to consider such lattices with their associated Lie groups being candidate grand unified groups. Such lattices occur only in dimensions which are multiples of eight, with one in eight dimensions, $\Lambda_R(E_8)$, two in sixteen, twenty-four in twenty-four but very many in dimensions which are higher multiples of eight[25].

In Lorentzian space such lattices occur only in dimensions two more than a multiple of eight and are unique in such dimensions, being denoted $II_{8n+1,1}$. The aforementioned Euclidean lattices of 8n dimensions can be found by considering the lattices "transverse" to a light like point δ of $II_{8n+1,1}$, i.e. the points of $II_{8n+1,1}$ orthogonal to δ and identified modulo δ, just as the polarisation was transverse in (12). It is striking that the critical dimensions of string theory, 10 and 26, have the form 8n+2.

By considering the action of the Weyl group generated by reflections in points of $II_{8n+1,1}$ with $a^2=2$, acting on the positive sheet of a mass shell hyperboloid it is possible to identify Weyl chambers and hence Dynkin diagrams with points corresponding to the bounding hyperplanes[14].

The Dynkin diagram for $II_{9,1}$ is E_{10}

$E_{10} = \Pi_{9,1}$

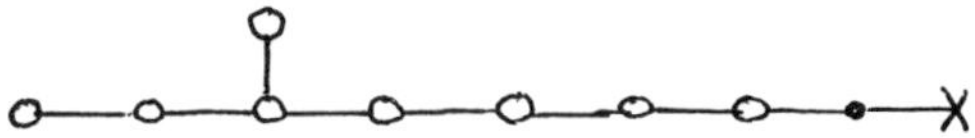

Deletion of the crossed point yields E_9, the diagram of $\hat{E}_8$, while the further deletion of the solid point yields E_8.

The Dynkin diagram for $II_{17,1}$ has 19 rather than 18

points:

$II_{17,1}$ 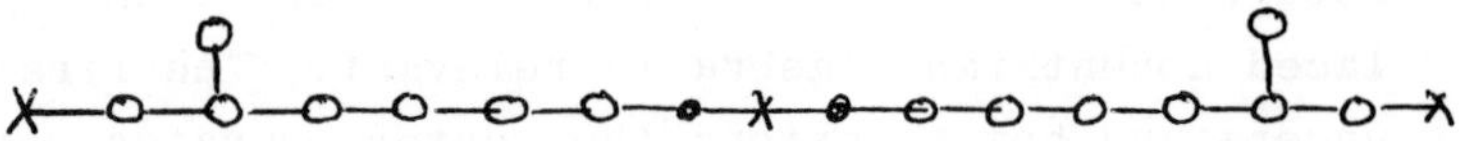

Deletion of the middle crossed point yields the diagram for $\hat{E}_8 \times \hat{E}_8$. Further deletion of the two solid points yields $E_8 \times E_8$. On the other hand, deletion of the two crossed points at the ends yields $\hat{D}_{16}$ while the further deletion of one of the remaining tip points yields D_{16}. These algebras D_{16} and $E_8 \times E_8$ are the ones associated with the two even self-dual sixteen dimensional lattices. Indeed the above diagram can be constructed by reversing the above procedure. It was speculated on this basis by Goddard and myself[14)] that these groups, D_{16}, $E_8 \times E_8$ (as well as E_8) might be particularly suitable choices of gauge group in extended supersymmetric gauge theories. Subsequently Green and Schwarz[13)] demonstrated that D_{16} (and later $E_8 \times E_8$) uniquely lead to anomaly cancellation. The incorporation of these gauge groups into string theory via the heterotic construction[26)] made use of the above lattice constructions. The modularity we have mentioned is apparently related to the cancellation of global anomalies.

The Dynkin diagram of $II_{25,1}$ has an infinite number of points and so cannot be drawn, yet its algebra seems to be particularly related to the bosonic string theory which shares the dimension 26. $II_{33,1}$ and higher lattices seem not to be interesting because of the occurrence of ghosts in their representations.

Despite being ten-dimensional the $II_{9,1} = E_{10}$ lattice does not seem appropriate to the fermionic string theory in 10 dimensions. That theory possesses two tachyons, with m^2

= -1 and -2, and although the latter decouples its Regge recurrences do not. We therefore suspect that a non-simply laced Lorentzian algebra is relevant. The first step is to understand how to extend the vertex operator construction in order to obtain level one representations of the affine untwisted Kac-Moody algebra $\hat{g}$ when g is no longer simply laced. We mentioned the solution to this[16,18] at the beginning of this section and now the motivation is clear.

We think that there are many exciting developments to come in this direction which has been based on speculative physical ideas and yet has already been vindicated by some of the recent developments in string theory. Mathematically this direction is uncharted and any physical intuition is particularly valuable.

REFERENCES

1) Virasoro M., Phys. Rev. D1, 2933 (1970).
2) Lovelace C., Phys. Lett. 34B, 500 (1971).
3) See Olive D. in Proceedings of the XVIIth International Conference on High Energy Physics, London 1974, ed. by J.R. Smith, p1-269 (Rutherford Lab).
4) Brower R., Phys. Rev. D6, 1655 (1972).
Goddard P. and Thorn C., Phys. Lett. 40B, 235 (1972).
5) Ramond P., Phys. Rev. D3, 2415 (1971).
Neveu A. and Schwarz J.H., Nucl. Phys. B31, 86 (1971), Phys. Rev. D4, 1109 (1971).
6) Brink L., Olive D., Rebbi C. and Scherk J., Phys. Lett. 45B, 379 (1973).
7) Gervais J-L. and Sakita B., Nucl. Phys. B34, 632 (1971).
8) Gliozzi F., Olive D. and Scherk J., Phys. Lett. 65B, 282 (1976)., Nucl. Phys. B122, 253 (1977).
9) Green M. and Schwarz J.H., Nucl. Phys. B181, 502 (1981).

10) Kac V.G., Functs. Anal. Prilozhen 1, 82 (1967); Moody R.V., Bull. Amer. Math. Soc. 73, 217 (1967): See Kac V.G. "Infinite-dimensional Lie Algebras - An Introduction" (Birkhauser 1983, second edition Cambridge 1985).

11) Frenkel I.B. and Kac V.G., Inv. Math 62, 23 (1980).

12) Segal G., Comm. Math. Phys. 80, 301 (1981).

13) Green M. and Schwarz J.H., Phys. Lett. 149B, 117 (1984).

14) Goddard P. and Olive D., in "Vertex Operators in Mathematics and Physics", MSRI Publication 3, p51, (Springer 1984).

15) Goddard P. and Olive D., "Kac-Moody and Virasoro Algebras in Relation to Quantum Physics", Int. J. Mod. Phys. A., To appear.

16) Goddard P., Olive D. and Schwimmer A., "Vertex Operators for Non-simply Laced Algebras", DAMTP preprint, Cambridge. 17) Waterson G., Phys. Lett. 171B, 77 (1986).

18) Bernard D. and Thierry-Mieg J., to appear.

19) Frenkel I.B., in "Proceedings of meeting on "Applications of Group Theory in Physics and Mathematical Physics"., Chicago 1982, Lectures in Applied Maths. 21, 355 (1985).

20) Goddard P., Nuyts J. and Olive D., Nucl. Phys. B125, 1 (1977).

21) Del Giudice E., Di Vecchia P. and Fubini S., Annals of Phys. 70, 378 (1972).

22) Winnberg E.B., Iszvestija AN USSR (ser. mat.) 35, 1083 (1971).

23) See Olive D. "Magnetic monopoles and electromagnetic duality conjectures" in "Monopoles in Quantum Field Theory", ed. by Craigie N., Goddard P. and Nahm W., p157 (World Scientific, Singapore 1982).

24) Olive D. and West P., Nucl. Phys. B217, 1 (1983).
25) Serre J-P., "A Course in Arithmetic" (Springer, New York 1973).
26) Gross D., Harvey J.A., Martinec E. and Rohm R., Phys. Rev. Lett. 54, 502 (1985), Nucl. Phys. B256, 253 (1985).

CRITICAL DIMENSIONS AND SUPERSYMMETRIC MEASURES

P. van Nieuwenhuizen*

Institute for Theoretical Physics
P.O. Box 80.006, 3508 TA Utrecht
The Netherlands

CONTENTS

* Visiting F.C. Donders professor of physics, on leave from the Institute for Theoretical Physics, State University of New York at Stony Brook, Stony Brook, N.Y. 11794, USA.

1. INTRODUCTION

As advocated by Fujikawa[1], anomalies arise in quantum field theory when the quantum action has a symmetry, but the measure of the path-integral is not invariant under that symmetry. The anomaly is then the product of the Jacobians at all points. In order to regulate this infinite product one uses a positive definite regulator R/M^2 with eigenvalues λ_m^2/M^2 and considers the limit $M \to \infty$. In this way one obtains the correct axial anomalies[2], trace anomalies[3], and critical dimensions[4], [5], [6] of strings, for ordinary field theories as well as for higher derivative field theories[7]. One may also define the anomaly as the product of the Jacobians at all points if the original action is not invariant, for example in the case of trace anomalies for ordinary (nonimproved) scalars.

It would be desirable to decide upon the correct measure by requiring that divergences in loop graphs cancel, in the same way as in nonlinear σ-models the $\delta^4(o)$ measure cancels quartic divergences. That would give the measure a direct meaning in terms of Feynman diagrams. This program we intend to execute in the future, but for the time being another route has been followed, namely requiring naive coordinate invariance (see below).

In order to fix the measure one requires that it is invariant under some preferred symmetries; these are then the symmetries without anomalies. Having fixed the measure in this way, one may then proceed to calculate the anomalies for other symmetries. Different choices of the measure (i.e., different choices of preferred symmetries) will in general lead to different anomalies for the non-preferred symmetries. The choice of preferred symmetries is a matter of physical prejudice and is analogous to the requirement of vector conservation in Yang-Mills-theories which then leads to axial vector anomalies.

For systems coupled to an external gravitational field, one chooses a measure which is invariant under reparametrizations[8]. This is analogous to the case of external Yang-Mills fields coupled to fermions with a measure which is invariant under vector transformations. As is well-known[9], one can always remove a reparametrization anomaly by shifting it to a local Lorentz anomaly, just as one can shift the vector anomaly in Yang-Mills theories to an axial vector anomaly.

When the gravitational field itself is quantized, the natural symmetry to impose[8] is BRST coordinate symmetry, which is the residual rigid symmetry of the quantum action which remains after the local classical gauge invariance of reparametrizations has been fixed and coordinate ghosts have been introduced. Since the coordinate ghosts and antighosts serve the purpose of removing the unphysical degrees of freedom of the graviton at the quantum level, their measure should be treated simultaneously with that of the graviton.

In theories with fermions, the gravitational field is usually described by the vielbein field $e_\mu^m(x)$. In addition to local coordinate invariance, one has then at the classical level also another gauge invariance, namely local Lorentz invariance. One should also add for this symmetry a gauge fixing term and an ensuing ghost action, and one should again treat the measures of the Lorentz ghosts and antighosts simultaneously with that of the vielbein and coordinate ghosts and antighosts. The natural gauge fixing term for local Lorentz invariance is the square of the antisymmetric part of the vielbein[10]

$$\mathcal{L}(\text{Lorentz fix}) = \alpha\left(e_{\mu m} - \delta_\mu^n \delta_m^\nu e_{n\nu}\right)^2 \tag{1.1}$$

with α a constant. The Lorentz ghosts and antighosts are then non propagating.

Instead of starting with local Lorentz invariance at the classical level and then treating it on a par with other local symmetries by means of the covariant quantization techniques, one could begin by choosing a suitable "unitary" gauge in the classical theory. The most natural such gauge is the "symmetric gauge" in which the vielbein is symmetric. Other local classical symmetries, such as reparametrizations, will in general break this symmetry and one has to add field-dependent compensating local Lorentz transformations which bring one back in the symmetric gauge. It has been shown in ref. (11) that quantization and classical gauge fixing are commuting operations: Integrating out the Lorentz ghosts and antighosts, the Lorentz ghosts cannot just be dropped, but they become equal to a certain combination of other fields and lead in this way to extra interactions. These same extra interactions are obtained if one applies the standard covariant quantization methods to the "corrected" classical symmetries (corrected in the sense that they maintain the symmetric gauge). In what follows we shall not work in the symmetric gauge, and treat local Lorentz symmetry on the same footing as a local Yang-Mills symmetry.

We now give a brief discussion of the Becchi-Rouet-Stora-Tyutin formalism. The BRST symmetry for gauge theories is a quantum extension of classical gauge symmetry. The classical gauge action plus gauge-fixing terms plus Faddeev-Popov ghost action has a rigid symmetry with a constant anticommuting parameter Λ, which is an extension of the classical local gauge invariance of the classical gauge action.

Let the classical gauge fields be generically denoted by ϕ^i, and let the classical gauge transformations be given by

$$\delta\phi^i = R^i{}_\alpha(\phi)\xi^\alpha \tag{1.2}$$

The indices i and α run over internal as well as spacetime variables. For example, in Yang-Mills theory $\delta W^a_\mu(x) = (D_\mu\Lambda)^a$ where $\{a, \mu, x\}$ constitute i, and $\{a,x\}$ correspond to α. To each local parameter ξ^α one associates a ghost field C^α with opposite statistics. Then the BRST transformation rule of the classical gauge fields is

$$\delta\phi^i = R^i{}_\alpha(\phi)C^\alpha\Lambda \tag{1.3}$$

We shall assume that the classical gauge algebra closes. That means that the commutator of two local gauge transformations is a sum of local gauge transformations

$$\begin{aligned}[\delta(\eta), \delta(\xi)]\phi^i &= R^i{}_{\alpha,j}R^j{}_\beta\eta^\beta\xi^\alpha - \xi \leftrightarrow \eta \\ &= R^i{}_\gamma f^\gamma{}_{\alpha\beta}\eta^\beta\xi^\alpha \end{aligned} \tag{1.4}$$

The symbol $R^i_{\alpha,j}$ denotes the right derivative of R^i_α w.r.t. ϕ^j. This formula defines the structure constants,which may be (and are in the case of supergravity theories) field dependent. The BRST transformation laws of the classical gauge fields ϕ^i are then nilpotent if the ghost fields C^α transform under BRST transformations as follows

$$\delta C^\alpha = -\tfrac{1}{2} f^\alpha{}_{\beta\gamma} C^\gamma \Lambda C^\beta \tag{1.5}$$

In this result and other results below we shall use equations which are valid for bosonic and fermionic local symmetries simultaneously[12].

The nilpotency of BRST transformations on gauge fields ϕ^i follows from the closure of the classical gauge algebra. Nilpotency on the ghosts follows from the Jacobi identities for three local gauge transformations[12]. This is the geometrical part of BRST transformations.

To complete the BRST transformations, one adds auxiliary fields d^α which are BRST inert

$$\delta d^\alpha = 0 \tag{1.6}$$

The antighosts transform per definition into the auxiliary fields

$$\delta C^{*\alpha} = -d^\alpha \Lambda \tag{1.7}$$

Obviously, the complete BRST transformation rules are still nilpotent as long as the fields commute or anticommute (i.e., neglecting normal ordering effects).

The quantum action is given by [13]

$$\mathcal{L}(\text{quantum}) = \mathcal{L}(\text{class},\phi^i) - (-)^\alpha d^\alpha \gamma_{\alpha\beta}(F^\beta + \tfrac{1}{2} d^\beta) + C^{*\alpha}\gamma_{\alpha\beta} F^\beta{}_{,\gamma} C^\gamma \tag{1.8}$$

if $\gamma_{\alpha\beta}$ is field-independent.

(There is no sign in the last term since C^α and $C^{*\alpha}$ always commute with F_α). The symbol $F^\beta,{}_\gamma\zeta^\gamma$ means the gauge-variation of F^β with parameter ζ. Sometimes one uses field-dependent $\gamma_{\alpha\beta}$; in that case

$$\mathcal{L}(\text{quantum}) = \mathcal{L}(\text{class},\phi^i) + \delta/\delta\Lambda\big(C^{*\alpha}\gamma_{\alpha\beta}(F^\beta+\tfrac{1}{2}d^\beta)\big) \tag{1.9}$$

where $\delta/\delta\Lambda$ indicates first making a BRST variation and then removing Λ from the right.

One can even construct a theory of local BRST invariance in superspace[14], but that seems (?) not of use for string theories.

Let us now first discuss how the measure is determined by requiring naive coordinate invariance.

2. THE CORRECT INTEGRATION VARIABLES

Consider a scalar field $S(x)$. In the path-integral, one should use as integration variable $\tilde{S}(x) = S(x)\,(\det e)^k$, where $\det e$ is the determinant of the vielbein field $e_\mu^{\ m}$ and the constant k is to be determined by requiring that the measure $D\tilde{S}$ be naively invariant under BRST coordinate transformations.

Expand $\tilde{S}$ into a complete set of orthonormal eigenfunctions φ^m with eigenvalues λ_m of some hermitian operator O

$$\tilde{S} = \sum a_m \varphi^m \tag{2.1}$$

Under a BRST coordinate transformation, $\tilde{S}$ changes as follows

$$\delta\tilde{S} = C^\mu\Lambda\;\partial_\mu\tilde{S} + k(\partial_\mu C^\mu\Lambda)\tilde{S} \tag{2.2}$$

Hence $\tilde{S} + \delta\tilde{S} = \sum\,(a_m+\delta a_m)\varphi^m$ where due to the orthonormality

$$\delta a_m = \int d^2x\;\varphi^m(x)\sum_n\,[C^\mu\Lambda\partial_\mu\varphi^n+k(\partial_\mu C^\mu\Lambda)\varphi^n]a_n \tag{2.3}$$

The measure $D\tilde{S}$ can also be written as $\Pi(Da_n)$, since the change of basis from $\tilde{S}(x)$ to a_n is (presumably) unitary. Hence the Jacobian for an infinitesimal change of the integration variable $\tilde{S}$ reads

$$J = 1 + \sum_m \partial\delta a_m/\partial a_m = 1 + \sum_m \int d^2x\ \varphi^m[C^\mu\Lambda\partial_\mu\varphi^m + k\partial_\mu C^\mu\Lambda\varphi^m] \tag{2.4}$$

For $k = \frac{1}{2}$ the deviation J-1 is a total derivative

$$(J-1) = \int d^2x\ \tfrac{1}{2}\partial_\mu(\sum_m \varphi^m C^\mu \Lambda\varphi^m) \tag{2.5}$$

To regulate the sum over m, one uses the fact that λ_m are real and defines for a general operator A

$$\text{tr}A = \sum_m \varphi^m A\varphi^m \equiv \lim_{M\to\infty} \varphi^m A e^{-\lambda_m^2/M^2}\varphi^m = \lim_{M\to\infty} \text{trace } Ae^{-O^+O/M^2} \tag{2.6}$$

The trace is in practice always evaluated on a basis of plane waves.

If we forget about regularization, we obtain naive invariance: the total derivative gives no contribution and the Jacobian for an infinitesimal (and hence for a finite) BRST coordinate transformation is unity. Hence, the correct integration variable for a scalar field is

$$\tilde{S} = S\ (\det e)^{\frac{1}{2}}. \tag{2.7}$$

For spinors one finds the same result, because they transform under general coordinate transformations, and thus under BRST coordinate transformations, in the same way as scalars. Hence

$$\tilde{\lambda}^a = \lambda^a\ (\det e)^{\frac{1}{2}} \tag{2.8}$$

For antighosts one has already the correct integration variables, because antighosts transform under BRST transformations into the auxiliary BRST fields d^α, so that their Jacobian is always unity. The coordinate antighosts $C^{*\mu}$ and the supersymmetry antighosts $\sum^a$ and the Lorentz antighosts C^{*mn} thus satisfy

$$\tilde{C}^{*\mu} = C^{*\mu},\ \tilde{\textstyle\sum}^a = \textstyle\sum^a,\ \tilde{C}^{*mn} = C^{*mn} \tag{2.9}$$

Also the BRST auxiliary fields d^α are already the correct integration variables, because they are inert under any BRST transformation.

We turn now to the ghost fields. The supersymmetry ghost fields S^a and Lorentz ghost fields C^{mn} transform under BRST coordinate transformations the same way as scalars; hence

$$\tilde{S}^a = S^a\,(\det e)^{\frac{1}{2}}\ ,\ \tilde{C}^{mn} = C^{mn}\,(\det e)^{\frac{1}{2}} \tag{2.10}$$

However for the coordinate ghost field the issue is less clear.

For a general covariant and contravariant vector field one has naive invariance under reparametrizations if

$$(D\tilde{B}^\mu) = D\left(B^\mu\ \det e^{\left(\frac{d+2}{2d}\right)}\right)$$

$$(D\tilde{A}_\mu) = D\left(A_\mu\ \det e^{\left(\frac{d-2}{2d}\right)}\right) \tag{2.11}$$

We shall assume that the same results hold for the vielbein and coordinate ghosts. For a discussion of this critical point see ref. (15). Hence

$$\tilde{e}_\mu^{\ m} = e_\mu^{\ m}(\det e)^{(d-2)/2d}\ ;\ \tilde{C}^\mu = C^\mu(\det e)^{(d+2)/2d} \tag{2.12}$$

For the gravitino, we have of course no problem; it is treated like $\tilde{A}_\mu$

$$\tilde{\psi}_\mu^{\ a} = \psi_\mu^{\ a}(\det e)^{(d-2)/2d} \tag{2.13}$$

Thus, in d=2 the vielbein and gravitino field are unmodified.

3. CRITICAL DIMENSIONS OF THE SPINNING STRINGS

There are three classes of spinning string models, namely the d=2 supergravities with N=1, N=2 or N=4 local supersymmetries.

The N=1 supergravity model [16) can be extended to a nonlinear sigma model[17] coupled to supergravity[18], and one can add a Wess-Zumino (WZ) term[19] also coupled to supergravity[20]. The fields are

$$e_\mu^m \, , \, \psi_\mu^a \, , \, \phi^I \, , \, \lambda^I \; (I=0,\ldots,d-1) \tag{3.1}$$

One may add a supergravity auxiliary field S, and a matter auxiliary field F^I, although they will not contribute to the critical dimension, because they appear in the action as $(\tilde{S})^2$ and $(\tilde{F}^I)^2$ and thus decouple from gravity.

The N=2 model has as matter fields d complex ϕ^I and d complex λ^I. The supergravity fields are e_μ^m, a complex ψ_μ^a and a real vector field B_μ which gauges the group U(1) [21]. It describes two ordinary spinning strings, coupled to each other by B_μ [22]. One can extend it to a nonlinear sigma model with a WZ term, coupled to supergravity[23]. The rigid N=2 nonlinear sigma model with a WZ term was given, both in x-space and in superspace, in ref. (24).

The N=4 model as a nonlinear sigma model coupled to supergravity was recently constructed[25]. This supergravity model can be extended to contain a WZ term[23]. This is then the most general sigma model with WZ term. A surprising result found in ref. (25) is that in the rigid case one has always a hyper-Kähler geometry but in the local case not only a quatermionic geometry but also a hyper-Kähler geometry is possible. The fields in this model are: e_μ^m, four $\psi_\mu^{a,I}$, a nonpropagating SU(2) gauge field $\vec{B}_\mu$, together with 4n scalars ϕ^I and 4n Majorana spinors λ^I.

Consider the simplest model, the N=1 linear sigma model without WZ term. Choosing the usual superconformal gauge for reparametrizations, local Lorentz symmetry and local supersymmetry

$$F_\alpha = \left\{ \begin{matrix} \tilde{e}_{11} + \tilde{e}_{22} = 0 \\ \tilde{e}_{12} + \tilde{e}_{21} = 0 \end{matrix} \; ; \; \tilde{e}_{12} - \tilde{e}_{21} = 0 \; ; \; \begin{matrix} \gamma_1 \tilde{\psi}_1 + \gamma_2 \tilde{\psi}_2 = 0 \\ \text{all indices flat} \end{matrix} \right\} \tag{3.2}$$

the vielbein is given by $e_\mu^m = \delta_\mu^m \sqrt{\rho}$ and the gravitino by $\psi_\mu = \gamma_\mu \varphi$. (As explained in section 2, the twiddles on vielbein and gravitino are redundant.) The complete quantum action becomes

$$\mathcal{L}(\text{quantum}) = \mathcal{L}(\text{class}) + \mathcal{L}(\text{fix}) + \mathcal{L}(\text{ghost})$$

$$\mathcal{L}(\text{class}) = -\tfrac{1}{2}(\partial_\mu\phi\partial_\nu\phi\eta^{\mu\nu}) - \tfrac{1}{2}(\bar{\lambda}\rho^{\frac{1}{4}})\not{\partial}(\rho^{\frac{1}{4}}\lambda)$$

$$\mathcal{L}(\text{fix}) = d^\alpha\ F_\alpha + \text{"extra terms" times } F_\alpha$$

$$\mathcal{L}(\text{ghost}) = -i(\vec{C}^*\sqrt{\rho})(\tau_3\partial_1-\gamma_o\partial_o)\vec{C}$$

$$-i(\bar{\Sigma}\ \rho^{-\frac{1}{4}})(\gamma_1\partial_1+\gamma_o\partial_o)(\rho^{-\frac{1}{4}}S)$$

$$+i(C^*_{12}\sqrt{\rho})(2C_{12}+\tfrac{1}{2}(\bar{\psi}_1\gamma_o-\bar{\psi}_o\gamma_1)S + \partial_o C^1 + \partial_1 C^o) \tag{3.3}$$

The "extra terms" are due to inserting into $\mathcal{L}$(class) and $\mathcal{L}$(ghost) the gauge conditions; these extra terms are clearly proportional to F_α and only shift d^α. The coordinate ghosts C^μ have been written as vectors $\vec{C}$ and the bar on the supersymmetry antighost field Σ is the Dirac bar. A derivation of this $\mathcal{L}$(quantum) was first given in (6). For an extended set of lectures on this derivation and the BRST formalism for strings, see (26).

We next replace all fields in $\mathcal{L}$(quantum) by twiddled fields times the corresponding powers of dete = ρ. Then we remove the total ρ dependence by rescaling each twiddled field. This rescaling leads to Jacobians whose product equals unity in the critical dimension.

Consider first ϕ. Its action is given by

$$-\tfrac{1}{2}\partial_\mu(\tilde{\phi}e^{-\sigma})\ \partial_\nu(\tilde{\phi}e^{-\sigma})\ \eta^{\mu\nu}\ \ ;\ \rho \equiv e^{2\sigma} \tag{3.4}$$

since dete = ρ. We scale $\tilde{\phi}$ from $\tilde{\phi}$ to $\tilde{\phi}e^\sigma$ but in little steps. Suppose we have reached a point where we have $\rho_t = e^{2\sigma_t} \equiv e^{2\sigma(1-t)}$. The next rescaling is thus $\tilde{\phi} \to e^{\sigma dt}\ \tilde{\phi} = (1+\sigma dt)\tilde{\phi}$. As regulator we use

$$R_\varphi = [e^{-\sigma_t}\ \Box\ e^{-\sigma_t}] \tag{3.5}$$

which is hermitian and positive definite (in Euclidean space). The

Jacobian is then

$$J = 1 + \int d^2x\, \sigma(x) dt \int \frac{d^2k}{(2\pi)^2} e^{-ikx} e^{R_\varphi M^{-2}} e^{ikx} \tag{3.6}$$

Pulling the plane wave exp ikx to the left, results in replacing ∂_μ by $\partial_\mu - ik_\mu$. The relevant k-integrals all are covered by the following formula[5,6)]

$$\int \frac{d^2k}{(2\pi)^2} e^{-ikx} e^{\rho_t^{\frac{1}{\alpha}} \not\partial \rho_t^{\beta} \not\partial \rho_t^{\frac{1}{\alpha}}} e^{ikx} =$$

$$\frac{1}{4\pi} \left(M^2 \rho_t + \frac{(\alpha+\beta)}{3} (-\Box\, \ell n \rho_t) + \mathcal{O}\frac{1}{M} \right) \tag{3.7}$$

For example, for the scalars $\tilde{S}$ one has $\beta = 0$ and $\alpha = \frac{1}{2}$.

The final Jacobian for a given field is then

$$J = \exp \frac{-1}{12\pi} c \int d^2x\, \sigma \Box \sigma \tag{3.8}$$

where c is a constant depending on the number of field components, their statistics and their weight of its rescaling.

The coefficients c in front of the Jacobians for the various fields can be read off from $\mathcal{L}$(quantum). One has

Field	Value of c	
one scalar field ϕ	$\frac{1}{2}$	ref. (4)
two real coordinate ghosts C^μ	-8	ref. (4)
a Majorana fermion λ	$\frac{1}{4}$	ref. (6)
two real coordinate antighosts $C^{*\mu}$	-5	ref. (6)
two anti-Majorana* susy ghosts S^a	15/4	ref. (6)
two Majorana susy antighosts Σ^a	7/4	ref. (6)

* The coordinate ghosts and supersymmetry ghosts always have opposite reality properties, see (26).

one real scalar Maxwell ghost -1 } ref. (27)

one real scalar Maxwell antighost 0 }

We can now at once determine the critical dimensions from the field content of the various models.

For the bosonic string the critical dimension is thus given by

$$d\tfrac{1}{2} - 8 - 5 = 0 \quad \rightarrow \quad d=26 \tag{3.9}$$

For the N=1 linear σ-model one has

$$d\tfrac{1}{2} + d\tfrac{1}{4} - 8 - 5 + \frac{15}{4} + \frac{7}{4} = 0 \quad \rightarrow \quad d=10 \tag{3.10}$$

For the N=2 linear σ-model one has

$$d + d\tfrac{1}{2} - 8 - 5 + 2\,\frac{15}{4} + 2\,\frac{7}{4} - 1 = 0 \quad \rightarrow \quad d=2 \tag{3.11}$$

The nonpropagating SO(2) field B_μ does not contribute, see next section.

For the N=4 linear σ-model one has

$$2d + d - 8 - 5 + 4\,\frac{15}{4} + 4\,\frac{7}{4} - 3 = 0 \quad \rightarrow \quad d=-2 \tag{3.12}$$

The nonpropagating SU(2) gauge field B_μ does not contribute, see next section.

Since in two dimensions there are no transversal modes and negative dimensions have no physical significance (?), only the cases d=26 and d=10 remain.

The divergent M^2 terms cancel in supersymmetric models. Their contribution is equal to the sum of the numbers of bosonic field components minus the number of fermionic field components, each weighted with the weight of the rescaling[6)]. This sum cancels, but note that this sum is not the same sum as just the difference of the number of bosonic and fermionic field components[28)].

For the N=2 vector fields B_μ, the abelian U(1) gauge invariance is fixed by the gauge fixing term $\partial_\mu B_\nu \eta^{\mu\nu} = 0$. Note that in d=2, B_μ equals $\tilde{B}_\mu$ so that the gauge condition is linear in twiddled fields[6]. The Maxwell ghost action becomes

$$C^*(\partial_\mu \partial^\mu)(\tilde{C}\ \mathrm{det}e^{-\frac{1}{2}}) \tag{3.13}$$

where C^* and $C = \tilde{C}\ \mathrm{det}e^{-\frac{1}{2}}$ are the Maxwell antighost and ghost. As hermitian regulator for the Maxwell ghost we choose

$$R = (\mathrm{det}e)^{-\frac{1}{2}} \Box\Box\ (\mathrm{det}e)^{-\frac{1}{2}} \tag{3.14}$$

Note that we must rescale $\tilde{C}$ by $\tilde{C} \rightarrow \tilde{C}\ e^{+\sigma}$ but C^* is not rescaled. Doing the k-integral, one finds c = −1 and c = 0. (In an earlier study[6]), a hermitian ghost action $C^*(\mathrm{det}e)^{-\frac{1}{2}} \Box (\mathrm{det}e)^{-\frac{1}{2}} \tilde{C}$ was obtained which yielded $c = -\frac{1}{2}$ and $c = -\frac{1}{2}$ instead of c = −1 and c = 0. A detailed analysis shows that indeed for Maxwell ghosts the c for $\Box\Box$ is twice the c for $\Box$, but we consider the argument used to obtain the hermitian ghost action no longer convincing.)

4. CRITICAL DIMENSIONS OF NONLINEAR σ-MODELS

We refer the reader to a recent publication[27]), but make here some comments.

It is clear that for a nonlinear σ-model one cannot use the linearized action as regulator, because in that case one would follow the same steps as for the linear σ-model, and one would end up with d=10, instead of the lower critical dimension

$$d(\mathrm{crit}) = 10 - \frac{1}{3}\dim G - \frac{2}{3}\dim G/\left[1+\tfrac{1}{2}C_2(G)k^{-1}\right] \tag{4.1}$$

where the group manifold on which the nonlinear σ-model is defined has dim G coordinates, Casimir operator $C_2(G)$ and winding number k.

The full nonlinear action, on the other hand, leads to

complicated k-integrals, too difficult to perform exactly. (These integrals correspond in Feynman language to infinitely many loops of scalar modes.) However, it is here that nonabelian bosonization comes to the rescue: certain nonlinear bosonic path-integrals are equal to corresponding more tractable (often linear) fermionic path-integrals, and vice-versa. However, for our applications one needs such equivalences for models coupled to gravity, and not many of such equivalences have rigorously been proven. We refer the reader to ref. (27).

A particular detail involves a nonpropagating Yang-Mills field coupled to fermions. In flat space, integrating out the fermions, one obtains a two-dimensional QCD determinant. For our purposes we must extend these results to nonpropagating Yang-Mills fields coupled to fermions in an external gravitational field, and the corresponding computations have, to our knowledge, not yet been performed. Work is under way to do this. We expect the results to be such that in our computations of critical dimensions we can altogether forget about these Yang-Mills fields (but not about their ghosts, see previous section). In that case we may indeed omit the U(1) gauge field of the N=2 model, and the SU(2) gauge field from the N=4 model.

5. SUPERSYMMETRY OF THE MEASURE IN x-SPACE

It seems natural to require for supersymmetric theories that the measure be also invariant under BRST supersymmetry transformations. In x-space, the measure was already fixed by naive BRST coordinate invariance, and hence one would expect that this measure is also invariant under BRST supersymmetry transformations. In ref. (29,6) it was shown that the x-space measure is naively supersymmetric, meaning that the BRST variation of the product of the (dete) factors of all integration variables cancels the Jacobians for the supersymmetry variations of all integration variables. We call this naive invariance, because no regularization is involved.

For example, the integration variables of the d=2 Poincaré multiplet are

$$(De^m_\mu)(D\psi^a_\mu)D(Sg^{\frac{1}{4}})D(C^\mu g^{\frac{1}{2}})D(C^{mn}g^{\frac{1}{4}})D(S^a g^{\frac{1}{4}}) \tag{5.1}$$

where S is an auxiliary field, C^μ the coordinate ghosts, C^{mn} the Lorentz ghost and S^a (a=1,2) the supersymmetry ghosts. (We have not written explicitly the antighosts and BRST auxiliary fields since they do not contribute to the Jacobians and are not multiplied by factors of g.) The total product of g factors is thus $g^{-\frac{1}{2}}$, which yields a naive BRST supersymmetry variation of $+\frac{1}{2}\bar{\psi}.\gamma S$. (Recall that δ (detg) = $\bar{\epsilon}\gamma.\psi$ detg) . For the vielbein variation the Jacobian is unity (since $\delta e^m_\mu = \frac{1}{2}\bar{\epsilon}\gamma^m\psi_\mu$ is independent of e^m_μ), but the other fields yield $(-\frac{1}{4},+\frac{1}{4},0,0,-\frac{1}{2})$ times $\bar{\psi}.\gamma S$, respectively (6). Clearly, the measure is naively supersymmetric in x-space.

In general the Jacobian for auxiliary fields cancels the variation of the g-factor for that same auxiliary field. This is for example the case for the field S above. Hence, one may omit auxiliary fields without destroying supersymmetry invariance of the measure. As is well-known[30)] this will introduce quartic ghost couplings in the quantum action.

In conformal supergravities one cannot drop the measure for the dilaton field, which is an independent field over which one should integrate even though it cancels from the classical action[31)] and also from the quantum action[31)]. Let us elaborate somewhat on this point. (The nonconformal reader may skip to the end of this section.)

Under ordinary (Q) supersymmetry $\delta b_\mu = -\frac{1}{2}\bar{\epsilon}\phi_\mu$. The S-supersymmetry gauge field ϕ_μ is not independent, but follows from the constraint $\gamma^\mu R_{\mu\nu}(Q) = 0$, where

$$R_{\mu\nu}(Q) = \mathcal{D}_\mu\psi_\nu - \mathcal{D}_\nu\psi_\mu + \gamma_\nu\varphi_\mu - \gamma_\mu\varphi_\nu \tag{5.2}$$

and $\mathcal{D}_\mu$ is given by

$$\mathcal{D}_\mu\psi_\nu = \partial_\mu\psi_\nu + \tfrac{1}{4}\omega_\nu^{mn}\gamma_{mn}\psi_\nu + \tfrac{1}{2}b_\mu\psi_\nu + \text{"more"} \tag{5.3}$$

where "more" denotes for example an axial term in d=4. The spin connection ω_μ^{mn} is found from the constraint $R(P)^m_{\mu\nu} = 0$, and contains the dilaton

$$\omega_\mu^{mn} = \omega_\mu^{mn}(e,\psi) - \left(b^m e_\mu^n - b^n e_\mu^m\right) \tag{5.4}$$

One finds from $\gamma^\mu R\,(Q)_{\mu\nu} = 0$ the following result for ϕ_μ in d dimensions (d=3 or d=4 are only relevant)

$$\phi_\mu = -\frac{1}{d-2}\gamma^\nu \left(\mathcal{D}_\mu\psi_\nu - \mathcal{D}_\nu\psi_\mu\right) + \frac{1}{(d-1)(d-2)}\gamma_\mu\gamma^{\rho\sigma}\mathcal{D}_\rho\psi_\sigma \tag{5.5}$$

Thus, the b_μ terms in δb_μ are given by

$$-\tfrac{1}{2}\bar{\varepsilon}[-\frac{1}{d-2}\gamma^\nu\{\tfrac{1}{2}b_\mu\psi_\nu - \tfrac{1}{2}b_\nu\psi_\mu - \tfrac{1}{4}2b^m e_\mu^n\gamma_{mn}\psi_\nu + \tfrac{1}{4}2b^m e_\nu^n\gamma_{mn}\psi_\mu\} +$$

$$\frac{1}{(d-1)(d-2)}\gamma_\mu\gamma^{\rho\sigma}\{\tfrac{1}{2}b_\rho\psi_\sigma - \tfrac{1}{4}2b^m e_\rho^n\gamma_{mn}\psi_\sigma\}] \tag{5.6}$$

The Jacobian is thus

$$\partial b_\mu/\partial b_\mu = -\tfrac{1}{2}\bar{\varepsilon}[-\frac{1}{d-2}\gamma^\nu\{\frac{d}{2}\gamma_\nu - \tfrac{1}{2}\psi_\nu + \tfrac{1}{4}\,2\gamma_{\mu\nu}\psi_\mu\} +$$

$$\frac{1}{(d-1)(d-2)}\gamma_\mu\gamma^{\rho\sigma}\{\tfrac{1}{2}\delta_\rho^\mu\psi_\sigma - \tfrac{1}{4}\,2\gamma_{\mu\rho}\psi_\sigma\}] \tag{5.7}$$

After some elementary gamma matrix algebra one finds

$$\delta b_\mu/\delta b_\mu = (\bar{\varepsilon}\gamma\cdot\psi)\,[\tfrac{1}{4}(\frac{d-1}{d-2}) - \tfrac{1}{4}(\frac{d-1}{d-2})$$

$$-\frac{1}{(d-1)(d-2)}\{\tfrac{1}{4}(d-1) + \tfrac{1}{4}(d-3)(d-1)\}] = -\tfrac{1}{4}\bar{\varepsilon}\gamma\cdot\psi \tag{5.8}$$

The g factor for a covariant vector field is in d-dimensions $D\left(b_\mu g^{(d-2)/(4d)}\right)$, hence one finds a factor $g^{(d-2)/4}$. Clearly, the variation of this d-dependent factor does not cancel the d-independent Jacobian. Thus one should not omit the dilaton field from the measure, even though it cancels from the action.

6. SUPERSYMMETRY OF THE MEASURE IN SUPERSPACE

As a step towards proving non-naive invariance of the x-space measure under supersymmetry, we first analyze the coupling of a scalar superfield S to N=1 d=2 supergravity in superspace, see ref. (32,33). The classical action is given by

$$\mathcal{L} = \mathrm{sdet}E\ (\mathcal{D}_b S)(\mathcal{D}_a S)\varepsilon^{ab} \tag{6.1}$$

where $D_a S = E_a{}^{\Lambda}\,\partial_\Lambda S$ and $z^\Lambda = \{x^\mu, \theta^\alpha\}$. The index a is a flat spinor index. One chooses the superconformal gauge in superspace

$$E_a{}^{\Lambda}\,\partial_\Lambda = e^{\psi}(D_a) \tag{6.2}$$

where D_a is the covariant derivative of rigid supersymmetry

$$D_a = \partial/\partial\theta^a + i\gamma^m_{ab}\,\theta^b\partial_m \tag{6.3}$$

In this gauge, $\mathrm{sdet}E_\Lambda^M = e^{-2\psi}$ so that all ψ dependence cancels from the classical action, similar to what happens in the x-space case. The ghost action is obtained by varying the five gauge conditions which fix the general supercoordinate invariance and the local Lorentz invariance. Defining the superfields $\mathcal{E}_M{}^{\Lambda}(x,\theta)$ by

$$E_M{}^{\Lambda}\,\partial_\Lambda = \mathcal{E}_M{}^{N}\,D_N \quad , \quad D_N = \{D_a, \partial_\mu\} \tag{6.4}$$

the gauge conditions are

$$\mathcal{E}_a{}^{m} = 0\ ,\ \mathcal{E}_{a=1}{}^{b=1} - \mathcal{E}_{a=2}{}^{b=2} = 0 \tag{6.5}$$

In spinor notation a=1 corresponds to + chirality, and a=2 corresponds to − chirality. Vector indices are rewritten in spinor language as $v^{ab} = v^m\gamma_m{}^{ab}$, and only by v^{++} and v^{--} are nonvanishing since $\gamma_m{}^{ab}$ is diagonal for m=0 and m=1. Thus the gauge conditions read

$$E_{\pm}{}^{++} = 0 \ , \ E_{\pm}{}^{--} = 0 \ , \ E_{+}{}^{+} - E_{-}{}^{-} = 0 \tag{6.6}$$

Note that the superscripts denote indices which are contracted with the flat indices of rigid supersymmetry.

The super vielbein is not totally arbitrary, but must satisfy certain constraints which define the kinematics of superspace. A thorough discussion of these issues can be found in ref. (32,33) and in a set of lectures on d=2 superspace[34], but here we shall only mention that as a consequence of these constraints, the above conditions also imply that

$$E_{+}{}^{-} = E_{-}{}^{+} = 0 \tag{6.7}$$

Thus $E_{\pm}{}^{M} = \delta_{\pm}^{M} e^{\psi}$.

From here on we follow reference (33). The corresponding ghost action is obtained by varying the gauge conditions. From

$$\delta \mathcal{D}_M = [\mathcal{D}_M, K] \ ; \ M = \{m, a\} = \text{flat superindex}$$

$$\mathcal{D}_M = E_M{}^N \, D_N + \phi_M \, L \ ; \ L \text{ is Lorentz generator}$$

$$K = K^{\Lambda} \, D_{\Lambda} + \Lambda \, L \ ; \ \Lambda \text{ is local Lorentz parameter} \tag{6.8}$$

we find $\delta E_{\pm}{}^{\substack{++\\--}}$ and $\delta(E_{+}{}^{+} - E_{-}{}^{-})$. In the result we substitute again the gauge conditions (for a discussion of this point see below) and we thus obtain

$$\begin{aligned}
\mathcal{L}(\text{ghost}) = {} & \bar{C}_{--}{}^{+}(e^{\psi} \, D_{+} \, C^{--}) \\
& + \bar{C}_{++}{}^{-} \, (e^{\psi} \, D_{-} \, C^{++}) \\
& + \bar{C}_{++}{}^{+} \, (e^{\psi} \, D_{+} \, C^{++} + 2i \, e^{\psi} \, C^{+}) \\
& + \bar{C}_{--}{}^{-} \, (e^{\psi} \, D_{-} \, C^{--} - 2i \, e^{\psi} \, C^{-})
\end{aligned}$$

$$+ \bar{C} \left(e^{\psi} D_{+} C^{+} - e^{\psi} D_{-} C^{-} - e^{\psi} C\right) \tag{6.9}$$

The two terms with a factor 2i are due to the rigid superspace anti-commutator $\{D_a, D_b\} = 2i\gamma^m_{ab} D_m$.

The correct integration variables are not C^{Λ} and C (or equivalently C^{++}, C^{--}, C^{+}, C^{-} and C) but rather $\tilde{C}^{\Lambda}$ and $\tilde{C}$, where $\tilde{C}^{\Lambda}$ and $\tilde{C}$ are the product of C^{Λ} and C with factors of $(\mathrm{sdet}E_{\Lambda}{}^{M})$. We fix $\tilde{C}^{\Lambda}$ and $\tilde{C}$ by requiring that the measure $D\tilde{C}^{\Lambda}D\tilde{C}$ be naively <u>super</u>-reparametrization invariant. This is the most interesting aspect of this problem, but for reasons of coherence we will first give the answer and complete the determination of the critical dimension. In the next section we will discuss the derivation of the relations between $\tilde{C}^{\Lambda}$, $\tilde{C}$ and C^{Λ}, C.

As integration variables in the path integral one should take

$$\tilde{S} = S(\mathrm{sdet}E)^{\frac{1}{2}},\ \tilde{C}^{\mu} = C^{\mu}(\mathrm{sdet}E)^{3/2},\ \tilde{C}^{\alpha} = C^{\alpha}(\mathrm{sdet}E)$$

$$\tilde{C} = C(\mathrm{sdet}E)^{\frac{1}{2}}\ ;\ \tilde{E}_M{}^{\mu} = E_M{}^{\mu}(\mathrm{sdet}E)^{3/2}\ ;\ \tilde{E}_M^{\alpha} = E_M{}^{\alpha}(\mathrm{sdet}E) \tag{6.10}$$

The antighosts need not be modified by factors of sdetE since their Jacobian under BRST transformations is unity (antighosts transform only into BRST auxiliary fields). Also the auxiliary BRST fields need not be modified (they are inert).

As gauge conditions we take conditions <u>linear</u> in the correct integration variables[6)]. Hence, as gauge conditions we should have taken

$$e^{-3\psi}\ E_{\pm}{}^{\pm\pm}\ ;\ \ e^{-2\psi}\left(E_{+}{}^{+} - E_{-}{}^{-}\right) \tag{6.11}$$

In the ghost action this merely leads to extra factors of $e^{-\psi}$ since in the unweighted gauge one need not vary ψ (since the result would anyhow vanish due to the unweighted gauge). With these modifications the ghost action becomes

$$\mathcal{L}(\mathrm{ghost}) = \bar{C}_{\pm\pm}{}^{\mp}\ e^{-2\psi}\ D_{\mp}(e^{3\psi}\tilde{C}^{\pm\pm}) + \bar{C}_{\pm\pm}{}^{\pm}[e^{-2\psi}\ D_{\pm}(e^{3\psi}\tilde{C}^{\pm\pm}) \pm 2i\ \tilde{C}^{\pm}]$$

$$+ \quad \bar{C}\, e^{-\psi} \left[D_+ (e^{\psi}\, \tilde{C}^+) - D_- (e^{\psi} \tilde{C}^-) - e^{\psi}\, \tilde{C} \right] \qquad (6.12)$$

We now first integrate over $\bar{C}$ and $\tilde{C}$. The net result is that one may omit the last line in $\mathcal{L}$(ghost) because the Jacobian, proportional to e^{ψ}, belongs to a nonpropagating field ($\tilde{C}$) and thus equals unity. (In the expansion into a complete set of modes, only the mode which is identically zero remains). In fact, the Jacobian is even ψ - independent.

Next we integrate over $\bar{C}_{++}{}^{+}$, $\bar{C}_{--}{}^{-}$, C^+ and C^-. The net result is now that we may omit the complete second line in $\mathcal{L}$(ghost). There is now not even a Jacobian. The ghost action thus reduces to

$$\mathcal{L}(\text{ghost}) = \bar{C}_{--}{}^{+}\, e^{-2\psi}\, D_+ (e^{3\psi}\, \tilde{C}^{--}) + \bar{C}_{++}{}^{-}\, e^{-2\psi} D_- (e^{3\psi}\, \tilde{C}^{++}) \qquad (6.13)$$

The complete quantum action is thus given by

$$\mathcal{L}(\text{quantum}) = \varepsilon^{ab} (D_b\, \tilde{S} e^{\psi}) D_a (\tilde{S} e^{\psi}) + \mathcal{L}(\text{ghost}) + \hat{d}^{\alpha}\, F_{\alpha} \qquad (6.14)$$

where F_{α} are the five gauge conditions given before. The field $\hat{d}^{\alpha}$ is equal to the BRST auxiliary field d^{α}, plus extra terms due to using the gauge condition $F_{\alpha} = 0$ in $\mathcal{L}$(ghost) and $\mathcal{L}$(classical). It follows that if we integrate over d^{α} in the path-integral, we find a product of delta functions, whose arguments are linear in the variables $\tilde{E}$. Subsequent integration over the five $\tilde{E}$ variables then yields unity, without a further Jacobian. Hence, the $\hat{d}^{\alpha}\, F_{\alpha}$ may be omitted, and the only supergravitational dependence is through e^{ψ}. We now decouple $\tilde{S}$ and the (anti)ghost fields from e^{ψ} by rescaling these integration variables.

These rescalings will lead to Jacobians, and the net result will be a free field theory for $\tilde{S}$, $\bar{C}_{\mp\mp}{}^{\pm}$ and $\tilde{C}^{\pm\pm}$, plus a term only dependent on e^{ψ} which is due to summing the infinitesimal Jacobians. In the critical dimension these ψ-contributions cancel.

For $\tilde{S}$ this rescaling is given by

$$\tilde{S} \to \tilde{S} e^{-\psi} \qquad (6.15)$$

For the scalar superfield $\tilde{S}$ we have a hermitian action: $\tilde{S}\, e^{\psi}\, D^a D_a\, e^{\psi}\, \tilde{S}$. However, $D^a D_a$ contains only one spacetime derivative, whereas we need a Dalembertian to make the integration over plane waves convergent (see below). Hence we use as regulator the square of the action

$$R_{\tilde{S}} = e^{\psi}\, D^a D_a\, e^{2\psi}\, D^b D_b\, e^{\psi} \tag{6.16}$$

The antighosts and ghosts we rescale as follows

$$\bar{C}_{\mp\mp}{}^{\pm} \to \bar{C}_{\mp\mp}{}^{\pm}\, e^{2\psi};\; \tilde{C}^{\pm\pm} \to e^{-3\psi}\, \tilde{C}^{\pm\pm} \tag{6.17}$$

As regulator we seek again a hermitian positive definite operator. Now the operator $e^{-2\psi}\, D_+\, e^{3\psi}$ maps from $\tilde{C}^{--}$ space into the dual of $\bar{C}_{--}{}^{+}$ space(which is $\bar{C}_{++}{}^{-}$ space), and $e^{3\psi}\, D_-\, e^{-2\psi}$ maps from $\bar{C}_{++}{}^{-}$ space to the dual of $\tilde{C}^{++}$ space. Thus $e^{3\psi}\, D_-\, e^{-4\psi}\, D_+\, e^{3\psi}$ maps from $\tilde{C}^{--}$ space into itself. As regulators we cannot yet choose the operators O appearing in

$$\tilde{C}^{++}(e^{3\psi}\, D_-\, e^{-4\psi}\, D_+\, e^{3\psi})\tilde{C}^{--}$$

$$\bar{C}_{--}{}^{+}(e^{-2\psi}\, D_+\, e^{6\psi}\, D_- e^{-2\psi})\bar{C}_{++}{}^{-} \tag{6.18}$$

because these operators O are not hermitian. (Recall that all ghost fields, included $\bar{C}_{\pm\pm}{}^{\mp}$ are real, and $(D_+)^{\dagger} = D_+$.) Instead of O we may use $O^{\dagger}O$ or $OO^{\dagger}$. In this way we find the following regulators

$$\text{for } \tilde{C}^{\mp\mp}:\; R_1(\psi) = e^{3\psi}\, D_{\pm}\, e^{-4\psi}\, D_{\mp}\, e^{6\psi}\, D_{\mp}\, e^{-4\psi}\, D_{\pm}\, e^{3\psi}$$

$$\text{for } \bar{C}_{\pm\pm}{}^{\mp}:\; R_2(\psi) = e^{-2\psi}\, D_{\mp}\, e^{6\psi}\, D_{\pm}\, e^{-4\psi}\, D_{\pm}\, e^{6\psi}\, D_{\mp}\, e^{-2\psi} \tag{6.19}$$

We do the rescaling again in little steps. Suppose one has come to a point where ψ has decreased to $\psi(1-t)$. The next rescaling is $\tilde{S} \to \tilde{S}(1-\psi dt)$; $\bar{C}_{\mp\mp}{}^{\pm} \to \bar{C}_{\mp\mp}{}^{\pm}(1+2\psi dt)$; $\tilde{C}^{\pm\pm} \to \tilde{C}^{\pm\pm}(1-3\psi dt)$. The Jacobians are thus $(1-\psi dt)$, twice $(1+2\psi dt)$ and twice $(1+3\psi dt)$, respectively. (A sign changes since one needs super determinants.) Each Jacobian has to be regularized with the appropriate regulator. We obtain the following infinitesimal contributions to the Jacobian

$$\lim_{M\to\infty} \int d^2x\, d^2\theta (q_i \psi dt) \int \frac{d^2k}{(2\pi)^2} d^2\chi\, e^{-iz.K} e^{R_i(\psi_t)/M^2} e^{iz.K} \quad (6.20)$$

where $iz.K = ik.x + \bar{\chi}\theta$ (with $\bar{\chi}\theta \equiv \chi^\alpha \varepsilon_{\alpha\beta}\theta^\beta$ imaginary) and where $q_i = -d$ for $\tilde{S}$, -1 for $C^{-}{}_{--}{}^{+}$ and for $\bar{C}_{++}{}^{-}$, and $+2$ for $\tilde{C}^{++}$ and $\tilde{C}^{--}$. The regulators R_i were given before. Afterwards we must integrate dt from 0 to 1.

We can always write each R_i as

$$R = g^{\Lambda\Pi} \partial_\Pi \partial_\Lambda + V^\Lambda \partial_\Lambda + W$$

$$= \hat{g}^{MN} D_N D_M + \hat{V}^M D_M + \hat{W} \quad (6.21)$$

because the product of three spinorial derivatives is equal to a spinorial and a vector derivative

$$D^2 D_a = - D_a D^2 = 2i\, \gamma^m{}_{ab}\, D^b\, \partial_m \;;\; D^2 \equiv D^a D_a \quad (6.22)$$

(To prove this, use $\varepsilon^{ab}\,(D_a D_b D_c + D_b D_c D_a + D_c D_a D_b) = 0$ and bring D_c to the far left or the far right).

We will use the following lemma:

Lemma I: $$\lim_{M\to\infty} \int \frac{d^2k}{(2\pi)^2} d^2\zeta\, e^{-iz.K} e^{R_i M^{-2}} e^{iz.K} = \frac{i}{2\pi} \left(\text{sdet} g^{\Lambda\Pi}_{R_i}\right)^{-\frac{1}{2}} \quad (6.23)$$

Proof: pulling the superplane wave exp iz.K from right to left, and rescaling $K \to KM$ the superJacobian for this rescaling is unity (because the M^2 for d^2k cancels the M^{-2} for $d^2\zeta$). Hence, only the term $-g^{\Lambda\Pi} K_\Pi K_\Lambda$ remains in the exponent. For $\zeta = 0$ the k_μ integral gives $\frac{1}{4\pi}(\det g^{\mu\nu})^{-\frac{1}{2}}$. For $k = 0$ the ζ-integral gives

$$\int d^2\zeta \; e^{+g^{\alpha\beta}\zeta_\beta\zeta_\alpha} = 2ig^{12} = 2i(\det g^{\alpha\beta})^{\frac{1}{2}} \qquad (6.24)$$

(which is real). This proves (6.23). (The superplane waves are orthonormal due to Berezinian integration. We define $d^2\zeta = id\zeta^1\, d\zeta^2$ and $\bar{\zeta}\theta = \zeta^1\theta^2 - \zeta^2\theta^1$ in order that

$$\int \frac{d^2k}{(2\pi)^2} d^2\zeta \; e^{-ik.x-\bar{\zeta}\theta} e^{iky+\bar{\zeta}\eta} = \delta^2(x-y)\,\delta^2(\theta-\eta) \qquad (6.25)$$

where $\delta^2(\theta-\eta) = -i(\theta^2-\eta^2)\,(\theta^1-\eta^1)$ and $\int d\theta^1\theta^1 = 1$.) Note that only with this real $d^2\zeta$ one obtains orthonormality).

For applications a second useful lemma is

Lemma II: $\operatorname{sdet} g^{\Lambda\Pi} = \operatorname{sdet} \hat{g}^{MN}$

Proof: the difference between $g^{\Lambda\Pi}\, K_\Pi K_\Lambda$ and the corresponding expression with $\hat{g}^{MN}$ but with $\hat{K}_\Pi = (k_a + i\bar{\theta}\not{k}_a, k_m)$ amounts to the shift $k_a \to k_a + i\bar{\theta}\not{k}_a$, which is unimodular.

If we now pick out of R_i the terms with $\hat{g}^{MN}_{R_i}\, D_N D_M$, the infinitesimal Jacobian is then given by

$$\frac{i}{2\pi}\int d^2x\, d^2\theta\, \left(q_i\psi\delta t\right)\left(\operatorname{sdet} \hat{g}^{MN}_{R_i}\right)^{-\frac{1}{2}} \qquad (6.26)$$

In d=2 superspace, the superdeterminant of $g^{\Lambda\Pi}$ can be written in the following simple form

$$\operatorname{sdet} g^{\Lambda\Pi} = (\det g^{\mu\nu})\varepsilon_{\alpha\beta}(g^{\alpha\beta} - g^{\alpha\mu}g_{\mu\nu}g^{\nu\beta}) \qquad (6.27)$$

where $g_{\mu\nu}$ is the inverse of $g^{\mu\nu}$. Thus we have:

Lemma III: If $R = A + B^\alpha D_\alpha + CD^\alpha D_\alpha + E^{\alpha\beta}\gamma^m_{\alpha\beta}\,\partial_m + F^\alpha\gamma^m_{\alpha\beta}\, D^\beta\partial_m + G\,\Box$ (6.28)

then

$$\left(\operatorname{sdet} \hat{g}^{MN}\right)^{-\frac{1}{2}} = \left(G^{-1}C - \tfrac{1}{4}\, G^{-2}F^\alpha F_\alpha\right) \qquad (6.29)$$

The total Jacobian is then given by

$$J = \exp \sum_j \int_o^1 dt \, \frac{i}{2\pi} \int d^2x \, d^2\theta \, (q_j\psi)[G_j^{-1}C_j - \tfrac{1}{4} G_j^{-2}F_j^\alpha F_{j\alpha}] \quad (6.30)$$

The actual evaluation of G, C and F^α for $R_{\tilde{S}}$ in (6.16) and R_1, R_2 in (6.19) requires some interesting manipulations. In what follows we will only retain terms which contribute to G, C, or F^α. We begin with $R_{\tilde{S}}$ and note that if in $e^\psi D^a D_a e^{2\psi} D^b D_b e^\psi$ the $D^a D_a$ both act to the right of $e^{2\psi}$ one obtains $G = -4e^{4\psi}$ according to (6.22). If only one D_a acts to the right of $e^{2\psi}$ one finds from (6.22) that $F^a = -8ie^{4\psi}D^a\psi$. If both $D^a D_a$ act on $e^{2\psi}$ one finds $C = e^{4\psi}[2(D^2\psi) + 4(D^a\psi)(D_a\psi)]$. Hence

$$\left(\text{sdet } g_{\tilde{S}}^{MN}\right)^{-\frac{1}{2}} = -\tfrac{1}{2} D^a D_a \psi \quad (6.31)$$

Next we consider the regulator $R_1 = e^{3\psi} D_+ (e^{-4\psi} D_- e^{6\psi} D_- e^{-4\psi}) D_+ e^{3\psi}$ and note that the expression between parentheses yields two terms: $e^{-2\psi} i\gamma^m_{--}\partial_m + 6e^{2\psi}(D_-\psi)D_- e^{-4\psi}$. The first term yields then $G=e^{4\psi}$, since $D_+D_+ = i\gamma^n{}_{++}\partial_n$. In the second term the free D_- cannot act on $e^{-4\psi}$, as $(D_-\psi)(D_-\psi) = 0$, hence it yields $6e^{\psi} D_+(D_-\psi)D_- D_+ e^{3\psi}$. If the D_+ on the left acts on $(D_-\psi)$ one finds with $(D_+D_-\psi) = -\tfrac{1}{2}(D^2\psi)$ that $C = -3/2$, whereas if D_+ on the left acts past $(D_-\psi)$ one finds a term $D_+D_-D_+ = -iD_-\gamma^m_{++}\partial_m$. This term does not contribute since it corresponds to F^+ whereas there is no corresponding F^-. Hence we find

$$\left(\text{sdet } \hat{g}_{R_1}^{MN}\right)^{-\frac{1}{2}} = -3/2 \, (D^a D_a \psi) \quad (6.32)$$

Finally we consider the regulator $R_2 = e^{-2\psi} D_- (e^{6\psi} D_+ e^{-4\psi} D_+ e^{6\psi}) D_- e^{-2\psi}$. The expression between parentheses yields

$$(e^{2\psi} i\gamma^m_{++}\partial_m e^{6\psi} - 4e^{2\psi}(D_+\psi)D_+ e^{6\psi}) \quad (6.33)$$

The derivatives ∂_m and D_+ must act past the $e^{6\psi}$. Acting with D_- on the first term in (6.33) yields with $\gamma^m_{++}\partial_m\gamma^n_{--}\partial_n = -\partial^m\partial_m$

$$e^{4\psi}\partial^m\partial_m + e^{4\psi} 8i(D_-\psi)\gamma^m_{++}\partial_m D^+ \quad (6.34)$$

where we used $D_- = D^+$. Acting with D_- on the second term in (6.33) yields

$$-4e^{6\psi}[8(D_-\psi)(D_+\psi) + (D_-D_+\psi) - (D_+\psi)D_-]D_+D_-e^{-2\psi}$$

$$= e^{4\psi}[8(D^a\psi)(D_a\psi)D^2 + (D^2\psi)D^2 + 4i(D_+\psi)D^-\gamma^m_{--}\partial_m]$$

where we used $D_+ = -D^-$ and $D_+D_- = -D_-D_+ = -\frac{1}{2}D^2$. Hence in this case $G = e^{4\psi}$, $C = \{8(D^a\psi)(D_a\psi) + (D^2\psi)\}e^{4\psi}$ while $F^+ = 8i\, e^{4\psi}(D_-\psi)$ and $F^- = 4ie^{4\psi}(D_+\psi)$. Therefore, with $F^2 = -2F^+F^-$, we get

$$\left(\text{sdet}\ \hat{g}^{MN}_{R_2}\right)^{-\frac{1}{2}} = (D^2\psi) \tag{6.35}$$

The total anomaly in (6.28) is thus proportional to the sum of (6.31), (6.32) and (6.35), weighted with $q_i = \{-d, +3, +3, +2, +2\}$, respectively. One obtains

$$\frac{d}{2} + 2\times 3\times\left(\frac{-3}{2}\right) + 2\times 2\times(1) = \frac{1}{2}(d-10) \tag{6.36}$$

Hence, in d=10 the supergravitational modes decouple completely, and the trace anomaly is proportional to the action $\int d^2x d^2\theta (D^a\psi)(D_a\psi)$. Note that trilinear terms proportional to $\psi(D^a\psi)(D_a\psi)$ did cancel for each regulator separately.

7. CORRECT INTEGRATION VARIABLES IN SUPERSPACE?

Just as in x-space, we determine the correct integration variables for the supervielbein and supercoordinate ghosts by requiring their measure to be invariant under BRST supercoordinate transformations. However, we hit upon two surprises (ref. (33))

(i) the measures of the vector and spinor parts of a supervector need different powers of sdet $E_\Lambda{}^M$

(ii) these different powers are not uniquely fixed by requiring BRST supercoordinate invariance.

We shall fix these powers uniquely by the additional requirement that the measures with flat indices are the same as for scalars [35]. There is no compelling reason for this additional requirement, although it does lead to the correct critical dimension, see last section. The fact that supercoordinate invariance alone does not fix the measure entirely may be an indication that the quantum action has further local symmetries; requiring these extra gauge invariances to be of the preferred type would then fix the measure uniquely. What these extra symmetries are is not clear to us at this moment.

Let us now give the details. A contravariant supervector transforms under general supercoordinate transformations as follows

$$\delta V^{\Lambda} = - K^{\Pi}\partial_{\Pi}V^{\Lambda} + V^{\Pi}\partial_{\Pi}K^{\Lambda} \tag{7.1}$$

(Expanding $V = V^{\Lambda}\partial_{\Lambda} = \hat{V}^{M}D_{M}$, one would find in $\delta\hat{V}^{M}$ an extra term $V^{a}K^{b}\gamma^{m}_{ab}\delta^{M}_{m}$ due to $\{D_{a},D_{b}\} \sim \gamma^{m}_{ab}\partial_{m}$). Suppose the correct integration variable were

$$\tilde{V}^{\Lambda} = V^{\Lambda}\ (\mathrm{sdet}E)^{k} \tag{7.2}$$

Then

$$\delta\tilde{V}^{\Lambda} = \tilde{V}^{\Pi}\partial_{\Pi}K^{\Lambda} - K^{\Pi}\partial_{\Pi}\tilde{V}^{\Lambda} - k\tilde{V}^{\Lambda}(\partial_{\Pi}K^{\Pi}(-)^{\Pi}) \tag{7.3}$$

In the super Jacobian, one would find a supertrace

$$\partial(\delta\tilde{V}^{\Lambda})/\partial\tilde{V}^{\Lambda}\ (-)^{\Lambda} \tag{7.4}$$

Since there are equal numbers of bosonic and fermionic components of $\tilde{V}^{\Lambda}$ (namely, 2 plus 2), the last two terms of $\delta\tilde{V}^{\Lambda}$ do not contribute, leaving only the first term. (The same conclusion holds for $\hat{V}^{M}$ since the extra term with δ^{M}_{m} does not contribute to the supertrace.) The first term in $\delta\tilde{V}^{\Lambda}$ contributes

$$\partial_{\Lambda}K^{\Lambda}(-)^{\Lambda} \tag{7.5}$$

to the supertrace, a nonvanishing result which is, moreover, k-independent. Thus for no k does $V^{\Lambda}(\mathrm{sdet}E)^{k}$ yield an invariant measure.

One possible way out is to treat the bose and fermi parts of V^Λ differently and to consider as integration variables

$$\tilde{V}^\mu = V^\mu(\text{sdet}E)^{k(\text{bos})} \tag{7.6}$$

$$\tilde{V}^\alpha = V^\alpha(\text{sdet}E)^{k(\text{fer})} \tag{7.7}$$

where k(bos) (k(fer)) is a constant for the vector (spinor) parts of $\tilde{V}^\Lambda$. Now

$$\delta\tilde{V}^\mu = \tilde{V}^\Lambda\partial_\Lambda K^\mu - K^\Lambda\partial_\Lambda\tilde{V}^\mu - k(\text{bos})\tilde{V}^\mu\left(\partial_\Pi K^\Pi(-)^\Pi\right) \tag{7.8}$$

$$\delta\tilde{V}^\alpha = \tilde{V}^\Lambda\partial_\Lambda K^\alpha - K^\Lambda\partial_\Lambda\tilde{V}^\alpha - k(\text{fer})\tilde{V}^\alpha\left(\partial_\Pi K^\Pi(-)^\Pi\right) \tag{7.9}$$

Thus, expanding $\tilde{V}^\Lambda$ on a complete basis as $a^\Lambda_m\phi^m$, one finds

$$\partial\delta\tilde{V}^\mu/\partial\tilde{V}^\mu = \sum_m \int d^2x\, d^2\theta\, \phi^m$$

$$[\partial_\mu K^\mu - 2K^\mu\partial_\mu - 2K^\alpha\partial_\alpha - 2k(\text{bos})(\partial_\mu K^\mu - \partial_\alpha K^\alpha)]\phi^m \tag{7.10}$$

and

$$\partial\delta\tilde{V}^\alpha/\delta\tilde{V}^\alpha = \sum_m \int d^2x\, d^2\theta\, \phi^m$$

$$[\partial_\alpha K^\alpha - 2K^\mu\partial_\mu - 2K^\alpha\partial_\alpha - 2k(\text{fer})(\partial_\mu K^\mu - \partial_\alpha K^\alpha)]\phi^m \tag{7.11}$$

The factors 2 are due to the vector and spinor traces.

Requiring that

$$\partial\delta\tilde{V}^\Lambda/\partial\tilde{V}^\Lambda = \partial\delta\tilde{V}^\mu/\partial\tilde{V}^\mu - \partial\delta\tilde{V}^\alpha/\partial\tilde{V}^\alpha \tag{7.12}$$

be a total derivative yields

$$\partial_\mu K^\mu\{1-2k(\text{bos}) + 2k(\text{fer})\}$$

$$+ \partial_\alpha K^\alpha\{+2k(bos) - 1 - 2k(fer)\} = 0 \qquad (7.13)$$

Hence

$$k(bos) - k(fer) = +\tfrac{1}{2} \qquad (7.14)$$

So, in d=2, the condition of a unit Jacobian does not fix the measures completely.

If the measure for a vector with flat indices is the same as that for a scalar, then $\tilde{V}^M = V^M = V^M \ \text{sdet}E^{1/2}$. Then $V^m = \tilde{V}^\mu \ e^{-3\psi}$ and $\tilde{V}^a = \tilde{V}^\alpha e^{-2\psi}$ because $E_\Lambda{}^M$ equals $e^{-2\psi}$ in the bose-bose sector and $e^{-\psi}$ in the fermi-fermi sector. Hence the correct integration variables are [35]

$$D(V^\mu \ \text{sdet}E^{3/2}) \ D(V^\alpha \ \ \text{sdet}E) \qquad (7.15)$$

ACKNOWLEDGEMENT:

The results discussed above were obtained in collaboration with P. Bouwknecht, with M. Rocek and C.S. Zhang, and with A. Eastough, L. Mezincescu and E. Sezgin. It is a pleasure to thank them for their cooperation.

REFERENCES

1. K. Fujikawa, Phys. Rev. Lett. 42, 1195 (1979).
2. K. Fujikawa, Phys. Rev. D21, 2848 (1980); Phys. Rev. D22, 1499 (1981), (erratum), Phys. Rev. D29, 285 (1984).
3. K. Fujikawa, Phys. Rev. Lett. 44, 1733 (1980); Phys. Lett. 108B, 33 (1982); Phys. Rev. D23, 2262 (1981).
4. A.M. Polyakov, Phys. Lett. 103B, 211 (1981).
5. K. Fujikawa, Phys. Rev. D25, 2584 (1982).
6. P. Bouwknecht and P. van Nieuwenhuizen, Class. Quantum Grav. 3, 207 (1986).

7. P.H. Frampton, D.R.T. Jones, S.C. Zhang and P. van Nieuwenhuizen, contribution to Festschrift for E.S. Fradkin.
8. K. Fujikawa, Nucl. Phys. B226, 437 (1983),
K. Fujikawa and O. Yasuda, Nucl. Phys. B245, 436 (1984).
9. W.A. Bardeen and B. Zumino, Nucl. Phys. B244, 421 (1984).
10. S. Deser and P. van Nieuwenhuizen, Phys. Rev. D10, 411 (1974).
11. P. van Nieuwenhuizen, Phys. Rev. D24, 3315 (1981).
12. P. van Nieuwenhuizen, Physics Reports 68, section 2, (1981).
13. N.K. Nielsen, Phys. Lett. 103B, 197 (1981),
F.R. Ore and P. van Nieuwenhuizen, Phys. Lett. 112B, 364 (1982).
14. F.R. Ore and P. van Nieuwenhuizen, Nucl. Phys. B204, 317 (1982).
15. P. van Nieuwenhuizen, Proceedings 1985 Scottish Universities Summer School.
16. L. Brink, P. di Vecchia and P. Howe, Phys. Lett. 65B, 471 (1976),
S. Deser and B. Zumino, Phys. Lett. 65B, 369 (1976).
17. T.L. Curtright and C.K. Zachos, Phys. Rev. Lett. 53, 1799 (1984).
18. L. Alvarez-Gaumé and D.Z. Freedman, C.M.P. 80, 443 (1981).
19. E. Witten, C.M.P. 92, 455 (1983).
20. E. Bergshoeff, S. Randjbar-Daemi, A. Salam, H. Sarmadi and E. Sezgin, Nucl. Phys. B269, 389 (1986).
21. L. Brink and J.H. Schwarz, Nucl. Phys. B121, 285 (1977).
22. E.S. Fradkin and A.A. Tseythin, Phys. Lett. 106B, 63 (1981)
23. B. de Wit and P. van Nieuwenhuizen, to be published.
24. S.J. Gates, C.M. Hull and M. Rocek, Nucl. Phys. B248, 157 (1984).
25. M. Pernici and P. van Nieuwenhuizen, Phys. Lett. 169B, 381 (1986).
26. P. van Nieuwenhuizen, lectures at the 1986 Physics School at Dubrovnik, Utrecht preprint.
27. A. Easthough, L. Mezincescu, E. Sezgin and P. van Nieuwenhuizen, Stony Brook preprint January 1986.
28. Ref. (22), below eq. (24).
29. M.K. Fung, D.R.T. Jones and P. van Nieuwenhuizen, Phys. Rev. D22, 2995, (1980).
30. E.S. Fradkin and M.A. Vassiliev, Phys. Lett. 72B, 70 (1977),
G. Sterman, P.K. Townsend and P. van Nieuwenhuizen, Phys. Rev. D17, 1501 (1978),

R.E. Kallosh, Nucl. Phys. B141, 141 (1978).

31. For a review of conformal supergravities, see ref. 12, chapter 4.
32. P.S. Howe, J. Phys. A 12, 393 (1979).
33. S.C. Zhang, M. Rocek and P. van Nieuwenhuizen, Ann. of Physics, to be published.
34. P. van Nieuwenhuizen, Teyler's lectures, Leiden preprint 1986.
35. These and related issues will be discussed in a forthcoming publication with N. K. Nielsen.

Gauge Covariant String Field Theory

Peter West

King's College, Strand, London WC2.

Abstract

A review of gauge covariant string field theory is given. It contains a pedagogical discussion of the free and interacting theories.

1. INTRODUCTION

Trying to learn string theory is more difficult than many other areas of theoretical physics. The situation is unlike, say, general relativity. There one first encounters the symmetry of general co-ordinate transformations. One then constructs invariants and in particular the Einstein action. Finally the S-matrix is calculated by using this action to weight the sum over paths in the Feynman path integral. The subjects of supergravity and Yang-Mills theories are developed in a similar linear systematic manner starting from a symmetry principle.

In complete contrast string theories arose from enforcing requirements such as unitarity and duality. The latter property was thought to be a desirable feature of strong interaction physics. In fact, rather than starting from a symmetry principle and computing the S-matrix, the first step in string theory was made by Veneziano [1] who literally guessed the four-string scattering amplitude at the tree level that possessed the above requirements. Despite the extensive development [2] that followed, no real attempt was made to deduce the enormous symmetry group which clearly underlies string theory. Reading this literature, one often obtains veiled glimpses of this symmetry which manifested itself in the many miraculous cancellations that were observed in string theory.

The recent effort to find a gauge-invariant theory of strings is in fact the quest to identify the symmetry group of string theory. Hopefully its identification will lead to many further developments in string theory. In this approach string theory can be seen to be a natural extension of many of the concepts that have become important to modern physics.

Since string theory involves an infinite number of spins, one might think that a covariant formulation which necessarily contains fields for each spin may be rather complicated. As we shall see, however, the results are very simple and elegant once cast in the correct formalism. This formalism is, however, so powerful and to some extent unlike previous developments that it can be difficult to appreciate what is really going on and in particular to recognise familiar manipulations. Consequently, I will begin by giving an introductory and rather pedestrian approach, before utilizing the full power of the formalism.

The results presented in this article were found in collaboration with André Neveu, while some of the results of the free theory were also found in collaboration with Hermann Nicolai and John Schwarz.

2. FREE GAUGE COVARIANT STRING THEORY

The task before us is to second-quantize an extended object, namely the string. Before doing this we will discuss the well-known case of the point particle, which, despite its simplicity, has many features in common with that of the string theory.

2.1 The Point Particle

The trajectory of a classical point particle in space time is parameterized by its proper time τ and is given by $x^{\mu}(\tau)$. The path it takes is so as to be an extremum of the action

$$A = -m\int d\tau \sqrt{-\dot{x}^{\mu}\dot{x}^{\nu}\eta_{\mu\nu}} = \frac{1}{2}\int d\tau \{V^{-1}\dot{x}^{\mu}\dot{x}^{\nu}\eta_{\mu\nu} - mV\} \tag{2.1}$$

where $\dot{x}^{\mu} = \frac{dx^{\mu}}{d\tau}$. This action is invariant under reparameterizations of the proper time $\tau \to \tau + f(\tau)$ with the transformation of x^{μ} being $\delta x^{\mu} = f(\tau)\dot{x}^{\mu}$.

Let us now give a Hamiltonian treatment of the first action of equation (1), although the same results can be found from the second action. The canonical momentum is given by

$$p_{\mu} = \frac{\delta A}{\delta \dot{x}^{\mu}(\tau)} = \frac{m\dot{x}_{\mu}}{\sqrt{-\dot{x}^{\mu}\dot{x}^{\nu}\eta_{\mu\nu}}} \tag{2.2}$$

and the equation of motion is of the form

$$\partial_{\tau} p^{\mu} = 0 \tag{2.3}$$

Due to the reparameterization invariance, the system is constrained by

$$p^{\mu} p_{\mu} + m^2 = 0 \tag{2.4}$$

and the Hamiltonian vanishes:

$$H = p^{\mu} \dot{x}_{\mu} - L \tag{2.5}$$

The method of dealing with such a system was given in reference [3] and we now apply this method in outline to the point particle.

The reader who wishes to read further details is encouraged to consult reference [4]. We take the Hamiltonian to be proportional to the constraints, i.e.

$$H = v(\tau)(p_{\mu}^2 + m^2) \tag{2.6}$$

where $v(\tau)$ is an arbitrary function of τ. One may verify that in this case there are no further constraints and that H generates time translations or reparameterizations in the sense that

$$\frac{dx^{\mu}}{d\tau} = \{x^{\mu}, H\} = 2v(\tau)p^{\mu} \tag{2.7}$$

The fundamental Poisson brackets vanish except for

$$\{x^{\mu}, p^{\nu}\} = \eta^{\mu\nu}. \tag{2.8}$$

To quantize the theory we make the usual transition from Poisson brackets to commutators, with an appropriate factor of $i\hbar$. These commutators are represented by the replacements

$$x^{\mu} \rightarrow x^{\mu} \quad ; \; p^{\mu} \longrightarrow -i\hbar \frac{\partial}{\partial x^{\mu}} \tag{2.9}$$

The constraints then become

$$\hat{\phi} = (-\partial^2 + m^2) \tag{2.10}$$

To find the second-quantized field theory, we consider the state to be described by a field $\psi(x^\mu, \tau)$ and we impose the constraint

$$\hat{\phi}\psi = (-\partial^2 + m^2)\psi = 0 \tag{2.11}$$

We also impose the Schrödinger equation

$$i\hbar \frac{\partial \psi}{\partial \tau} = H\psi \tag{2.12}$$

The right-hand side of this equation vanishes and we find that ψ is independent of τ. In the second-quantized theory, there is in any case more than one particle and so the concept of more than one proper time is problematical.

The action that leads to the above Klein-Gordon equation is

$$A = \int d^4x \quad \psi(-\partial^2 + m^2)\psi \tag{2.13}$$

and we may use it to weight the Feynman path integral which can then be used to find the Green's function of the second-quantized theory.

We note that the original reparameterization invariance of the proper time which was so important for determining the form of the classical action is absent in the second quantized theory; its only remnant being the field equation itself. Since we performed a Hamiltonian quantization with respect to the proper time this is only to be expected, but one might consider if one could second-quantize in such a way as to maintain this invariance. It is the above steps that we now wish to repeat for the string.

2.2 Introduction to the Bosonic String

The bosonic string, whose length is parameterized by $(0 \leqslant \sigma \leqslant \pi)$ sweeps out in time τ, a two-dimensional surface parameterized $\xi^{\alpha} = (\tau, \sigma)$, in a space-time x^{μ} according to the function $x^{\mu}(\xi)$. This trajectory is in such a way as to sweep out an extremal area and so its action [5] is given by

$$A = -\frac{1}{2\pi\alpha'} \int d^2\xi \sqrt{-\det \partial_\alpha x^\mu \partial_\beta x^\nu \eta_{\mu\nu}} \tag{2.14}$$

where α' has the dimensions of $(\text{mass})^{-2}$. It is invariant under arbitrary reparametrizations of the two-dimensional surface

$$\xi^\alpha \to \xi^\alpha + f^\alpha(\xi) \quad ; \quad x^\mu \to x^\mu + \xi^\alpha \partial_\alpha x^\mu \tag{2.15}$$

as well as invariant under the Poincaré group transformations acting on the space-time co-ordinate x^{μ}.

The canonical momentum is given by

$$P_\mu = \frac{\delta A}{\delta\left(\frac{\partial x^\mu}{\partial \tau}\right)} = \frac{\partial_\tau x_\mu (x^{\nu\prime})^2 - x'_\mu \partial_\tau (x^\nu x'_\nu)}{2\pi\alpha' \sqrt{-\det(\partial_\alpha x^\mu \partial_\beta x^\nu \eta_{\mu\nu})}} \tag{2.16}$$

where $x^{\mu\prime} = \frac{\partial x^\mu}{\partial \sigma}$.

Due to the invariance mentioned above, we have the constraints

$$P_\mu^2 + \frac{1}{(2\pi\alpha')^2} (x'^\mu)^2 = 0 \tag{2.17}$$

$$x'^\mu P_\mu = 0 \tag{2.18}$$

It is convenient to extend mathematically the range of σ to lie between $-\pi$ to π by requiring

$$x^\mu(\sigma) = \begin{cases} x^\mu(\sigma) & 0 \le \sigma \le \pi \\ x^\mu(-\sigma) & -\pi \le \sigma \le 0 \end{cases} \tag{2.19}$$

that is, $x^\mu(\sigma) = x^\mu(-\sigma)$. Using this extension, the above constraints can then be written as

$$(\mathcal{P}_\mu)^2 = \left(p^\mu - \frac{1}{2\pi\alpha'} x'^\mu\right)^2 \tag{2.20}$$

It will be advantageous to take the Fourier transform of these constraints and so we define

$$L_n = \frac{\pi\alpha'}{2}\int_{-\pi}^{\pi} d\sigma (\mathcal{P}^\mu)^2 e^{-in\sigma} \tag{2.21}$$

One finds that the L_n's obey the algebra [6].

$$\{L_n, L_m\} = -i(n-m)L_{n+m} \tag{2.22}$$

Since the Hamiltonian vanishes we take it to be proportional to the constraints, i.e.

$$H = \sum_{n=-\infty}^{\infty} c_n L_n \tag{2.23}$$

One may verify that the L_n's are the generators of two-dimensional conformal transformations on the world sheet of the string. Generally, the conformal group has only a finite number of generators, but for two dimensions only, it is an infinite dimensional algebra which corresponds to making an arbitrary analytic transformation in $z = \tau + i\sigma$. Clearly, after suitable Euclidean rotation, the two-dimensional flat metric which can be written as $dzd\bar{z}$ scales under $z \to f(z)$. The emergence of the two-dimensional conformal group rather than the original two-dimensional general co-ordinate group is related to the choice of τ as the time to be used in the Hamailtonian approach. As we shall see, the Virasoro algebra and hence the two-dimensional conformal group play an important role in the second-quantized gauge covariant theory.

The fundamental Poisson brackets of the theory are

$$\{x^\mu(\sigma), P^\nu(\sigma')\} = \eta^{\mu\nu}\,\delta(\sigma-\sigma')$$
$$\{x^\mu(\sigma), x^\nu(\sigma')\} = 0 = \{P^\mu(\sigma), P^\nu(\sigma')\} \tag{2.24}$$

To quantize the theory we replace these relations by commutator relations which are represented by the changes

$$x^\mu(\sigma) \to x^\mu(\sigma)\ ;\ P^\mu(\sigma) \to -i\hbar\frac{\delta}{\delta x_\mu(\sigma)} \tag{2.25}$$

Making these replacements in the generators of the constraints, we find that

$$L_n = \frac{\pi\alpha'}{2}\int_{-\pi}^{\pi} d\sigma\, e^{-in\sigma}\,\hat{P}^\mu(\sigma)\,\hat{P}^\nu(\sigma)\,\eta_{\mu\nu} \tag{2.26}$$

where

$$\hat{P}^\mu(\sigma) = -i\hbar\frac{\delta}{\delta x_\mu(\sigma)} - \frac{1}{2\pi\alpha'}\frac{\partial x^\mu}{\partial\sigma} \tag{2.27}$$

We now impose the following constraints on the functional $\Psi[(x^\mu(\sigma)]$, which, like the particle, is τ-independent.

$$\begin{aligned}(L_0-1)\Psi &= 0\\ L_n\Psi &= 0 \qquad n\geq 1\end{aligned} \tag{2.28}$$

These are not entirely what we might naively expect. The -1 in the first equation corresponds to the possibility of their being, due to L_0 not being uniquely defined by the clasical theory, a normal ordering constant. We shall see that this is fixed to be -1 by requiring a ghost-free spectrum of on-shell states. In the second equation we do not require all the L_n's to vanish on ψ, since this would imply that ψ itself wculd vanish due to the central term in the Virasoro algebra which we will discuss shortly. However, the latter equation implies that

$$(\psi, L_n \psi) = (L_{-n} \psi, \psi) \; \forall n \tag{2.29}$$

as $L_n^+ = L_{-n}$. In this way one recovers in the classical limit that all the L_n's vanish in accord with equation (20). This procedure is the same in the Gupta-Bleuther formulations of quantum electrodynamics.

The necessity of the constraints of equation (2.28) is guaranteed by the following theorem.

Theorem [7]

Equations (2.28) describe a ghost-free set of on-shell states provided the dimension D of space-time is less than or equal to 26.

In fact for $D < 26$ there are other problems and for the remainder of this contribution we will take $D = 26$. It may be helpful to recall the distinction between a ghost and a tachyon: for a scalar with action

$$A = \int dx \left(- c(\partial_\mu A)^2 - d A^2\right). \tag{2.30}$$

we say it is a ghost if $c < 0$ and a tachyon if $d < 0$. We will see that the open bosonic string does indeed possess a tachyon.

We are now in a position to specify what requirements a second-quantized gauge covariant formulation of strings must satisfy. We must demand that there be an action whose equations of motion imply equations (2.28). Of course, we may have to make some gauge choices and we expect $L_n\psi = 0$ $n \geq 1$ may be the result of the gauge choices, while $(L_0 - 1)\psi = 0$ is the remaining equation of motion. We further expect the action to be local in the sense that the free theory should contain no more than two space-time derivatives. The actions we will obtain will contain the fields of Yang-Mills and gravity for the open and closed bosonic strings respectively; they will therefore possess these corresponding gauge invariances and so we must find that the gauge symmetries of the string will contain these particular symmetries.

To quantize this system we will use this action after appropriate gauge fixing and ghosts to weight a Feynman path integral. The vacuum-to-vacuum amplitude is given generally by

$$\int \mathcal{D}\psi \exp \frac{i}{\hbar} A \tag{2.31}$$

where the action A is of the generic form

$$A = \int \mathcal{D}x^\mu(\sigma) f \;\psi[x^\mu(\sigma)] \tag{2.32}$$

+ gauge fixing + ghost + source terms.

Up until recently, the second-quantization of strings has either been performed with constraints being present or carried out in a given gauge, such as the light-cone gauge [8], where the constraints have been solved. Quantization using BRS techniques of the linearized theory in a given gauge has been discussed in reference [9]. It is possible that one can, in principle, discover all the properties of a theory by quantizing in a given gauge. However, without the ability to use all the wisdom acquired with second-quantization, this may be difficult. Also, the whole subject of the non-perturbative semi-classical phenomena and spontaneous symmetry breaking has up to now only been developed in the gauge-invariant framework. It is also possible that the finiteness properties of strings may become particularly apparent in a covariant formulation, as they did in the case of supersymmetric theories.

Another advantage of obtaining a second-quantized field theory of strings is that it will help us to understand what strings are. One of the remarkable developments of modern physics is that the theories relevant to nature are almost entirely determined by symmetries. For example, a theory possessing local gauge invariance realized on a vector potentials A_μ and which must be no more than second order **in derivatives in the** action, can only be on the Yang-Mills action. Similarly, Einstein's action is determined uniquely by general co-ordinate transformations realized on the metric and supergravity is controlled by local supersymmetry. The string theories are more or less unique up to distinctions about being open or closed, supersymmetric or bosonic and they also contain the local symmetries mentioned above. It is natural to suppose that the string is also completely determined by a symmetry. A knowledge of this symmetry would explain the many wondrous cancellations found in string theory, as well as, hopefully, lead to many more. Clearly, possessing an invariant action under a set of transforming fields whose algebra is known would make it much easier to guess the principle which underlies this symmetry, and hence string theory itself.

2.3 Oscillator Formalism

In order to analyze the Virasoro conditions of equation (2.28), it is useful to re-express the quantities discussed above in terms of creation and annihilation oscillators. We may write

$$x^\mu(\sigma) = \sum_{n=-\infty}^{\infty} x_n^\mu \, e^{in\sigma} \tag{2.33}$$

where $x_0^\mu = x^\mu$ and $x_{-n}^\mu = (x_n^\mu) = x_n^\mu$ as a result of equation (2.19) and the reality of $x^\mu(\sigma)$. The form of the above expansion for $x^\mu(\sigma)$ is such as to obey the boundary conditions for the open string

$$\frac{\partial x^\mu}{\partial \sigma}\bigg|_{\sigma=0} = \frac{\partial x^\mu}{\partial \sigma}\bigg|_{\sigma=\pi} = 0 \tag{2.34}$$

We may use the chain rule to rewrite the functional derivatives:-

$$\frac{\delta}{\delta x^\mu(\sigma)} = \sum_{n=-\infty}^{\infty} \frac{\partial x_n^\nu}{\partial x^\mu(\sigma)} \frac{\partial}{\partial x_n^\nu} = \frac{1}{2\pi} \sum_{n=-\infty}^{\infty} e^{in\sigma} \frac{\partial}{\partial x_n^\mu} \tag{2.35}$$

Let us define

$$\hat{P}^\mu(\sigma) = -\frac{1}{\sqrt{2\alpha'}\,\pi} \sum_{n=-\infty}^{\infty} \alpha_n^\mu \, e^{in\sigma} \tag{2.36}$$

Using equation (2.33) we find that

$$\alpha_n^\mu = i\left(\left(\frac{\alpha'}{2}\right)^{\frac{1}{2}} \frac{\partial}{\partial x_n^\mu} + \frac{n}{\sqrt{2\alpha'}} x_n^\mu\right). \tag{2.37}$$

From the reality of $p^\mu(\sigma)$ we find that $\alpha_{-n}^\mu = a_n^{\mu +}$. The α's commutation relations are

$$[\alpha_n^\mu, \alpha_m^\nu] = 0 \; ; \; [\alpha_n^\mu, \alpha_m^{\nu +}] = n\,\delta_{n,m}\,\eta^{\mu\nu} \tag{2.38}$$

for n,m > 1. The Virasoro operators [6] can be expressed in terms of the α_n's by

$$L_n = \frac{1}{2} : \sum_{m=-\infty}^{\infty} \alpha^\mu_m \alpha^\nu_{n-m} \eta_{\mu\nu} : \qquad (2.39)$$

where L_0 is understood to be normal ordered. They obey the modified algebra

$$[L_n , L_m] = (n-m) L_{n+m} + \frac{D}{12} n(n^2-1) \delta_{n,-m} \qquad (2.40)$$

The latter term is called the central term and it can most easily be computed by taking vacuum expectation values of this equation. We find in particular that

$$L_0 = \frac{1}{2} \alpha^\mu_0 \alpha_{0\mu} + \sum_{m=1}^{\infty} \alpha^{\mu+}_m \alpha_{m,\mu} \qquad (2.41)$$

$$L_1 = L^+_{-1} = \alpha^\mu_0 \alpha_{1\mu} + \sum_{m=1}^{\infty} \alpha^{\mu+}_m \alpha_{m+1,\mu} \qquad (2.42)$$

where

$$\alpha^\mu_0 = i \sqrt{\frac{\alpha'}{2}} \frac{\partial}{\partial x^\mu}$$

We may write the state of the string in occupation number basis in terms of the creation operators $\alpha^{\mu+}_n$ by

$$\psi[x^\mu(\sigma)] = \{ \phi(x) + i A^1_\mu(x) \alpha_1^{\mu+} + i A^2_\mu(x) \alpha^{\mu+}_2 + h_{\mu\nu} \alpha^{\mu+}_1 \alpha^{\nu+}_1 + \cdots \} \langle x^\mu(\sigma) | 0 \rangle \qquad (2.43)$$

The vacuum satisfies the equation

$$\alpha^\mu_n \langle x^\mu(\sigma) | 0 \rangle = 0 \qquad n \geq 1 \qquad (2.44)$$

The vacuum of equation (2.44) is of the form

$$\langle x^\mu(\sigma) | 0 \rangle = \prod_{n=1}^{\infty} c_n \exp\left(-\frac{n}{2\alpha'} x^\mu_n x_{n\mu}\right) \qquad (2.45)$$

The action of the $\alpha_n^{\mu +}$ on $\langle x^\mu(\sigma)|0\rangle$ produces the well-known complete set of Hermite polynomials. In terms of component fields, we find (2.28) has as consequences

$$\left(\partial^2+\frac{4}{\alpha'}\right)\phi(x) = 0 = \left(\partial^2-(\ell-1)\frac{4}{\alpha'}\right)A_\mu^\ell \qquad (2.46)$$
$$= \left(\partial^2-\frac{4}{\alpha'}\right)h_{\mu\nu} = \cdot\cdot\cdot\cdot\cdot$$

as well as

$$\partial^\mu A_\mu^1 = 0 = \sqrt{2}\,\partial^\mu A_\mu^2 + h_\nu^{\ \nu} \qquad (2.47)$$
$$= \frac{1}{\sqrt{2}}\partial^\mu h_{\mu\nu} + A_\nu^2 = \cdot\cdot\cdot\cdot$$

The theorem of reference [7] is remarkable in the sense that the presence of a_n^{0+} leads to many ghost states which are all excluded by the Virasoro constraints of equation (2.28).

2.4 The Gauge Covariant Theory at Low Levels

We now wish to find an action which is built from unconstrained fields and instead possesses gauge invariances that allow the constraints of equation (2.28) to arise as a gauge choice upon the equation of motion. Consequently, we expect an infinite number of gauge invariances. This can be achieved mass level by mass level by successively releasing the constraints on Ψ.

At the first level, we release the constraint $L_1\Psi = 0$ but Ψ is still subject to

$$0 = L_2\Psi = L_1^2\Psi = L_3\Psi = L_2L_1\Psi \ldots \qquad (2.48)$$

Consider now the transformation of Ψ

$$\delta\Psi = L_{-1}\Lambda^1 \qquad (2.49)$$

where Λ^1 is subject to

$$0 = L_1 \Lambda^1 = L_2 \Lambda^1 = \cdots \tag{2.50}$$

corresponding to the constraints of equation (2.48) on Ψ. Using the form of L_{-1} given in equation (2.42), we find that this invariance contains the transformation $\delta A^1 = \partial_\mu \Lambda_1(x)$ which is the Abelian transformation expected for a linearized Yang-Mills theory.

An action invariant under $\delta\Psi = L_{-1}\Lambda_1$ is given by

$$\frac{1}{2}\left(\Psi, \left(L_0 - 1 - \frac{1}{2} L_{-1} L_1\right)\Psi\right) \tag{2.51}$$

The equation of motion is given by

$$\left(L_0 - 1 - \frac{1}{2} L_{-1} L_1\right)\Psi = 0 \tag{2.52}$$

We note that this equation at the first level is easily seen to contain the equation $\partial^2 A_\mu - \partial_\mu \partial^\nu A_\nu = 0$. Explicitly testing the invariance, we find that

$$0 = \left(L_0 - 1 - \frac{1}{2} L_{-1} L_1\right)\Lambda^1 = \left(L_0 L_{-1} - L_{-1} - L_{-1} L_0\right)\Lambda_1 \tag{2.53}$$

since $L_1 \Lambda_1 = 0$.

Equation (2.52) was probably known to a few people in the old heyday of string theory, but has been rediscovered more recently [10],[11].

The projector P of a string field onto the physical states of equation (2.28) was given in reference [12]. The projector is an object that has the property $PL_{-n} = 0$ and at lowest order it is given by

$$P = 1 - \frac{1}{2} L_{-1} \frac{1}{L_0} L_1 + \cdots \tag{2.54}$$

We note that <u>at this level</u> the equation of motion is given by

$$(L_0 - 1)\, P\psi = 0 \tag{2.55}$$

Equation (2.55) is not the correct equation of motion of all levels of the string, despite the above coincidence for the spin one at the first level. The projector, P, has been formally computed for all levels, and (2.55) is explicitly non-local at the second level and more and more so for higher levels. One's suspicions are further aroused by the fact that P can be constructed in an arbitrary space-time dimension and that D = 26 is not particularly favoured.

In fact, a gauge-covariant theory of strings cannot be built with ψ alone. Let us momentarily consider the closed string, repeating the above procedure, but now with the boundary conditions $x^\mu(\sigma) = x^\mu(2\pi)$ where σ takes values from 0 to 2π; we find twice as many oscillators ($\alpha^\mu, \tilde{\alpha}^\mu$) and so twice as many Virasoro generators (L_n and $\tilde{L}_n$). The Virasoro conditions are

$$\begin{aligned}(L_0 + \tilde{L}_0 - 2)\,\psi &= 0\\ L_n\psi = \tilde{L}_n\psi &= 0 \qquad n \geq 1\end{aligned} \tag{2.56}$$

and ψ is taken to be subject to $(L_0 - \tilde{L}_0)\psi = 0$. At the first level these constraints imply that

$$\partial^2 h_{\mu\nu} = \partial^\mu h_{\mu\nu} = 0 \tag{2.57}$$

where $h_{\mu\nu} = h_{\nu\mu}$ is the only field contained in ψ at this level. The on-shell states described by these equations are one massless spin 2 and one massless spin 0 (i.e. $h^\mu_\mu = 0$) at this level.

Writing down the most general equation of the form $\partial\partial h = 0$ one soon discovers [13] that it can never lead to $\partial^\nu h_{\nu\mu} = 0$ and $\partial^2 h_{\mu\nu} = 0$.

The solution to this problem is well known since this system is just that of Einstein plus a spin 0. We describe the gravity by $h_{\mu\nu}$ and the spin 0 by an additional scalar

ϕ. Since we are dealing with strings we must work with spin functionals, so ϕ is the first component of a string functional $\phi^1[x^u(\sigma)]$:

$$\phi^1[x^\mu(\sigma)] = [\phi(x) + \cdots \quad]\langle x^\mu(\sigma)|0\rangle$$

In fact, quite generally, the equations for spin 0, $\frac{1}{2}$ and 1 are exceptional in that they do not require supplementary fields. The equations above spin 1 do. This is reflected in the fact that projectors of higher spin equations are not rendered local by an application of ∂^2. For example, the closed string projector for the first level states is given by

$$R_\mu{}^\rho R_\nu{}^\lambda h_{\rho\lambda} = 0$$

where (2.58)

$$R_\mu{}^\rho = (\delta_\mu{}^\rho - \partial_\mu \partial^\rho / \partial^2)$$

We observe it contains a $\partial_\mu \partial^\kappa \partial_\nu \partial^\lambda / (\partial^2)^2$ term and so multiplication by ∂^2 does not render it local.

Let us now examine how many supplementary fields are required at the second level of the bosonic open string [14]. In order to count the number of on-shell states at this level we could examine the Virasoro contraints at this level. The Virasoro constraints, however, possess on-shell gauge invariance that must be chosen before the on-shell states become apparent. This is clear even at the first level for which the conditions are $\partial^2 A_\mu = 0$ and $\partial^\mu A_\mu = 0$. As is well known, it is the additional on-shell invariance $\delta A_\mu = \partial_\mu \Lambda$ where $\partial^2 \Lambda = 0$ which allows the reduction to two rather than three on-shell states. A much faster method is to use the fact that the constraints are solved in the light-cone gauge and so one only has the on-shell states corresponding to oscillators α_n^i.i = 1 to 24. In this case Ψ has the expansion

$$\{ \phi + i\alpha_1^{i\dagger} A_i^1 + i\alpha_2^{i\dagger} A_i^2 + h_{ij}\alpha_1^{i\dagger}\alpha_1^{j\dagger} \qquad (2.59)$$
$$+ \cdots \} \langle x^i(\sigma)|0\rangle$$

At the first level we have (D-2) states corresponding to the (D-2) states contained in the vector representation of SO(D-2), which is the little group for massless particles in D dimension. Since the massive states must belong to representations of the little group SO(D-1), this demonstrates that the normal-ordering constant in equation (2.28) was chosen correctly. Any other choice would violate Lorentz invariance in D dimensions since (D-2) states can not carry a representation of SO(D-1).

At the second level we have

$$\frac{1}{2}(D-2)(D-1) + (D-2) = \frac{D(D-1)}{2} - 1 \qquad (2.60)$$

on-shell states. These can only be identified with the second-rank traceless symmetric representation of SO(D-1). We shall refer to this as "pure spin two".

"Pure spin two" is described on-shell by the field $h_{\mu\nu} = h_{\nu\mu}$ subject to the equations

$$\partial^\mu h_{\mu\nu} = h_\mu{}^\mu = (\partial^2 - m^2) h_{\mu\nu} \qquad (2.61)$$

The projector is well known. It involves terms of the form $\partial_\mu\partial_\lambda\partial_\rho\partial_\sigma/(\partial^2)^2$ and hence multiplication (∂^2) does not lead to a local field equation. In fact, there is no way to describe in a Lorentz-covariant way only a massive spin-two particle in terms of only $h_{\mu\nu} = h_{\nu\mu}$ subject to $h_\mu^\mu = 0$. The correct equations of motion involve the introduction of a supplementary field ϕ and are given by

$$(-\partial^2 + m^2) h_{\mu\nu} + (\partial_\mu \partial^\rho h_{\rho\nu} + \partial_\nu \partial^\rho h_{\rho\mu})$$
$$- \frac{2}{D}\eta_{\mu\nu}\partial^\rho\partial^\lambda h_{\rho\lambda} = \frac{D-2}{D-1}\left(\partial_\mu\partial_\nu\phi - \frac{\eta_{\mu\nu}}{D}\partial^2\phi\right)$$

$$\partial^\mu \partial^\nu h_{\mu\nu} = (\partial^2 - \frac{D}{D-2} m^2)\phi \tag{2.62}$$

where D is the dimension of space-time.

Indeed, these coupled equations lead to the desired result, namely

$$0 = \partial^\mu h_{\mu\nu} = (\partial^2 - m^2) h_{\mu\nu} \tag{2.63}$$

The above equations illustrate a more general method [15] of introducing additional fields in order to propagate higher spin fields. At the second level the string contains the fields $h_{\mu\nu}$ and A^2_μ, however A^2_μ is gauge away leaving $h_{\mu\nu}$ traceless and so from the above discussion we require one extra supplementary field ϕ to implement the massive spin-two field equation. This field ϕ will be the lowest component of a supplementary string field

$$\chi^2[x^\mu(\sigma)] = [\phi(x) + \cdots \quad]\langle x^\mu(\sigma)|0\rangle \tag{2.64}$$

We will now find the gauge-covariant action at the next level [14]. Starting from the action of equation (2.51) and releasing the constraints of equation (2.48) of the first level, we subject ψ and χ to

$$\begin{aligned} L_3\psi &= L_2 L_1 \psi = L_1^3 \psi = 0 = \cdots \\ L_1 \chi^2 &= L_2 \chi^2 = 0 = \cdots \end{aligned} \tag{2.65}$$

The most general expression of the correct order is

$$(L_0 - 1 - \tfrac{1}{2} L_{-1} L_1 - \tfrac{1}{4}\gamma L_{-2} L_2)\psi + (L_{-1}^2 + \tfrac{3}{2}\beta L_{-2})\chi^2 = 0 \tag{2.66}$$

$$(L_1^2 + \tfrac{3}{2}\beta L_2)\psi = (aL_0+b)\chi^{(2)} \tag{2.67}$$

The use of the same β in (2.65) and (2.66) is required by demanding that the equations of motion follow from an action. An alternative way of searching for a gauge invariance is to demand that L_1 on equation (2.66) should vanish when we use equation (2.67). Carrying this out and using constraints of equation (2.65) we find

$$\begin{aligned} -\tfrac{1}{2} L_{-1}(L_1^2 + \tfrac{3}{2}\gamma L_2)\psi &+ (4L_{-1}L_0 + 2L_{-1})\chi^{(2)} \\ &+ \tfrac{9}{2}\beta L_{-1}\chi^{(2)} = 0 \end{aligned} \tag{2.68}$$

To eliminate ψ by equation (2.67) requires $\gamma = \beta$ and we find

$$-\tfrac{1}{2} L_{-1}(aL_0+b)\chi^{(2)} + L_{-1}(4L_0 + 2 + \tfrac{9}{2}\beta)\chi^{(2)} = 0 \tag{2.69}$$

Hence we conclude that $a = 8$ and $b = 4+9\beta$. Applying L_2 in a similar way fixes $\beta = 1$, and we find the equations of motion

$$\begin{aligned} &(L_0 - 1 - \tfrac{1}{2}\sum_{n=1}^{2} \tfrac{1}{n} L_{-n}L_n)\psi + (L_{-1}^2 + \tfrac{3}{2}L_{-2})\chi^{(2)} = 0 \\ &(L_1^2 + \tfrac{3}{2}L_2)\psi = (8L_0 + 13)\chi^{(2)}. \end{aligned} \tag{2.70}$$

This system of equations is in fact invariant under the gauge transformations

$$\begin{aligned} &\delta_1\psi = L_{-1}\Lambda^1, \quad \delta_1\chi^{(2)} = \tfrac{1}{2}L_1\Lambda^1, \quad \delta_2\psi = L_{-2}\Lambda^2 \\ &\delta_2\chi^{(2)} = \tfrac{3}{2}\Lambda^2 \end{aligned} \tag{2.71}$$

with

$$L_2\Lambda^1 = L_1^2\Lambda^1 = 0 = \ldots$$

$$L_1 \Lambda^2 = L_2 \Lambda^2 = 0 = \dots \tag{2.72}$$

We stress that equations (2.70) are Λ^2 invariant only for D = 26. It may be possible with the introduction of further supplementary fields, to relax this condition. The corresponding action is given by

$$\frac{1}{2}\left(\Psi,\left(L_0-1-\frac{1}{2}L_{-1}L_1-\frac{1}{4}L_{-2}L_2\right)\Psi\right)+\left(\Psi,\left(L_{-1}^2+\tfrac{3}{2}L_{-2}\right)\chi^{(2)}\right)$$
$$-\frac{1}{2}\left(\chi^{(2)},(8L_0+13)\chi^{(2)}\right) \tag{2.73}$$

This completes the second level.

Before constructing the action to all orders, it will be instructive to rewrite the second-level result in a kind of first-order form. Completing the squares in the $L_1\Psi$ and $L_2\Psi$ terms in the action, we may rewrite equation (2.73) as

$$\frac{1}{2}\left(\Psi,(L_0-1)\Psi\right)-\frac{1}{4}\left(L_1\Psi-2L_{-1}\chi^{(2)},L_1-2L_{-1}\chi^{(2)}\right)$$
$$-\frac{1}{8}\left(L_2\Psi-6\chi^{(2)},L_2\Psi-6\chi^{(2)}\right)-2\left(\chi^{(2)},(L_0+1)\chi^{(2)}\right) \tag{2.74}$$

where we have used the fact that at this level $L_1\chi^{(2)} = 0$. If we now introduce the auxiliary fields $\phi^{(1)}$ and $\phi^{(2)}$ we may rewrite the action as

$$\frac{1}{2}\left(\Psi,(L_0-1)\Psi\right)+\left(\phi^1, L_1\Psi+L_{-1}\xi^1_1\right)$$
$$+\left(\phi^2, L_2\Psi+3\xi^1_1\right)+\left(\phi^1,\phi^1\right)$$
$$+2\left(\phi^2,\phi^2\right)-\frac{1}{2}\left(\xi^1_1,(L_0+1)\xi^1_1\right) \tag{2.75}$$

where $\xi^1{}_1 = -\frac{1}{2}\chi^{(2)}$.

The construction up to the sixth level involving the systematic introduction of supplementary fields was given in [14].

2.5 The Finite Set

We now find an action which is gauge-invariant to all levels. We require the gauge invariance

$$\delta\Psi = L_{-1}\Lambda^1 + L_{-2}\Lambda^2 \tag{2.76}$$

This was seen to be a consequence of our lowest-order construction, but it can also be seen from the on-shell gauge invariances, which now become off-shell, of the Virasoro constraints or from the associated projector. We note that since $L_{-3} = -[L_{-2},L_{-1}]$ we will also have the gauge invariance $\delta\Psi = L_{-3}\Lambda_3$ and by an extension $\delta\Psi = L_{-n}\Lambda_n$ for any $n \geq 1$.

Given an action A invariant under (2.75) we can deduce that

$$0 = \left(\Lambda^1, L_1 \frac{\delta A}{\delta\Psi}\right) + \sum_i \left(\delta\phi^i, \frac{\delta A}{\delta\phi^i}\right) \tag{2.77}$$

The ϕ^i represent all the other string functionals that occur in A. If we enforce all the field equations except that of Ψ we find that

$$L_1 \frac{\delta A}{\delta\Psi} = 0 \tag{2.78}$$

and by a similar argument $L_2 \frac{\delta A}{\delta\Psi} = 0$. The Ψ equation must be of the form $(L_0-1)\Psi + \ldots = 0$. Applying L_1 and L_2 we require a knowledge of $L_1\Psi$ and $L_2\Psi$ in order to get zero. This requires the introduction of two fields ϕ^1 and ϕ^2 in order to specify $L_1\Psi$ and $L_2\Psi$. As such, we require a term $(\phi^1,L_1\Psi) + (\phi^2,L_2\Psi)$ in the action. Consequently ϕ^1 and ϕ^2 must occur in the Ψ equation as

$$(L_0-1)\Psi + L_{-1}\phi^1 + L_{-2}\phi^2 = 0 \tag{2.79}$$

Having introduced these new terms we must now specify $L_1\phi^1$, $L_2\phi^1$ and $L_1\phi^2$ and $L_2\phi^2$ which are determined by the field equations of the new fields $\zeta^1{}_1$, $\zeta^2{}_1$, $\zeta^1{}_2$ and $\zeta^2{}_2$ respectively. We now write these field equations down and add in, with arbitrary coefficients, any other possible terms of the correct level. For example in the ϕ^2 equation which begins $L_2\psi$ we may add a term $c\zeta^1{}_1$, which is also of level 2. We then determine the arbitrary coefficients by applying L_1 and L_2 on $\frac{\delta A}{\delta\psi}$ and demanding zero. We find the equations of reference [16]. These field equations are:

field	field equation
ψ	$(L_0-1)\psi + L_{-1}\phi^1 + L_{-2}\phi^2 = 0$
ϕ^1	$L_1\psi = -2\phi^1 - L_{-1}\zeta^1{}_1 - L_{-2}\zeta^2{}_1$
ϕ^2	$L_2\psi = -4\phi^2 - L_{-1}\zeta^1{}_2 - L_{-2}\zeta^2{}_2 - 3\zeta^1{}_1$
$\zeta^1{}_1$	$L_1\phi^1 = (L_0+1)\zeta^1{}_1 - 3\phi^2$
$\zeta^2{}_1$	$L_2\phi^1 = (L_0+2)\zeta^1{}_2$
$\zeta^1{}_2$	$L_1\phi^2 = (L_0+2)\zeta^2{}_1$
$\zeta^2{}_2$	$L_2\phi^2 = (L_0+3)\zeta^2{}_2$

(2.80)

We leave the application of L_1 to the reader and consider the application of L_2 in more detail: we find that

$$(L_0+1)L_2\psi + 3L_1\phi^1 + L_{-1}L_2\phi^1 + \left(4L_0 + \frac{D}{2}\right)\phi^2 + L_{-2}L_2\phi^2 = 0 \qquad (2.81)$$

Substituting for $L_2\psi$, $L_1\phi^{(1)}$, $L_2\phi^{(1)}$ and $L_2\phi^{(2)}$, we find that

$$(D-26)\phi^2 = 0 \qquad (2.82)$$

Consequently, we discover that this system only exists in the critical dimension D = 26.

Given the field equations, we can search for the full gauge invariance. These are entirely determined from $\delta\psi = L_{-1}\Lambda^1 + L_{-2}\Lambda^2$ and one may easily check that they are given by reference [16].

$$\delta\psi = \sum_{n=1}^{2} L_{-n}\Lambda^n \quad ; \quad \delta\phi^n = -(L_0 + n - 1)\Lambda^n \qquad (2.83)$$

$$\delta\zeta^n{}_m = -L_m \Lambda^n - (2m+n)\Lambda^{n+m} \qquad (2.84)$$

for $n, m = 1, 2$.

It is straightforward to write down an action from which the above equation follows:

$$\tfrac{1}{2}(\psi,(L_0-1)\psi) + \sum_{n=1}^{2}(\phi^n, L_n \psi) + \sum_{n,m=1}^{2}(L_n \phi^m, \zeta^n{}_m)$$
$$+ \sum_{n=1}^{2} n(\phi^n, \phi^n) - \tfrac{1}{2}\sum_{n,m=1}^{2}(\zeta^n{}_m, (L_0 + n + m - 1)\zeta^m{}_n) \qquad (2.85)$$
$$+ \sum_{n,m=1}^{2}(2n+m)(\phi^{n+m}, \zeta^n{}_m)$$

We will refer to this system as the "finite set".

In fact, one can find free-gauge-covariant formulation of all known strings in this way; the open and closed bosonic string, and the open and closed superstring theories and the hetoric string. We refer the reader to reference [16] for these other formulations.

2.6 The Infinite Set

One can extend the result for the open bosonic string given above so that it contains an infinite number of supplementary fields [17][18]. This will have the advantage that the generators of the Virasoro algebra will appear on a more equal footing. This is achieved by introducing the fields

$$\psi, \phi^n, \zeta^n{}_m \quad ; \quad n, m = 1, 2, \ldots \infty \quad (2.86)$$

and the action is the same as that of equation (2.85) except now all the sums run from 1 to ∞. Remarkably this action is invariant under the transformations of equation (2.83) except the sums also run from 1 to ∞. This requires the identity

$$-\sum_{m=1}^{n-1}(2n-m)(n+m) + \frac{D}{12}n(n^2-1) - 2n(n-1) = 0 \quad (2.87)$$

which only works if D = 26.

It will be useful to write out the equations of motion of this set for future use. They are

$$(L_0 - 1)\psi + \sum_{n=1}^{\infty} L_{-n}\phi^n = 0$$

$$L_n \psi + \sum_{m=1}^{\infty} L_{-m}\zeta^m{}_n + \sum_{n,m=1}^{\infty} \delta(m+p-n)(2p+m)\zeta^p{}_m \quad (2.88)$$

$$+ 2n\phi^n = 0$$

$$L_n \phi^m - (L_0 - n + m - 1)\zeta^m{}_n + (2n+m)\phi^{n+m} = 0$$

Of course, one must not only find a gauge-invariant action; one must also find an action which gives the correct count of on-shell states. One way to demonstrate this would be to show that after appropriate gauge choices we recover the Virasoro constraints. It is known that the "finite set" and the ∞ set lead to the correct count of states up to the tenth and sixth levels respectively [19].

Recently [20], some incomplete arguments have been advanced to suggest that they fail at these levels. Whether this is the case or not will emerge when the count of states is explicitly carried out at these levels.

Even at the third and sixth levels in the infinite and finite sets respectively one finds [19] the existence of complicated additional symmetries that are required for the counting of states and, for example, building the interacting theory.

We would, of course, like a formulation in which all the symmetries are manifest. This formulation is easily found by extending the ∞ set by the same process by which it was found. We require essentially that L_n vanish on all equations of motion. For the ϕ^n equation this requires a knowledge of $L_n\, \zeta^m{}_p$ and so we require a new field $\phi^{mn}{}_p$ which in turn requires a field $\psi^{mn}{}_{pq}$. Repeating the process indefinitely we find the so-called master theory, which has the field content

$$\psi^k_k \;, \quad \phi^{k+1}_k$$

where the index k indicates the number of indices. We will find that indices on a given level are completely symmetrized.

In fact, with some technology, the transition from the ∞ to the master set is very straightforward, and we will see how this is achieved later. At this point we will discuss some of the technology which enables a simple derivation of the master set.

One might hope to find the finite and infinite sets by a gauge choice from the master set. In this process one would find compensating transformations to preserve the choice of gauge. A preliminary examination of the complicated symmetries of the smaller set does indeed indicate that they are of this type. Later, we will show that the master set does have the correct on-shell count. Consequently, if one could recover the smaller sets through a gauge choice, this would show that the smaller sets also give the correct on-shell count.

2.7 The Master Set

The master set was found in reference [19] using the techniques outline in this article. It was also independently found in reference [21] by extending the gauge-covariant free theory given in references [18],[22],[23] and it was mentioned in reference [24].

It will prove useful to introduce an extension of space-time to include anticommuting co-ordinates. We consider the space which is parametrized by the co-ordinates $x^u(\sigma)$, $c(\sigma)$, $\bar{c}(\sigma)$. The co-ordinates $x^u(\sigma)$ are the usual 26 bosonic co-ordinates while $c(\sigma)$ and $\bar{c}(\sigma)$ are fermionic co-ordinates. We take these co-ordinates to satisfy the boundary conditions

$$\partial_\sigma x^\mu(\sigma) = \partial_\sigma c(\sigma) = \bar{c}(\sigma) \tag{2.89}$$

All vanish at $\sigma = 0$ and π. In fact, this 28-dimensional space emerges naturally in the context of string theory when one BRST-quantizes the Nambu action for the string [25],[26]. The fields $c(\sigma)$ and $\bar{c}(\sigma)$ correspond to the ghosts one must introduce when one fixes the two-dimensional reparametrization symmetry of the world sheet of the string. The reader should be clear that we are going to find a gauge-invariant second-quantized field theory of strings and not a BRST formulation. The latter can be found from the former by the usual method of gauge fixing and the introduction of corresponding ghosts.

However, it has emerged that many of the tools introduced in the first-quantized string action have a natural role in the gauge-invariant second-quantized theory.

While $x^u(\sigma)$ has the usual Fourier expansion, the Fourier expansions for c and $\bar{c}$ correponding to the above boundary conditions are

$$c(\sigma) = c^0 + 2\sum_{n=1}^{\infty} c_n \cos n\sigma$$

$$\bar{c}(\sigma) = 2\sum_{n=1}^{\infty} \bar{c}_n \sin n\sigma \qquad (2.90)$$

In analogy with $x^u(\sigma)$ we introduce fermionic annihilation and creation operators:-

$$\bar{\beta}(\sigma) = \frac{\delta}{\delta c(\sigma)} - \frac{1}{2\pi} c(\sigma) = \frac{1}{\pi}\frac{1}{\sqrt{2}}\sum_{n=-\infty}^{\infty} \bar{\beta}_n e^{in\sigma}$$

$$\beta(\sigma) = -\frac{\delta}{\delta c(\sigma)} + \frac{1}{2\pi} c(\sigma) = \frac{1}{\pi}\frac{1}{\sqrt{2}}\sum_{n=-\infty}^{\infty} \beta_n e^{in\sigma} \qquad (2.91)$$

Equipped with the obvious scalar product we find that

$$\bar{\beta}(\sigma)^{+} = \bar{\beta}(\sigma) \quad ; \quad \beta(\sigma)^{+} = \beta(\sigma). \qquad (2.92)$$

Note that unlike for $\frac{\delta}{\delta x^u(\sigma)}$ we do not require an i for hermiticity since we are dealing with anticommuting quantities and

$$\left(\frac{\delta}{\delta c(\sigma)}\right)^{+} = \frac{\delta}{\delta c(\sigma)}. \qquad (2.93)$$

As a result we find that

$$\bar{\beta}_n^{+} = \bar{\beta}_{-n} \quad ; \quad \beta_n^{+} = \beta_{-n} \qquad (2.94)$$

and in particular

$$\bar{\beta}_0^{\dagger} = \bar{\beta}_0 \quad ; \quad \beta_0^{\dagger} = \beta_0 \tag{2.95}$$

It is easily seen that these fermionic oscillators obey the relations

$$\begin{aligned} \{\bar{\beta}_n^{\dagger}, \beta_m\} &= \{\bar{\beta}_n, \beta_m^{\dagger}\} = \delta_{n,m} \\ \{\beta_n, \beta_m\} &= 0 = \{\bar{\beta}_n, \bar{\beta}_m\} \\ \{\beta_n, \beta_m^{\dagger}\} &= 0 = \{\bar{\beta}_n, \bar{\beta}_m^{\dagger}\} \\ & n, m \geq 0 \end{aligned} \tag{2.96}$$

We now define a vacuum with respect to these oscillators. Clearly we can take

$$\beta_n |\rangle = \bar{\beta}_n |\rangle = 0 \qquad n \geq 1 \tag{2.97}$$

as well as the usual condition for the bosonic α_n^u oscillators. The action of the zero modes on the vacuum, however, require more care [25]. We can define a vacua $|+\rangle$ by

$$\beta_0 |+\rangle = 0 \tag{2.98}$$

then under $\bar{\beta}_0$ we find a new vacuum

$$\bar{\beta}_0 |+\rangle = |-\rangle \tag{2.99}$$

Since $\bar{\beta}_0^2 = 0$, we find that $\bar{\beta}_0 |-\rangle = 0$. From the relation $\{\beta_0, \bar{\beta}_0\} = 1$ we also find $\beta_0 |-\rangle = |+\rangle$. We note that

$$\langle + | + \rangle = \langle - | \beta_0 \beta_0 | - \rangle = 0 \tag{2.100}$$

and similarly for $\langle -|-\rangle = 0$. However, we have the relations

$$\langle +|-\rangle = \langle -|\beta_0 \bar{\beta}_0|+\rangle = \langle -|+\rangle \tag{2.101}$$

and we choose $\langle +|-\rangle = 1$. We take that $|-\rangle$ vacuum to be odd and so the $|+\rangle$ vacuum is even as β_0 is an odd object.

Let us consider the most general functional χ of $x^u(\sigma)$, $c(\sigma)$, $\bar{c}(\sigma)$. In the oscillator basis it may be written as

$$\begin{aligned} |\chi\rangle &\equiv \psi\,|-\rangle + \varphi\,|+\rangle \equiv \\ &\sum_{\{n\}\{m\}} \beta^{n_1\dagger}\cdots\beta^{n_b\dagger}\,\bar{\beta}_{m_1}^{\dagger}\cdots\bar{\beta}_{m_a}^{\dagger}\,\psi_{n_1\ldots n_b}{}^{m_1\ldots m_a}[x^{\mu}(\sigma)]\,|-\rangle \\ &+\sum_{\{n\}\{m\}} \beta^{n_1\dagger}\cdots\beta^{n_b\dagger}\,\bar{\beta}_{m_1}^{\dagger}\cdots\bar{\beta}_{m_{a+1}}^{\dagger}\,\varphi_{n_1\ldots n_b}{}^{m_1\ldots m_{a+1}}[x^{\mu}(\sigma)]\,|+\rangle \end{aligned} \tag{2.102}$$

In fact, these are the string functionals that occur in covariant string field theory. The ghost co-ordinates, as we shall see, correctly encode the supplementary fields. This is similar to the superspace of sypersymmetric theories which enables one to encode in a systematic way the component fields of a supermultiplet.

We note that $\psi_{n_1\ldots n_b}{}^{m_1\ldots m_a} = \psi_{[n_1\ldots n_b]}{}^{[m_1\ldots m_a]}$ and that if $a + b$ is an odd integer then ψ^a_b is an anti-commuting field. Since we want to describe a gauge invariant string field theory and not a BRST invariant theory which has ghost fields, we require χ to be subject to a constraint. Note that the previously discussed master theory had a field content ψ^k_k , ϕ^{k+1}_k. This will be the content of $|\chi\rangle$ if we impose the constraint

$$\sum_{n=1}^{\infty} (\beta^{m\dagger}\bar{\beta}_m - \bar{\beta}^{\dagger}_m \beta^m + \beta_0\bar{\beta}_0)\,|\chi\rangle = 0$$

$$\text{or} \quad N|\chi\rangle = 0 \tag{2.103}$$

where N is defined to be the operator in the first equation. We now adopt this equation. Note that, due to the fermionic assignments of the vacuum, if $|\ \rangle$ is odd then the component fields ψ_k^k and ϕ_k^{k+1} are even. An extremely useful operator is the Virasoro charge which is constructed as follows. The Virasoro generators L_n classically obey the algebra $\{L_n, L_m\} = -i(n-m)L_{n+m} = -if_{nm}{}^p L_p$. The corresponding BRST charge is

$$Q' = \sum_{n=-\infty}^{\infty} \beta_{-n} L_n - \frac{1}{2} \sum_{n,m,p=-\infty}^{\infty} \bar{\beta}_p f_{nm}{}^p \beta_{-n} \beta_{-m} \tag{2.104}$$

This type of construction applied to first-quantized Hamiltonian theories can be found in reference [27]. The above object is not well defined unless it is normal ordered; in which case one may find a normal ordering constant. Hence we consider the object

$$Q = :\sum_{n=-\infty}^{\infty} \beta_{-n} L_n - \frac{1}{2} \sum_{n,m,p=-\infty}^{\infty} \bar{\beta}_p f_{nm}{}^p \beta_{-n} \beta_{-m} - a\beta_0 : \tag{2.105}$$

Note that for anticommuting quantities we must assign minus signs when normal ordering; for example

$$:\beta_n \bar{\beta}_{-m}: = -\bar{\beta}_{-m} \beta_n \quad \text{for } n,m \geq 1 \tag{2.106}$$

From its definition almost, one easily finds that $Q'^2 = 0$. However, the well-defined charge Q only satisfies $Q^2 = 0$ provided D = 26 and the intercept is 1 (i.e. a=1) [25].

We may rewrite Q in various ways that we will shortly need

$$Q = \beta_o K - 2\bar{\beta}_o M + d + D \tag{2.107}$$

where K, M, d and D do not involve any zero modes, i.e., any β_o and $\bar{\beta}_o$. We find that

$$K = L_o - 1 + \sum_{n=1}^{\infty} (n\beta_n^+ \beta_n + n\bar{\beta}_n^+ \beta_n)$$

$$M = -\sum_{n=1}^{\infty} n\beta_n^+ \beta_n$$

$$d = \beta^{n+} (L_n + f_{m,-n}{}^{p} \bar{\beta}_p^+ \beta^m + \tfrac{1}{2} \beta^{m+} f_{mn}{}^{p} \bar{\beta}_p)$$

$$D = \beta^n (L_{-n} + f_{m,-n}{}^{p} \beta^{m+} \bar{\beta}_p + \tfrac{1}{2} \bar{\beta}_p^+ f_{mn}{}^{p} \beta^m$$

(2.108)

all sums implied with n, m, p $\geqslant$ 1

We note that $D = d^\dagger$ and they satisfy the relations

$$d^2 = D^2 = 0 \; ; [K, d] = 0 = [K, D] = [K, M]$$

$$[M, d] = [M, D] = 0 \tag{2.109}$$

$$\{d, D\} - 2MK = 0 \tag{2.110}$$

as a consequence of which

$$Q^2 = \{d, D\} - 2MK = 0 \tag{2.111}$$

Let us now study the action of Q on an arbitrary functional.

If $|\chi\rangle$ is odd we find that

$$Q|\chi\rangle = (2M\phi + (d+D)\psi)|-\rangle + (K\psi + (d+D)\phi)|+\rangle \tag{2.112}$$

while if $|\chi\rangle$ is even

$$\varphi|\chi\rangle = (-2M\phi + (d+D)\psi)|-\rangle + (-K\psi + (d+D)\phi)|+\rangle. \tag{2.113}$$

where K,M, d and D are now defined by acting on $\psi_{n_1\ldots n_b}^{m_1\ldots m_a}$.

$$(K\psi)_{n_1\ldots n_b}^{m_1\ldots m_a} = (L_0 - 1 + m + n)\,\psi_{n_1\ldots n_b}^{m_1\ldots m_a}$$

$$(M\psi)_{n_1\ldots n_{b+1}}^{m_1\ldots m_{a-1}} = (-1)^b a\,\eta_{p[n_1}\psi^{p\,m_1\ldots m_{a-1}}_{n_2\ldots n_{b+1}]}$$

$$(d\psi)_{n_1\ldots n_{b+1}}^{m_1\ldots m_a} = L_{[n_1}\psi^{m_1\ldots m_a}_{n_2\ldots n_{b+1}]} + a f^{[m_1}_{p[n_1}\psi^{p\,m_2\ldots m_a]}_{n_2\ldots n_{b+1}]} - \frac{1}{2} b\, f^{\;\;p}_{[n_1 n_2}\psi_p{}^{m_1\ldots m_a}_{n_3\ldots n_{b+1}]}$$

$$D\psi_{n_1\ldots n_b}^{m_1\ldots m_{a-1}} = (-1)^b a\,[L_{-p}\psi^{p\,m_1\ldots m_{a-1}}_{n_1\ldots n_b} + b\, f^{\;\;q}_{[n_1,-p}\psi^{p\,m_1\ldots m_{a-1}}_{q\,n_2\ldots n_b]} - \frac{1}{2}(a-1) f^{m_1}_{k\ell}\psi^{k\ell m_2\ldots m_{a-1}}_{n_1\ldots n_b}$$

where $m = \sum_{i=1}^{a} m_i$, $n = \sum_{i=1}^{b} n_j$. These definitions differ from the definitions given in the literature by trivial factors. The difference in signs between $|\chi\rangle$ being even and odd comes from pushing the zero modes through the oscillators and fields onto the vacuum.

For example, we find that if

$$\psi^0_0 = \psi \;;\quad \phi^1_0 = \phi^n$$

$$(d\psi^0_0) = L_n\psi \quad ,(D\phi^1_0) = \sum_{n=-\infty}^{\infty} L_{-n}\phi^n$$

$$(d\phi^1_0) = L_m\,\phi^n + (2m+n)\phi^{n+m} \tag{2.114}$$

$$(M\phi^1_0) = n\,\phi^n \quad ; \quad (K\phi^1_0) = (L_0 - 1 + n)\phi^n$$

We now realize that the above operators are particularly suited to a discussion of the equations of motion (2.88) of the infinite set. These equations may be rewritten in the form

$$\begin{aligned} K\psi^0_0 + D\,\phi^1_0 &= 0 \\ d\,\psi^0_0 + D\,\psi^1_1 + 2m\,\phi^1_0 &= 0 \\ K\psi^1_1 + d\,\phi^1_0 &= 0 \end{aligned} \tag{2.115}$$

where $\psi^1_1 = \{-\zeta^n{}_m : n,m = 1,2,\ldots,\infty\}$ and the gauge invariances are

$$\begin{aligned} \delta\psi^0_0 &= D\,\Lambda^1{}_0 \quad ; \quad \delta\phi^1{}_0 = -K\Lambda^1{}_0 \\ \delta\psi^1_1 &= d\,\Lambda^1{}_0 \end{aligned} \tag{2.116}$$

The invariance of equations (2.115) under the gauge invariances of equation (2.116) is now easily shown. The generalization of the infinite set to the master set is obvious. The general equations are [19][21]

$$\begin{aligned} K\psi^a_a + D\phi^{a+1}_a + d\,\phi^a_{a-1} &= 0 \\ d\,\psi^a_a + D\,\psi^{a+1}_{a+1} + 2m\,\phi^{a+1}_a &= 0 \end{aligned} \tag{2.117}$$

while the gauge invariances are

$$\delta\psi^a_a = d\,\Lambda^a_{a-1} + D\,\Lambda^{a+1}_a - 2m\,\Omega^{a+1}_{a-1}$$

$$\delta\phi_a^{a+1} = -\kappa\Lambda_a^{a+1} + d\Omega_{a-1}^{a+1} + D\Omega_a^{a+2} \tag{2.118}$$

where Ω is a new symmetry. Some traces of this symmetry can be found in the infinite set, although due to gauge compensation it occurs in a rather complicated form.

The full use of the formalism given above gives an extremely simple description of the master set. Before giving this, however, let us momentarily return to the point particle [9]. Here, we only have one constraint and so

$$Q = \beta_0(-\partial^2 + m^2) \tag{2.119}$$

since the classical algebra of the constraints is Abelian. The well-known Klein-Gordon action can be written as

$$\int d^4x\, \psi(-\partial^2+m^2)\psi = \langle \chi | Q | \chi \rangle \tag{2.120}$$

where $|\chi\rangle = \psi|-\rangle + \phi|+\rangle$ is the most general functional of x^μ and c, and $\bar{c}$. Clearly, $Q^2 = 0$ and so the action is invariant under

$$\delta|\chi\rangle = Q|\Lambda\rangle \tag{2.121}$$

In components this reads

$$\delta\psi = 0 \quad , \quad \delta\phi = (-\partial^2+m^2)\Lambda \tag{2.122}$$

where $|\Lambda\rangle = \Lambda|-\rangle + \Omega'|+\rangle$ and so the symmetry is trivially realized.

Returning to the string, the action for the master set is in fact none other than [19][21][24]

$$\frac{1}{2} \langle \chi | Q | \chi \rangle \tag{2.123}$$

and the gauge invariance is

$$\delta|\chi\rangle = Q|\Lambda\rangle. \tag{2.124}$$

we recall that $|\chi\rangle$ is subject to the algebraic constraint $N|\chi\rangle = 0$ and this reflects itself, using $[\phi,N] = -N$ to give the constraint

$$(N+1)|\Lambda\rangle = 0 \tag{2.125}$$

on $|\Lambda\rangle$. The functional $|\Lambda\rangle = \Lambda|-\rangle+\Omega|+\rangle$ then contains the following functionals of $x^u(\sigma)$

$$\Omega^{K+1}_K, \ \Omega^{K+2}_K \quad \text{for all K.} \tag{2.126}$$

The action may be expressed in functional form:

$$\frac{1}{2}\int \mathcal{D}x^\mu(\sigma)\,\mathcal{D}c(\sigma)\,\mathcal{D}\bar{c}(\sigma)\,\chi[x^\mu(\sigma),c(\sigma),\bar{c}(\sigma)]$$
$$\cdot Q\ \chi[x^\mu(\sigma),c(\sigma),\bar{c}(\sigma)] \tag{2.127}$$

where Q in functional form is given by

$$Q = \frac{i\pi}{2} : \oint_{z=0} \frac{dz}{z}\,[\pi\alpha'\sqrt{2}\,(\hat{P}^{\mu 2}) - \frac{1}{\pi}z + z\frac{d}{dz}(\beta\bar{\beta})$$
$$+ z\frac{d}{dz}\bar{\beta}]\,\beta : \tag{2.128}$$

where $z = e^{-i\sigma}$. Note that χ and Q are odd, but so is $\mathcal{D}c(\sigma)\mathcal{D}\bar{c}(\sigma) = \prod_{n=0}^{\infty} dc_n \prod_{n=1}^{\infty} d\bar{c}_n$ due to the fact that $c(\sigma)$ has one more zero mode than $\bar{c}(\sigma)$.

The equation of motion is given by

$$Q|\chi\rangle = 0 \tag{2.129}$$

which, using equation (2.112), can be written in terms of component fields ψ_k^k and ϕ_k^{k+1} as

$$\begin{aligned} K\psi + d\phi + D\phi &= 0 \\ d\psi + D\psi + 2m\phi &= 0 \end{aligned} \qquad (2.130)$$

The invariance in terms of the component fields Λ_k^{k+1} Ω_k^{k+2} is given by

$$\begin{aligned} \delta\psi &= (d+D)\Lambda - 2M\Omega \\ \delta\phi &= -K\Lambda + (d+D)\Omega. \end{aligned} \qquad (2.131)$$

These are of course identical to equation (2.117) where the index structure is explicitly shown.

The action $\langle\chi|Q|\chi\rangle$ becomes in component fields

$$\begin{aligned} &\tfrac{1}{2}(\psi, K\psi) + (\psi, d\phi) + (D\phi, \psi) \\ &+ (\phi, M\phi) \end{aligned} \qquad (2.132)$$

In deriving this result we have taken account of the vacuum properties of equations (2.98) and (2.99) and used the fact that $(\psi d\phi) = (D\psi, \phi)$ for any ψ and ϕ.

We note that $|\Lambda\rangle = Q|\Lambda'\rangle$ leaves $|\chi\rangle$ inert and is a so-called hidden invariance. This automatically leads to the phenomenon of ghost for ghosts since the gauge-fixing term must involve $|\chi\rangle$ alone, and so the corrresponding ghost action will inevitably be invariant under this invariance. We will discuss this shortly.

Unlike the finite and infinite sets it is straightforward to show that the master set leads to the correct on-shell states for the string.

2.8 The On-shell Spectrum of the Master Set [19]

The string in D = 26 has the same number of on-shell degrees of freedom as there are fields in a functional of $x^i(\sigma)$ i = 1 to 24 and x^-. That is because in the light-cone gauge, where only $x^i(\sigma)$ and the x^- centre of mass co-ordinate remain, all degrees of freedom are physical. Hence, the number of on-shell degrees of freedom at each level N is p(N) where p(N) is given by

$$\sum_{n=0}^{\infty} p(N)\, x^N = \prod_{n=1}^{\infty} (1-x^n)^{-24}$$
$$= 1 + 24x + \ldots \tag{2.133}$$

Note that the partition function $\prod_{n=1}^{\infty}(1=x^n)^{-D}$ just records the number of degrees of freedom in a D-dimensional functional.

We must show this is the number of on-shell degrees of freedom predicted by $\langle x|Q|x\rangle$. We could do this in three ways.

(i) By counting the degrees of freedom classically, but Lorentz covariantly. This is non-trivial even for the photon.

(ii) Show we can go to the light-cone formalism by a gauge choice.

(iii) Quantize the system, that is, gauge fix and add ghosts and then show that the bosonic minus the fermionic degrees is p(N) at level N.

We now carry out route (iii). We begin with the classical theory whose component field content is ψ_k^k, ϕ_k^{k+1}. Examining equation (2.118) let us fix Λ_k^{k+1} by setting $\phi_k^{k+1} = 0$. Hence we insert $\delta(\phi_k^{k+1})$ in the functional integral and incur the following ghost term: $\bar{\Lambda}_{k+1}^{k}(-K\Lambda_k^{k+1} +D\Omega_k^{k+2}+d\Omega_{k-1}^{k+1})$. We are using the same notation for ghost fields as the symmetry from which they arise. This ghost action, however, possesses an invariance corresponding to the "hidden" invariance $|\Lambda\rangle = Q|\Lambda'\rangle$. Since $[Q,N] = -N$ we find that $N|\chi\rangle = 0$ implies that $(N+2)|\Lambda'\rangle = 0$, and so it has the field content $\{\Lambda'^{k+1}_{k-1}, \Omega'^{k+2}_{k-1}\}$. We fix this invariance by setting $\Omega^{k+1}_{k-1} = 0$, that is, we insert $\delta(\Omega^{k+1}_{k-1})$ in the functional integral and add the ghost term

$$\bar{\Lambda}'^{k-1}_{k+1}\left(K\Lambda'^{k+1}_{k-1} + d\Lambda'^{k+1}_{k-2} + D\Lambda'^{k+2}_{k-1}\right) \qquad (2.134)$$

This term again has an invariance corresponding to $|\Lambda'\rangle = Q|\Lambda''\rangle$ requiring ghost for ghost for ghost. Clearly this process continues indefinitely.

The net result is the field content

$$\psi^{k}_{k}, \quad \Lambda^{k+1}_{k}, \quad \Lambda'^{k+1}_{k-1}, \quad \Lambda''^{k+2}_{k-1}$$

$$\text{and} \quad \bar{\Lambda}^{k}_{k+1}, \quad \bar{\Lambda}'^{k-1}_{k+1}, \quad \bar{\Lambda}''^{k-1}_{k+2}. \qquad (2.135)$$

and they occur in the action with the kinetic operator K. We observe that in this set there is one and only one tensor of the type $\Lambda^{m_1\ldots m_a}_{n_1\ldots n_b}$ for each a and b. These fields may be neatly fitted into a functional χ which is completely general except it has no $|+\rangle$ vacuum, i.e., it satisfies

$\bar{\beta}_0|\chi\rangle = 0$ and so

$$|\chi\rangle = \sum_{\{n\}\{m\}} \beta^{n_1\dagger}\cdots\beta^{n_b\dagger}\bar{\beta}^{\dagger}_{m_1}\cdots\bar{\beta}^{\dagger}_{m_a}\Lambda^{m_1\ldots m_a}_{n_1\ldots n_b}|-\rangle \qquad (2.136)$$

The free BRST action is given by

$$\langle\chi|\beta_0 K|\chi\rangle = \langle\chi|[\beta_0\bar{\beta}_0, Q]|\chi\rangle \qquad (2.137)$$

Of course, it is no longer gauge-invariant but is BRST-invariant, namely

$$\delta|\chi\rangle = \lambda\,\varphi|\chi\rangle \tag{2.138}$$

where λ is the anticommuting BRST parameter.

We will now demonstrate that this gauge-fixed action originally found by Siegel [9] by another approach has the correct number of degrees of freedom. Let us write the tensor Λ^{a+c}_{b+c} in the form

$$\Lambda^{m_1\cdots m_a;\,p_1\cdots p_c}_{\;\;n_1\cdots n_b;\,p_1\cdots p_c} \tag{2.139}$$

indicating that it has c indices in common.

This tensor first contributes at level $M = \sum_{i=1}^{a} m_i + \sum_{i=1}^{b} n_i + 2\sum_{i=1}^{c} p_i$ and is commuting or anti-commuting according to whether $\sum_i m_i + \sum_i n_i$ is even or odd. the net (bose-fermi) number of tensors at level M is c(m) where

$$\sum_{m=0}^{\infty} c(m)x^m = \prod_{n=1}^{\infty}(1-2x^n+x^{2n})$$
$$= \prod_{n=1}^{\infty}(1-x^n)^2 . \tag{2.140}$$

Since each such tensor contributes $T^{26}(N-M)$, the net total number of degrees of freedom at level N is given by

$$p(N) = \sum_{m=0}^{N} c(m)\,T^{26}(N-m)$$

and so

$$\sum_{N=0}^{\infty} p(N)x^N = \sum_{N=0}^{\infty}\sum_{m=0}^{\infty} c(m)\,T^{26}(N-m)\,x^{N-m}x^m$$
$$= \prod_{n=1}^{\infty}(1-x^n)^2 \prod_{n=1}^{\infty}\frac{1}{(1-x^n)^{26}} \tag{2.141}$$
$$= \prod_{n=1}^{\infty}(1-x^n)^{-24} = \sum_{n=0}^{\infty} T^{24}(N)x^N$$

Hence we recover the light-cone count and so confirm that $\langle x|Q|x\rangle$ does describe the spectrum of string theory.

A much shorter proof is as follows [19]

$$\begin{aligned}
&\sum_N n_B(N) - n_F(N)\, x^N \\
&= \mathrm{STr}\; x^{\sum_{n=1}^{\infty}(\alpha_n^{\mu}\alpha_n^{\mu} + \beta_n^{+}\beta_n + \bar{\beta}_n^{+}\bar{\beta}_n)} \\
&= \prod_{n=1}^{\infty} (1-x^n)^2 \prod_{n=1}^{\infty} \frac{1}{(1-x^n)^{26}} \qquad (2.142) \\
&= \prod_{n=1}^{\infty} \frac{1}{(1-x^n)^{24}}.
\end{aligned}$$

where we have made use of the fermionic partition function.

This concludes our discussion of the open bosonic string. The systematic method of construction given for this case can be easily used to find all the other free gauge-covariant string theories. One first finds the finite set, then generalizes this to the infinite set and then to the master set. This one can put in convenient form using extra co-ordinates. The count of states is given in the same way as above. See appendix (A) where this is carried out for the closed bosonic string and appendix (B) for a summary of the results for the open superstring and a discussion of the other superstrings.

3. A REVIEW OF USEFUL TOPICS IN INTERACTING STRING THEORY

There is a substantial literature on interacting string theory [2]. In this section we will summarize some parts of this old literature which are useful for covariant string theory. A substantial amount of this literature is concerned with the dual model, which was the original approach to string theory.

3.1 Dual Model

Using duality and crossing symmetry Veneziano [1] guessed the four-point amplitude at the tree level for tachyonic external states. This result, for the open bosonic string, was later generalized to the n-point function. Unitarity was then used to factorize the n-point amplitude and obtain the on-shell 3-string scattering amplitude, that is the scattering amplitude for arbitrary external states. This vertex [28] was written in the oscillator formalism and after a slight modification to make it cyclic was found to be [29]

$$V^{CSV} = \langle 0_{123} | \exp\left\{ - \sum_{r=1}^{3} \sum_{\substack{m=0 \\ n=1}}^{\infty} \frac{\alpha_m^{\mu r} \alpha_n^{\mu r+1} (-1)^m \Gamma(n)}{\Gamma(1+m)\Gamma(1+n-m)} \right. . \tag{3.1}$$

where

$$\langle 0_{123} | = {}_1\langle 0 | \; {}_2\langle 0 | \; {}_3\langle 0 |$$

The 3-string scattering was found by considering

$$V^{CSV} |\psi\rangle_1 \, |\psi\rangle_2 \, |\psi\rangle_3 \tag{3.2}$$

where $|\psi\rangle$ has the expansion given in equation (2.43) labelled according to the relevant external string. To obtain the actual expression for the scattering of any three external states only requires us to carry out the oscillator algebra for each string. The reader may wish to verify that the 3-Yang-Mills amplitude scattering agrees with the usual result on-shell, i.e., up to a gauge transformation. (For a discussion of this point see references [2] and [14].)

The dual model consisted in taking V^{CSV} and constructing graphs using the propagator $(L_0-1)^{-1}$. However, the combinatoric rules in the dual model, which were determined by unitarity, are not those of field theory. The rules of the latter are uniquely specified by the Feynman path integral. In particular, since V^{CSV} is automatically cyclic and indeed so are the higher-point functions constructed from it, one does not include a sum over graphs which are related by cyclic permutation. In fact, one must sum over all distinct topologies with unit weight. This gives $\frac{(n-1)!}{2}$ diagrams for the n-point tree diagram.

The reader will note that the dual model is manifestly Lorentz invariant and, although it was constructed to satisfy duality, this is not particularly apparent from the above vertex. The situation with loops, however, is not so straightforward. Clearly, the manifest Lorentz symmetry implies we have, for example A_μ $\mu = 0,\ldots,25,$; however, we know from usual field theory that ghost particles are required for unitarity. These particles were not present in the old dual-model formalism. To some extent this was overcome by introducing physical state projection operators into the loops.

By studying the dual model it was deduced that the dual model corresponded to the scattering of strings and led to the sum over world surfaces description of strings [30],[31],[32].

One aspect of the second-quantized approach to string field theory which has been known for some time was the one where one chooses the light-cone gauge. In this approach one can, if one wishes, regain the sum over world surfaces. The interaction in this formulation consisted of the splitting and joining of strings at their end points.

3.2 Relation between overlap δ functions and Oscillator Vertices

Although strings are extended objects we would like their interactions to be local in space-time. The most natural way for a string to achieve this is for it to break at a point into two strings. In this picture the interaction is a kind of infinitesimal event. The corresponding world sheet of the strings is given in fig.1.

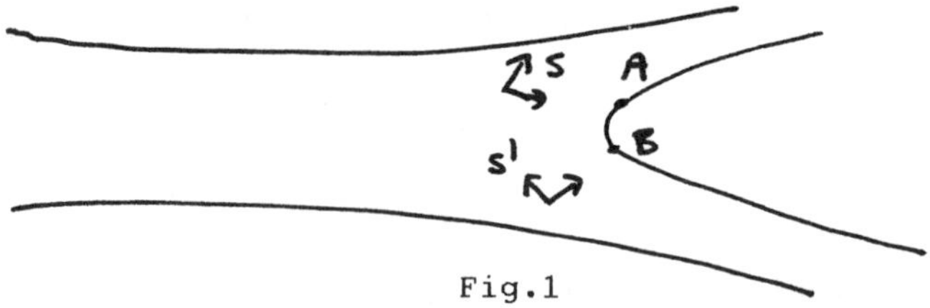

Fig.1

We also observe that the point at which the interaction takes place is observer-dependent. As shown in fig.1 the observers of the two frames of reference consider the interaction to have taken place at different space-time points marked A and B. In terms of mathematics, the above pictures mean that at the moment of the interaction, the incoming string can be identified, in the appropriate regions, with the two outgoing strings. By time reversal we can, of course, have two strings becoming one by joining at their end points. The most natural parameterization for such a process is to assign the strings a length. For the rth string of length $\alpha_r\pi$, we have

$$0 \leq \sigma_r \leq \alpha_r \pi \tag{3.3}$$

The lengths in the interaction will be preserved i.e.

$$\sum_{r=1}^{3} \alpha_r = 0 \tag{3.4}$$

In fact, in the light-cone gauge this is the way the second-quantized strings interact. In this case the string lengths are p^+ and are therefore automatically conserved. For gauge-covariant string theory, as we shall see, we will also have string lengths which appear as parameters in the theory and whose conservation is imposed by a δ function. The interactions will correspond to the splitting and joining of strings.

Let us now show[33],[34],[35],[36] how the above considerations can lead to a detailed vertex for string theory.

We consider the case of 2 strings joining to form a third and parametrize this process as a strip (fig.2).

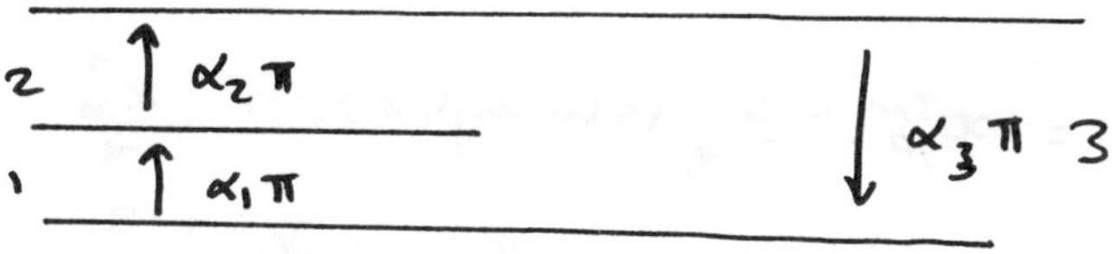

Fig.2.

Rather than use σ_r we use $\eta_r = \dfrac{\sigma_r\pi}{|\alpha_r|}$. The interaction is given by $\eta_3 = \beta_2\pi$ since at the top of string 2 $\alpha_3\eta_3 = \alpha_2\pi$ where $\beta_2 = \dfrac{\alpha_2}{\alpha_1+\alpha_2}$, $\beta_1 = \dfrac{\alpha_1}{\alpha_1+\alpha_2}$. The interaction in co-ordinate space is given by

$$V(x) \equiv V|x\rangle = \delta\left(x_3^\mu(\eta_3) - \theta(-\eta_3+\beta_2\pi)\, x_2^\mu(\eta_2) - \theta(\eta_3-\beta_2\pi)\, x_1^\mu(\eta_1)\right) \tag{3.5}$$

For the second-quantized light-cone theory $\mu = 1$ to 24 while for gauge-covariant string theory $\mu = 0$ to 25. In the latter case this is only part of the vertex, the remainder being constructed from c and $\bar{c}$. At the moment they touch the δ function implies that they overlap, that is

$$x^3(\eta_3) = \begin{cases} x_1(\eta_1) & \beta_2\pi < \eta_3 < \pi \\ x_2(\eta_2) & 0 < \eta_3 < \beta_2\pi \end{cases} \tag{3.6}$$

Now, the Fourier expansion of $x^{r\mu}(\eta_r)$ is given by

$$x^{r\mu}(\eta_r) = x_0^{r\mu} + 2\sum_{n=1}^{\infty} \cos n\eta_r\, x_n^{r\mu} \tag{3.7}$$

and using the overlap condition, we can relate the x_n^r for the 3 strings. Matching string 3 and string 1 in the region $0 \leq \eta_1 \leq \pi$ we find

$$\frac{1}{\pi}\int_0^\pi d\eta_1\, x^{1\mu}(\eta_1)\cos n\eta_1 = \frac{1}{\pi}\int_0^\pi d\eta_1\, \cos n\eta_1\, x^{3\mu}(\eta_3) \quad (3.8)$$

which implies that

$$x_0^{1\mu} = x_0^{3\mu} + \frac{2}{m\beta_1}(\sin m\beta_1\pi)(-1)^m \frac{x_m^{3\mu}}{\pi}$$

$$x_n^{1\mu} = \frac{(-1)^n}{\pi}\left[\frac{-2m\beta_1}{n^2-\beta_1^2 m^2}\right](\sin n\beta_1\pi)(-1)^m x_m^{3\mu} \quad (3.9)$$

Similarly for string 2 we find the equations

$$x_0^{2\mu} = x_0^{3\mu} + \frac{2}{m\beta_2}\sin m\pi\beta_2 \frac{x_m^{3\mu}}{\pi}. \quad (3.10)$$

$$x_n^{2\mu} = \frac{2m\beta_2}{n^2-m^2\beta_2^2}\sin m\pi\beta_2 \frac{x_m^{3\mu}}{\pi}$$

These results may be summarized in the generic form

$$x_1 = \hat{A}^1 x_3 \qquad ; \quad x^2 = \hat{A}^2 x_3$$

$$\sum_{r=1}^{3}\alpha^r x_0^r = 0 \qquad x_0^1 - x_0^2 = \alpha_3 \hat{B} x^3 \quad (3.11)$$

where A^1_{nm}, A^2_{nm} and B_n $n,m = 1,2,\ldots,\infty$ are read off from the preceding equation. Consequently the overlap δ function can be rewritten as

$$V|x\rangle = \delta(x_n^{1\mu} - \hat{A}^1_{nm}x_m^{3\mu})\,\delta(x_n^{2\mu} - \hat{A}^2_{nm}x_m^{3\mu})$$

$$\delta\left(\sum_{r=1}^{3}\alpha^r x_0^{r\mu}\right)\delta(x_0^{1\mu} - x_0^{2\mu} - \alpha_3 B_n x_n^{3\mu}) \quad (3.12)$$

We could have matched the strings from $0 \leq \eta_3 \leq \pi$ and obtained a different-looking, but equivalent result.

By a basis change we find the vertex in momentum space to be given by

$$\begin{aligned}\upsilon(p) \equiv \upsilon|p\rangle &= \int \prod_{s,n} dx_n^{s\mu}\ \upsilon|x_n^{s\mu}\rangle\langle x_n^{s\mu}|p_n^s\rangle \\ &= \int \prod_{s,n} dx_n^{s\mu}\ \exp\left(i\, p_{s\mu}^n\, x_n^{s\mu}\right)\ \upsilon\ |x_n^{s\mu}\rangle \\ &= \delta\left(p_0^{1\mu}+p_0^{2\mu}+p_0^{3\mu}\right)\delta\left(p^{3\mu}+A^2p^{2\mu}+A^1p^{1\mu}+B\,\mathbb{P}^\mu\right)\end{aligned} \tag{3.13}$$

where

$$\mathbb{P}^\mu = \alpha_1 p_0^{2\mu} - \alpha_2 p_0^{1\mu} = \alpha_2 p_0^{3\mu} - \alpha_3 p_0^{2\mu} = \alpha_3 p_0^{1\mu} - \alpha_1 p_0^{3\mu}$$

Before transforming to the oscillator basis we will discuss some elementary properties of coherent states.

Consider a single harmonic oscillator a, which satisfies $[a,a^+] = 1$. We define the coherent state by $|z\rangle = e^{za^+}|\rangle$ where z is complex and the vacuum satisfies $a|\rangle = 0$. We observe that $|z\rangle$ satisfies the relations

$$a|z\rangle = z|z\rangle \qquad , a^+|z\rangle = \frac{d}{dz}|z\rangle \tag{3.14}$$

The overlap function between momentum and coherent states is given by

$$\langle p|z\rangle = \exp\left(-\frac{p^2}{2} - \frac{z^2}{2} + \sqrt{2}\, p\cdot z\right) \tag{3.15}$$

This is easily found by inserting $p = \frac{1}{\sqrt{2}}(a+a^+)$ and $x = \frac{i}{\sqrt{2}}(a-a^+)$ and using the above relations. We also find that

$$\langle z'|z\rangle = \exp z^{*\prime} z \tag{3.16}$$

and the completeness relation is given by

$$\mathbb{1} = \int dz\,dz^*\ \exp(-z^*z) \tag{3.17}$$

The generalization to the string oscillators only requires factors of n in the appropriate places.

As a preliminary step we now find the vertex in the oscillator basis

$$V(z) \equiv V|z\rangle = \int \prod_{r,n} dp_{rn}\, V|p^r_n\rangle\langle p^r_n|z\rangle$$
$$= \int \prod_{r,n} dp_{rn}\, \delta\left(\sum_{r=1}^{3} p_0^r\right) \delta(p^3 + A^2 p^2 + A^1 p^1 + B\mathbb{P})$$
$$\exp\left\{-\frac{1}{2}\hat{p}^r C^{-1} p^r - \frac{1}{2}\hat{z}^r C^{-1} z^r + \sqrt{2}\,\hat{p}^r C^{-1} z^r\right\} \tag{3.18}$$

where $C_{mn} = n\delta_{mn}$. We do not integrate over p_0 and so $\delta(\sum_{r=1}^{\infty} p_0^r)$ remains and although present, we will not indicate it explicitly. The μ indices will also be suppressed until we find the final result. Rewriting the δ function as an integral we find

$$V|z\rangle = \int dp\, du\, \exp\left\{-\frac{1}{2}\hat{p}^r C^{-1} p^r - \frac{1}{2}\hat{z}^r C^{-1} z^r + \sqrt{2}\,\hat{p}^r C^{-1} z^r\right\} \exp iu\left(\sum_{r=1}^{3} A^r p^r + B\mathbb{P}\right)$$
$$= \int dp\, du\, \exp\left\{-\frac{1}{2}(\hat{p}^r - \sqrt{2}\,\hat{z}^r - i\hat{u}A^r C)C^{-1}(p^r - \sqrt{2}\,z^r - iC\hat{A}^r u) + \frac{1}{2}(\sqrt{2}\,\hat{z}^r + i\hat{u}A^r C)C^{-1}(\sqrt{2}\,z^r + iC\hat{A}^r u) + i\hat{u}B\mathbb{P} - \frac{1}{2}\hat{z}^r C^{-1} z^r\right\}$$
$$= \int du\, \exp\left\{\frac{1}{2}\hat{z}^r C^{-1} z^r - \frac{1}{2}\hat{u}\Gamma u + i\hat{u}W\right\} \tag{3.19}$$

where

$$\Gamma = \sum_{r=1}^{3} A^r C \hat{A}^r, \quad \hat{W} = \sqrt{2}\,\hat{z}^r \hat{A}^r + \hat{B}\mathbb{P} \tag{3.20}$$
$$A^3_{nm} = \delta_{n,m}$$

Doing the final integral we obtain the result

$$
\begin{aligned}
V|z\rangle &= \exp\left\{\tfrac{1}{2}\tilde{z}^{r} C^{-1} z^{r} - \tfrac{1}{2}\tilde{W}\,\Gamma^{-1} W\right\} \\
&\equiv \exp\left\{\tfrac{1}{2} z^{r}_{n} N^{rs}_{nm} z^{s}_{m} + N^{r}_{m} z^{r}_{m}\, \mathbb{P} \right. \\
&\qquad \left. + K\,\mathbb{P}^{2}\right\}
\end{aligned}
\tag{3.21}
$$

where

$$
\begin{aligned}
N^{rs}_{nm} &= \delta^{rs}(C^{-1})_{nm} - 2\,(\hat{A}^{r}\,\Gamma^{-1} A^{s})_{nm} \\
N^{r}_{m} &= -\sqrt{2}\,(\hat{A}^{r}\,\Gamma^{-1} B)_{m} \\
K &= -\tfrac{1}{2}\,\hat{B}\,\Gamma^{-1} B
\end{aligned}
\tag{3.22}
$$

Hence the vertex in terms of oscillators is given by the expression

$$
\begin{aligned}
&{}_{1}\langle 0|\,{}_{2}\langle 0|\,{}_{3}\langle 0|\, \exp\left\{\tfrac{1}{2}\alpha^{\mu r}_{n} N^{rs}_{nm} \alpha^{\mu s}_{m} \right. \\
&\left. + N^{r}_{m}\alpha^{\mu r}_{m}\,\mathbb{P}^{\mu} + K(\mathbb{P}^{\mu})^{2}\right\}
\end{aligned}
\tag{3.23}
$$

by virtue of equation (3.14).

It can be shown that [36][37]

$$
N^{rs}_{nm} = -\frac{\alpha_1\alpha_2\alpha_3}{\alpha_r\alpha_s}\,\frac{mn}{n\alpha_s + m\alpha_r}\, f_n(\gamma_r)\, f_m(\gamma_s)
\tag{3.24}
$$

$$
\gamma_r = -\frac{\alpha_{r+1}}{\alpha_r}\,;\quad f_n(\gamma_r) = \frac{1}{n\gamma_r}\binom{n\gamma_r}{n} = \frac{(n\gamma_r - 1)\cdots(n\gamma_r - n + 1)}{n!}
$$

while

$$
N^{r}_{m} = \frac{1}{\alpha_r} f_m(\gamma_r)\,, \quad K = -\frac{\tau_b}{2\alpha_1\alpha_2\alpha_3}
\tag{3.25}
$$

An indirect derivation of this result will be a consequence of the discussion of the next section.

3.3 The Mandelstam Map

Although the string interaction is not well defined in x-space, it has a simple interpretation in terms of the splitting and joining of strings. After making some basis transformations on the co-ordinate space vertex we found an oscillator vertex which is well defined, but has a rather complicated form involving the f_m and $N_m{}^r{}_n{}^s$ functions.

From the point of view of the sum over the world surfaces of the string these complications can be seen to arise as follows. In this approach, one is required to find the Green's function for point sources on the strip, i.e.,

$$\left(\frac{\partial^2}{\partial\tau^2} - \frac{\partial^2}{\partial\sigma^2}\right) G(\sigma,\tau;\sigma',\tau') = 2\pi\,\delta(\sigma-\sigma')\,\delta(\tau-\tau') \tag{3.26}$$

subject to $\frac{\partial G}{\partial\eta(\sigma,\tau)}(\sigma,\tau;\sigma'\tau') = f(\sigma,\tau)$ where $\frac{\partial}{\partial\eta}$ is the normal derivative at the boundary and f is an arbitrary function. This is identical to a problem in two-dimensional electrostatics and there we know that it is most easily solved by transforming the appropriate region to a more simple one. In this case we should transform from the strip to the upper half plane by a Schwartz-Christoffel map. The Green's function on the two-dimensional upper half plane is rather simple; namely, for one charge it is $\ln(z-z')$. Upon transforming this simple function back to the strip to find the Green's function, we encounter the $f_m{}^r$ and $N_m{}^r{}_n{}^s$ functions as the Fourier components of this transformed Green's function.

Even though the approach of gauge-covariant field theory is somewhat different we will find that the map from the strip to the upper half plane plays an important role. (In fact, it is the vertices of the dual model that live on the upper half plane, and the above transformation takes these to those of gauge-covariant string theory on the strip.)

We now discuss in detail the map from the strip to the upper half plane [38]. Consider 3-string scattering shown in the figure below

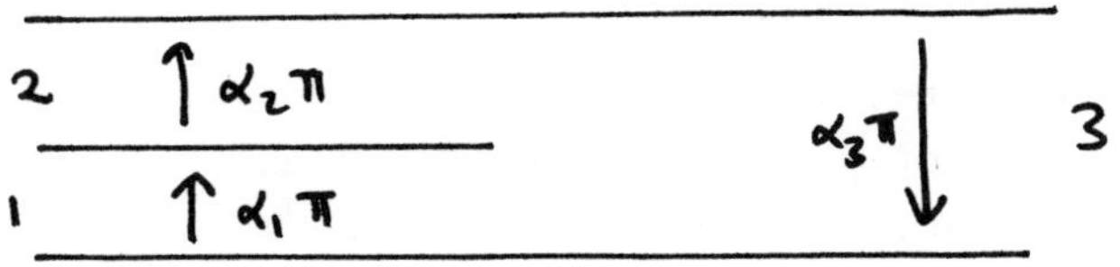

The strip is described by $\rho = \tau + i\sigma$; $0 \leqslant \sigma \leqslant \alpha_3\pi$; $-\infty < \tau < +\infty$, the interaction being at $\tau = 0$. The individual strings have the strip co-ordinates

$$\rho_1 = \alpha_1(\xi_1 + i\eta_1) \quad \text{for string 1}$$

$$\rho_2 = \alpha_2(\xi_2 + i\eta_2) + i\alpha_1\pi \quad \text{for string 2}$$

$$\rho_3 = \alpha_3(\xi_3 + i\eta_3) - i\alpha_3\pi \quad \text{for string 3}$$

(3.27)

where $0 \leqslant \eta_r \leqslant \pi$; $\tau_r = \alpha_r\xi_r$ and $-\infty \leqslant \xi_r \leqslant 0$.

We now map the strip onto the upper half plane as shown below

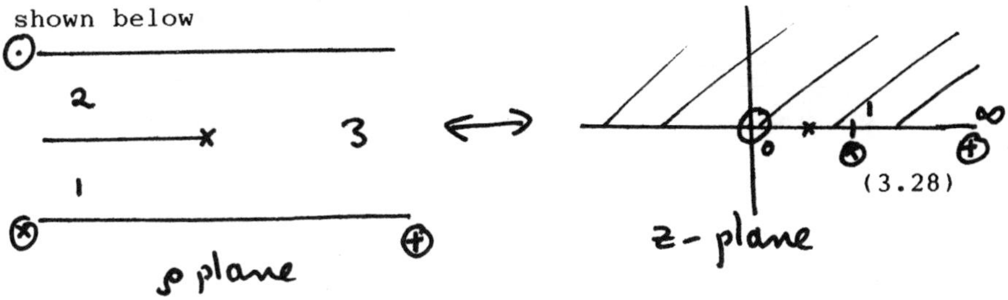

(3.28)

by the transformation [38],[39]

$$\rho' \equiv \rho + \tau_0 = \alpha_1 \ln(z-1) + \alpha_2 \ln z \tag{3.29}$$

One finds in particular that the real axis of the upper half plane is mapped onto the boundary of the string. The three-end points of the strings are mapped onto the three points of the upper half plane as shown. Starting at $z = +\infty$, i.e., at point ⊕ and moving along the real axis, we move along the strip boundary to the end of string 1 (i.e., point ⊗ at $z = 1$). Here, $(z-1)$ changes argument by π and so we jump to the top of the first string and move towards the interaction point. When we arrive at this point we then travel along the bottom of string 2 towards the end of string 2 etc.. Since the derivatives at the interaction point are discontinuous the map is not invertible at this point, i.e.,

$$\frac{d\rho'}{dz} = 0 \rightsquigarrow z_b = -\frac{\alpha_2}{\alpha_3} \tag{3.30}$$

and

$$\tau_b = \alpha_1 \ln\left(-\frac{\alpha_1}{\alpha_3}\right) + \alpha_2 \ln\left(-\frac{\alpha_2}{\alpha_3}\right) \tag{3.31}$$

The time evolution of the string on the upper half plane is demonstrated in the diagram below.

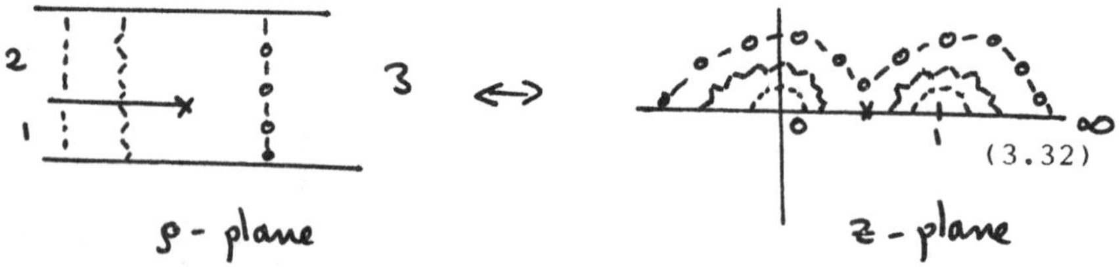

(3.32)

We will require the inverse of this map, that is the map from the upper half plane to the strip. To this end let us define the object $y(\gamma, x)$ by

$$y \equiv \gamma \ln(1 + x e^{y})$$
$$= \gamma\left(x e^{y} - \tfrac{1}{2}(x e^{y})^2 + \ldots\right)$$

$$= \gamma x + \frac{\gamma}{2}(2\gamma-1)x^2 + \cdots \qquad (3.33)$$

$$\equiv \gamma \sum_{n=1}^{\infty} f_n(\gamma)\, x^n$$

It can be shown that [38]

$$f_n(\gamma) = \frac{1}{n\gamma}\binom{n\gamma}{n} = \frac{1}{n!}(n\gamma-1)(n\gamma-2)\ldots(n\gamma-n+1) \qquad (3.34)$$

The inverse map from the half plane to the strip is given by [33]

$$z = \begin{cases} -\frac{1}{\zeta_3}\, e^{-y_3(\zeta_3)} & \text{for string } 3 \\ e^{-\frac{\alpha_1}{\alpha_2} y_1(\zeta_1)} & \text{for string } 1 \\ \zeta_2\, e^{-\frac{\alpha_1}{\alpha_3} y_2(\zeta_2)} & \text{for string } 2 \end{cases} \qquad (3.35)$$

where

$$\zeta_r = e^{\tau_b + \rho_r} \; ; \; y_r \equiv \gamma_r \ln(1 + \zeta_r e^{y_r})$$

$$\gamma_r = \frac{-\alpha_{r+1}}{\alpha_r} \qquad (\alpha_{3+1} \equiv \alpha_1) \qquad (3.36)$$

One can easily verify by substituting in equation (3.29) that this is the correct inverse map.

In particular we will be concerned with the map at the interaction time $\tau = 0$. In this case $\zeta_r = e^{-i\eta_r} e^{\tau_b}$. Although the e^{τ_b} factors are essential, we will not in what follows explicitly indicate them in order to keep the formulae simple.

4. INTERACTING GAUGE-COVARIANT OPEN BOSONIC STRINGS

The work reported in this section can be found in references [40] and [41]. The gauge-covariant vertex is in agreement with the independent work of reference [42]. For alternative approaches the reader is referred to references [21] and [43].

We saw that the free theory had a rather simple formulation in terms of functionals on the space $x^{\mu}(\sigma)$, $c(\sigma)$, $\bar{c}(\sigma)$, the free action being simply $\langle X|Q|X\rangle$. We will also use this formalism to construct the interacting theory.

The one second-quantized field theory of strings which has been known for some time is the light-cone gauge-fixed theory [33][35][38]. Although this is a very specific gauge one expects some of its features to be generic to any second-quantized field theory. In particular, one finds in the light-cone gauge the beautiful picture of the splitting and joining of strings as described in the previous section. This allows one to relate the second-quantized string field theory to the first-quantized sum over world sheet approach.

We wish to carry over the picture of splitting and joining of strings to covariant string theory. As a result we will take the strings to have lengths $\alpha_r\pi$. We will regard α's as parameters in the theory and impose their conservation by including in the interaction the factor

$$\int \prod_{r=1}^{3} d\alpha_r \ \delta\left(\sum_{r=1}^{3} \alpha_r\right) \tag{4.1}$$

If this factor is not written explicitly it is to be understood to be present. In fact, an additional motivation for introducing lengths was to solve the so-called duality problem which is discussed later. As we shall see the string lengths do not appear in on-shell quantities. As one may suspect, the interaction will be an overlap delta function in the space $x^u(\sigma)$, $c(\sigma)$, $\bar{c}(\sigma)$ with one important modification which we will discuss below.

Derivation of the Gauge Covariant Vertex

Consider three-string scattering which is governed by the cubic vertex in the action. We adopt the strip parametrization of equation (3.37). We recall that the relation between the η_r's is of the form

$$\eta_3 = \begin{cases} -\beta_2\eta_2 + \beta_2\pi & \text{for } 0 \le \eta_3 \le \beta_2\pi \\ -\beta_1\eta_1 + \pi & \text{for } \beta_2\pi \le \eta_3 \le \pi \end{cases} \tag{4.2}$$

If the vertex contains an overlap δ function in co-ordinate space, then the string co-ordinates denoted generically by $Z^r(\eta_r)$ will obey the relations

$$Z^3(\eta_3) = \begin{cases} Z^1(\eta_1) & \text{for } 0 \le \eta_3 \le \beta_2\pi \\ Z^2(\eta_2) & \text{for } \beta_2\pi \le \eta_3 \le \pi \end{cases} \tag{4.3}$$

It is important to stress that the vertex may contain more than the overlap δ function but we assume, at this point, that it is generically of the form

$$V = N\,\delta\big(Z^3(\eta_3) - \theta(-\eta_3+\beta_2\pi)Z^2(\eta_2) - \theta(\eta_3-\beta_2\pi)Z^1(\eta_1)\big) \tag{4.4}$$

where N is some function of Z's. In fact, we will see that N contains ghost co-ordinates evaluated at the interaction point. In particular, for the open bosonic string we have

$$Z^r(\eta_r) = \left(x_\mu^r(\eta_r),\, c^r(\eta_r),\, \frac{\bar{c}^r}{\alpha^r}(\eta_r)\right) \tag{4.5}$$

Due to the fact that the interaction contains an overlap δ function we find an identity, the generic form of which is given by

$$V \sum_{r=1}^{3} \int_0^{\pi} d\eta_r \; f(\eta_r) \; S \tag{4.6}$$

where V is the vertex and S is a function of Z^r and $\delta/\delta Z^r$ and derivatives of these objects with respect to η_r. Under the change of variable from η_3 to η_1 and η_2 in the appropriate region, we assume that S is adjusted so as not to scale.

The identity (4.6) is rather trivial, as is apparent when one examines a particular example. Let us take $S = \alpha^r \alpha^r$; then the above identity reads

$$v \left\{ \sum_{r=1}^{3} \int_0^{\pi} d\eta_r \, f(\eta_r) \, \alpha^r \, c^r(\eta_r) \right\} = 0 \tag{4.7}$$

Examining the term involving $c^3(\eta^3)$ and using equation (4.4) we find that

$$\begin{aligned} v \int_0^{\pi} d\eta_3 f(\eta_3) \alpha^r c^3(\eta_3) &= v \int_0^{\pi} \beta_2 d\eta_2 \, f(\eta_3(\eta_2)) \alpha^3 c^3(\eta_3 \\ &+ \int_0^{\pi} \beta_1 d\eta_1 \, f(\eta_3(\eta_1)) \, \alpha^3 \, c^3(\eta_3) \Big\} \\ &= - v \Big\{ \int_0^{\pi} d\eta_2 \, f(\eta_3(\eta_2)) \, \alpha^2 c^2(\eta_2) + \int_0^{\pi} d\eta_1 \, f(\eta_3(\eta_1)) \alpha_1 c^1(\eta_1) \Big\} \end{aligned} \tag{4.8}$$

Thus the identity is none other than $x\delta(x) = 0$.

We cannot however take any function f of η in equation (4.8) since it must lead to a convergent sum in all three expansions. Since the expansions involve a sum over coefficients times, in general, operators, we must insist that they have finite matrix elements. This for example is not true for expressions of the form

$$\sum_{n=1}^{\infty} e_n \alpha^{\mu}_{-n} \quad \text{or} \quad \sum_{n=1}^{\infty} d_n L_{-n} \tag{4.9}$$

Unless we pick a particular form for the allowed functions f, we will generically find such terms. The correct procedure is to utilize the Mandelstam map of equation (3.29) discussed in the previous section. We recall that the inverse of this map was given in equation (3.45). We take the f's to be arbitrary functions of z which itself is a function of the strip variable ζ.

While one could take any polynomial in z, it is most convenient to take one which has only a single z^{-n} pole in z. Such functions are of the form

$$F_m^r(z) = \sum_{p=0}^{n-1} \binom{n\gamma_r}{p} (-z)^{n-p} \tag{4.10}$$

To derive the gauge identities, we need to know the expansion of $F_m(z)$ in terms of z_r for each r; this is given by the following theorem:

<u>Theorem</u>

$$F_m^3(z) = \begin{cases} \dfrac{1}{\zeta^m} - m\gamma_3 f_m(\zeta_3) - m \displaystyle\sum_{n=1}^{\infty} N_{mn}^{33} \zeta_3^n & \text{for string 3} \\ \displaystyle\sum_{p=0}^{m-1} \binom{m\gamma_3}{p} (-1)^{m-p} - m \sum_{n=1}^{\infty} N_{mn}^{31} \zeta_1^n & \text{for string 3} \\ -m \displaystyle\sum_{n=1}^{\infty} N_{mn}^{32} \zeta_2^n & \text{for string 2} \end{cases} \tag{4.11}$$

where

$$N_{mn}^{rs} = -\frac{\alpha_1\alpha_2\alpha_3}{n\alpha_r + m\alpha_s} \frac{mn}{\alpha_r\alpha_s} f_m(\gamma_r) f_n(\gamma_s) \tag{4.12}$$

The expansion coefficients N_{mn}^{rs} are none other than the Fourier transform of the Neumann functions which play such a central role in string vertices. The reader will recall that we have suppressed the presence of τ_b.

For a proof of this theorem see reference [41].

As explained in ref.[41], one may obtain cycled identities by making the replacements

$$z \to \frac{1}{1-z} \quad \text{and} \quad z \to 1-\frac{1}{z} \tag{4.13}$$

in $F_m(z)$.

Finally, we mention a subtlety associated with the gauge identities which concerns the treatment of the interaction points. For reasons that will be explained for each case, it is sometimes necessary to consider functions $f(z)$ which vanish at the interaction point, in which case we consider the polynomials

$$\bar{F}_m^r(z) = F_m^r(z) - F_m^r(z^{int}) \tag{4.14}$$

We note that

$$z_{int} = \frac{1}{\gamma_2} = 1-\gamma_3 = \frac{\gamma_1}{\gamma_1 - 1} \tag{4.15}$$

The above discussion is quite general and may be applied to any string or superstring theory. We now illustrate it for the gauge-covariant formulation of the bosonic open string which has a cubic vertex denoted by V. This vertex can be written in the form

$$V = V^B V^F \tag{4.16}$$

where V^B involves only the bosonic co-ordinates $x^\mu(\eta)$, and V^F involves only the fermionic co-ordinates $c(\eta)$ or $\bar{c}(\eta)$. In the oscillator basis, they are functions of $\alpha_n^{\mu r}$ and β_n^r, $\bar{\beta}_n^r$ respectively.

Let us first take S to be

$$P_\mu^r(\eta_r) = \frac{1}{\pi\sqrt{2\alpha'}}\left[-i\hbar\frac{\delta}{\delta x_r^\mu(\eta_r)} - \frac{1}{2\pi\alpha'}\frac{\partial x_r^\mu}{\partial \eta_r}\right] \tag{4.17}$$

In terms of $\alpha_n^{\mu r}$ oscillators, we have

$$P_\mu^r(\eta_r) = -\frac{1}{\pi}\frac{1}{\sqrt{2\alpha'}}\sum_{n=-\infty}^{\infty} \alpha_n^{\mu r} \zeta_r^{-n} \tag{4.18}$$

we may verify that P_μ^r does scale appropriately, and since there is no need in this case to subtract at the interaction point, the corresponding identity when written in terms of ζ_r rather than η_r is of the form

$$V\sum_{r=1}^{3}\oint_{\zeta_r=0}\frac{d\zeta_r}{\zeta_r} F_m(z) P_\mu^r(\zeta_r) = 0 \tag{4.19}$$

One easily finds that the α's obey the identity

$$\begin{aligned} V[\alpha_{-n}^{\mu,r} - m\gamma_r f_m(\gamma_r)\alpha_0^{\mu,r} - m f_m(\gamma_r)\alpha_0^{\mu,r+1} \\ - m\sum_{n=1}^{\infty}\sum_{s=1}^{3} N_{mn}^{rs}\alpha_n^{\mu s}] = 0 \end{aligned} \tag{4.20}$$

The vertex which satisfies this equation is easily seen to be

$$\begin{aligned} V^B = \langle 0_{123}| \exp\{\tfrac{1}{2}\alpha_n^{\mu r} N_{nm}^{rs}\alpha_m^{\mu s} + N_m^{rs}\alpha_m^{\mu r}\alpha_0^{\mu s} \\ \exp \tau_b \sum_{r=1}^{3}\left(\frac{L_0^r - 1}{\alpha^r}\right) \end{aligned} \tag{4.21}$$

where

$$N_m^{rs} = \frac{1}{\alpha_r} f_m(\gamma_r)(\alpha_r \delta_{r+1,s} - \alpha_{r+1}\delta_{r,s}) \tag{4.22}$$

$$\alpha_0^\mu = p^\mu\sqrt{2\alpha'}\ , \quad \langle 0_{123}| = \underset{1}{\langle 0|}\underset{2}{\langle 0|}\underset{3}{\langle 0|}$$

$$\tau_b = \alpha_1 \ln\left(-\frac{\alpha_1}{\alpha_3}\right) + \alpha_2 \ln\left(-\frac{\alpha_2}{\alpha_3}\right) \tag{4.23}$$

Although we find that the factor $\exp \tau \sum_t (L_0^t - 1/\alpha_t)$ is required to obtain the correct four-point functions, we will take the effects of this factor in gauge identities and vertices to be understood in order not to overcomplicate expressions. We could also apply the above strategy to the light-cone formulation; in this case we find the above vertex, the only difference being that $\mu = 1,\ldots,23$.

In the later sections we will need the action of

$$L_n = :\frac{\alpha'}{2}\oint_{z=0}\frac{dz}{z}\, z^n\, \hat{P}_\mu^{\,2} \tag{4.24}$$

on V. This identity may be deduced directly, or from Eq.(4.20). Care must be taken, as L is normal-ordered. In this case, one must use the functions $\bar{F}_m(z)$, which vanish at the interaction point. This is because $P^{\mu 2}$ is badly defined at the interaction point due to the occurrence of the step functions and derivatives in P^μ. The L_n's do not scale with unity but $(L_n^{(r)}/\alpha^r)$ does. Applying the above reasoning, we find the identity

$$V\Big\{\frac{L_m^r}{\alpha^r} - m\gamma_r f_m(\gamma_r)\frac{L_0^r}{\alpha^r} - m f_m(\gamma_r)\frac{L_0^{r+1}}{\alpha^{r+1}} - m\sum_{\substack{n=1\\ s=1,2,3}}^{\infty} N_{mn}^{rs}\frac{L_n^s}{\alpha^s} - S_m^r\sum_{s=1}^{3}\frac{L_0^s}{\alpha^s} - \frac{D}{\alpha^r}\sum_{n=1}^{m-1}\frac{n(m-n)}{2}N_{n,\,m-n}^{r,\,r}\Big\} = 0 \tag{4.25}$$

where

$$S_m^r = \sum_{p=0}^{m-1}\binom{m\gamma_r}{p}(\gamma_r-1)^{m-p} \tag{4.26}$$

The constant term comes about since L is normal-ordered and its value may easily be checked by taking $|0\rangle_{123}$ on the left-hand side of (4.25).

We now turn to the determination of the V^F part of the vertex by considering the $\beta,\bar{\beta}$ identities

$$\beta(\eta) = \frac{1}{\pi\sqrt{2}}\sum_{n=-\infty}^{\infty}\beta_n z^{-n} = -\frac{\delta}{\delta\bar{c}(\eta)} + \frac{1}{2\pi}c(\eta) \tag{4.27}$$

$$\bar{\beta}(\eta) = \frac{1}{\pi}\frac{1}{\sqrt{2}}\sum_{n=-\infty}^{\infty}\bar{\beta}_n z^{-n} = \frac{\delta}{\delta c(\eta)} - \frac{1}{2\pi}\bar{c}(\eta) \tag{4.28}$$

As explained in reference [40], the part of the vertex involving c and $\bar{c}$ must contain three c ghosts at the vertex in order to recover the Lorentz covariant generalization of the light-cone result. Due to the $\delta/\delta c$ derivatives in $\bar{\beta}$, we must insist that it satisfy an interaction-point-subtracted identity. The β identity, on the other hand, is unsubtracted. It is easily seen that $\bar{\beta}^n$ and $\alpha^r\beta^r$ scale with unity, and hence we have the identities

$$V\{\bar{\beta}^r_{-m} - m\gamma_r f_m(\gamma_r)\bar{\beta}^r_0 - m f_m(\gamma_r)\bar{\beta}^{r+1}_0 - m\sum_{\substack{n=1\\ s=1,2,3}}^{\infty} N^{rs}_{mn}\bar{\beta}^s_n - S^r_m \sum_{s=1}^{3}\bar{\beta}^r_0\} = 0 \tag{4.29}$$

$$V\{\alpha_r\beta^r_{-m} - m\gamma_r f_m(\gamma_r)\alpha_r\beta^r_0 - m f_m(\gamma_r)\alpha_{r+1}\beta^{r+1}_0 - m\sum_{n=1}^{\infty}\sum_{s=1}^{3} N^{rs}_{mn}\alpha_s\beta^s_n\} = 0 \tag{4.30}$$

as well as

$$V\sum_{r=1}^{3}\alpha_r\beta^r_0 = 0 \tag{4.31}$$

From Eq.(4.29) we can deduce the vertex; it is found to be

$$V^F = \langle t_{123}|\exp\{-\beta^r_n(n N^{rs}_{nm})\bar{\beta}^s_m + R^{rs}_n\beta^r_n\bar{\beta}^s_0\} \tag{4.32}$$

where

$$R^{rs}_m = -m f_m(\gamma_r)(\gamma_r\delta_{r,s} + \delta_{r+1,s}) - S^m_r$$

$$\langle t_{123}| = \langle t_1|\langle t_2|\langle t_3| \tag{4.33}$$

It is easily verified that this vertex does satisfy Eqs.(4.30) and (4.31).

Consequently, the vertex for the open bosonic string is given by Eqs.(4.32) and (4.21). These have been derived under the assumption that the vertex contains in co-ordinate space an overlap δ function. Important in the derivation has been whether the identity for a given object is subtracted at the interaction point. It is not clear that this is equivalent to the vertex required in an invariant theory; however, in Section 4.3 we will show that this is indeed the case. We recall from reference [40] that the above vertex does indeed yield the correct interacting theory at lowest order.

4.2 On-shell Three-String Scattering, Duality and String Lengths

The on-shell scattering amplitude for the scattering of three strings [29] which was deduced from unitarity and duality is known to be described by the oscillator vertex [30] of equation (3.1). Any proposed gauge-covariant theory of strings must give the same scattering amplitude as the above vertex for on-shell states. We recall that an on-shell state satisfies the constraints

$$(L_0 - 1)\,|\psi\rangle = 0$$

$$L_n\,|\psi\rangle = 0 \qquad n \geq 1$$

By examining explicitly the way the $\alpha_n^{\mu r}$ oscillators are transmitted and reflected by V^{CSV}, one can show that there exists a conformal transformation which takes one from V^{CSV} to another vertex which satisfies the simple overlap conditions given in equation (4.20). Since the overlap conditions uniquely fix the vertex, this other vertex is none other than the covariant vertex of equation (4.21). The map is none other than the conformal map from the upper half plane to the strip given in equation (3.35). The reader is referred to reference [41] for more details.

The net result is that

$$V^{CSV} \exp\left\{\sum_{r=1}^{3}\sum_{n=1}^{\infty} A_n^r L_n^r\right\} = V_B$$

where $\exp \sum_{r=1}^{3} \sum_{n=1}^{\infty} A_n^r L_n^r$ is the generator of the conformal map from the upper half plane to the strip. In this sense the V^{csv} lives on the upper half plane.

Consquently, for on-shell states $|\psi\rangle$

$$V^{CSV} |\psi\rangle_1 |\psi\rangle_2 |\psi\rangle_3 = V |\psi\rangle_1 |\psi\rangle_2 |\psi\rangle_3$$

and we have shown that the gauge-covariant theory does reproduce the known result for three-string scattering.

Since V^{csv} is independent of string lengths α we realize that the α's come about as a type of conformal squashing of the strings. As such the string lengths do not occur in on-shell quantities for three-string scattering. Some thought shows the same result is very likely to hold for four-string scattering and it would seem likely that on-shell the string length disappears for all on-shell quantities.

The presence of parameters which do not contribute to on-shell quantities is perhaps not desirable, and it would be better to find a larger formalism in which the string lengths emerge as a kind of gauge choice, as indeed they do in the light-cone formalism. We note that the possibility of different string lengths is related to the fact, discussed in section 3, that different observers see the interaction point occur at different space-time points.

It is perhaps appropriate to comment on the motivation for introducing string lengths. Since the vertex V^{csv} of equation (3.1) gives the correct on-shell three-string result one might, at first sight, wonder if one could use this vertex in the gauge covariant theory. This however would not give the correct answer for any of the higher-point tree graphs. The reason is that V^{vcs} does give the correct higher-point tree graphs provided one uses it in the context of dual theory. Here the combinatoric rule is that one includes all graphs of a given topology with unit weight. That is, one does not include graphs which are related by a cycling of their external legs. The reason for this is that graphs constructed in the dual model possess duality which is summarized by the property

$$G(1,2,\ldots n) = G(n,1,2\ldots)$$

This implies that a given graph will contain poles from more than one channel. For example, the dual graph given below

contains both s and t channel poles. For the n-point tree graph we have $\frac{(n-1)!}{2}$ graphs.

In gauge covariant string field theory, on the other hand, we work from a Feynman path integral formulation and so the combinatoric rules must be those of field theory. At the level of the four-point function, we must, in the gauge-covariant field theory, include the graphs below

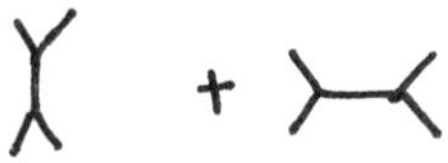

In the dual model these graphs are numerically equal. Hence if we were to use V^{CSV} in the gauge-covariant field theory we would overcount as this vertex gives the correct answer in the dual model. For the four-point function we would overcount by a factor of 2 and it gets successively worse for the higher-point functions.

This problem was of course solved in the context of second-quantized light-cone field theory [35]. Here the strings have a length proportional to p^+ and one uses the vertex of equation (4.21) but with $i = 1,\ldots,D-2$. One finds that the combinatoric rules of field theory do indeed give the correct result for the n-point functions. In fact, it was discovered by computing the four-point function that one required a quartic interaction in the action.

In covariant string theory we assign the string lengths, but we find that if we use the vertex of equation (4.21 and 4.32) we do recover the same combinatoric mechanism as for light-cone field theory.

4.3 Invariance and Closure

In reference [14], the generic form of the action and transformation laws in a gauge-covariant string theory were given. The action must be of the form

$$A = \frac{1}{2} \langle x | Q | x \rangle + \frac{1}{3} g V |x\rangle_1 |x\rangle_2 |x\rangle_3 \tag{4.34}$$

and the transformation law of the generic form

$$\delta \underset{3}{\langle} x| = \frac{1}{g} \underset{3}{\langle} \Lambda | Q^3 + V'(|x\rangle_1 |\Lambda\rangle_2 - |\Lambda\rangle_1 |x\rangle_2) \tag{4.35}$$

The question arises as to whether the vertex V' in the transformation law is the same as the vertex V in the action. In fact, the vertices V and V' can be deduced from a knowledge of the bosonic part V^B of V alone. Using this procedure, in ref.[40] the expressions for $\delta \langle x|$ are given at lowest order. The reader may readily verify that these transformation laws are reproduced by the equation

$$\delta \underset{3}{\langle} x| = \frac{1}{g} \underset{3}{\langle} \Lambda | Q^3 + V\left(\frac{\alpha_2}{\alpha_3} |x\rangle_1 |\Lambda\rangle_2 - \frac{\alpha_1}{\alpha_3} |\Lambda\rangle_1 |x\rangle_2\right) \tag{4.36}$$

This result uses the fact that we assigned the $|-\rangle$ vacuum to be odd in the sense of anticommuting. The $|+\rangle$ vacuum must then be even. We adopt this to be the correct transformation rule and examine the criterion for invariance. The variation of the action of Eq.(4.34) at order g^0 under Eq.(4.36) is

$$\delta A = \frac{1}{3} V\left[\frac{\alpha_2}{\alpha_3} |x\rangle_1 |\Lambda\rangle_2 \varphi^3 |x\rangle_3 - \frac{\alpha_1}{\alpha_3} |\Lambda\rangle_1 |x\rangle_2 \varphi^3 |x\rangle_3 + \text{cyclic}\right]$$
$$+\frac{1}{3} V(\varphi^1 |\Lambda'\rangle |x\rangle_2 |x\rangle_3 + |x\rangle_1 \varphi^2 |\Lambda\rangle_2 |x\rangle_3$$
$$+ |x\rangle_1 |x\rangle_2 \varphi^3 |\Lambda\rangle_3) \tag{4.37}$$

As a result, the necessary and sufficient condition for invariance is

$$V \sum_{t=1}^{3} \frac{\varphi^t}{\alpha^t} = 0 \tag{4.38}$$

We can also consider the criterion for closure. Carrying out the commutator of two transformations on $\langle x|$, we find at order g^{-1}

$$[\delta_{\Lambda'}, \delta_\Lambda] \langle x|_3 = \frac{1}{g} V\left[\frac{\alpha_2}{\alpha_3} \varphi^1 |\Lambda'\rangle_1 |\Lambda\rangle_2 - \frac{\alpha_1}{\alpha_3} \varphi^2 |\Lambda\rangle_1 |\Lambda'\rangle_2\right]$$
$$- (\Lambda \leftrightarrow \Lambda') \tag{4.39}$$

From the lowest-order results, one finds that the closure at this order is consistent with the result

$$[\delta_{\Lambda'}, \delta_\Lambda] \langle x|_3 = -\frac{\alpha_1 \alpha_2}{g \alpha_3^2} V\left(|\Lambda'\rangle_1 |\Lambda\rangle_2 - (\Lambda \leftrightarrow \Lambda')\right) \varphi^3 \tag{4.40}$$

Taking this to be the correct result to all levels, we find that the condition for closure at order g^{-1} is Eq.(4.38). Thus the conditions for invariance and closure are the same, and this confirms Eqs.(4.36) and (4.40) to all levels.

It now only remains to show that V does indeed satisfy this condition. We recall that

$$\varphi = :\sum_{n=-\infty}^{\infty} \beta_n L_n - \sum_{n,m,p=-\infty}^{\infty} \frac{1}{2} \bar{\beta}_p f_{nm}{}^p \beta_{-n} \beta_{-m} - \beta_0 :$$
$$\equiv :\sum_{n=-\infty}^{\infty} \beta_n \mathcal{L}_{-n} - \beta_0 : \tag{4.41}$$

where

$$\mathcal{L}_n = L_n + \frac{1}{2} L_n^{\beta}$$

and

$$L_n^{\beta} = - : \sum_m (n+m) \, \beta_m \, \bar{\beta}_{n-m} :$$

We note that Q is a normal-ordered version of the BRS charge corresponding to the Virasoro algebra, and that it satisfies $Q^2 = 0$ in $D = 26$. In functional language, Q is given in Eq.(2.128). Clearly this form does satisfy Eq.(4.38) in functional language, but this ignores all questions of normal ordering, as indeed does Eq.(2.128) itself.

We now consider the proof of Eq.(4.38) in the oscillator formalism. We must show that

$$V \sum_{\substack{n=-\infty \\ r=1,2,3}}^{\infty} : \left(\beta_{-n}^{r} \frac{\mathcal{L}_n^{r}}{\alpha^r} - \frac{\beta_0^{r}}{\alpha^r} \right) : = 0 \qquad (4.42)$$

Now L_n^{β} obeys a similar relation to L_n, namely Eq.(4.25), except that the constant due to L_n^{β} being normal-ordered is different; this identity reads

$$V \Big[\frac{L_{-m}^{\beta r}}{\alpha^r} - m \gamma_r f_m(\gamma_r) \frac{L_0^{\beta r}}{\alpha^r} - m f_m(\gamma_r) \frac{L_0^{\beta r+1}}{\alpha^{r+1}} - \frac{m}{\alpha^r} R_m^{rr}$$
$$+ \frac{1}{\alpha^r} \sum_{n=1}^{m-1} (n+m)(m-n) N_{n,m-n}^{rr} - m \sum_{n=1}^{\infty} N_{mn}^{rs} \frac{L_n^{\beta s}}{\alpha^s} - S_m^{r} \sum_{s=1}^{3} \frac{L_0^{\beta s}}{\alpha^s} = 0 \qquad (4.43)$$

In deriving this equation, we have normal-ordered expressions involving anticommuting variables. Normal-ordering Q before we take it through V, Eq.(4.42) becomes

$$V \sum_{r=1}^{3} \left(\sum_{m=1} \beta_{-m}^{r} \frac{\mathcal{L}_m^{r}}{\alpha^r} + \frac{\beta_0^{r} \mathcal{L}_0^{r}}{\alpha^r} + \sum_{m=1}^{\infty} \frac{\mathcal{L}_{-m}^{r}}{\alpha^r} \beta_m^{r} - \frac{\beta_0^{r}}{\alpha^r} \right) = 0 \qquad (4.44)$$

Using Eqs.(4.25) and (4.43) to replace L^r_{-m} and dragging β^r_{-m} and $B_0{}^r$ explicitly through V, we find the following four terms:

$$V\Big\{+\Big[-\beta^s_n(n N^{sr}_{nm})\frac{\mathcal{L}^r_m}{\alpha^r}\Big]+\Big[R_n{}^{rs}\beta^r_n\frac{\mathcal{L}^s_0}{\alpha^s}\Big]$$

$$+\Big[m\gamma_r f_m(\gamma_r)\frac{\mathcal{L}^r_0}{\alpha^r}+m f_m(\gamma_r)\frac{\mathcal{L}^{r+1}_0}{\alpha^{r+1}}+m N^{rs}_{mn}\frac{\mathcal{L}^s_n}{\alpha^s}$$

$$+S^m_r\sum_{s=1}^{3}\frac{\mathcal{L}^s_0}{\alpha^s}+\frac{D}{2\alpha_r}\sum_{p=1}^{m-1}N^{rr}_{p,m-p}\,p(m-p)$$

$$-\frac{1}{2\alpha^r}\sum_{p=1}^{m-1}N^{r\,r}_{p,m-p}(m-p)(m+p)+\frac{m}{2\alpha^r}R^{rr}_m\Big]\beta^r_m$$

$$+\Big[m f_m(\gamma_r)\Big(\gamma_r-\frac{1}{\gamma_r}\Big)+\alpha_r S^r_m\sum_{s=1}^{3}\frac{1}{\alpha^s}\Big]\frac{\beta^r_m}{\alpha^r}\Big\}=0 \qquad (4.45)$$

Examining this expression, in which summation signs are to be understood, we find that the terms in would cancel if the β^s_n's were always on the same side of L_n. Using the relation

$$[\beta^r_n,\mathcal{L}^s_m]=\frac{1}{2}(2m+n)\,\beta^r_{n+m}\,\delta_{r,s} \qquad (4.46)$$

we find that

$$V\Big\{\sum_{\substack{n=1\\ r=1,2,3}}^{\infty}\frac{\beta^r_n}{\alpha^r}\Big[\sum_{m=1}^{n-1}\frac{D}{2}m(n-m)N^{rr}_{m,n-m}-\sum_{m=1}^{n-1}(n+m)N^{rr}_{m,n-m}$$

$$-\gamma_r n^2 f_n(\gamma_r)-n S^r_n+n f_n(\gamma_r)\Big(\gamma_r-\frac{1}{\gamma_r}\Big)$$

$$+\alpha_r S^r_n\sum_{s=1}^{3}\frac{1}{\alpha_s}\Big]\Big\}=0 \qquad (4.47)$$

Upon using the identity

$$S^r_n=\sum_{m=1}^{n-1}n\,N^{rr}_{m,n-m}+(\gamma_r-1)\,n\,f_n(\gamma_r) \qquad (4.48)$$

the above equation can be written in the form

$$\sum_{\substack{n=1\\ r=1,2,3}}^{\infty} V\Big\{\frac{\beta_n^r}{\alpha^r}\Big[\sum_{m=1}^{n-1} N_{m,n-m}^{r\ r}\big(m^2(1-\tfrac{D}{2})+n^2(\tfrac{D}{4}-2)\big)$$

$$+n\alpha_r\sum_{t=1}^{3}\frac{1}{\alpha_t}+n(n-1)(1-2\gamma_r)f_n(\gamma_r)\Big]\Big\}=0 \quad (4.49)$$

Recalling that we may write

$$N_{n-m,m}^{r\ r}=\gamma_r(\gamma_r-1)\frac{m(n-m)}{n}f_{n-m}(\gamma_r)f_m(\gamma_r) \quad (4.50)$$

one may verify that Eq.(4.49) is satisfied provided D = 26. This completes the proof of the invariance of the action and the closure at this level.

We now demonstrate that the cubic vertex given in reference [40] and re-examined in reference [41] is indeed the same as that conjectured in ref.[42]. Using Eq.(4.48), the fermonic part of our vertex V^F given in Eq.(4.32) can be rewritten in the form

$$V^F=\langle +_{123}|\exp\Big\{-\sum_{\substack{n,m\\ r,s}}\beta_n(nN_{nm}^{rs})\bar{\beta}_m^s$$

$$+\sum_{r,s,n}\bar{\beta}_0^r\beta_n^s\, n f_n(\gamma_s)\Big[\frac{\alpha_{r-1}-\alpha_{r+1}}{\alpha_r}\delta_{r,s}+\frac{\alpha_r}{\alpha_{r+1}}\delta_{r+1,s}$$

$$-\frac{\alpha_r}{\alpha_{r-1}}\delta_{r-1,s}\Big]\Big\} \quad (4.51)$$

which can be seen to agree with that of ref.[42].

The only point that remains to finish the construction of the open bosonic string is to complete the closure and invariance to all orders in g. As explained in ref.[40], we expect the occurrence of a four-point function but no higher-point function in the action. This can be seen, for example, from demanding that the four-point function possess duality. The latter method would follow closely the light-cone calculation of refs.[34] and [38], which deduced the existence of the light-cone four-point function for the open string. Many features of this calculation are the same, except that now we must also compute the contribution from the anticommuting oscillators. It is apparent that the integration regions will not match and a quartic interaction will be required. This calculation will be reported elsewhere.

For the algebra, one may suspect that

$$\delta \underset{3}{<} \chi | = \frac{1}{g} \underset{3}{<} \Lambda | \Phi^3 + V(1,2,3) \left(\frac{\alpha_2}{\alpha_3} |\chi\rangle_1 |\Lambda\rangle_2 - \frac{\alpha_1}{\alpha_3} |\chi\rangle_2 |\Lambda\rangle_1 \right) \tag{4.52}$$

holds to all orders in g and that the commutator of two Λ transformations is given to all orders by

$$[\delta_{\Lambda'}, \delta_{\Lambda}] = \delta_{\bar{\Lambda}} \tag{4.53}$$

where

$$\underset{3}{<} \bar{\Lambda} | = - \frac{\alpha_1 \alpha_2}{\alpha_3} V(1,2,3) \left[|\Lambda'\rangle_1 |\Lambda\rangle_2 - (\Lambda \leftrightarrow \Lambda') \right] \tag{4.54}$$

This equation defines the symmetry group underlying the string. The lowest component of this equation is none other than the group law for Yang-Mills theory; however, $|\Lambda\rangle$ contains as many gauge parameters as there are fields in the theory. The higher-level group relations can be deduced from Eq.(4.54), and the structure functions are related to the N's. Thus Eq.(4.54) represents an enormous generalization of Lie group theory which deserves further study. It is straightforward to evaluate the necessary condition that Eq.(4.54) hold; carrying out the commutator of two Λ transformations on $\langle\chi|$ and neglecting the g^{-1} term which was discussed earlier, we find that

$$[\delta_{\Lambda'}, \delta_{\Lambda}] \underset{3}{<} \chi | = V(1,2,3) \Big\{ \frac{\alpha_2}{\alpha_3} \bar{V}(1',2',1) \Big[\frac{\alpha_{2'}}{\alpha_1} |\chi\rangle_{1'} |\Lambda'\rangle_{2'}$$

$$- \frac{\alpha_{1'}}{\alpha_1} |\Lambda'\rangle_1 |\chi\rangle_{2'} \Big] |\Lambda\rangle_2 - \frac{\alpha_1}{\alpha_3} |\Lambda\rangle_1 \bar{V}(1',2',2) \Big[\frac{\alpha_{2'}}{\alpha_2} |\chi\rangle_{1'} |\Lambda'\rangle_{2'} \tag{4.55}$$

$$- \frac{\alpha_1'}{\alpha_2} |\Lambda'\rangle_{1'} |\chi\rangle_{2'} \Big\} - (\Lambda \leftrightarrow \Lambda')$$

In order that this should agree with the g^{-1} term, it must be of the form

$$- V(1,2,3) \Big[\frac{\alpha_2}{\alpha_3} |\chi\rangle_1 \bar{V}(1',2',2) \frac{\alpha_{1'}\alpha_{2'}}{\alpha_2^2} |\Lambda'\rangle_{1'} |\Lambda\rangle_{2'}$$

$$+ \frac{\alpha_1}{\alpha_3} \bar{V}(1',2',1) \frac{\alpha_{1'}\alpha_{2'}}{\alpha_1^2} |\Lambda'\rangle_{1'} |\Lambda\rangle_{2'} |\chi\rangle_2 \Big\} - (\Lambda \leftrightarrow \Lambda') \tag{4.56}$$

In the above equation

$$V(1,2,3)|\chi\rangle_3 \equiv {}_3\langle\chi| \bar{V}(1,2,3) \tag{4.57}$$

Defining

$$V(1,2,3) \frac{1}{\alpha_1} \bar{V}(1',2',1) = V(2,3;1',2') \tag{4.58}$$

we find that the necessary and sufficient condition for closure to all orders in g is

$$\Big[\big[V(2,3;1',2') - V(3,2;1',2') - (1' \leftrightarrow 2') \big]$$

$$- \big[V(1'3;2,2') - V(3,1';2',2) \big] \Big] + (2 \leftrightarrow 2') = 0 \tag{4.59}$$

This is a generalized Jacobi identity. As explained in ref.[40], we must, in order to ensure the correct field content at every level, add the Chan-Paton reversed vertex. In the above closure discussion, it is to be understood that this has been done; in fact, it is required in order for Eq.(4.59) to hold. This calculation will be given elsewhere.

It is straightforward to find the oscillator vertex of equations (4.21) and (4.32) in coordinate space using the same techniques as in section 3.2. We refer the reader to reference [41], but the final result is

$$V_F(c,\bar{c}) = \delta(\bar{c}^1 - \hat{\tilde{A}}^1 \bar{c}^3)\delta(\bar{c}^2 - \hat{\tilde{A}}^2 \bar{c}^3)\delta(c^1 - \hat{\tilde{A}}^1 c^3)$$
$$\delta(c^2 - \hat{\tilde{A}}^2 c^3)\,\delta\Big(\sum_{s=1}^{3} \alpha_s c_0^s\Big)\,\delta(c_0^1 - c_0^2 - \tilde{B} c^3 \alpha_3)$$
$$\delta\Big(c_0^3 + \tilde{c}^3\Big(\frac{1}{\sqrt{2}}\Gamma C^{-1} S_3 + B\alpha_1\Big)\Big) \tag{4.60}$$

where

$$\hat{\tilde{A}}^r = C\hat{A}^r C^{-1} \tag{4.61}$$

Let us consider what part of this result is due to overlap δ functions. In particular, let us suppose that

$$c^3(\eta_3) = \begin{cases} c^2(\eta_2) & \text{for } 0 \le \eta_3 \le \beta_2\pi \\ c^1(\eta_1) & \text{for } \beta_2\pi \le \eta_3 \le \pi \end{cases} \tag{4.62}$$

Take Fourier transforms in the regions $0 \leqslant \eta_1 \leqslant \pi$ and $0 \leqslant \eta_2 \leqslant \pi$ of the expressions

$$c^r(\eta_r) = c_0^r + 2\sum_{n=1}^{\infty} c_n^r \cos n\eta_r \tag{4.63}$$

we find, as for $x_\mu^r(\eta)$, that

$$c^1 = \tilde{A}^1 c^3,\quad c^2 = \tilde{A}^2 c^3,\quad \sum_r \alpha_r c_0^r = 0 \tag{4.64}$$

and $c_0^1 - c_0^2 = \alpha_3 c^3 B$ where

$$(\tilde{A}_1)_{nm} = \frac{(-1)^n}{\pi}\;\frac{-2m\beta_1}{(n^2-\beta_1^2 m^2)}(-1)^m \sin m\beta_1\pi$$

$$(\tilde{A}_2)_{nm} = \frac{2m\beta_2}{n^2 - m^2\beta_2^2} \cdot \frac{\sin m\beta_2\pi}{\pi}$$

$$\alpha_3 B_m = -\frac{4}{m\beta_1\beta_2} \sin m\beta_2\pi \tag{4.65}$$

For $\bar{c}$ we take the overlap condition

$$\frac{\bar{c}^3(\eta_3)}{\alpha_3} = \begin{cases} \dfrac{\bar{c}^2(\eta_2)}{\alpha_2} & \text{for } 0 \le \eta_3 \le \beta_2\pi \\ \dfrac{\bar{c}^1(\eta_1)}{\alpha_1} & \text{for } \beta_2\pi \le \eta_3 \le \pi \end{cases} \tag{4.66}$$

The fact that $\bar{c}$ vanishes as $\eta = 0$ and $\eta = \pi$ means we should take Fourier tranforms of the equations

$$\bar{c}^r(\eta_r) = 2\sum_{n=1}^{\infty} \bar{c}_n^r \sin n\eta_r \tag{4.67}$$

One finds that

$$\bar{c}^1 = \hat{\tilde{A}}\bar{c}^3 \quad , \quad \bar{c}^2 = \hat{\tilde{A}}\, c^3 \tag{4.68}$$

Consequently, the vertex consists of the overlap δ functions for c and $\bar{c}$ as well as an additional factor of

$$c_0^3 + \tilde{c}_3\left(\frac{1}{\sqrt{2}}\Gamma C^{-1} S_3 + B\alpha_1\right). \tag{4.69}$$

APPENDIX A

In preparation for giving the interacting closed bosonic string, we now present the most useful free action for this theory. The method of construction is very similar to that for the open bosonic string given in section 2 it begins with the formulation of the closed bosonic string in terms of the infinite set [7] which contains the fields ψ, ζ^m_n, $\bar{\zeta}^m_n$, $\bar{\chi}^m_n$, χ^m_n and ϕ^n, $\bar{\phi}^n$.
Let us identify these fields as

$$\psi^{00}_{00} = \psi \quad ; \zeta^m{}_n = -\psi^{1\,0}_{1\,0} \; ; \psi^{1\,0}_{0\,1} = \bar{\chi}^m{}_n$$
$$\psi^{0\,1}_{1\,0} = -\chi^m{}_n ; \psi^{0\,1}_{0\,1} = \bar{\zeta}^m{}_n \quad , \phi^{10}_{00} = \phi^n \tag{A.1}$$
$$\phi^{01}_{00} = \bar{\phi}^n$$

With this notation, the equations of motion of ref.[17] become

$$(\kappa+\hat{\kappa})\,\psi^{00}_{00} + D\,\phi^{10}_{00} + \hat{D}\,\phi^{01}_{00} = 0$$
$$d\,\psi^{00}_{00} + D\,\psi^{1\,0}_{1\,0} + \hat{D}\,\psi^{0\,1}_{0\,1} + \eta\,\phi^{10}_{00} = 0$$
$$\hat{d}\,\psi^{00}_{00} + D\,\psi^{1\,0}_{0\,1} + \hat{D}\,\psi^{0\,1}_{0\,1} + \hat{\eta}\,\phi^{01}_{00} = 0$$
$$d\,\phi^{10}_{00} + (\kappa+\hat{\kappa})\,\psi^{1\,0}_{1\,0} = 0$$
$$-d\,\phi^{10}_{00} + (\kappa+\hat{\kappa})\,\psi^{1\,0}_{0\,1} = 0 \tag{A.2}$$
$$d\,\phi^{01}_{00} + (\kappa+\hat{\kappa})\,\psi^{0\,1}_{1\,0} = 0$$
$$\hat{d}\,\phi^{01}_{00} + (\kappa+\hat{\kappa})\,\psi^{0\,1}_{0\,1} = 0$$

and the transformation laws are

$$\delta\,\psi^{00}_{00} = D\,\Lambda^{10}_{00} + \hat{D}\,\Lambda^{01}_{00}$$
$$\delta\,\phi^{10}_{00} = -(\kappa+\hat{\kappa})\,\Lambda^{10}_{00} \; ; \; \delta\,\phi^{01}_{00} = -(\kappa+\hat{\kappa})\,\Lambda^{01}_{00}$$

$$\delta\psi^{10}_{10} = d\,\Lambda^{10}_{00} \quad ; \quad \delta\psi^{10}_{01} = -\tilde{d}\,\Lambda^{10}_{00}$$

$$\delta\psi^{01}_{10} = d\,\Lambda^{01}_{00} \quad ; \quad \delta\psi^{01}_{01} = \tilde{d}\,\Lambda^{01}_{00} \qquad \text{(A.3)}$$

The symbols d, D, K and η are given by

$$(K\psi)^{m_1\ldots m_a\, p_1\ldots p_c}_{n_1\ldots n_b\, q_1\ldots q_d} = (L_0 + m + n - 1)\,\psi^{m_1\ldots m_a\, p_1\ldots p_c}_{n_1\ldots n_b\, q_1\ldots q_d}$$

$$m = \sum_i m_i \quad , \quad n = \sum_i n_i \, .$$

$$(d\psi)^{m_1\ldots m_a\, p_1\ldots p_c}_{n_1\ldots n_{b+1}\, q_1\ldots q_d} = L_{[n_1}\,\psi^{m_1\ldots m_a\, p_1\ldots p_c}_{n_2\ldots n_{b+1}]\, q_1\ldots q_d}$$

$$+\, a\, f^{[m_1}_{p,[n_1}\,\psi^{p\ldots m_a],\, p_1\ldots p_c}_{n_2\ldots n_{b+1}\, q_1\ldots q_d} - \frac{1}{2}\, b\, f^{\;p}_{[n_1 n_2}\,\psi^{m_1\ldots m_a\, p_1\ldots p_c}_{p\, n_3\ldots n_{b+1}]\, q_1\ldots q_d}$$

$$(D\psi)^{m_1\ldots m_{a-1}\, p_1\ldots p_c}_{n_1\ldots n_b\, q_1\ldots q_d} = (-1)^b a \Big\{ L_{-p}\,\psi^{p\, m_1\ldots m_{a-1}\, p_1\ldots p_c}_{n_1\ldots n_b\, q_1\ldots q_d}$$

$$+\, b\, f_{n-p}^{\;\;q}\,\psi_q{}^{p\, m_1\ldots m_{a-1}\, p_1\ldots p_c}_{n_2\ldots n_b\, q_1\ldots q_d} - \frac{1}{2}(a-1)\, f^{[m_1}_{kl}\,\psi^{kl\, m_2\ldots m_{a-1}]\, p_1\ldots p_c}_{n_1\ldots n_b,\, q_1\ldots q_d}$$

$$(\eta\psi)^{m_1\ldots m_{a-1}\, p_1\ldots p_c}_{n_1\ldots n_{b+1}\, q_1\ldots q_d} = (-1)^b a\, \eta_{q[n_1}\,\psi^{q\, m_1\ldots m_{a-1}\, p_1\ldots p_c}_{n_2\ldots n_{b+1}]\, q_1\ldots q_d} \qquad \text{(A.4)}$$

The symbols d, D, K and η are the same except that they operate on $p_1\ldots p_c$ and $q_1\ldots q_d$, leaving $m_1\ldots m_a$ and $n_1\ldots n_b$ alone. The objects obey the relations

$$d^2 = D^2 = 0 \quad ; \quad [K, d] = 0 = [K, D] = [K, M]$$

$$[M, d) = 0 = [M, D] \quad ; \quad \{d, D\} - 2MK = 0 \qquad \text{(A.5)}$$

as well as similar relations for the tilda's operators.

We now wish to generalize this system to the master set which contains the fields

$$\psi^{k_1 k_2}_{k_3 k_4} \quad \text{with} \quad k_1 + k_2 - k_3 - k_4 = 0$$

and

$$\psi^{k_1 k_2}_{k_3 k_4} \quad \text{with} \quad k_1 + k_2 - k_3 - k_4 = 1 \tag{A.6}$$

All these fields are subject to

$$(\kappa - \hat{\kappa})\,\psi = (\kappa - \hat{\kappa})\,\phi = 0 \tag{A.7}$$

as were those of the finite and infinite system. One soon discovers this generalization of the above equation of motion to those of the master set to be [41]

$$(\kappa + \hat{\kappa})\,\psi^{k_1 k_2}_{k_3 k_4} + d\,\phi^{k_1 k_2}_{k_3 - 1\ k_4} + \hat{d}\,\phi^{k_1 k_2}_{k_3\ k_4 - 1}(-1)^{k_1 + k_3}$$
$$+ D\,\phi^{k_1 + 1\ k_2}_{k_3\ k_4} + \hat{D}\,\phi^{k_1\ k_2 + 1}_{k_3\ k_4}(-1)^{k_1 + k_3} = 0 \tag{A.8}$$

$$d\,\psi^{k_1 k_2}_{k_3 k_4} + \hat{d}\,\psi^{k_1 k_2}_{k_3 + 1\ k_4 - 1}(-1)^{k_1 + k_3 + 1}$$
$$+ D\,\psi^{k_1 + 1\ k_2}_{k_3 + 1\ k_4} + \hat{D}\,\psi^{k_1\ k_2 + 1}_{k_3 + 1\ k_4}(-1)^{k_1 + k_3 + 1}$$
$$+ \eta\,\phi^{k_1 + 1\ k_2}_{k_3\ k_4} + \hat{\eta}\,\phi^{k_1\ k_2 + 1}_{k_3 + 1\ k_4 - 1} = 0 \tag{A.9}$$

and the invariance to be [41]

$$\delta\psi^{k_1 k_2}_{k_3 k_4} = D\,\Lambda^{k_1 + 1\ k_2}_{k_3\ k_4} + \hat{D}\,\Lambda^{k_1\ k_2 + 1}_{k_3\ k_4}(-1)^{k_1 + k_3}$$
$$+ d\,\Lambda^{k_1\ k_2}_{k_3 - 1\ k_4} + \hat{d}\,\Lambda^{k_1 k_2}_{k_3\ k_4 - 1}(-1)^{k_1 + k_3}$$

$$\delta\phi^{k_1 k_2}_{k_3 k_4} = -(\kappa + \hat{\kappa})\,\Lambda^{k_1 k_2}_{k_3 k_4} \tag{A.10}$$

These transformations contain extra and hidden symmetries which, when taken into account, lead to the correct count of states. The pattern is so similar to that for the open string that we shall not go through it here.

We now wish to place this generalized system in a more convenient formulation. Let us take our base space to contain $x^{\mu}(\sigma)$, $c(\sigma)$ and $\bar{c}(\sigma)$, where c and $\bar{c}$ are anti-commuting. All these co-ordinates obey the same periodic boundary conditions, which for $\bar{c}(\sigma)$ read

$$\bar{c}(-\pi) = \bar{c}(\pi) \tag{A.11}$$

They have a similar Fourier expansion and for c this is

$$c(\sigma) = \sum_{n=-\infty}^{\infty} c_n e^{in\sigma} \tag{A.12}$$

We now define creation and annihilation operators by

$$\frac{\delta}{\delta c(\sigma)} - \frac{1}{2\pi}\bar{c}(\sigma) = \frac{1}{\pi}\frac{1}{\sqrt{2}}\sum_{n=-\infty}^{\infty}\bar{\beta}_n e^{in\sigma}$$

$$-\frac{\delta}{\delta \bar{c}(\sigma)} + \frac{1}{2\pi}c(\sigma) = \frac{1}{\pi}\frac{1}{\sqrt{2}}\sum_{n=-\infty}^{\infty}\beta_n e^{in\sigma}$$

$$\frac{\delta}{\delta c(\sigma)} + \frac{1}{2\pi}\bar{c}(\sigma) = \frac{1}{\pi}\frac{1}{\sqrt{2}}\sum_{n=-\infty}^{\infty}\hat{\bar{\beta}}_n e^{in\sigma}$$

$$\frac{\delta}{\delta \bar{c}(\sigma)} + \frac{1}{2\pi}c(\sigma) = \frac{1}{\pi}\frac{1}{\sqrt{2}}\sum_{n=-\infty}^{\infty}\hat{\beta}_n e^{in\sigma} \tag{A.13}$$

The first-quantized BRS charge associated with the closed string Virasoro algebra is

$$Q = :\Big[\sum_n \beta_n L_{-n} - \frac{1}{2}\sum_{n,m,p}\bar{\beta}_p f_{nm}{}^{p}\beta_{-n}\beta_{-m} - \beta_0$$
$$+ \sum_n \hat{\beta}_n \hat{L}_{-n} - \frac{1}{2}\sum_{n,m,p}\hat{\bar{\beta}}_p f_{nm}{}^{p}\hat{\beta}_{-n}\hat{\beta}_{-m} - \hat{\beta}_0\Big]:$$
$$= \beta_0 K - 2\bar{\beta}_0\eta + d + D + \hat{\beta}_0\hat{K} - 2\hat{\bar{\beta}}\hat{\eta} + \hat{d} + \hat{D} \tag{A.14}$$

where η, d, D, $\tilde{\eta}$, etc., are as in Eq.(A.4) with an appropriate scattering of minus signs. This charge Q, however, is not quite the one we wish to consider. For the closed string, the measure $\int x^{\mu} \prod_{n=0}^{\infty} dc_n d\bar{c}_n$ is even, and so the action cannot be of the form $\langle x|Q|x\rangle$. Let us consider, however, the object

$$\begin{aligned}\hat{Q} &= Q - \delta_0(K - \hat{K}) + 2\hat{\bar{\delta}}_0(\eta - \hat{\eta}) \\ &= \gamma_0(K + \hat{K}) - 2\bar{\gamma}_0(\eta + \hat{\eta}) + Q'\end{aligned} \tag{A.15}$$

where

$$\gamma_0 = \frac{1}{2}(\beta_0 + \hat{\beta}_0) \quad , \quad \bar{\gamma}_0 = \frac{1}{2}(\bar{\beta}_0 + \hat{\bar{\beta}}_0)$$

$$\delta_0 = \frac{1}{2}(\beta_0 - \hat{\beta}_0) \quad , \quad \bar{\delta}_0 = \frac{1}{2}(\bar{\beta}_0 - \hat{\bar{\beta}}_0)$$

and

$$Q' = d + D + \hat{d} + \hat{D} \tag{A.16}$$

We note that

$$\hat{Q}^2 = -(\eta - \hat{\eta})(K - \hat{K})$$

and so vanishes for the fields of interest to us.

The most straightforward procedure is to consider a space in which β_0 and $\bar{\beta}_0$ do not occur. The vacuum in this space is annihilated by β_n, $\tilde{\beta}_n$, $\bar{\beta}_n$, $\tilde{\bar{\beta}}_n$ for $n \geq 1$ and the $|+\rangle$ vacuum satisfies

$$\gamma_0 |+\rangle = 0$$

while

$$|-\rangle = 2\bar{\gamma}_0|+\rangle. \tag{A.17}$$

The most general functional of $x^\mu(\sigma)$, $c(\sigma)$ and $\bar{c}(\sigma)$ is given by

$$|\chi\rangle = \sum_{\substack{n,m\\p,q}} \beta^+_{n_1}\cdots\bar{\beta}^+_{m_1}\cdots\tilde{\beta}^+_{p_1}\cdots\tilde{\bar{\beta}}^+_{q_1}\cdots\psi^{m_1\ldots m_b q_1\ldots q_d}_{n_1\ldots n_a p_1\ldots p_c}|-\rangle$$

$$+\sum_{\substack{n,m\\p,q}} \beta^+_{n_1}\cdots\bar{\beta}^+_{m_1}\cdots\tilde{\beta}^+_{p_1}\cdots\tilde{\bar{\beta}}^+_{q_1}\cdots\phi^{m_1\ldots m_b q_1\ldots q_d}_{n_1\ldots n_a p_1\ldots p_c}|+\rangle \tag{A.18}$$

For the gauge-invariant theory, we take only the bosonic fields in $|\chi\rangle$ which belong to the master set, and so it is subject to the constraint

$$\left[\sum_{n=1}^{\infty}(\beta^+_n\bar{\beta}^n - \bar{\beta}^+_n\beta_n) + [\beta\leftrightarrow\tilde{\beta},\bar{\beta}\leftrightarrow\tilde{\bar{\beta}}] + \tfrac{1}{2}\gamma_0\bar{\gamma}_0\right]|\chi\rangle = 0 \tag{A.19}$$

We also enforce the condition

$$(\kappa-\tilde{\kappa})|\chi\rangle = 0 \tag{A.20}$$

The action [41][44] which leads to the equation of motion of Eq.(A.10) is given by

$$\langle\chi|\hat{Q}|\chi\rangle \tag{A.21}$$

and it is invariant under

$$\delta|\chi\rangle = \hat{Q}|\Lambda\rangle \tag{A.22}$$

provided one uses Eq.(A.20).

We may also formulate the free action in an alternative way by reintroducing δ_0 and $\bar{\delta}_0$. The most general functional is then of the form

$$|x\rangle = \psi|-+\rangle + \phi|++\rangle + \tilde{\psi}|+-\rangle + \tilde{\phi}|--\rangle \tag{A.23}$$

We take a $|-\rangle$ vacuum to be odd while a $|+\rangle$ vacuum is even, and hence if $|x\rangle$ is odd, ψ and $\tilde{\psi}$ are even while ϕ and $\tilde{\phi}$ are odd. We note that $\delta_0|x\rangle = 0$ implies that $\tilde{\phi} = \tilde{\psi} = 0$ and we recover the previous field content. The above vacua are defined by the equations

$$\gamma_0|+,\pm\rangle = 0 = \delta_0|\pm,+\rangle$$

and

$$|-,+\rangle = 2\gamma_0|+,+\rangle \quad , \quad |+,-\rangle = 2\bar{\delta}_0|+,+\rangle$$

$$|-,-\rangle = 4\bar{\gamma}_0\bar{\delta}_0|+,+\rangle \tag{A.24}$$

We note that

$$\langle --|++\rangle = \langle ++|--\rangle^{+} = 4\langle ++|\bar{\delta}_0\bar{\gamma}_0|++\rangle$$

$$= -\langle ++|--\rangle \tag{A.25}$$

and as a result we may choose

$$\langle --|++\rangle = i \tag{A.26}$$

By further manipulations one finds that

$$\langle +-|-+\rangle = i = -\langle -+|+-\rangle \tag{A.27}$$

As a consequence of these scalar products, one finds that $\langle x|x\rangle = 0$, for example.

We impose the following algebraic constaints on $|x\rangle$:

$$(K + \tilde{K})|x\rangle = 0 \tag{A.28}$$

and

$$\left[\sum_{n=1}^{\infty}(\beta_n^{+}\bar{\beta}_n-\bar{\beta}_n^{+}\beta_n)+(\beta\leftrightarrow\hat{\beta},\ \bar{\beta}\leftrightarrow\hat{\bar{\beta}})-2\bar{\delta}_0\delta_0+2\gamma_0\bar{\gamma}_0\right]|\chi\rangle=0 \tag{A.29}$$

This last equation ensures that we have the field content of Eq.(A.6) as well as $\tilde{\phi}^n$ and $\tilde{\Psi}$, which have the reversed index structure to ϕ and Ψ respectively.

As noted above, the gauge-covariant free action cannot be $\langle\chi|Q|\chi\rangle$, and the correct action [41] is

$$i\,\langle\chi|\,\bar{\delta}_0\,Q\,|\chi\rangle \tag{A.30}$$

We note that we can replace Q by $\hat{Q}$ since $\bar{\delta}_0(Q-\hat{Q})$ vanishes on $|\chi\rangle$. This action is Hermitian as a result of the fact that $i\,\langle\chi|\{\bar{\delta}_0,Q\}|\chi\rangle$ vanishes due to Eq.(A.24). Evaluating this action, we find we can recover an action which leads to the correct field equations of Eqs.(A.8) and (A.9). In fact, if we remove the constraint $(K-\tilde{K})|\chi\rangle = 0$, we find that it is imposed on Ψ and ϕ by the still-Hermitian action of Eq.(A.30), since $\tilde{\Psi}$ and $\tilde{\phi}$ act as Langrange multipliers. The action of Eq.(A.30) is invariant under

$$\delta|\chi\rangle = Q|\Lambda\rangle \tag{A.31}$$

and this reproduces the transformation laws of Eq.(A.10).

We now turn to the construction of the interacting closed bosonic string. This can be found by a straightforward application of the methods discussed in the context of the open string in this paper. We will only briefly give the results. The homogeneous term in the transformation law of $|\chi\rangle$ can be seen to require an additional anticommuting term. Since the closed string is somewhat like the product of two open strings, we expect the vertex [41] to be

$$V^{closed} = \langle 0_{1,2,3} | V_F(\beta, \bar{\beta}, \bar{\beta}_0) V_F(\hat{\beta}, \hat{\bar{\beta}}, \hat{\bar{\beta}}_0) \cdot V_B(\alpha) V_B(\hat{\alpha}) P \tag{A.32}$$

where P is the projector

$$P = \oint_{z=0} \frac{dz}{z} z^{(L_0 - \hat{L}_0)}$$

We take the vacuum to be

$$\langle 0_{123} | = {}_1\langle + - | \, {}_2\langle + - | \, {}_3\langle + - | \tag{A.33}$$

We therefore postulate the transformation law to be

$$\delta \, {}_3\langle \chi | = \frac{1}{g} \, {}_3\langle \Lambda | \, \Phi^3 + 2V^{closed} \delta_0^3 \left(\frac{\alpha_2}{\alpha_3} |\chi\rangle_1 |\Lambda\rangle_2 - \frac{\alpha_1}{\alpha_3} |\chi\rangle_2 |\Lambda\rangle_1 \right) \tag{A.34}$$

while the interaction term is

$$\frac{2ig}{3} V^{closed} |\chi\rangle_1 |\chi\rangle_2 |\chi\rangle_3 \tag{A.35}$$

We find the condition for closure and invariance of the action [41] is

$$V^{closed} \sum_{r=1}^{3} \frac{\Phi^r}{\alpha^r} = 0 \tag{A.36}$$

The proof of this equation is very similar to that for the open bosonic string and is straightforward provided one utilizes the constraint of Eq.(A.28), and realizes that the $\bar{\delta}_0$ terms in V^{closed} die due to the choice of the vacuum in V.

For the same reasons as discussed above as to why the open string requires a four-point function, we do not expect any more terms for the closed string, and we expect the above action and transformation laws to be complete.

APPENDIX B

In this appendix we will give the free gauge-covariant action for the open superstring. We employ the same method as for the open bosonic string explained in detail in the main text.

The generators of the superconformal algebra are denoted L_A. In the Ramond sector $L_A = \{L_n, F_m;\ n,m = 0,\pm 1,.\}$ and in the Neveu-Schwarz sector they are $L_A = \{L_n,\ G_r,\ n = 0,\pm 1,\ldots \quad r = \pm\frac{1}{2}\pm\frac{3}{2},\ldots\}$. Their algebra is generically of the form

$$[L_A, L_B\} = f_{AB}{}^{C} L_C \tag{B.1}$$

and in detail these relations are:

$$\{F_m, F_n\} = 2L_{n+m} + D/_2\, m^2\, \delta_{m+n,0}$$
$$[L_m, F_n] = (\tfrac{m}{2} - n)\, F_{n+m} \tag{B.2}$$
$$[L_m, L_n] = (m-n)\, L_{n+m} + \frac{D}{8} m^3\, \delta_{m+n,0}$$

and

$$\{G_r, G_s\} = 2L_{r+s} + D/_2 (r^2 - \tfrac{1}{4})\, \delta_{r+s,0}$$
$$[L_m, G_r] = (\tfrac{m}{2} - r)\, G_{r+m}$$
$$[L_m, L_n] = (m-n)\, L_{m+n} + \frac{D}{8} m(m^2-1)\, \delta_{m+n,0} \tag{B.3}$$

The reader will have little trouble finding the "finite" set which was given in reference [16]. It is then easily generalized to the "infinite" set [40] which is as follows.

The field equations in the Neveu-Schwarz sectors are

$$(L_0 - \tfrac{1}{2})\, \psi + \sum_{r=\frac{1}{2}}^{\infty} G_{-r}\, \phi^r + \sum_{n=1}^{\infty} L_{-n}\, \chi^n = 0$$

$$L_m \chi^n - (L_0 + m + n - \tfrac{1}{2})\, \xi^n{}_m + (2m+n)\, \chi^{m+n} = 0$$

$$G_r \Psi + 2\phi^r + \sum_{s=\frac{1}{2}}^{\infty} G_{-s} \zeta^s{}_r + \sum_{n=1}^{\infty} L_{-n} \zeta^n{}_r$$
$$+ \sum_{s,n} \{ (s+\tfrac{3n}{2}) \zeta^n{}_s - 2 \zeta^s{}_n) \, \delta(n+s-r) = 0$$
$$L_n \Psi + 2n \chi^n + \sum_{s=\frac{1}{2}}^{\infty} G_{-s} \zeta^s{}_n + \sum_{m=1}^{\infty} L_{-m} \zeta^m{}_n \qquad \text{(B.4)}$$
$$- \sum_{s,r=\frac{1}{2}}^{\infty} (\tfrac{3r+s}{2}) \zeta^r{}_s \, \delta(r+s-n) + \sum_{p,m=1}^{\infty} (2p+m) \zeta^p{}_m \, \delta(n-p-m) = 0$$
$$G_r \phi^s + (L_0 + r + s - \tfrac{1}{2}) \zeta^s{}_r - (\tfrac{3r+s}{2}) \chi^{r+s} = 0 ; \; L_n \phi^s + (s+\tfrac{3n}{2}) \phi^{n+s}$$
$$- (L_0 + n + s - \tfrac{1}{2}) \zeta^s{}_n = 0 ; \; G_s \chi^n - (L_0 + s + n - \tfrac{1}{2}) \zeta^n{}_s + 2\phi^{n+s} = 0$$

and they are invariant under the transformations

$$\delta \Psi = \sum_{s=\frac{1}{2}}^{\infty} G_{-s} \Lambda^s + \sum_{n=1}^{\infty} L_{-n} \Omega^n$$
$$\delta \phi^r = -(L_0 + r - \tfrac{1}{2}) \Lambda^r \; ; \; \delta \chi^n = -(L_0 + n - \tfrac{1}{2}) \Omega^n$$
$$\delta \zeta^r{}_s = G_s \Lambda^r - (\tfrac{3s+r}{2}) \Omega^{r+s} ; \; \delta \zeta^r{}_n = -L_n \Lambda^r - (r + \tfrac{3n}{2}) \Lambda^{r+n} \qquad \text{(B.5)}$$
$$\delta \zeta^n{}_r = -G_r \Omega^n - 2 \Lambda^{r+n} ; \; \delta \zeta^n{}_m = -L_m \Omega^n - (2m+n) \Omega^{n+m}$$

In the Ramond sector we have the field equations

$$F_0 \Psi + \sum_{n=1}^{\infty} (F_{-n} \phi^n + L_{-n} \chi^n) = 0 ; \; F_n \phi^m + F_0 \zeta^m{}_n$$
$$- \tfrac{m}{2} \lambda^m{}_n + 2\mu^m{}_n - (\tfrac{m+3n}{2}) \chi^{m+n} = 0 ; \; F_n \chi^m + F_0 \lambda^m{}_n$$
$$- 2\tau^m{}_n + 2\zeta^m{}_n + 2\phi^{m+n} = 0 ; \; L_n \phi^m + F_0 \mu^m{}_n - \tfrac{n}{2} \zeta^m{}_n$$
$$- m \tau^m{}_n + (\tfrac{3n}{2} + m) \phi^{n+m} = 0 ; \; L_n \chi^m - F_0 \tau^m{}_n - \tfrac{n}{2} \lambda^m{}_n$$
$$- 2\mu^m{}_n + (2n+m) \chi^{n+m} = 0 ; \; F_n \Psi + \sum_{m=1}^{\infty} (F_{-m} \zeta^m{}_n + L_{-m} \lambda^m{}_n)$$
$$+ \sum_{p,m=1}^{\infty} (2\mu^p{}_m + (\tfrac{3}{2}p + m) \lambda^p{}_m) \delta(p+m-n) - 2F_0 \phi^n + \tfrac{5n}{2} \chi^n = 0 \qquad \text{(B.6)}$$
$$L_n \Psi + \sum_{m=1}^{\infty} (F_{-m} \mu^m{}_n + L_{-m} \tau^m{}_n) + 2n F_0 \chi^n + \tfrac{5n}{2} \phi^n$$
$$+ \sum_{p,m=1}^{\infty} \{ (2p+m) \tau^p{}_m - (m + \tfrac{3p}{2}) \zeta^p{}_m \} \, \delta(p+m-n) = 0$$

and they are invariant under the transformations

$$\delta\psi = \sum_{n=1}^{\infty}(F_{-n}\Lambda^n + L_{-n}\Omega^n)$$

$$\delta\phi^n = F_0\Lambda^n - \frac{n}{2}\Omega^n \;;\; \delta\chi^n = -F_0\Omega^n - 2\Lambda^n$$

$$\delta\zeta^n{}_m = F_m\Lambda^n - \left(\frac{n+3m}{2}\right)\Omega^{m+n} ; \; \delta\lambda^n{}_m = -F_m\Omega^n - 2\Lambda^{n+m}$$

$$\delta\mu^n{}_m = -L_m\Lambda^n - \left(\frac{3m}{2}+n\right)\Lambda^{n+m} , \; \delta\tau^n{}_m = -L_m\Omega^n - (2m+n)\Omega^{n+m} \tag{B.7}$$

We now find the analogues of K, η, d and D of the bosonic model in order to generalize the "infinite" set to the master set. To do this we examine the Virasoro charge. In the Neveu-Schwarz sector we define the bosonic annihilation and creation operators denoted g_r and $\bar{g}_r$. They obey the relations

$$[g_r, \bar{g}_s] = \delta_{r+s,0}$$
$$[g_r, g_s] = 0 = [\bar{g}_r, \bar{g}_s] \tag{B.8}$$

They have the Hermiticity properties

$$g_r^\dagger = g_{-r} \quad ; \quad \bar{g}_r^\dagger = \bar{g}_{-r}$$

A general functional of the α's and these oscillators is of the form

$$|\chi\rangle_{N-S} = \sum_{\substack{n,m\\r,s}} \beta^{n_1\dagger}\ldots\bar{\beta}_{m_1}^\dagger\ldots g^{r_1\dagger}\ldots\bar{g}_{s_1}^\dagger\ldots\psi^{m_1\ldots s_1\ldots}_{n_1\ldots r_1\ldots}|-\rangle$$

$$+\sum_{\substack{n,m\\r,s}} \beta^{n_1\dagger}\ldots\bar{\beta}_{m_1}^\dagger\ldots g^{r_1\dagger}\ldots\bar{g}_{s_1}^\dagger\ldots\phi^{m_1\ldots s_1\ldots}_{n_1\ldots r_1\ldots}|+\rangle \tag{B.9}$$

where the vacuum obeys the equations

$$g_r|\pm\rangle = \beta_n|\pm\rangle = \bar{\beta}_m|\pm\rangle$$
$$= \bar{g}_s|\pm\rangle = 0 \qquad n,m,r,s > 0$$

and

$$\beta_0|+\rangle = 0 \qquad ; |-\rangle = \bar{\beta}_0|+\rangle \tag{B.10}$$

The field content of the master set is clearly of the form

$$\psi^{K}_{K} , \ \phi^{K+1}_{K}.$$

where the indices can be either n or r. In order to restrict $|x\rangle_{N-S}$ to have this field content we impose the constraint

$$N_{N-S}|\chi\rangle_{N-S} = \left(N - \sum_{r=\frac{1}{2}}^{\infty}\left(g^{r\dagger}\bar{g}^{r} - \bar{g}^{r\dagger}g^{r}\right)\right)|\chi\rangle_{N-S} = 0 \tag{B.11}$$

The BRST charge corresponding to the superconformal group is

$$Q = :\sum_{A}\beta_A L_{-A} - \frac{1}{2}\sum_{A,B,C}\bar{\beta}_C f_{AB}{}^{C}\beta_{-A}\beta_{-B} - \frac{1}{2}\beta_0: \tag{B.12}$$

where

$$\beta_A = (\beta_n, g_r) \quad ; \ \bar{\beta}_A = (\bar{\beta}_n, \bar{g}_r)$$

It is shown in reference [46] that $Q^2 = 0$ in ten dimensions. We may rewrite this super-Virasoro charge as

$$Q^{N-S} = \beta_0 K - 2\bar{\beta}_0 M + D + d \tag{B.13}$$

where K, M, D and d are not involved. β_0 and $\bar{\beta}_0$ satisfy D = d and are defined by the same procedure as in the bosonic case. They obey the relations

$$d^2 = D^2 = 0 \ ; [d, K] = [D, K] = [M, d] = [M, D] = 0 \qquad ; \{d, D\} = 2MK \tag{B.14}$$

We then realize that the generalizations of equation (B.4) to the infinite set are given by [47]

$$K\psi + (D+d)\phi = 0$$
$$d\psi + D\psi + 2M\phi = 0 \qquad (B.15)$$

Equation (B.12) is invariant under the transformations [47]

$$\delta\psi = d\Lambda + D\Lambda - 2M\Omega.$$
$$\delta\phi = -K\Lambda + d\Omega + D\Omega \qquad (B.16)$$

The Neveu-Schwarz sector is almost identical to the open bosonic string. The equations of motion of the master set follow from the action [47][44]

$${}_{N-S}\langle\chi|\,Q^{N-S}\,|\chi\rangle_{N-S} \qquad (B.17)$$

and this action is obviously invariant under

$$\delta|\chi\rangle_{N-S} = Q^{N-S}\,|\Lambda\rangle_{N-S} \qquad (B.18)$$

We now repeat the procedure in the Ramond sector where life is a little more complicated. We introduce the oscillators f_n and $\bar{f}_n$ which obey the relations [46].

$$[f_n, \bar{f}_m] = \delta_{n+m,0}$$
$$[f_n, f_m] = 0 = [\bar{f}_n, \bar{f}_m] \qquad (B.19)$$

They obey the hermiticity properties $f_n^+ = f_n$, $\bar{f}_n^+ = -\bar{f}_{-n}$.

The super-Virasoro charge is given by

$$Q^R = :\sum_A f_A L_A - \frac{1}{2}\sum_{A,B,C} \bar{\beta}_C f_{AB}{}^C \beta_{-A}\beta_{-B} \qquad (B.20)$$

where $\beta_A = \{\beta_n, f_n\}$ and similarly for $\bar{\beta}_A$ in this sector. In ten dimensions it has been shown [46] that $Q^2 = 0$. We may rewrite Q as

$$Q^R = :f_0 F + \beta_0 K - 2\bar{\beta}_0 M + D + d + \bar{f}_0 J - \bar{\beta}_0 f_0^2 \tag{B.21}$$

We note that F, D, d, M, K and J obey the relations

$$\{d, D\} = 2KM + FJ \quad ; \quad F^2 = K$$
$$J = [M, F] \quad ; \quad d^2 = D^2 = [M, K] = 0 \text{ etc} \tag{B.22}$$

and all other commutators, and anticommutators vanish.

The master set contains the x-space functionals

$$\psi^K_K \; ; \; \phi^{K+1}_K . \tag{B.23}$$

Here K denotes the number of indices of either type. In terms of the action of the above operators on the component fields, the field equation of the master set may be written as [47]

$$F\psi + (D + d)\phi = 0$$
$$-(MF + FM)\phi + (D + d)\psi = 0 \tag{B.24}$$

and their gauge invariances [47] are

$$\delta\psi = (D + d)\Lambda - (MF + FM)\Omega$$
$$\delta\phi = F\Lambda + (D + d)\Omega . \tag{B.25}$$

We now wish to cast the above results in more compact form using Q^R. However, we observe that f_0 is a bosonic object and so no power of it vanishes. Clearly, in the master set there do not correspond the infinite set of fields that would be contained in an arbitrary expansion in f_0. We therefore consider an expansion of the form [47][48]

$$|\chi\rangle_R = (\psi + f_0 \phi)|-\rangle + F\phi|+\rangle \qquad \text{(B26)}$$

where

$$\psi = \beta^{+n_1} \cdots \bar{\beta}^{+}_{m_1} \cdots f^{p_1 +} \cdots \bar{f}^{+}_{q_1} \cdots \psi^{m_1 \cdots m_a \, q_1 \cdots q_d}_{n_1 \cdots n_b \, p_1 \cdots p_c} \qquad \text{(B.27)}$$

and similarly for ϕ.

The vacuua obey the conditions

$$0 = \beta_n|\pm\rangle = \bar{\beta}_n|\pm\rangle = f_n|\pm\rangle = \bar{f}_n|\pm\rangle \qquad n \geq 1$$

and (B.28)

$$\bar{f}_0|\pm\rangle = 0 \,, \; \beta_0|+\rangle = 0 \,; \; \bar{\beta}_0|+\rangle = |-\rangle$$

In order that $|\chi\rangle_R$ only contain the master set we must also impose the constraint

$$0 = N_R|\chi\rangle_R \equiv \Big(N - \sum_{n=1}^{\infty}(f^{n+}\bar{f}_n - \bar{f}_n^{+} f^n) - f_0 \bar{f}_0\Big)|\chi\rangle_R \qquad \text{(B.29)}$$

We can now examine the action of Q^R on $|\chi\rangle_R$. One finds

$$Q_R|\chi\rangle_R = (\hat{\psi} + f_0 \hat{\phi})|-\rangle + F\hat{\phi}|+\rangle \qquad \text{(B.30)}$$

where

$$\hat{\psi} = (D+d)\psi - (MF + FM)\phi$$

and

$$\hat{\phi} = F\psi + (D+d)\phi \qquad \text{(B.31)}$$

We observe that the action of Q maintains the form of the expansion.

Clearly the field equations (B.24) are given by

$$\beta_0 \varphi_R |\chi\rangle_R = 0 \tag{B.32}$$

and we now wish to find an action from which it follows. There are may ways of doing this; however, one method which was also useful in establishing the supersymmetry [47][48] of the open superstrings is as follows. Let us introduce

$$\chi_R = \psi + f_0 \phi + F \phi \beta_0 \tag{B.33}$$

We note that

$$|\chi\rangle_R = \chi_R |-\rangle. \tag{B.34}$$

We then introduce a new type of vacuum $|\tilde{\pm}\rangle$ which satisfies all the conditions that $|\pm\rangle$ does with the exception of the condition $\bar{f}_0 |\pm\rangle = 0$, which we now replace by

$$f_0 |\tilde{\pm}\rangle = 0 \tag{B.35}$$

Since under Hermitian conjugation

$$\begin{aligned} (|\pm\rangle)^\dagger &= \langle \tilde{\pm}| \\ (|\tilde{\pm}\rangle)^\dagger &= \langle \pm | \end{aligned} \tag{B.36}$$

we have the equation

$$\langle -| f_0 = 0 \tag{B.37}$$

The action [47][48] that leads to the equation is (B.32) is given by

$$-\frac{1}{2} \langle -| \bar{f}_0 \chi_R^\dagger \beta_0 \varphi_R \chi_R |-\rangle. \tag{B.38}$$

The reader is referred to references [44] and [49] for some alternative methods of formulating the master set of reference [47].

The counting of on-shell states goes the same way as for the open bosonic string. A little thought shows that one does indeed recover the light-cone count.

REFERENCES

[1] G. Veneziano, Nuovo Cimento 57A, 190 (1968).

[2] Some reviews of this literature are:-
V. Alessandrini, D. Amati, M. Le Bellac and D.I. Olive, Phys. Reports 1C, 170 (1971).
P. Frampton, Dual Resonance Models (Benjamin, 1974).
S. Mandelstam, Phys. Reports 13C, 259 (1974).
C. Rebbi, Phys. Reports 12C, 1 (1974).
J. Scherk, Rev. Mod. Phys. 47, 1213 (1975).
J.H. Schwarz, Phys. Reports 8C, 269 (1973).
G. Veneziano, Phys. Reports 9C, 199 (1974).
J.H. Schwarz, Phys. Reports 89, 223 (1982).
M. Green, in Surveys of High Energy Physics 3, 127 (1983).

[3] P.A.M. Dirac, Lectures in Quantum Mechanics. (Belfer Graduate School of Science, Yeshiva University; New York 1964).

[4] P. Goddard, J. Goldstone, C. Rebbi and C. Thorn, Nucl. Phys. B56, 109 (1973).

[5] Y. Nambu, Proc. Int. Conf. on Symmetries and Quark Modes, Detroit 1969, (Gordon and Breach, New York 1970).

[6] M. Virasoro, Phys. Rev. D1, 2933 (1970).

[7] R.C. Brower, Phys. Rev. D6, 1655 (1972);
P. Goddard and C.B. Thorn, Phys. Lett. 40B, 235 (1972).

[8] E. Cremmer and J.-L. Gervais, Nucl. Phys. B90, 410 (1975);

M. Kaku and K. Kikkawa, Phys. Rev. D10, 1110, 1823 (1974).

[9] W. Siegel, Phys. Lett. 148B, 556 (1984); 149B, 157 (1984).

[10] S. Raby and P.C. West, unpublished.

[11] T. Banks and M. Peskin, Proceedings of the Symposium on Anomalies, Geometry and Topology, Argonne Nat. Lab., March 1985;
M. Kaku and J. Lykken, ibid.

[12] C. Brower and C.B. Thorn, Nucl. Phys. B31, 163 (1971).

[13] P. van Nieuwenhuizen, Nucl. Phys. B60, 478 (1973).

[14] A. Neveu and P. West, Nucl. Phys. B268, 125 (1986).

[15] L.P.S. Singh and C.R. Hagen, Phys. Rev. D9, 898 (1974).

[16] A. Neveu, H. Nicolai and P. West, Nucl. Phys. B264, 573 (1986).

[17] A. Neveu, J. Schwarz and P. West, Phys. Lett. 164B, 51 (1985).

[18] T. Banks and M. Peskin, SLAC preprint 3740 (1985).

[19] A. Neveu, H. Nicolai and P. West, Phys. Lett. 167B, 307 (1986).

[20] Y. Meurice, C.E.R.N. preprints, CERN TH4356/86.

[21] E. Witten, "Non-commutative Geometry and String Field Theory, Princeton preprint (1985).

[22] W. Siegel and B. Zweibach, Berkeley preprint UC12-PTH-85130 (1985).

[23] K. Itoh, T. Kugo, H. Kunitomo and H. Ooguri, Kyoto preprint.

[24] A. Restuccia and J. Taylor, unpublished.

[25] K. Kato and K. Ogawa, Nucl. Phys. B212, 443 (1983). These co-ordinates were also utilized in reference [9].

[26] For a discussion of this point see also the lectures of Tom Banks in this volume.

[27] E. Fradkin and G. Vilkovisky, Phys. Lett. 55B, 224 (1975).

I.A. Batalin and G. Vilkovisky, Phys. Lett. 69B, 309 (1977).

E.S. Fradkin and T.E. Fradkina, Phys. Lett. 72B, 343 (1978).

[28] S. Scinto, Lett. Nuovo Cimento 2, 411 (1969).

[29] L. Caneschi, A. Schwimmer and G. Veneziano, Phys. Lett. 30B, 351 (1969)

[30] J.L. Gervais and B. Sakita, Phys. Rev. Lett. 30, 719 (1973).

S. Mandelstam, Nucl. Phys. B64, 205 (1973), B69, 77 (1974).

[31] A.M. Polyakov, Phys. Lett. 703B, 207 (1961), 103B, 211 (1961).

[32] For further discussion see the lectures of L. Alvarez-Gaumé in this volume.

[33] E. Cremmer and J. Gervais, Nucl. Phys. B76, 209 (1974).

[34] E. Cremmer and J. Gervais, Nucl. Phys. B90, 410 (1975).

[35] M. Kaku and K. Kikawa, Phys. Rev. D10, 110 1823 (1974).

M. Kaku, Phys. Rev. D10, 3943 (1974),

[36] J.F.L. Hopkinson, R.W. Tucker and P.A. Collins, Phys. Rev. D12, 1653 (1975).

[37] M. Green and J. Schwarz, Nucl. Phys. B218, 43 (1983).

[38] S. Mandelstam, Nucl. Phys. B64, 205 (1973).

[39] S. Mandelstam, Nucl. Phys. B69, 77 (1974).

[40] A. Neveu and P. West, Phys. Lett. 168B, 192 (1985).

[41] A. Neveu and P. West, "Symmetries of the Interacting Gauge-Covariant Bosonic String", CERN preprint TH4358/86 (1986).

[42] H. Hata, K. Itoh, T. Kugo, H. Kunitomo and K. Ogawa, Kyoto preprints.

[43] M. Awada, Cambridge University preprints (1985).

[44] T. Banks, M.E. Peskin, C.R. Preitschopf, D. Friedan and E. Martinec, SLAC Preprint 3853 (1985).

[45] J.H. Schwarz, Caltech preprint 68-1304 (1985).

[46] Y. Kazama, A. Neveu, H. Nicolai and P. West, Symmetry Structures of Superstring Field Theories, CERN-TH4301/85, (1985) Nucl. Phys. B to be published.

[47] Y. Kazama, A. Neveu, H. Nicolai and P. West, "Space-time Supersymmetry of the Covariant Superstring", CERN preprint TH.4418/86 (1986).

[48] J. Araytn and A.H. Zimerman, Rev. J. Mod. Phys. A to be published;
H. Terao and J. Distler, Hiroshima preprint RRK-86-3 (1985);
A. le Clair and J. Distler, Harvard preprint HUTY-86/A008 (1986);
G.D. Daté, M. Gunaydin, M. Pernici, K. Pilch and P. van Nieuwenhuizen, Stony Brook preprint ITP-SB-86 (1986);
S.P. de Alvis and N. Ohta, Texas preprint UTTG-06-86 (1986).

GAUGE INVARIANT ACTIONS FOR STRING MODELS*

THOMAS BANKS

Stanford Linear Accelerator Center
Stanford University, Stanford, California 94305
and
Tel Aviv University Ramat Aviv, Israel

ABSTRACT

String models of unified interactions are elegant sets of Feynman rules for the scattering of gravitons, gauge bosons, and a host of massive excitations. The purpose of these lectures is to describe progress towards a nonperturbative formulation of the theory. Such a formulation should make the geometrical meaning of string theory manifest and explain the many "miracles" exhibited by the string Feynman rules. As yet only partial success has been achieved in realizing this goal. Most of the material presented here already appears in the published literature but there are some new results on gauge invariant observables, on the cosmological constant, and on the symmetries of interacting string field theory.

1. INTRODUCTION (WITH APOLOGIES TO E. AMBLER)[1]

A Frenchman named Chamfort, who should have known better, once said that chance was a nickname for Providence.

It is one of those convenient, question begging aphorisms coined to discredit the unpleasant truth that chance plays an important, if not predominant, part in human affairs. Yet it was not entirely inexcusable. Chance does occasionally operate with a sort of fumbling coherence readily mistakable for the workings of a self conscious Providence.

The history of the string model is an example of this. The fact that an S matrix theorist like Veneziano[2] should discover the key to the quantum theory of gravity by looking for solutions of hadronic finite energy sum rules is alone grotesque. That Wess and Zumino[3] should discover spacetime supersymmetry by generalizing the two dimensional supersymmetry of the Neveu-Schwarz-Ramond model, that Schwarz and Green[4] should spend years they could ill afford probing into the shadowy structure of the superstring, that the theory of Everything should be discovered because Harvey[5] and Thierry-Mieg[6] noticed an odd identity for Casimir operators ... these facts are breathtaking in their absurdity.

---

*Work supported by the Department of Energy, contract DE–AC03–76SF00515.

Yet, when these events are seen side by side with the other facts in the case[7] it is difficult not to become lost in superstitious awe. Their very absurdity seems to prohibit the use of the words "chance" and "coincidence". For the sceptic there remains only one consolation: if there should be such a thing as superhuman Law, it is administered with subhuman efficiency. The choice of the dual model of hadrons as an instrument for revealing the superstring to the world could have been made only by an idiot.

More than most physical theories then, the string model cries out for an ahistorical presentation which starts from simple and deep physical principles and derives the dual amplitudes in a logical manner. My attempt to present such a description of string theories will be hampered by the fact that a true Theory of Strings does not yet exist. I will therefore begin with a brief exposition of the most coherent description of string theory now available; the covariant Feynman rules for string perturbation theory. I will then describe the framework within which all current attempts to derive these rules are based: gauge invariant functional field theory (GIFFT) I am not at all sure that the ultimate description of string theory will fall within this framework. Nonetheless, I believe that GIFFT is a correct description of string theories which is based on conceptually simple generalizations of things that we know. Even if it turns out to be a bit clumsy (as it seems at present) it appears to be the likeliest avenue of access to a more satisfying description of strings.

There has been an explosion of interest in string field theory this year. I have been unable to refer to every paper on the subject in the text. Reference 49 is a long (but probably still incomplete) list of papers on string field theory. The approach to string field theory presented here is a combination of ideas developed in collaboration with Michael Peskin and Christian Preitschopf and the elegant formalism invented by Ed Witten. I would like to thank them and Dan Friedan and Emil Martinec for numerous conversations about string theory. Finally, I would like to remind the reader that the initial development of covariant string field theory was solely due to Warren Siegel (ref. 18). His papers have been a continuing source of inspiration for myself and my collaborators.

2. STRINGY FEYNMAN RULES

In ordinary field theory, the perturbation expansion may be expressed in terms of path integrals for a single spacetime degree of freedom $x^\mu(\tau)$, which propagates with the standard action of a relativistic particle. The propagation may be visualized as taking place on a net in spacetime whose topological properties depend on the field theory Lagrangian and the order of perturbation theory. The endless series of possibilities for the topology of such a net is responsible for the complexity of higher order Feynman graphs and at least partially responsible for the wide range of possible Lagrangians.

If on the other hand we consider a string propagating in space time we see a very different picture. The string world sheet is a two dimensional manifold, and the simple requirement that it be smooth considerably restricts its topology. If the manifold is to represent the propagation of a closed oriented string for example, it is completely classified by giving the number of handles and the number of boundaries (corresponding to incoming and outgoing strings). This leads one to suspect that a perturbation expansion of a theory of interacting strings would have only one diagram in each order. This is indeed correct, though as we shall see, these unique dual diagrams are related to ordinary Feynman diagrams in a manner reminiscent of the relation between the latter and the diagrams of old fashioned time ordered perturbation theory.

At the present time, the only complete derivation of the dual Feynman rules which I am about to present is in the so called light cone gauge. A theory of interacting strings is constructed in infinite momentum frame Hamiltonian formalism. The perturbation series for this Hamiltonian is worked out and then one sees how to add up the various terms in a given order to give a single dual diagram. To be frank, a complete and rigorous derivation of the dual diagrams has only been constructed (including such details as numerical factors) through one loop order. Lorentz invariance becomes obvious only at the end of the calculation, and other fabulous properties of string models such as duality and general coordinate invariance, are not obvious at all. It is tempting to regard the infinite momentum frame string theory as a gauge fixed version of a beautiful action which manifests all of these wondrous properties at a glance. In later chapters we will review the attempts that have been made to realize this dream.

The string Feynman rules are written in terms of a conformally invariant two dimensional field theory, which is also invariant under two dimensional general coordinate transformations. For the expansion around flat ten dimensional space this field theory is given by a Lagrangian first written by Brink, DiVecchia, Howe,[8] Deser and Zumino[9] and explored by Polyakov.[10]

$$\mathcal{L} = \sqrt{g}\, g^{\alpha\beta} \frac{\partial x^\mu}{\partial \xi^\alpha} \frac{\partial x^\nu}{\partial \xi^\beta} \tag{1}$$

String S matrix elements are given in terms of expectation values of certain generally coordinate invariant, conformally invariant operators (called vertex operators) in this field theory. This gives us the tree level S matrix. As usual the expectation values can be written as Euclidean functional integrals. The fields in these functional integrals are defined on the plane, or equivalently, the Riemann sphere. L loop corrections to the amplitudes are given by the same functional integral on a surface with L handles. (These are the rules for closed oriented strings. Open and non-orientable strings have slightly more complicated rules.)

In order to actually compute these expectation values we have to choose a gauge for two dimensional coordinate transformations and Weyl transformations and introduce the corresponding Faddeev-Popov ghosts. We also have to make sure that there are no anomalies in the classical invariances of the action. This restricts the dimension of space time to be 26 for the bosonic string. If we are expanding around a curved space time, it also restricts the geometry to be such that the corresponding non-linear sigma model has vanishing beta function. It has been argued[7] that this restriction is equivalent to the classical equations of motion for the string field theory. We will derive the restriction to d=26 in another way below.

The ghosts for local conformal invariance are trivial, they have no kinetic energy. Those of two dimensional reparametrizations have a Lagrangian

$$\mathcal{L} = b\partial_{\bar{z}}c + \bar{b}\partial_z\bar{c} \; ; \qquad \begin{array}{l} z = \xi_1 + i\xi_2 \\ \bar{z} = \xi_1 - i\xi_2 \end{array} . \tag{2}$$

The total Lagrangian of the system is

$$\mathcal{L} = \frac{\partial x^\mu}{\partial z}\frac{\partial x^\mu}{\partial \bar{z}} + b\partial_{\bar{z}}c + \bar{b}\partial_z\bar{c} \; . \tag{3}$$

It is invariant under the Becchi-Rouet-Stora-Tyutin (BRST) transformation:

$$\begin{aligned} \delta x^\mu &= \epsilon(c\partial_{\bar{z}}x^\mu + \bar{c}\partial_z x^\mu) \qquad \delta b = \epsilon\left(\bar{c}\partial_z b + c\partial_{\bar{z}} b + 2\partial_{\bar{z}}cb + \frac{\partial x}{\partial z}\frac{\partial x}{\partial z}\right) \\ \delta c &= \epsilon(c\partial_{\bar{z}}c + \bar{c}\partial_z c) \\ \delta \bar{c} &= \epsilon(c\partial_{\bar{z}}\bar{c} + \bar{c}\partial_z\bar{c}) \qquad \delta\bar{b} = \epsilon\left(\bar{c}\partial_z\bar{b} + c\partial_{\bar{z}}\bar{b} + 2\partial_z\bar{c}b + \frac{\partial x}{\partial \bar{z}}\frac{\partial x}{\partial \bar{z}}\right) . \end{aligned} \tag{4}$$

At the level of classical manipulations, the BRST transformation is nilpotent. However, this breaks down at the quantum level. The commutation relations for the conformal gauge variables of the string are given by (we specialize to open strings where $\bar{b}$ and $\bar{c}$ are fixed in terms of b, c by the boundary conditions)

$$x^\mu(z) = \hat{x}^\mu + \frac{i}{2}\sum_{n>0}\frac{1}{n}\left(\alpha^\mu_n - \alpha^\mu_{-n}\right)\left(z^n + \bar{z}^n\right)$$

$$\begin{aligned} b(z) &= \sum z^{-n-2} b_n ; \\ c(z) &= \sum z^{-n+1} c_n ; \end{aligned}$$

$$[\alpha_n^\mu, \alpha_m^\nu] = n\delta(n+m)\eta^{\mu\nu}\,; \quad [b_n, c_m]_+ = \delta(n+m)\ . \tag{5}$$

The vacuum state is defined by

$$\alpha_n^\mu\,|\Omega\rangle = b_n\,|\Omega\rangle = c_n\,|\Omega\rangle = 0\,; \quad n > 0\ . \tag{6}$$

We will normal order all operators with respect to this state.

Formally, if T_a are the generators of the residual gauge group of a gauge theory:

$$[T_a, T_b] = f_{ab}{}^c T_c \tag{7}$$

then the BRST generator is given by

$$Q = c^a(T_a + \tfrac{1}{2}\,T_a^{gh}) \tag{8}$$

where $T_a^{gh}(\equiv\ c^b f_{ab}{}^c\ \frac{\partial}{\partial c^c})$ are the generators of residual gauge transformations on the ghosts. Simple manipulations involving the Jacobi identity prove that $Q^2 = 0$.

For the string the residual gauge transformations are the conformal transformations generated by the Virasoro operators:

$$\begin{aligned} L_n^x &= \tfrac{1}{2}\ :\sum \alpha_{n+k}^\mu \alpha_{-k}^\mu\ : \\ L_n^{gh} &=: \oint \frac{dx}{(2\pi i)}\, z^{n+1}(c\partial_z b + 2(\partial_z c)b)\ : -1\ \delta(n)\ . \end{aligned} \tag{9}$$

These operators have anomalous commutation laws

$$\begin{aligned} [L_n^x, L_m^x] &= (n-m)\,L_{n+m}^x + \frac{d}{12}\,(n^3-n)\delta(m+n) \\ [L_n^{gh}, L_m^{gh}] &= (n-m)\,L_{n+m}^{gh} - \frac{13}{6}\,(n^3-n)\delta(m+n)\ . \end{aligned} \tag{10}$$

The operator L_0^{gh} which appears in ((9)-(10)) is not normal ordered with respect to the vacuum. Rather, its ordering is chosen so that L_0, L_{-1}, L_1 generates an SL(2) subalgebra of the Virasoro algebra. For the Virasoro generators of the x^μ variables, this requirement is equivalent to normal ordering with respect to the vacuum. In other words, we normal order the Virasoro generators with respect to an SL(2) invariant state (it is unique). The vacuum defined above is related to this SL(2) invariant state by the action of c_{-1}. It therefore satisfies

$$|\Omega\rangle = c_1 |0\rangle$$
$$(L_0 + 1) |\Omega\rangle = 0 . \tag{11}$$

The anomalous commutation rules of the L_n and the necessity of normal ordering the BRST charge, ruin the proof that $Q^2 = 0$. These difficulties cancel when the spacetime dimension, d, equals 26. Notice that it is at precisely this point that the full conformal generators $L_n + L_n^{gh}$ have no commutator anomaly. It can be proven in general (as long as the ghosts are free fields) that the BRST charge is nilpotent if and only if the full commutator anomaly vanishes.[11]

Given a nilpotent BRST charge, string Feynman rules are constructed in the following way: we identify a set of local vertex operators whose integrals over the world sheet are BRST invariant. These are all dimension two conformal fields.[11] If the spacetime background which defines the conformal field theory we are studying has some isometries, the vertex operators can be classified by their transformation laws under the isometry group. In particular, if spacetime has some flat, Euclidean, dimensions, the vertex operators can be chosen to carry fixed energy and momentum. Scattering amplitudes are then defined as the vacuum expectation values of time ordered products of integrated vertex operators in the two dimensional conformal field theory. Higher loop corrections are calculated by expressing the expectation values as Euclidean functional integrals, and then performing the same functional integral on world sheets of arbitrary topology.

In order to complete our description of the machinery necessary for constructing these rules we must show how to compute expectation values in the theory at hand. The space of states in the BRST invariant theory is a tensor product with many factors. Of particular interest is the factor describing the ghost zero modes. These are fermionic operators obeying:

$$[b_0, c_0]_+ = 1 .$$

The factor space in which they act is two dimensional. Define the state $|\downarrow\rangle$ by

$$b_0 |\downarrow\rangle = 0 . \tag{12}$$

Then the space is spanned by

$$|\downarrow\rangle \; ; \quad c_0 |\downarrow\rangle . \tag{13}$$

These states each have zero scalar product with themselves but their overlap is one. The piece of the vacuum $|\Omega\rangle$ which is in the zero mode factor, is the state $|\downarrow\rangle$. Thus operators which have nonzero expectation value in this state must carry precisely one factor of the zero mode c_0.

It should be emphasized that the general framework for string perturbation theory presented above is not tied to the flat space backgrounds in which string theory is traditionally presented. Any representation of the Virasoro algebra with appropriate central charge can be the basis for a sensible string perturbation theory (actually there are further restrictions which appear only at one loop). However since we do not yet know very much about the representations of the conformal group associated with curved manifolds, practical calculations are restricted to generalized toroidal backgrounds.

Let us now introduce a useful operator: the ghost number g, by

$$g \equiv \sum_{n>0}(b_{-n}c_n + c_{-n}b_n) + \tfrac{1}{2}\left[c_0, b_0\right]$$

g is antihermitian and takes the values $-1/2$ and $-3/2$ respectively on the states $|\Omega\rangle$ and $|0\rangle$. The ghost number of any other state is determined by noting that $c(z)$ carries ghost number 1 and $b(z)$, -1. It follows that $[g, Q] = Q$.

The basic equation of the first quantized string theory that we have described is $QA = 0$. Obviously solutions are determined only up to BRST exact forms $A = Q\epsilon$. It it thus of interest to determine what the nontrivial BRST invariant states look like. This has been determined by Kato and Ogawa.[12] A more rigorous mathematical discussion has been given by Frenkel et. al.[13] The methods of these authors are extremely powerful and elegant and should be useful for further investigations of the subject. The results of both of these groups are the following. The only nontrivial solutions of $QA = 0$ are states of ghost number $\pm 1/2$ and (but only for spacetime momentum equal to zero) $\pm 3/2$. The fact that we get both signs of the ghost number has to do with the ambiguity in the choice of the vacuum for the ghost zero mode. We resolve the ambiguity by choosing physical states to have ghost number -1/2. The general solution of the BRST equation with ghost number -1/2 is

$$A = \Psi \otimes \Omega \tag{14}$$

where Ω is the ghost vacuum annihilated by b_0 and Ψ is a physical state:

$$L_n^x \Psi = 0; \quad n > 0; \quad (L_0^x - 1)\Psi = 0 \tag{15}$$

Note that the minus one in this equation is determined in the following way. We define L_0 for any conformal field theory as the commutator of L_{-1} and L_1

divided by two. This resolves the normal ordering ambiguity. One then finds that L_0 annihilates the vacuum in the x^μ sector while it has eigenvalue -1 on the ghost vacuum, Ω.

The nontrivial BRST cohomology with nonzero momentum thus consists of states of ghost number -1/2 which are in one to one correspondence with the physical states of the dual resonance model. The ghost number -3/2 cohomology is one dimensional and consists of the state which is invariant under the SL(2) subgroup of the conformal group.

This ends our brief survey of string Feynman rules. In the next chapter we will exploit the technology that we have described here to construct a gauge invariant, free string field theory.

3. FIELD THEORY OF OPEN BOSONIC STRINGS

The basic equation of the first quantized string theory that we described in the previous section is the BRST equation

$$QA = 0\ . \tag{16}$$

The program of string field theory is to derive this equation as the equation of motion of a functional action $S[A]$. Since the solutions of (16) are determined only up to a BRST exact form, $\delta A = Q\epsilon$, the action we will write down has a natural gauge invariance. One may wonder why a similar gauge invariance does not arise in the transition from the relativistic particle action to the Klein-Gordon equation. In fact it does,[14] but the gauge transformations are trivial. The usual scalar field is a gauge invariant field strength, and since it has a local action, there is no need to keep the gauge structure.

The equation Q=0 does not completely determine the class of allowable string states. Physical states of the string are BRST invariant equivalence classes with ghost number $-1/2$. Generically they take the form

$$\Psi \otimes \Omega \qquad L_n\Psi = (L_0 - 1)\Psi = 0\,; \quad n > 0 \tag{17}$$

though there are still some states of the form $Q\epsilon$ within this class. An action of the form $\langle A|Q|A\rangle$ will obviously have $QA = 0$ as its equation of motion. Furthermore, this action will be nonzero only if the ghost number of A is $-1/2$. Thus we make the natural assumption that: *The classical fields of string field theory are the states of the first quantized string with ghost number* $-1/2$. The action is then

$$\langle A|Q|A\rangle \tag{18}$$

which is nonvanishing.

As promised, this action has the gauge invariance $\delta A = Q\epsilon$, where ϵ is a general state with ghost number -3/2. To understand the content of this invariance we expand A, ϵ, and the action as:

$$A = \int d^d x \{\phi(x) - i\alpha^{\mu}_{-1} A_{\mu}(x) + b_{-1} c_0 \xi(x) + \dots\} |\Omega, x\rangle$$

$$\epsilon = \int d^d x \{b_{-1}\Lambda(x) |\Omega, x\rangle + \dots\}; \quad \hat{x}^{\mu} |\Omega, x\rangle = x^{\mu} |\Omega, x\rangle \ ; \quad p_{\mu} = \frac{1}{i}\frac{\partial}{\partial x^{\mu}}$$

$$Q = c^0 \left(\tfrac{1}{2} p^2 - 1 + \sum_{n>0} \left[\alpha^{\mu}_{-n}\alpha^{\mu}_{n} + n(c_{-n}b_n + b_{-n}c_n) \right] \right)$$

$$+ c_1(p_{\mu}\alpha^{\mu}_{-1} + \tfrac{1}{2} c_{-1} b_0 + \dots) + \dots \qquad (3.4)a$$

$$\begin{aligned} &\Delta A = Q\epsilon \\ &\Delta\phi(x) = 0 \\ &\Delta A_{\mu}(x) = \partial_{\mu}\Lambda(x) \\ &\Delta\xi(x) = \tfrac{1}{2} p^2 \Lambda(x) \end{aligned} \qquad (3.4)b$$

$$S = \tfrac{1}{2} \int \phi(p^2 - 2)\phi + \tfrac{1}{2} \int A^{\mu} p^2 A_{\mu} + \tfrac{1}{2} \int \xi^2 + i \int \xi p^{\mu} A_{\mu}$$

$$\to \tfrac{1}{2} \int \phi(p^2 - 2)\phi - \tfrac{1}{4} \int (\partial_{\mu} A_{\nu} - \partial_{\nu} A_{\mu})^2 \qquad (19)c$$

Note that the scalar field ϕ is a tachyon and is gauge invariant (remember our comment about the relativistic particle and the Klein-Gordon equation). The string gauge invariance reduces to ordinary Maxwell gauge invariance for the massless vector field. At higher levels we find massive spinning fields described by Stueckelbergian[15] gauge invariant Lagrangians. Siegel and Zweibach[16] have conjectured that these Lagrangians can all be obtained by dimensionally reducing massless spinning Lagrangians in one higher dimension. They have carried out the expansion of the string action through particles of spin five. These explicit computations depend of course on the detailed structure of the Lagrangian describing single string propagation in flat space time. It should be emphasized that nothing else that we will say in these lectures depends on these details.

Rather, it involves only the structure of the Virasoro algebra and its superconformal generalizations. Thus most of our discussion is applicable to the description of small string fluctuations about an arbitrary background spacetime whose associated sigma model is conformally invariant. As mentioned in the previous section, these are the classical solutions of string theory.[7]

It is important to point out that the classical gauge invariance of the string action depended on the fact that the BRST charge is nilpotent, which is only true in 26 dimensions. Classical string theory is not a gauge invariant system in any other dimension. There are probably further violations of string gauge invariance in string loop amplitudes. Eliminating them further restricts the structure of the theory. (There is a good argument, which I will not reproduce here, that all such anomalies vanish if the integrands of the loop amplitudes are invariant under modular transformations)

We now have a gauge invariant action which obviously reproduces the correct kinematics for the low lying levels of the string. In order to see that it gives a complete description of the string spectrum we must somehow fix the gauge and show that the quantum mechanics of our system is equivalent to the light cone functional field theory of strings invented by Kaku and Kikkawa.[17] In order to show this we will first gauge fix our action to the covariant gauge discovered by Siegel.[18] Then we will use an argument of Parisi and Sourlas[19] to show that Siegel's formalism is equivalent to that of Kaku and Kikkawa. A direct derivation of the descent to the light cone gauge can be found in Ref. 20.

The gauge fixing condition for Siegel's gauge is simply $b_0 A = 0$. It is easy to see that it is always possible to choose the gauge parameter ϵ so that this equation is satisfied. Just expand the BRST operator in powers of the ghost zero modes:

$$Q = c_0 K + d + \delta - 2b_0 \Downarrow \ . \tag{20}$$

Explicit expressions for the coefficients of c_0 and b_0 are:

$$\begin{aligned} K &= \tfrac{1}{2} p^2 + \sum_n : \alpha_n^\mu \alpha_{-n}^\mu : + \sum_n : n b_n c_{-n} : -1 \\ \Downarrow &= \sum_n : n c_n c_{-n} : \ . \end{aligned} \tag{21}$$

Nilpotency of Q implies:

$$\left[K, \begin{matrix} d \\ \delta \\ \Downarrow \end{matrix} \right] = 0 \quad \left[\Downarrow, \begin{matrix} d \\ \delta \end{matrix} \right] = 0 \quad d\delta + \delta d = 2K \Downarrow \quad d^2 = \delta^2 = 0 \ . \tag{22}$$

We can also expand the state A and the gauge parameter ϵ as:

$$\begin{aligned} A &= \Phi + c_0 \chi \\ \epsilon &= \epsilon_0 + c_0 \epsilon_1 \ . \end{aligned} \tag{23}$$

The gauge transformations are then:

$$\begin{aligned} \Delta\Phi &= (d+\delta)\epsilon_0 - 2 \Downarrow \epsilon_1 \\ \Delta\chi &= K\epsilon_0 - (d+\delta)\epsilon_1 \ . \end{aligned} \tag{24}$$

It is clear that as long as the operator K is invertible we can use ϵ_0 to eliminate χ. This is similar to the elimination of the longitudinal components of the vector potential in Landau gauge electrodynamics. There is no more off shell gauge invariance. Transformations which preserve the condition $b_0 A = 0$ do not effect the Φ component, except for states with $K = 0$.

The Faddeev-Popov determinant for this gauge fixing condition is the determinant of the operator $b_0 Q$, which is then the operator which appears in the Faddeev Popov ghost action. Since $Q^2 = 0$, this operator is singular and the ghost action has its own gauge invariance $\Delta G = QE$, with E an arbitrary state of ghost number $-5/2$. There is no similar gauge invariance for the antighost field. The operator b_0 projects out the piece of the antighost field (which is a state of ghost number $1/2$) that does not contain the operator c^0, but since this projection is completely nondynamical, we do not have to include a Faddeev-Popov determinant to compensate for it. The ghost system thus bears some resemblance to that encountered in the theory of p-form gauge fields[21] (there are ghosts for ghosts) but is somewhat simpler (there are no hidden ghosts[22] and the ghost counting is trivial). Baulieu and Ouvry[23] have shown how to introduce extra auxiliary fields to make the analogy with p form gauge fields exact. It is not clear that this is useful for studying the interacting theory.

It is clear that we can gauge fix the ghost system with the same gauge condition that we employed for the classical fields. The ghosts of ghosts will be bosons, their action is gauge invariant under a transformation identical to that of the classical field and the ghost (except that the gauge parameter has ghost number $-7/2$) and the procedure obviously continues indefinitely. The upshot of all of this is that our gauge fixed system consists of *all* states of the Hilbert space of string fields and ghosts which are annihilated by the antighost zero mode. The only piece of the BRST charge which has non-zero matrix elements between such states is the term proportional to K. Our gauge fixed action thus has the form

$$\langle \Phi | c_0 K | \Phi \rangle \ ; \quad b_0 \, |\Phi\rangle = 0 \tag{25}$$

which is the action originally proposed by Siegel.

We next want to show that Siegel's action is equivalent to the light cone gauge action of Kaku and Kikkawa. It is important to remember that the expansion coefficients of odd numbers of two dimensional ghost fields in the functional field $\Phi(x, b, c)$ are Grassmann variables since they are odd order ghosts in the spacetime sense. This rule can be easily remembered, for it is simply the requirement that the functional field have fixed statistics. This is a remarkable connection between spacetime and world sheet statistics. We will see more of it when we deal with the superstring.

Let us assume that we can choose the conformal field theory which defines our "matter" representation of the Virasoro algebra so that it is a free field theory in two space time dimensions (the two light cone directions) plus an operator $K_\perp$ which only refers to the transverse dimensions. If we think of L_0 as the Hamiltonian of a nonlinear sigma model, we are taking the background spacetime metric to be in light cone gauge. The string field theory action now takes the form

$$\left\langle \Phi \middle| c_0 \left[\tfrac{1}{2}(p_+ p_-) + L_0^{gh} - 1 + : \sum_n \alpha_n^+ \alpha_{-n}^- : + K_\perp \right] \middle| \Phi \right\rangle . \tag{26}$$

Since

$$[\alpha_n^+, \alpha_m^-] = n\delta(n+m) \tag{27}$$

we can write this as: $(\alpha_n \equiv \alpha_n^-)$

$$\left\langle \Phi \middle| c_0 \left[\tfrac{1}{2}(p_+ p_-) + (K_\perp) + \sum_n n \left(\alpha_{-n} \frac{\partial}{\partial \alpha_{-n}} + c_{-n} \frac{\partial}{\partial c_{-n}} \right) \right] \middle| \Phi \right\rangle \tag{28}$$

which has an obvious (Parisi Sourlas) supersymmetry connecting α_n and c_n. Now consider an arbitrary "String Green's Function", of functional fields which do not depend on the ghosts or the longitudinal oscillators

$$\int d\Phi\, e^{iS[\Phi]} \Phi(p_+^1, p_-^1, p_\perp^1, x_\perp^1(\sigma)) \cdots \Phi(p_+^k, p_-^k, p_\perp^k x_\perp^k(\sigma)) . \tag{29}$$

Clearly, this will have the same value as the Kaku-Kikkawa Green's function:

$$\int d\Phi\, e^{iS_{KK}[\Phi]} \Phi(p_+^1, p_-^1, p_\perp^1, x_\perp^1(\sigma)) \cdots \Phi(p_+^k, p_-^k, p_\perp^k, x_\perp^k(\sigma))$$

$$S_{KK}[\Phi] = \left\langle \Phi \middle| \tfrac{1}{2}(p_+ p_-) + (K_\perp - 1) \middle| \Phi \right\rangle \tag{30}$$

if

$$\det(K_\perp + p^2) = s\det\left(K_\perp + \Sigma n\left(\alpha_n \frac{\partial}{\partial \alpha_n} + c_n \frac{\partial}{\partial c_n}\right) + p^2\right) . \tag{31}$$

Note that integration over the Siegel gauge functional fields gives a superdeterminant because the coefficients of odd numbers of ghost coordinates have Fermi statistics. The superdeterminant is given by:

$$\begin{aligned} s\det\left(K_\perp + \sum_n n\left(\alpha_n \frac{\partial}{\partial_{\alpha_n}} + c_n \frac{\partial}{c_n}\right) + p^2\right) \\ = \int_0^\infty \frac{dt}{t}\, tr\, e^{-t(K_\perp+p^2)} \prod_n \left[tr\, e^{-tn\alpha\partial_\alpha} str\, e^{-tnc\partial_c}\right] \\ = \int_0^\infty \frac{dt}{t}\, tr\, e^{-t(K_\perp+p^2)} \prod_n \left[\frac{1}{1-e^{-nt}}\, \left(1 - e^{-nt}\right)\right] \\ = \det(K_\perp + p^2) \end{aligned} \tag{32}$$

and the equivalence with the Kaku Kikkawa formalism is proven.

So there we have it. Free string theory is a gauge invariant functional field theory and the appearance of gauge bosons and gravitons is no longer a mystery. As in any gauge theory, it is interesting to enquire about the gauge invariant observables of the theory. This is particularly true for string theories, where at present the only sensible quantity one can compute is the S-matrix - an unacceptable situation. Indeed, most of the massive states of the string model are unstable, and their S-matrix is meaningless (they have widths of order the Planck mass). Among the massless states, only neutral ones like the graviton have an S-matrix which is free of infrared divergences. In Yang-Mills theory, the S-matrix of physical states is derived from gauge invariant Green's functions. Are there analogs of such Green's functions in string theory?

The equation of motion $QA = 0$ is a statement of the vanishing of the field strength of our string one form. Let us try to construct a general gauge invariant linear functional of A. It has the form $< G|A >$ and is non-zero only if G has ghost number 1/2. It is gauge invariant only if G is BRST invariant. If G is BRST exact, then $< G|A >$ vanishes when A satisfies the equations of motion. On the other hand we have seen that the only BRST invariant quantities which are not exact are on shell states which satisfy $K|G >= 0$. This would seem to lead to the disappointing conclusion that the only gauge invariant quantities in string theory are on shell. Before despairing however, let us compute the two point functions of $< G|A >$ for some simple G's. We will compute in the Siegel gauge and so we might as well take G's which are proportional to c_0. BRST invariance then implies

$$(d+\delta)G = \Downarrow G = 0 \tag{33}$$

Simple solutions of this are:

$$\begin{aligned} G &= c_0 \int d^d x \, J(x) \, |\Omega, x\rangle \\ G &= \int d^d x J_\mu(x) \, c_0 \alpha^\mu_{-1} \, |\Omega, x\rangle \;\; ; \partial^\mu J_\mu = 0 \; . \end{aligned} \tag{34}$$

The two point function of $< G|A >$ computed in the Siegel gauge is $\langle G | \, \frac{b_0}{K} \, | G \rangle$. For the first example, this just gives the two point function of the scalar tachyon field while for the second (if we write $J^\mu = \epsilon^{\mu\nu\lambda\kappa} \partial_\nu M_{\lambda\kappa}$) we get the two point function of the Maxwell field strength tensor. Thus our argument that gauge invariant off shell quantities vanish was fallacious. It is analogous to the claim that the off shell Green's function of a scalar field vanishes because any off shell source can be written as the Klein-Gordon operator acting on some other function. It will be a major challenge to generalize these gauge invariant quantities to the interacting theory that we will discuss in the last lecture.

4. OPEN SUPERSTRINGS

In the previous section we constructed an elegant gauge invariant action for bosonic open string theory. The basic objects from which the action was built were differential forms on the Virasoro group, otherwise known as Faddeev-Popov ghosts for reparametrizations. In order to repeat our performance for the superstring we will have to understand the corresponding ghosts for super-reparametrizations. Lack of time (and its conjugate variable, energy) force me to forgo a detailed discussion of the reparametrization invariant Lagrangian for the first quantized spinning string, its gauge fixing in the superconformal gauge and the derivation of the corresponding BRST formalism. I will refer you to the excellent discussions of these points in the existing literature[24] Thus, as in the case of the bosonic string, I will begin by writing down the algebra of residual gauge transformations in the superconformal gauge:

$$\begin{aligned} [L_n, L_m] &= (n-m)L_{n+m} + \frac{d}{8} \, (n^3 - n) \, \delta(m+n) \\ [L_n, F_{\dot m}] &= \left(\tfrac{1}{2} n - \dot m\right) \, F_{\dot m + n} \\ [F_{\dot n}, F_{\dot m}] &= 2L_{\dot m + \dot n} + \frac{d}{2} \, (\dot n^2 - \tfrac{1}{4} \, \delta(\dot m + \dot n) \; . \end{aligned} \tag{35}$$

The dotted indices take on integer values in the so called Ramond sector, and half integer values in the Neveu-Schwarz sector. These sectors arise because the fermion fields ψ^μ of the spinning string can satisfy two different sorts of boundary conditions on the world sheet. Both sectors are needed in order to construct a unitary string theory. The superconformal generators $F_{\dot{n}}$ (in the Neveu Schwarz sector they are usually denoted G) are built out of odd numbers of fermion fields and so carry integer or half integer values of L_0 depending on the moding of the fermions. The ground state energy (L_0) is zero in the Ramond sector and $-1/2$ in the Neveu-Schwarz sector, when we take the ghosts into account.

We introduce the usual ghosts b_n and c_n for the Virasoro generators and corresponding superghosts $\gamma_{\dot{n}}$ and $\beta_{\dot{n}} = -\frac{\partial}{\partial \gamma_{-\dot{n}}}$. The ghost fields are $\gamma(z) = \sum z^{\frac{1}{2}-\dot{n}}\gamma_{\dot{n}}$ and $\beta(z) = \sum z^{-\frac{3}{2}-\dot{n}}\beta_{\dot{n}}$. Since we are dealing with a superalgebra, the superghosts are commuting variables. The commutation relations *between* ordinary and superghosts are a matter of convention which can be changed by a Klein transformation. In[25] we chose them to anticommute so that they mimic the properties of superdifferential forms. However, it appears to be more convenient to follow the main body of literature on the subject and have superghosts and ghosts commute like ordinary fermi and bose fields.

There are many inequivalent representations of the superghost commutation relations. This is true for all Bose systems, but usually a positive definite Hamiltonian picks a unique representation based on its ground state. The superghost Lagrangian is however first order and no such principle applies. Indeed, the inequivalent representations are essential for understanding the covariant fermion vertex operator and supersymmetry.[26] We define the qth Bose sea vacuum by

$$\begin{aligned} \beta_n \left|q\right\rangle &= 0 \qquad n > -q - 3/2 \\ \gamma_n \left|q\right\rangle &= 0 \qquad n \geq q + 3/2 \end{aligned} \tag{36}$$

q is integer in the Neveu-Schwarz sector and half integer in the Ramond sector. If we represent the states of the superghost system as functions of the γ_n, the Bose sea vacua are given by:

$$\left|q\right\rangle = \prod_{n=q+3/2}^{\infty} \delta(\gamma_n) \ . \tag{37}$$

It is easy to see from this representation that we cannot transform one Bose sea vacuum into another by acting with polynomials in β and γ. In a moment we will construct operators that do this job.

We will define the scalar product in the superghost state space in such a way that $\beta_{\dot{n}} = -\beta^{\dagger}_{-\dot{n}}$ and $\gamma_{\dot{n}} = \gamma^{\dagger}_{-\dot{n}}$. Using these hermiticity properties and the commutation relations, it is easy to show that

$$\langle p|q\rangle = \delta(p+q+2)\ . \tag{38}$$

All other scalar products can be obtained from this one by using the commutation relations. Finally, we define the superghost number by

$$g_F = \sum_n [\beta_{\dot{n}}, \gamma_{-\dot{n}}]_+ + \tfrac{1}{2}\,[\beta_0, \gamma_0]_+\ . \tag{39}$$

Note that it is an antihermitian operator.

Superghost number is the integral of a conserved world sheet current

$$g_F = \int :\beta\gamma: \equiv \int j(z)\ . \tag{40}$$

We would like to use this current to obtain another representation of the superghost Hilbert space analogous to the bosonization of fermions.[27] Thus we introduce a scalar field $\phi(z) = \int^z j(z')$. The operator product expansions of the superghost energy momentum tensor with itself and the superghost number current are:

$$\begin{aligned} T(z)j(w) &\sim \frac{-2}{(z-w)^3} + \frac{j(z)}{(z-w)^3} \\ T(z)T(w) &\sim \frac{11}{2(z-w)^4} + \frac{2}{(z-w)^2}\,T(w) + \frac{1}{z-w}\,\partial_w T(w)\ . \end{aligned} \tag{41}$$

To reproduce the first of these we must write

$$T(z) = :\ \tfrac{1}{2}\,j^2(z) + \partial_z j(z)\ : \tag{42}$$

but then the c-number in the second OPE comes out 13 instead of 11. This indicates that "bosonization" does not work in a naive manner. Another indication of this comes when we try to write the β and γ fields as exponentials of ϕ. The exponentials are fermions rather than bosons.

In ref. 26, both of these problems are solved by introducing a free fermion system whose c number anomaly is -2. The fields are called ξ and η and they have dimensions 0 and 1 respectively. The energy momentum tensor is

$$T_{\xi\eta} = -\eta \partial_z \xi \tag{43}$$

and

$$[\eta_m, \xi_n]_+ = \delta(m+n) \ . \tag{44}$$

Fields with the properties of β and γ are constructed as

$$\begin{aligned} \gamma &= e^{\phi}\eta \\ \beta &= e^{-\phi}\partial_z \xi \ . \end{aligned} \tag{45}$$

The conserved $\xi - \eta$ charge is

$$g_{\xi\eta} = \sum : \xi_n \eta_{-n} : + \tfrac{1}{2}\left[\xi^0, \eta^0\right] . \tag{46}$$

It takes the value -1/2 on the vacuum state defined by

$$\begin{aligned} \eta_n \left|\Omega\right\rangle &= 0 \quad n \geq 0 \\ \xi_n \left|\Omega\right\rangle &= 0 \qquad n > 0 \ . \end{aligned} \tag{47}$$

The qth Bose sea vacuum is related to this state by

$$\left|q\right\rangle = e^{q\phi(0)} \left|\Omega\right\rangle \ . \tag{48}$$

Finally, by combining the superghost number g_F with the $\xi - \eta$ charge we get a quantity $B = g_F + g_{\xi\eta}$ which commutes with β and γ. B is thus the label for the inequivalent Bose sea Hilbert spaces: the qth sea has $B = q + \frac{1}{2}$.

The BRST charge for the spinning string is constructed in the same way as that of the bosonic string:

$$Q = \sum \left[c^{-n} \left(L_n + \tfrac{1}{2} L_n^{gh} \right) + \gamma^{-\dot{n}} \left(F_{\dot{n}} + \tfrac{1}{2} F_{\dot{n}}^{gh} \right) \right] . \tag{49}$$

$Q^2 = 0$ if the spacetime dimension is 10.

We would now like to build a spinning string field theory action in terms of the BRST charge. We must first choose the quantum numbers of physical states, the ghost number and Bose sea charge. In the bosonic case there was a unique non-trivial cohomology of the BRST operator with ghost number -1/2. Here this is no longer true because of the occurrence of Bose seas. We will make the conventional choice of Neveu Schwarz sector as the $q = -1$ Bose sea.

The vacuum state in this sector has total ghost number (ghost + superghost number) -1/2 and Bose sea charge -1/2 as well. We will restrict all classical Neveu-Schwartz string fields to have these quantum numbers. We do not make a separate restriction on ghost and super ghost number because the BRST charge does not have a homogeneous transformation law under these symmetries.

It is now clear that an expression like $\langle A|Q|A\rangle$ must vanish. It has zero ghost number (Q has ghost number 1) but Bose sea charge -1. Clearly we must find an operator with ghost number zero and Bose sea charge 1 to insert in the action. To see what is going on let us examine the zero mode subspace of the $\xi - \eta$ Hilbert space. It is isomorphic to the zero mode subspace of the bosonic ghosts. η_0 annihilates the vacuum, the vacuum expectation value of 1 is 0, and the vacuum expectation value of ξ_0 is 1. Clearly, then, the expectation value of Q vanished above because Q (which is built from β and γ) contains no ξ_0. ξ_0 is just the operator with $B = 1$ and $g = 0$ that we needed above. We thus write the action for the Neveu-Schwarz model as

$$\langle A|\xi_0 Q|A\rangle \tag{50}$$

If we restrict the A field by $\eta_0 A = 0$, this is gauge invariant under $A \to A + Q\epsilon$. To prove this we have to compute the anticommutator of Q with ξ_0. This comes only from the terms proportional to η_0 in Q and has no term proportional to ξ_0. Thus if we restrict A to be annihilated by η_0 (which anticommutes with Q) the anticommutator has zero expectation value and the action is gauge invariant.

Let us now expand the BRST charge in ghost zero modes, just as we did for the bosonic string:

$$Q = c_0 K + d + \delta - 2b_0 \Downarrow \tag{51}$$

We have used the same names for the expansion coefficients here, not because they are the same operators that we encountered for the bosonic string, but because they have the same properties. Indeed, the hermiticity properties and commutation relations of these operators are determined by the nilpotence and hermiticity of Q and by the properties of the zero modes.

Given this decomposition of Q we can now proceed to gauge fix the Neveu Schwarz string field theory by following exactly the manipulations we used for the bosonic string. Siegel gauge fixing is thus trivially carried out. When we attempt to fix the light cone gauge, two complications arise. First, we must decompose the fermionic partners of the string coordinates into transverse and longitudinal components. This is easily done. More importantly, we must discuss the statistics of the various space time fields that appear in our action. In the case of the bosonic string we recognized a remarkable correlation between space time and world sheet statistics. Coefficients of fermionic world sheet fields obeyed fermi statistics in space time. If we insist on retaining this principle for the spinning string we come to a remarkable conclusion: half of the fields in the model must be thrown out if we wish to preserve the spin statistics connection for space-time fields. All of the fields in the Neveu Schwarz model carry integer spin. Consider however, the classical sector with Q = B = -1/2. Since the ψ^μ are anticommuting, half of these fields will be fermions, and should be omitted from the action. Which half are fermions depends on the overall statistics we choose for the string field A. The quadratic form $\langle A|\xi_0 Q|A\rangle$ is antisymmetric since $(\xi_0 Q)^\dagger = Q\xi_0 = -Q\xi_0 + \{Q, \xi_0\}$ and the anticommutator has zero expectation value. Thus A must be an anticommuting variable in order to get non-zero action. Schwinger[28] invented a similar argument for Majorana spinors many years ago. If A is a fermion, then we must throw away all coefficients of even powers of ψ^μ if we do not wish to violate the spin statistics theorem. This is precisely the GSO projection!

One may object that such a classical argument should not be able to discriminate between bosons and fermions. Indeed, Schwinger's argument can be circumvented at the classical level by introducing extra degrees of freedom.[28] A quantum mechanical calculation is necessary to show that this leads to a violation of unitarity. It is not at all clear how the present argument is related to those based on modular invariance[29] or world sheet locality[30] which also show the necessity of the GSO projection for the spinning string. Note that we have not shown the necessity of including the Ramond sector in the action. We only show that the NS sector must be projected (we will show the same thing for the Ramond sector below). This ends our discussion of the Neveu-Schwarz sector of the spinning string.

The Ramond sector of the spinning string has created a lot of confusion in the literature on string field theory. Many papers[31] have been written about it, not all of which are correct. Several groups[32] found an action which could be gauge fixed to the correct set of light cone degrees of freedom, but the connection of this action to the BRST charge remained obscure. Yamron[33] found an action based on the BRST charge which can be partially gauge fixed to give the action of Ref. 32.[34] Finally, an elegant form of the Ramond action which generalizes easily to the interacting case, was found by Witten.[35] We will describe Witten's action and show how it is related to that of Ref. 32. The reader is asked to consult the literature for a discussion of how to gauge fix the action to light cone gauge. The gauge fixing is a simple generalization of what we have already done.

There are two obvious problems with any simple minded attempt to generalize what we have done to the case of Ramond superstrings. First of all the coefficient of c_0 in the BRST charge, which gives the diagonal term of the bosonic string action, is a second order differential operator. This is appropriate for a bosonic action but not for a fermionic one. The second problem has to do with ghost number. The simplest Bose sea for the Ramond sector is $q = -1/2$. The vacuum state in this sector has vanishing ghost number and Bose sea charge. We take all classical Ramond fields to have these quantum numbers.

As in the Neveu Schwarz sector we see that the BRST charge has the wrong quantum numbers to give a nonvanishing action. Anticipating a necessary factor of ξ_0 we write

$$S = \langle \Psi | \xi_0 Y Q | \Psi \rangle \tag{52}$$

where Y must have $B = -1 = g$. We have used Witten's conventions for the space time Dirac algebra. Henceforth we drop the factor ξ_0 to simplify the notation. The action will be gauge invariant if $[Q, Y] = 0$. The authors of Refs. 26 have found a BRST invariant local operator with $g = B = -1$. They call it the "inverse" picture changing operator. Witten has shown that it can be written as

$$Y = \left[\beta_0, c\left(\frac{\pi}{2}\right) e^{-\phi(\pi/2)}\right] . \tag{53}$$

Let us now expand Q and Ψ as

$$\begin{aligned}
Q &= c_0 K + J + b_0 M \\
M &= -\tfrac{1}{4}\gamma^{02} - 2 \Downarrow \\
J &= -\tfrac{1}{2}\gamma_0 F + (d+\delta) - 2\beta_0[F, \Downarrow] \\
F &= F_0^{(x)} + \sum_{m>0} [m(\beta_{-m} c^m - c^{-m}\beta_m) + b_{-m}\gamma^m + \gamma^{-m} b_m] \\
\Downarrow &= \sum_{m>0} [m c^{-m} c^m + \tfrac{1}{4}\gamma^{-m}\gamma^m] \\
\Psi &= \Psi_0 + c_0 \Psi_1 \\
Q\Psi &= c_0(K\Psi_0 - J\Psi_1) + J\Psi_0 + M\Psi_1 .
\end{aligned} \tag{54}$$

From these formula it is easy to prove that Ψ_0 can be gauge transformed to be annihilated by M and that the required gauge fixing is algebraic and does not produce a Faddeev-Popov determinant. Furthermore the Ψ_1 field equation

$$M\Psi_1 = -J\Psi_0 \tag{55}$$

is also algebraic (Ψ_1 is an auxiliary field) so we can use it to simplify the action.

The elimination of Ψ_1 and the gauge condition $M\Psi_0 = 0$ reduce the number of fields in Witten's action to the number employed in ref. 32. The gauge condition has two linearly independent functions of γ_0 in its solution. Therefore the field content is equivalent to two functions of the nonzero modes of the ghosts and superghosts. We will combine these two functions into a single state by introducing fermionic operators b and c with an algebra isomorphic to that of the reparametrization ghost zero modes. It should be emphasized that they have no conceptual connection with the ghost zero modes.The two states represent the two independent solutions of the gauge conditions.

The proof that Witten's action for these two fields reduces to the one we will present has not appeared in the literature. We use the new fermion operators b and c to define a nilpotent operator $\tilde{Q}$

$$\tilde{Q} = cF + d + \delta + b[F, \Downarrow]_+$$

We make the convention that b and c commute with all the other operators in the problem. d and δ are the BRST charges for the positive and negative frequency parts of the Ramond algebra. Nilpotence of $\tilde{Q}$ then follows from the fact that F anticommutes with d and δ and that $(d+\delta)^2 = F[F, Downarrow]_+$. Both of these are consequences of nilpotence of the full BRST charge.

From this point onwards, the algebra involved in gauge fixing the action to the light cone gauge is a straightforward generalization of that which we did in the Neveu-Schwarz sector.A notable feature of the covariant gauge fixing in this sector is that the complications usually associated with Nielsen-Kallosh ghosts for high spin fermionic gauge fields are completely bypassed. The discussion of spin and statistics parallels that in the Neveu-Schwarz sector. Since Q and Y commute, the Ψ field must still be anticommuting, but now, since the coefficient fields carry half integer spin, we must project out states with odd numbers of fermion operators. Again this coincides with the proper GSO projection. We therefore have a completely sensible picture of the free superstring in terms of gauge invariant functional field theory. Witten has extended this formalism to the interacting open superstring.

5. CLOSED STRINGS

It is easy to be fooled into thinking that the field theory of closed strings is a trivial generalization of that for open strings. The Hilbert space of the closed bosonic string is the direct product of two open string spaces (corresponding to left and right moving modes), except for the zero mode of the x^μ field. There are two copies of the Virasoro algebra and a BRST charge $Q = Q_L + Q_R$ for the whole system. The non-trivial cohomology of the BRST charge is given by physical states of the form

$$|\Phi\rangle \otimes |\Omega\rangle$$

where $|\Omega\rangle$ is the ghost vacuum and:

$$L_n |\Phi\rangle = \overline{L}_n |\Phi\rangle = 0; \quad n > 0$$

$$(L_0 + \overline{L}_0 - 2) |\Phi\rangle = (L_0 - \overline{L}_0) |\Phi\rangle = 0$$

These states have total (left moving plus right moving) ghost number equal to -1.

If we try to make a field theory action that reproduces these equations by using the BRST charge in the action we find that the ghost number does not add up correctly and the naive action is zero. The obvious cure is to insert an operator of ghost number one in the action, in analogy with what we did for the Neveu Schwarz superstring. This has been done for example by Lykken and Raby.[36] However, it gives a gauge invariant action only if the string functional is subjected to the a priori constraint $(L_0 - \overline{L}_0) |A\rangle = 0$. Thus one of the Virasoro conditions must be imposed as an off shell constraint rather than an equation of motion or a gauge condition. The resulting formalism can be described in a subspace of the closed string Hilbert space in which one of the zero modes of the ghost is discarded, and the states are subjected to the above constraint. If c_0 is the remaining zero mode, the operator: (barred and unbarred quantities refer to left and right movers respectively)

$$\widetilde{Q} = c_0 \left(\frac{K + \overline{K}}{2} \right) + d + \overline{d} + \delta + \overline{\delta} + b_0(\Downarrow + \overline{\Downarrow})$$

is nilpotent and can be used to form the gauge invariant action $\left\langle A|\widetilde{Q}|A \right\rangle$. The gauge invariance of this action includes linearized general coordinate transformations. Gauge fixing proceeds in precisely the same manner as it did for the open string (the algebra is completely identical in the constrained subspace) and we obtain the Kaku-Kikkawa light cone action for closed strings. Note that in this action the field is also constrained to be annihilated by $L_0 - \overline{L}_0$.

Although one of the Virasoro conditions is being treated in a very different manner from the others, there is no apparent inconsistency in the action we have written. Indeed in the light cone gauge one can write an interacting action with fields which satisfy the constraint. This action is the basis for the only manifestly unitary calculations of string scattering amplitudes and as far as one knows[37] gives results in agreement with the world sheet path integral. Nonetheless I feel a bit uneasy with the present formulation of the theory and I have the feeling that we have to do a lot more to truly understand the closed string. I will present two more pieces of evidence that something is wrong.

The first has to do with superstrings. A closed spinning type II string has four sectors, corresponding to the choice of Ramond or Neveu-Schwarz boundary conditions for left movers and right movers separately. The heterotic string has two sectors, the left movers can be described by a compactified bosonic string while the right movers are either the NS or R sectors of the spinning string. When describing Ramond sectors, we will use the formalism of Ref. 25 in which all zero modes of the superghosts are discarded. By throwing away one of the bosonic ghost zero modes as well, we can in every case except the Ramond-Ramond sector, construct a modified BRST operator which is nilpotent in the subspace of states with equal left moving and right moving energy. For the NS-NS sector of the type II string or the NS sector of the heterotic string, this operator has the same form as the one we constructed for the closed bosonic string. The d's and $\delta's$ must be reinterpreted to stand for BRST generators of the NS algebra, but their algebra remains the same. In the NS-R sectors of type II, or the R sector of the heterotic string we write:

$$\tilde{Q} = c_0 F + (-1)^{N_F}(d + \delta + \bar{d} + \bar{\delta}) - b_0(F \Downarrow + \Downarrow F + F\bar{\Downarrow} + \bar{\Downarrow} F) \ .$$

This operator is again nilpotent in the subspace of states with equal left and right moving energy.

In the Ramond-Ramond sector this analogy breaks down. It is tempting to write down a straightforward generalization of what has gone before, gluing together two Ramond open strings. A gauge invariant action can be constructed in this manner - but it has the wrong spectrum! Knowledgeable readers will be reminded of a well known difficulty with the field theory formulation of the massless level of the type IIB string. This sector contains a massless particle which can be described by a rank four antisymmetric tensor gauge field. The field equation for this gauge potential is the self duality of its (rank five) field strength. This is the ten dimensional analog of a right moving scalar field in two dimensions. No one has ever found a covariant bilinear lagrangian formulation of this system. If one simply introduces a Lagrange multiplier field, one doubles the number of degrees of freedom. This is precisely what occurs for the covariant action of the Ramond-Ramond sector described above. For both type I and type II systems it gives twice the correct number of degrees of freedom.

The only way that we have found to resolve this problem is extremely ugly. We write an action with no spacetime derivatives in its diagonal terms:

$$\begin{aligned} S &= \left\langle B|\widetilde{Q}|B\right\rangle \\ \widetilde{Q} &= c^0 + (d + \delta + \overline{d} + \overline{\delta}) - b_0(F\{F, \Downarrow\} + \overline{F}\{\overline{F}, \overline{\Downarrow}\}) \ . \end{aligned} \tag{56}$$

$\widetilde{Q}$ is nilpotent in the space of states satisfying:

$$(F - \overline{F})\,|\lambda\rangle\, 0 \tag{57}$$

so the action is gauge invariant. In (57) F and $\overline{F}$ are the world sheet supersymmetry generators for left and right movers. Unfortunately, this constraint is dynamical, unlike the $K = \overline{K}$ constraint we have encountered in other sectors. This is a consequence of the fact that the left moving and right moving fermions ψ^μ have independent zero modes.

In effect, a dynamical constraint is an equation of motion that does not follow from the Lagrangian. To get a sensible system which we can quantize, we must solve the constraint. This should be done without inverting any time derivatives, otherwise the resulting action will be non-local in time. The only simple way we have found to solve the constraint is to use light cone coordinates. The action can be gauge fixed to the light cone gauge by arguments exactly analogous to those already presented. The result is

$$S = \tfrac{1}{2}\, tr\, B^T\, \frac{p^2 + K_\perp}{2p^+}\, B \tag{58}$$

where B is an 0(8) bispinor. This is the correct light cone gauge Lagrangian for this sector and contains (after GSO projection) the correct spectrum of states.

This treatment of the Ramond-Ramond sector of the type II superstrings is a major disaster. It is not a manifestly covariant formulation of the theory. One gets the feeling (as one already did for the closed bosonic string) that we are leaving out some gauge invariance but the precise nature of the missing ingredients is unclear.

A further indication that something is wrong comes from the calculation of the one loop "cosmological constant" in string field theory. We will describe this computation below and find that our computation disagrees with that done by Polchinski in the world sheet path integral formalism[38] One should emphasize that the "cosmological constant", defined as the logarithm of the string field theory vacuum amplitude is not necessarily a physical object in string theory.

At present the only fully gauge invariant objects that we know are scattering amplitudes. In light cone string field theory, one can calculate one loop scattering amplitudes which "contain" the cosmological constant (as the coefficient of the dilaton tadpole) and find an answer which agrees with Polchinski's. However when one calculates the *vacuum amplitude* in the light cone field theory one gets the result we will present below. Thus it is not clear whether we should worry about the disagreement.

In any free field theory, the logarithm of the vacuum amplitude is the trace of the log of the kinetic operator K. For open strings this gives:

$$sTr\ell nK = \int_0^\infty \frac{dt}{t} sTr\exp(-tK) = \int_0^\infty \frac{dt}{t} Tr\exp(-tK_\perp)$$

In the second equality, $K_\perp$ is the Kaku-Kikkawa light cone kinetic energy which we obtained by the Parisi-Sourlas trick. This expression coincides with that derived from the world sheet formalism. t represents the modulus (ratio of circumference to length) of the world cylinder, and it is integrated over the correct domain.

An analogous calculation for closed strings gives

$$\int_0^\infty \int_0^{2\pi} \frac{dt}{t}\, d\theta Tr\exp\left(-t(K_\perp + \overline{K}_\perp) + i\theta(K_\perp - \overline{K}_\perp)\right)$$

The θ integral imposes the constraint that left moving and right moving energies match. It is easy to see that this result is simply the sum of the one loop vacuum energies of all of the physical particle states contained in the closed string. Polchinski has shown that this is the same, up to an infinite factor, as the vacuum bubble given by the world sheet path integral. The (t, θ) integration region is a strip of width 2π in the upper half plane. Polchinski shows that the integrand is invariant under modular transformations of the integration region. Thus, our result is the integral over a fundamental region of the modular group (which is the world sheet result) times the infinite order of that group. I have been unable to motivate dropping this infinite factor on the basis of the rules for the closed string field theory that we have formulated. It cannot be a consequence of some additional discrete gauge invariance of the string field which is not fixed by Siegel's condition. This would lead to an infinite factor multiplying the vacuum amplitude, while we have an infinite factor in the log of the vacuum amplitude. I do not know if this annoying discrepancy is a red herring or a profound clue to the construction of a correct closed string field theory.

6. INTERACTING STRINGS

Several proposals have appeared in the literature for generalizing the gauge invariant free string actions that we have discussed to include interactions. Peter West has described the Siegel-CERN- Kyoto[39] approach to this problem in his lectures at this school. I will describe the elegant approach invented by E. Witten. A gauge invariant string field theory action should satisfy at least three criteria. It should generalize non-abelian gauge invariance (and for closed strings, general coordinate invariance) in an interesting way, and provide an explanation for the ubiquitous occurrence of gauge bosons and gravitons in string theory. It should provide a *simple* derivation of the elegant dual Feynman rules, and it should be simple enough to allow us to understand non-perturbative effects in string theory. Witten's action satisfies the first criterion, at least for open string theories. It does not appear to satisfy the second. The third, which is obviously the most important, may or may not be satisfied. It is too early to tell.

The starting point of Witten's construction is the idea[40] that string fields are differential forms on the Virasoro group. More precisely, they are semi infinite differential forms. In physicists language, an ordinary differential form on the Virasoro group would be a functional of the ghost field $c(z)$. Alternatively, one could describe it as a state built on the empty Dirac sea of the ghosts. However, it is clear that in order to have a sensible action of the Virasoro algebra on the forms (in particular to have finite eigenvalues of L_0) we must fill the Dirac sea. Thus in effect, all forms with finite L_0 have infinite rank. As usual it is convenient to introduce a new concept of rank that measures the difference of the rank of a form from the rank of the "vacuum" form (the full Dirac sea). This is essentially the ghost number. In this language, the BRST charge Q is the exterior derivative on forms.

Our free action for string forms was a bilinear in the form A and its field strength or exterior derivative QA. Readers with the right kind of background will be immediately reminded of a Chern Simons form. Witten's formalism is a fleshing out of this analogy. Firstly, we clearly want to call A a one form. This means that the rank of a form is given by its ghost number plus $3/2$. The gauge parameter is then a zero form. In order to construct a topological invariant such as the Chern-Simons form, we have to find analogs of the wedge product and the integral of a differential form. Let us call the analog of the wedge product of two differential forms A and B

$$A * B \, .$$

It should take forms of rank n and m and multiply them together to give a form of rank $n+m$. Thus $A * B$ should have ghost number equal to $g_A + g_B - 3/2$. The exterior derivative Q should be a graded derivation of wedge multiplication:

$$Q(A * B) = QA * B + (-)^A A * QB$$

In this formula $(-1)^A$ is minus one if A is a form of odd rank and plus one if the rank of A is even.

The standard wedge product for forms turns the space of forms into a Grassman algebra. However, from Yang-Mills theory, we are familiar with forms that take values in matrix algebras, for which the wedge product is an associative but not anticommutative operation. Thus we should only require that $*$ be associative. Given a wedge product we are almost ready to define most of the usual operations of exterior calculus. What is lacking as yet is the notion of the integral of a differential form. This is a linear c-number valued function on forms which satisfies:

$$\int A * B = (-1)^{AB} \int B * A \qquad \int QA = 0\ .$$

Finally, we should require that $*$ and $\int$ are such that the abelian Chern-Simons invariant $\int A * QA$ is the free string action $< A|Q|A >$ that we constructed above.

If we can define such a product and integral then we can repeat many of the usual constructions of Yang-Mills theory. Define gauge transformations of a 1 form A by:

$$\delta A = Q\epsilon + A * \epsilon - \epsilon * A$$

Then the field strength:

$$F = QA + A * A$$

transforms covariantly:

$$\delta F = F * \epsilon - \epsilon * F$$

and we can make various gauge invariant objects by integrating products of F and A to make Chern-Simons invariants. Usually one can only form a single Chern-Simons invariant for a given dimension. The same will be true here if we insist that the integral operation carries a fixed ghost number. In particular, if we want the integral of $A * QA$ to be nonzero, the integral must vanish unless its integrand carries ghost number 3/2. With this restriction we are led to a unique gauge invariant action:

$$S(A) = \int A * QA + 2/3 A * A * A$$

Note that it is automatically trilinear, as expected for a string theory. The field equation which follows from $S(A)$ is $F = 0$. It is thus an integrability condition, and raises the exciting possibility that string theory is a completely integrable system. Witten[41] and Friedan and Shenker[42] have suggested this in the context of rather different considerations.

The introduction of a product and integral in the space of string forms thus offers us the possibility of building an elegant and unique interacting string action. Witten has found explicit definitions of $\int$ and $*$ which obey the rules that we have outlined above.[43] He has argued that it gives the familiar three point couplings of the Veneziano model. Giddings[44] has computed the four point amplitude from this formalism. We do not have time here to go into the details of the construction, but will content ourselves with making some general comments about invariance properties and uniqueness of Witten's definitions. (Recently Giddings, Martinec and Witten[45] have shown that the Feynman rules of this theory generate the world sheet form of the dual amplitudes to all orders in the loop expansion.)

The dual Feynman rules are invariant under reparametrizations of the string world sheet. In actual computations one must choose a gauge, but even the gauge fixed Feynman rules are invariant under conformal transformations. One might have expected the string field theory action to be conformally invariant. This is indeed the case for the kinetic energy term that we have constructed, since the BRST charge commutes with all the conformal generators. However, the interaction term of Witten's action does not have full conformal invariance. I believe that this was inevitable and that any string field theory will have this problem. The world sheet diagram for an elementary open string interaction is shown in figure 1. Any description of this interaction in terms of an action for fields which are functionals of $X(\sigma)$, must make a cut along the world sheet to identify the surface on which the strings interact (assuming that the interaction is local). Witten's cut is shown in figure 2. This identification obviously violates conformal invariance.

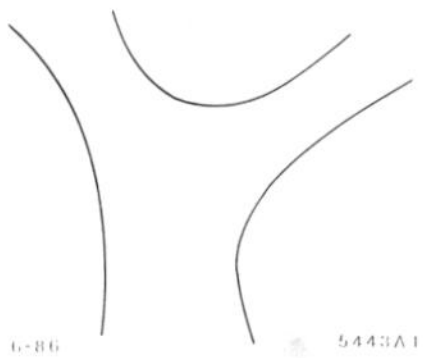

Fig. 1. The world sheet for 3-string scattering.

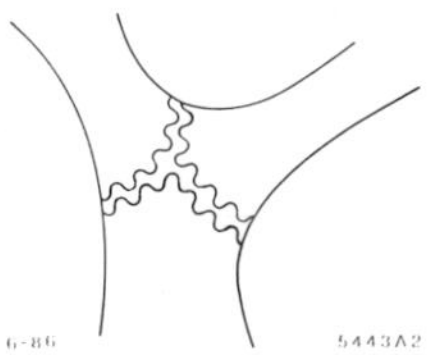

Fig. 2. The cut across the world sheet that defines Witten's interaction.

More formally, consider any action defined in terms of a $*$ product and an integral. The free action is conformally invariant by itself, so the interaction term must be as well. The only apparent way to enforce this is to insist that the conformal generators L_n be derivations of the $*$ algebra and that $\int L_n A = 0$ for all A. The integral is a linear functional on the space of string states and thus can be represented as the scalar product with some state $|I\rangle$. I must be a state of ghost number -3/2, and if the action is conformally invariant it must be annihilated by all the conformal generators. In particular, it must be annihilated by L_0. But there is only one state of ghost number $-3/2$ which is annihilated by L_0. It is the SL_2 invariant state $|0\rangle$, described in Section I. This state is not annihilated by L_n with $n < -2$. Thus there is no state annihilated by all the conformal generators. A given definition of $\int$ and $*$ is characterized by the subgroup of the conformal group which leaves it invariant. The integral is defined by a BRST invariant state I of ghost number -3/2. Thus $|I\rangle = |0\rangle + Q\,|B\rangle$ where $|0\rangle$ is the $SL(2)$ invariant state. I is probably completely specified by the Virasoro subalgebra which leaves it invariant. For example, if the subalgebra contains L_0 then I is the SL(2) invariant state. I will now argue that with a few assumptions, one can say the same about the product.

It is natural to assume that the Hilbert space contains an identity element for the star product. We will call this I, using the same symbol as that used for the integral for reasons which will become obvious in a moment. By definition:

$$A * I = I * A = A$$

for any A. Since Q and any symmetries of the action are (possibly graded) derivations, they must annihilate the identity. It is also clear that the identity has ghost number -3/2. If, as we conjectured above, the subalgebra of the conformal algebra which annihilates I completely determines it, then the identity state is the state which defines the integral. This would certainly be the case if the symmetry of the action contains L_0.

With one more assumption, we can show that the symmetry group of the action completely determines the * product. Any * product induces a correspondence between states and operators. The state A corresponds to the operator $\widehat{A}$ defined by

$$\widehat{A}(B) = A * B.$$

We can write the state A as $\widehat{A}(I)$, where I is the identity. If there were a unique operator that created A when applied to I we would be able to construct the product operation uniquely from the state I. Every state would be written as the corresponding operator acting on I, and the * product would just be the ordinary operator product. Note that this definition is associative, and has the right ghost number properties. The integral is defined as above by the scalar product with I. The integral of a BRST exact form then vanishes. The two properties of Witten's formalism that are not immediately apparent in this formulation are the graded commutativity of the integral of A*B and the fact that Q is a graded derivation.

Of course, the relation $A = \widehat{A}(I)$ for all states does not completely define the operators $\widehat{A}$. We can only make progress by putting further restriction on these operators. A natural restriction exists when I is the SL(2) invariant state. In that case, if we require that $\widehat{A}$ be the integral of a local density, then standard results of conformal field theory ensure that there is indeed a unique connection between states and operators. It is then fairly easy to prove the remaining Witten axioms. For example:

$$\begin{aligned} Q(A * B) &= Q\widehat{A}\widehat{B}(I) = [Q, \widehat{A}\widehat{B}](I) \\ &= ([Q, \widehat{A}]_{\pm}\widehat{B} + (-1)^A \widehat{A}[Q, \widehat{B}]_{\pm})(I) = Q(A) * B + (-1)^A A * Q(B) \end{aligned}$$

In this equation we have used the fact that if $\widehat{A}$ is the local operator which creates A from the SL(2) invariant vacuum then the local operator which creates Q(A) is the commutator or anticommutator of Q and A, according to whether A is bosonic or fermionic. Using the operator product expansion one can also prove the graded commutativity of the integral of A*B.

Thus, in the case where we choose the identity state to be the SL(2) invariant vacuum, there seems to be a unique definition of * product and integral that satisfies Witten's axioms. The Feynman rules for this formalism involve SL(2) invariant expectation values of vertex operators and seem to be rather directly related to the Koba-Nielsen formulas. (There are some troublesome points which I have not sorted out so I will not present the details here). Unfortunately, Witten's choice of identity is not the SL(2) invariant state. It is a state annihilated by the set of Virasoro generators which preserve a fixed time slice and preserve the midpoint of the string in that slice. These are the generators $K_n = L_n - (-1)^n L_{-n}$. I believe that the uniqueness theorems that hold for the SL(2) case are also valid here, so that Witten's formalism is completely defined by this invariance group.

We seem then to have a number of different gauge invariant actions for string field theory. What is the relation between them, and do they give equivalent results for scattering amplitudes? I can give only a partial answer to this question. Consider for definiteness, Witten's definition of the * product. Let U be a conformal transformation which is not an automorphism of * (e.g. e^{iaL_o}) Now define

$$A *_U B = U^{-1}(U(A) * U(B))$$

We must also define a new integral by

$$\int_U A = \int U(A)$$

It is easy to verify that this is an associative product which satisfies all of Witten's axioms. Thus, for example, we can go from Witten's product to one which is invariant under σ reparametrizations. This will define a different string field theory action. Using the fact that a conformal transformation of an on shell physical state is that same state plus a BRST exact form, it is easy to show that the S-matrices defined by these two actions are identical. Thus there is a large class of definitions of the * product which give an action different from but physically equivalent to Witten's. An action based on the SL(2) invariant state is not in this class. The symmetry subalgebra of Witten's formalism is not conjugate to the invariance algebra of the SL(2) vacuum. Nonetheless I suspect that a manifestly SL_2 invariant formalism exists and is equivalent to Witten's.

Recently, Giddings and Martinec[46] have described the connection between string field theory actions and triangularizations of Riemann surfaces. They also find that there should be a multitude of equivalent string field theory actions. It would be extremely interesting to relate the present discussion to theirs, for we seem to have an algebraic (according to invariance group) classification of possible actions whereas their classification is geometrical. Clearly both presentations are incomplete and much work remains to be done.

Although many of the details are cloudy, it seems clear that an elegant gauge invariant formulation of open string field theory exists. How should it be extended to the case of closed strings? We have seen already, that even free closed string field theory is not in the best of shape. A question of great importance, is whether we need to introduce separate closed string fields or whether closed strings are somehow already contained in the formalism at hand. Remember that closed string intermediate states automatically appear in the unitarization of covariant open string amplitudes.[47] In the light cone gauge, where closed strings are needed for covariance rather than unitarity, a separate closed string field is certainly needed. However, Witten has argued that the loop diagrams of his open string field theory already contain closed string contributions. He has made a beautiful proposal which suggests why this comes about.

Let us notice that the field equations of Witten's action can be rewritten in an interesting way by using the operators $\widehat{A}$ described above. It is easy to verify that the vanishing of the field strength is equivalent to the statement that:

$$(Q + d_A)^2 = 0.$$

$$d_A(B) \equiv A * B - (-1)^{AB} B * A$$

or in other words that $Q + d_A$ is itself a BRST operator. This is a realization of the suggestion of Friedan[48] that the equations of string field theory are the requirement that the background classical field define a two dimensional conformal field theory with the right value of the central charge. Closed strings fit very nicely into this framework. We have emphasized repeatedly that everything in our formalism goes through for any field theory of the x^μ field which is conformally invariant. Thus different consistent background space time geometries are equivalent to different choices of Q in the above formalism. Solutions of the open string equations correspond to different BRST operators which differ only by an inner derivation of the * algebra. Different consistent space time geometries (which we suspect correspond to different classical solutions of closed string field theory) correspond to BRST generators whose difference is not an inner derivation. Thus, according to Witten, we should think of a general state of the system of closed plus open strings as a general graded derivation of the * algebra. The classical equation of motion of this system is just $Q^2 = 0$. In order to complete this program we must find an action from which this equation follows and a suitable parametrization of the set of all derivations. I refer the reader to Witten's paper for further discussion.

7. CONCLUSIONS

It is clear that string field theory is still in a state of flux. The beautiful results obtained by Witten have pointed the way to a full understanding of the underlying structure of strings, but much work remains to be done. I believe that a proper understanding of all possible (equivalent?) realizations of Witten's axioms is necessary in order to make a clean connection with the dual Feynman rules and Riemann surfaces. The details of Witten's outline for closed string field theory must be filled in, in particular one must find an action and a useful parametrization of the space of all derivations of the * algebra. Witten has generalized his open string field theory to the superstring but the heterotic string still looks like a sport of nature. The ugly duckling has yet to be recognized as a swan. We must clearly find a better framework for describing it.

There remain many other obscure points and open problems. We must gain a much better understanding of the nature of the cosmological constant in string theory. We must find a Hamiltonian formulation of string theory analogous to the Wheeler-DeWitt formulation of quantum gravity. We must understand the "Wave Function of the Universe" and the tunneling processes between different classical vacua of string theory. We must find out whether we can attack the parts of string theory relevant to low energy physics by weak coupling and/or "effective field theory" techniques or whether there are some low energy questions which have "essentially stringy" answers. Most important of all, we must find some way to subject string theory to experimental tests, at least by calculating the parameters in an effective field theory which describes accessible physics. The developments in string field theory over the past year have made it seem that the answers to these questions may not lie asymptotically far in the future.

REFERENCES

1. The reference is to the delightful pseudo academic opening of Ambler's "A Coffin for Dmitrios", A. Knopf, London (1939). I first read this at the age of 14 and have been planning to get it into a physics paper ever since.
2. G. Veneziano, Nuovo Cimento 57A, 190, (1968); Phys. Lett. 30B, 351, (1969)
3. J.Wess,B.Zumino, Nucl. Phys. B70, 39, (1974)
4. J. Schwarz, Phys. Rep. 89C (1982) 223 (1974). M. B. Green, Surveys in High Energy Physics 3 (1982) 127.
5. J.Harvey private communication. Of course the discovery of anomaly cancellation for $E(8) \times E(8)$ was made by Schwarz and Green as well. I have left them out of the text for stylistic reasons.
6. J. Thierry-Mieg, Phys. Lett.156B, 199, (1985)
7. Other relevant miracles are Lovelace's discovery that bosonic strings are only consistent in 26 dimensions, and the connection between conformal invariance and classical equations of motion of the string. C. Lovelace, Lecture at Princeton University January 1985 and Rutgers preprint RU-85-51, January 1986; C. Callan, D. Friedan, E. Martinec, M. Perry; Nucl. Phys. B262, 593, (1985); A.Sen, Phys Rev. Lett. 55, 1846, (1985)
8. L. Brink, P. DiVecchia, P. Howe, Phys. Lett. 65B, 471, (1976)
9. S.Deser,B.Zumino, Phys. Lett. 65B, 369, (1976)
10. A. Polyakov, Phys. Lett. 103B, 207 (1981).
11. D. Friedan, E. Martinec, S. Shenker (Chicago preprint Print-86-0024 (Chicago), November 1985. T. Banks, D. Nemeschansky, A. Sen, SLAC-PUB-3885, February 1986.
12. M. Kato, K. Ogawa, Nucl. Phys. B212, 443, 1983.
13. I. Frenkel, L. Garland, G. Zuckerman, Yale Mathematics preprint 1986
14. W. Siegel, Phys. Lett. 149B, 157, (1984); Phys. Lett. 151B, 391 (1985)
15. E.C.G. Stueckelberg, Helv. Phys. Acta 11, 225 (1938)
16. W. Siegel, B. Zweibach, Nucl. Phys. B263, 105 (1986)
17. M. Kaku, K. Kikkawa, Phys. Rev. D10, 1110 (1974)
18. W. Siegel, Phys. Lett. 149B,162 (1984); 151B, 396 (1985);149B, 157 (1984); 151B, 391 (1985); 142B, 276 (1984)
19. G.Parisi,N.Sourlas, Phys. Rev. Lett.43, 744, (1979)

20. M. Peskin and C. Thorn, Nucl. Phys. B ; T. Banks, D. Friedan, E. Martinec, M. Peskin and C. Preitschopf SLAC-PUB-3853, Dec. 1985

21. M.Grisaru,S.Gates,M.Rocek,W.Siegel, *Superspace*, Benjamin (1982).

22. L.Baulieu,J.Thierry-Mieg, Nucl. Phys. B197,477, (1982) ; W.Siegel, Phys. Lett.93B, 170, (1980)

23. L.Baulieu,S.Ouvry, CERN-TH-4338/85, Dec.1985

24. E. Martinec, Phys. Rev. D28, 2604 (1983); A. Polyakov, Phys. Lett. 103B, 211 (1981).

25. T. Banks, D. Friedan, E. Martinec, M. Peskin, C. Preitschopf, SLAC-PUB-3853 Dec. 1985

26. D. Friedan, E. Martinec, S. Shenker, U. Chicago preprint 86-0024 Nov. 1985

27. S.Mandelstam, Phys. Rev. D11, 3026, (1975); Phys. Rep. 23C, 307, (1976)

28. J. Schwinger, *Particles Sources and Fields I*, p. 102, Addison-Wesley, Reading MA (1970).

29. N. Seiberg and E. Witten, Princeton preprint 1986

30. Friedan, Martinec, and Shenker, op. cit. Ref. 26

31. H. Ooguri, Kyoto preprint, KUNS-825, Feb. 1986; KUNS-819-Rev., Dec. 1985; S. deAlwis, N. Ohta, Austin preprint, UTTG-08-86, Apr. 1986; A.LeClair, Phys. Lett. 168B, 53 (1986). H. Terao, S. Uehara, Hiroshima preprint, RRK 86-14, Apr 1986; Phys. Lett. 168B, 70 (1986) A. Neveu, H. Nicolai, P. C. West, Nucl. Phys. B264, 573 (1986)

32. G. D. Date, M. Gunyadin, M. Pernici, K. Pilch, P. vanNieuwenhuizen, Stony Brook preprint, ITP-SB-86-3, Feb. 1986; T. Banks, M. Peskin, C. Preitschopf, D. Friedan, E. Martinec, SLAC-PUB-3853, Dec 1985 H. Terao, S. Uehara, Hiroshima preprint, RRK 86-8, Feb 1986. ;RRK 86-3, Jan 1986. A. LeClair, J. Distler, Harvard preprint HUTP-86/A008, Jan 1986.

33. J. P. Yamron, Berkeley preprint, UCB-PTH-86/3, Feb. 1986

34. C. Preitschopf and K. Pilch, private communication

35. E. Witten, Interacting Open Superstrings, Princeton preprint 1986

36. J. Lykken, S. Raby Los Alamos preprint 1986; P. West private communication.

37. S.Mandelstam, contribution to the Santa Barbara Superstring Workshop, July 1985; and ref. 49

38. J. Polchinski, Texas preprint, UTTG-13-85, Jun 1985

39. W. Siegel, Ref. 18, A. Neveu and P. West et al., Ref. 49, T. Kugo et al., Ref. 49.

40. T. Banks and M. Peskin Nucl. Phys. B264,513,(1986) ; I. Frenkel, L. Garland, G. Zuckerman Yale Mathematics preprint 1986

41. E. Witten, Nucl. Phys. B226, 245 (1986)

42. D. Friedan, S. Shenker, Chicago preprint, May 1986

43. E. Witten, Nuc. Phys. Bto appear

44. S. Giddings, Princeton preprint 1986

45. S. Giddings, E. Martinec, E. Witten, Princeton preprint, May (1986).

46. S. Giddings, E. Martinec, Princeton preprint 1986

47. C. Lovelace,Phys. Lett. 34B,500,(1971)

48. D. Friedan, Chicago preprint, EFI-87-27-CHICAGO, Apr. 1985; T. Yoneya, talk at RIFP "Workshop on Unified Theories", Dec. 1985; H. Hata, K. Itoh, T. Kugo, H. Kunitomo, K. Ogawa, Kyoto preprint, RIFP-656, Apr. 1986

49. Andre LeClair, Jacques Distler (Harvard U.), HUTP-86/A008 (1986). Andre LeClair, Phys. Lett. 168B, 53 (1986). H. Terao (Kyoto U.), S. Uehara (Hiroshima U.), RRK 86-14 (1986). H. Terao (Kyoto U.). S. Uehara, (Hiroshima U.), RRK 86-8 (1986). H. Terao (Kyoto U.) S. Uehara (Hiroshima U.), RRK 86-3 (1986). H. Terao and S. Uehara, Phys. Lett. 168B, 70 (1986). M. O. Katanayev, I. V. Volovich (Wroclaw U.), UWR-86/658 (1986). Steven B. Giddings, Emil Martinec (Princeton U.), Print-86-0415 (1986). Daniel Friedan, Stephen Shenker, EFI-86-18B-CHICAGO, (1986). Steven B. Giddings, Emil Martinec (Princeton U.), Print-86-0403 (1986). P. Ramond, V.G.J. Rodgers (Florida U.), Print-86-0396 (1986). Hiroyuki Hata (Kyoto U., RIFP), Katsumi Itoh, Taichiro Kugo, Hiroshi Kunitomo, Kaku Ogawa (Kyoto U.), RIFP-656 (1986). S. Mandelstam, Nucl. Phys. B64, 25 (1973). S. Mandelstam, Nucl. Phys. B69, 77 (1974). Joseph Lykken, Stuart Raby (Los Alamos), LA-UR-86-1334 (1986) submitted to Nucl. Phys. B. Neil Marcus, Augusto Sagnotti (UC, Berkeley & LBL, Berkeley), UCB-PTH-86/9 (1986). A. Cohen, G. Moore, P. Nelson, J. Polchinski, Harvard, HUTP-86/A028 (1986); G. West, LA-UR-86-1136, Los Alamos (1986); S. Samuel, CERN-TH-4365/86 (1986); Y. Meurice, CERN-TH-4398/86 (1986); S. de Alwis, N. Ohta, Texas preprint, UTTG-08-86 (1986); M. Awada, DAMPT preprint, Print-86-0294 (1986); J. L. Gervais, Ecole Normale preprint, LPTENS 86/1 (1986). Larry Carson, Yutaka Hosotani (Minnesota U.), UMN-TH-555/86 (1986). A. Neveu, P. West, CERN-TH-4358/86 (1986). C. G. Callan, I. R. Klebanov, M. J. Perry (Princeton U.), Print-86-0243 (1986). Steven B. Giddings (Princeton U.),

Print-86-0247 (1986). Jisuke Kubo, James G. McCarthy (SUNY, Stony Brook), ITP-SB-86-8 (1986). Hiroyuki Hata (Kyoto U., RIFP), Katsumi Itoh, Taichiro Kugo, Hiroshi Kunitomo, Kaku Ogawa (Kyoto U.), KUNS 829 (1986). Jonathan P. Yamron (UC, Berkeley), UCB-PTH-86/3 (1986). P. West (CERN), CERN-TH-4304/85 (1985). H. Aratyn, A. H. Zimerman (Hebrew U.), RI/86/10 (1986). Ngee-Pong Chang (CCNY), Han-ying Guo, Zongan Qiu, Ke Wu, CCNY-HEP-86/5 (1986). Emil Martinec (Princeton U.), DOE/ER/3072-32 (1986). I. Senda (Tokyo Inst. Tech.), TIT/HEP-90 (1986). G. D. Date, M. Gunaydin, M. Pernici, K. Pilch, P. van Nieuwenhuizen (SUNY, Stony Brook), ITP-SB-86-3 (1986). Hirosi Ooguri (Kyoto U.), KUNS-825 (1986). L. Baulieu, S. Ouvry, CERN-TH-4338/85 (1985). Yannick Meurice, CERN-TH-4356/86 (1986). Ingemar Bengtsson, CERN-TH-4348/86 (1986). Antal Jevicki, CERN-TH-4341/85 (1985). A. Neveu, P. West, Phys. Lett. 168B, 192 (1986). Edward Witten, Nucl. Phys. B268, 253 (1986). T. Jacobson, R. P. Woodard, N. C. Tsamis, UCSB-TH-10-1985 (1985). Hirosi Ooguri (Kyoto U.), KUNS-819-Rev. (1985) H. Aratyn, A. H. Zimerman (Hebrew U.), RIP/85/9 (1985). J. L. Gervais (Ecole Normale Superieure, LPT), LPTENS 85/35 (1985). Sumit R. Das (Tata Inst. & Fermilab), Spenta R. Wadia (Tata Inst.), FERMILAB-PUB-85/90-T (1985). Stanley Mandelstam (Ecole Normale Superieure, LPT), LPTENS 85/13 (1985). Charles B. Thorn, Phys. Lett. 159B, 107 (1985), Addendum-ibid. 160B, 430 (1985). Joseph Polchinski (Texas U.), UTTG-13-85 (1985). D. Friedan (Chicago U., EFI & Chicago U.), EFI-85-27-CHICAGO (1985). A. Neveu (CERN), P. West (King's Coll., London), CERN-TH-4358/86 (1986). Y. Kazama, A. Neveu, H. Nicolai, P. West (CERN), CERN-TH-4301/85 (1985). A. Neveu, P. West, Phys. Lett. 168B, 192 (1986). A. Neveu, H. Nicolai, P. West, Phys. Lett. 167, 307 (1986). A. Neveu, P. C. West, Phys. Lett. 165B, 63 (1985). A. Neveu, J. Schwarz, P. C. West, Phys. Lett. 164B, 51 (1985). A. Neveu, H. Nicolai, P. C. West, Nucl. Phys. B264, 573 (1986). A. Neveu, P. C. West, Nucl. Phys. B268, 125 (1986). Michael B. Green, John H. Schwarz, Phys. Lett. 109B, 444 (1982). Michael B. Green, John H. Schwarz, Nucl. Phys. B198, 441 (1982). Michael B. Green, John H. Schwarz, Nucl. Phys. B198, 252 (1982). S. Mandelstam, Nucl. Phys. B83, 413 (1974). Charles Marshall, P. Ramond, Nucl. Phys. B85, 375 (1975). Michio Kaku, K. Kikkawa, Phys. Rev. D10, 1110 (1974). Michio Kaku, K. Kikkawa, Phys. Rev. D10, 1823 (1974). S. P. de Alwis, N. Ohta (Texas U.), UTTG-09-86 (1986). M. Awada (Cambridge U., DAMTP), Print-86-0295 (1986). Peter Orland (Bohr Inst.), NBI-HE-86-09 (1986). D. Pfeffer, P. Ramond, V.G.J. Rodgers (Florida U.), UFTP-85-19 (1986). M.A. Awada (Cambridge U., DAMTP), Print-85-0988 (1985). Michio Kaku (CCNY), CCNY-HEP-85-11 (1985). Korkut Bardakci (LBL,

Berkeley & UC, Berkeley), UCB-PTH-85/33 (1985). Michio Kaku (Osaka U. & CCNY), OU-HET-79 (1985). Michio Kaku, Phys. Lett. 162B, 97 (1985). Yutaka Hosotani, Phys. Rev. Lett. 55, 1719 (1985). Michio Kaku, Nucl. Phys. B267, 125 (1986). Keiji Kikkawa (Osaka U.), OU-HET-76 (1985). C. Lovelace, Phys. Lett. 135B 75, (1984). S. Olariu, I. Popescu, Phys. Rev. D27, 383 (1983). A. M. Polyakov, Phys. Lett. 103B, 211 (1981). A. M. Polyakov, Phys. Lett. 103B, 207 (1981). N. V. Borisov, M. V. Ioffe, M. I. Eides, Yadern. Fiz. 21, 655 (1975). M. B. Green, Nucl. Phys. B124, 461 (1977). Tamiaki Yoneya, Prog. Theor. Phys. 56, 1310 (1976). E. Cremmer, J.-L. Gervais (Ecole Normale Superieure, LPT), PTENS 75/11 (1975). K. Kikkawa (Osaka U.), Print-75-0429 (1985). Michio Kaku, Phys. Rev. D10, 3943 (1974). E. Cremmer, J. L. Gervais, Nucl. Phys. B90, 410 (1975). Lay Nam Chang, Kenneth I. Macrae, Freydoon Mansouri, Phys. Rev. D13, 235 (1976). P. Ramond (Yale U.), YALE-3075-63, (1974).

RIEMANN SURFACES AND STRING THEORIES

Luis Alvarez-Gaumé

CERN Theory Division
CH-1211 Geneva 23
SWITZERLAND

Philip Nelson

Lyman Laboratory of Physics
Harvard University
Cambridge MA 02138
USA

CONTENTS

I. OVERVIEW

These lectures are intended to introduce some of the mathematical constructions needed to analyze the first-quantized theory of strings. In this overview we quickly summarize some of the issues which will arise in the subsequent sections.

In the first quantized formulation of string theory [1], one studies the propagation of a string on a fixed background space-time. Thus the computation of string partition and correlation functions requires that we solve a two-dimensional field theory on a surface. If for simplicity we only consider closed oriented strings, the world surface traced out by the string can be thought of as a Riemann surface. In the sequel, we will always assume that the Wick rotation has been performed both on the two-dimensional world sheet and in the target space-time where the string moves. In Polyakov's approach to string theory [2], one has to calculate objects of the form:

$$Z = \sum_g \lambda^g \int dX \int \frac{[d\gamma]}{\mathrm{Vol(Diff)\ Vol(Conf.)}} \, e^{-S[X,\gamma]} \quad . \tag{1.1}$$

$X(\sigma,\tau)$ is the field describing the embedding of the string in space-time. $\gamma(\sigma,\tau)$ describes the metric on a two-dimensional parameter space Σ, and $1/\mathrm{Vol(Diff)} \cdot \mathrm{Vol(Conf.)}$ formally denotes that one has to sum over inequivalent conformal structures on Σ. Finally g is an integer, the number of handles of the surface. Equation (1.1) represents the partition function for bosonic strings. (If one wanted to compute S-matrix amplitudes, one would have to include the insertion of vertex operators.)

On a two-dimensional surface, the metric tensor $\gamma_{ab}(\sigma,\tau)$ only has three independent components. Since an infinitesimal diffeomorphism has two components, and a conformal transformation depends on an arbitrary function, there are three gauge parameters, and one might think that it is always possible to transform any metric γ_{ab} into a standard one. This unfortunately is only true on the sphere. For surfaces with higher genus one is left with a finite number of parameters (the "moduli") which characterize conformally inequivalent deformations. Thus if the set of two-dimensional fields is chosen in such a way that the functional integral is conformally invariant, the integration over metrics in (1.1) is reduced to a finite dimensional integral over moduli space [3] (see Section IV). Therefore, to a large extent, the study of string perturbation theory is reduced to the behavior of the measure in (1.1) as a function of the moduli parameters. The parameters that describe the inequivalent conformal structures describe at the same time the inequivalent complex structures that one can construct on the topological surface Σ with g handles, i.e. the inequivalent ways of finding a set of complex coordinate charts (U_α, z_α) in Σ with holomorphic transition functions. The space of inequivalent complex structures on Σ is denoted by $\mathcal{M}_g$ (the moduli space of Riemann surfaces of genus g).

When one considers fermionic strings, the measure analogous to (1.1) can also be obtained [4], but now one has to "divide" by the supervolume of supercoordinate and super Weyl transformations of the theory, and $S[X,\gamma]$ describes a two-dimensional theory coupled to 2-dimensional supergravity. The main subtlety in the fermionic analog of (1.1) is that the action $S[X,\gamma]$ contains fermions. On a Riemann surface of genus g, there are 2^{2g} spin structures. As we will see below, the GSO [5] projection, necessary to obtain the physical spectrum, requires that we sum over all the possible spin structures for each fermion. This might at first sight seem to introduce a good deal of arbitrariness in the theory. This conclusion is erroneous, because the group of diffeomorphisms on Σ acts non-trivially on the spin structures. Therefore, if we want to have a theory invariant under diffeomorphisms, we have to pay particular attention to the invariance of the sum over spin structures under the group of disconnected diffeomorphisms on Σ. This is the question of modular invariance of central importance in string theory.

One of the most interesting recent developments is the proof of holomorphic factorization of the string integrand in moduli space [6].

The space $\mathcal{M}_g$ has many remarkable properties, but one of the most relevant to string theory is that $\mathcal{M}_g$ is a complex space. It's well known from elementary complex analysis that function theoretic arguments simplify considerably in the complex analytic category. In [6] it was shown that the string integrand is essentially the absolute value squared of a holomorphic function on moduli space. This result, combined with the complex analytic structure of $\mathcal{M}_g$ provides a very powerful framework to analyze string perturbation theory [7,8] and perhaps even to get non-perturbative information about the quantum theory of strings as suggested recently by Friedan and Shenker.

There is a vast literature on the theory of Riemann surfaces. In these lectures we would like to give a first introduction to the theory of Riemann surfaces in its application to string theory. Reference [9] also contains material both parallel and complementary to the present discussion.

The organization of these lectures is as follows: Section II gives an introduction to the geometry of Riemann surfaces, to the group of disconnected diffeomorphisms, and to its action on spin structures. In Section III we show explicitly how the modular group is related to the spectrum of string theories. In Section IV we write down the measure for multiloop diagrams in string theory, prove holomorphic factorization, and start our analysis of determinants of elliptic operators on Riemann surfaces. Section V studies in more detail the moduli space $\mathcal{M}_g$ and the behavior of determinants of Cauchy-Riemann operators as functions of the moduli. Here we will use some methods borrowed from algebraic geometry to obtain information about the string partition function.

II. GENERAL PROPERTIES OF RIEMANN SURFACES

A) Bases, Coordinates, and Modular Transformations

Throughout these notes Σ will denote a compact orientable surface with g handles. A surface is completely characterized topologically by g, which is called the genus of Σ. In particular the Euler number $\chi(\Sigma) = 2 - 2g$ as can be seen by triangulating Σ. Each surface Σ can be endowed with the structure of a complex manifold, and the complex structures on Σ are in 1-1 correspondence with conformal structures on Σ. To see this, let γ_{ab} be a given metric on Σ, and let $\{U_\alpha\}$ be a covering of Σ by simply-connected patches. On each patch we can choose Gaussian coordinates:

$$ds^2_{(\alpha)} = e^{2\sigma(\alpha)}(dx^2_{(\alpha)} + dy^2_{(\alpha)}) \quad .$$

Then since across coordinate patches $ds^2_{(\alpha)} = ds^2_{(\beta)}$ the coordinate transformation $z_{(\alpha)} = f_{\alpha\beta}(z_\beta)$, $z = x + iy$, is easily shown to be holomorphic or antiholomorphic. Since moreover Σ is orientable, we can choose every $f_{\alpha\beta}$ to be orientation-preserving, i.e. holomorphic. Hence Σ acquires a complex structure which clearly depends only on the conformal class of γ. Conversely, if we are given a complex structure on Σ we can on every coordinate patch consider the conformal class $ds^2_{(\alpha)} \propto dz_{(\alpha)} d\bar{z}_{(\alpha)}$. A two-dimensional complex manifold is called a Riemann surface.

In any geometrical construction it is always desirable to avoid insofar as possible introducing new data into the problem, since in general such data are not invariant under the symmetries of the original problem and so obscure the issue of whether those symmetries are preserved. Thus for example one strives to write expressions in forms which are manifestly independent of coordinate or gauge choices. Nevertheless such choices are sometimes necessary in order to have explicit formulae. In string theory, for example, the physical data include a Riemann surface, but we will often want to augment this by a specific set of curves on the surface, or more precisely by the homology classes of certain curves. Since the introduction of such curves is unphysical and breaks diffeomorphism invariance, one has to check by hand at the end of any construction that the answer is independent of the choice made. This is the problem of modular invariance mentioned earlier.

Specifically we will choose on Σ a basis of its first homology group $H_1(\Sigma, \mathbb{Z}) \cong \mathbb{Z}^{2g}$. Such a basis is shown in Fig. 2.1. The basis shown has the property that the intersection pairing of cycles satisfies

$$(a_i, a_j) = (b_i, b_j) = 0$$

$$(a_i, b_j) = -(b_i, a_j) = \delta_{ij} \quad . \tag{2.1}$$

Here the intersection pairing or intersection form is a quadratic form on $H_1(\Sigma, \mathbb{Z})$:

$$(\cdot,\cdot) : H_1(\Sigma, \mathbb{Z}) \times H_1(\Sigma, \mathbb{Z}) \to \mathbb{Z} \tag{2.2}$$

such that given any two 1-cycles γ and γ', (γ, γ') counts the number of intersections (including orientation).

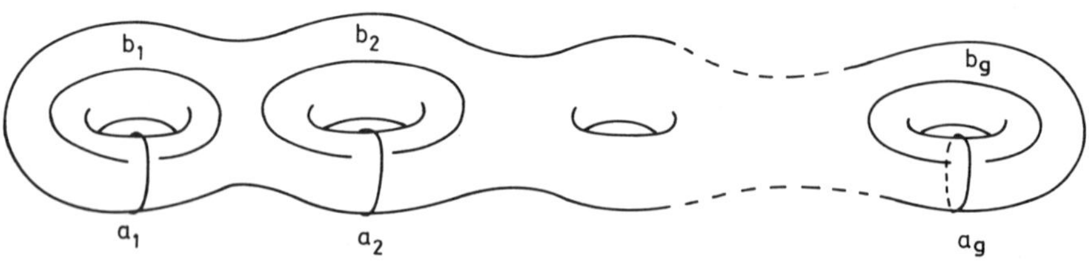

Fig. 2.1 A canonical homology basis on Σ.

Any basis satisfying (2.1) is called "canonical". We will always assume that such a basis has been chosen. In terms of it, any 1-cycle can be written as:

$$\gamma = n_i a_i + m_i b_i \quad , \quad n_i, m_i \in \mathbb{Z}$$

and therefore

$$(\gamma, \gamma') = (n, m)\begin{pmatrix} 0 & \mathbb{1} \\ -\mathbb{1} & 0 \end{pmatrix}\begin{pmatrix} n' \\ m' \end{pmatrix} \tag{2.3}$$

where $\mathbb{1}$ is the $g \times g$ unit matrix.

While the homology of Σ is rather easy to describe, its first homotopy group is a more complicated object. $\pi_1(\Sigma)$ is a non-abelian group with 2g generators and one relation. To exhibit the relation in more detail, we construct the cut surface Σ_c (see Fig. 2.2) as follows: Choose a point A on Σ, and draw curves starting at A and homotopic to the (a_i, b_i) curves of Fig. 2.1. If we cut the surface along these curves, and unfold it, we obtain a 4g-sided polygon. Each of the curves $\Gamma(a_i)$, $\Gamma(b_i)$, $1 \subseteq i \subseteq g$ generates a non-trivial class of $\pi_1(\Sigma)$. However one has that

$$[\Gamma(a_i)\Gamma(b_i)\Gamma(a_i)^{-1}\Gamma(b_i)^{-1} \ldots \Gamma(a_g)\Gamma(b_g)\Gamma(a_g)^{-1}\Gamma(b_g)^{-1}] \simeq [1] \tag{2.4}$$

is a trivial homotopy class. The deformation to the identity is obtained by noting that the boundary of Σ_c (Fig. 2.2) is a representative of (2.4), and now we only have to shrink the boundary continuously to any point in the interior of Σ. Later in this section we will describe $\pi_1(\Sigma)$ in terms of fuchsian groups (for surfaces with $g \geq 2$) in more detail.

The representation of Σ as a cut surface is very useful in proving a number of identities. For example, if θ, η are closed 1-forms, then [10]

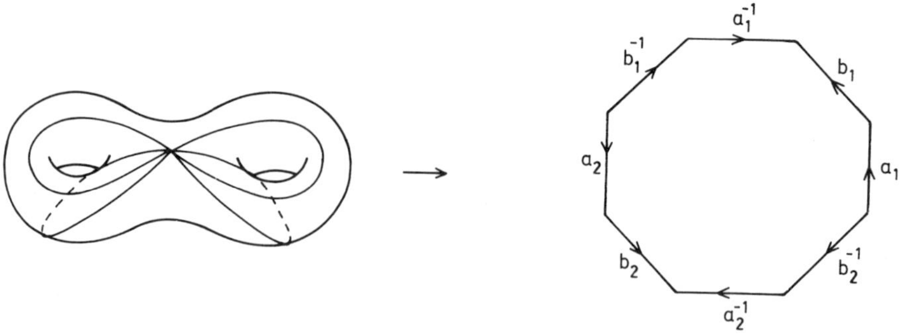

Fig. 2.2 The "cut" Riemann Surface Σ_c.

$$\int_\Sigma \theta \wedge \eta = \sum_{i=1}^{g} \left[\int_{a_i} \theta \int_{b_i} \eta - \int_{a_i} \eta \int_{b_i} \theta \right] . \tag{2.5}$$

By the Hodge-De Rham theorem one can also find a set of real harmonic 1-forms α_i, β_i, $1 \le i \le g$ dual to the homology basis:

$$\int_{a_i} \alpha_j = \int_{b_i} \beta_j = \delta_{ij}$$

$$\int_{a_i} \beta_j = \int_{b_i} \alpha_j = 0 \tag{2.6}$$

Thus far we have made no use of the complex structure on Σ. We can now combine the closed forms α_i, β_i into g holomorphic and g antiholomorphic closed 1-forms. These are known as the "abelian differentials" ω_i, $\bar{\omega}_i$; like the α's and β's they are determined once a canonical homology basis has been chosen. A standard way of normalizing the ω_i's is to require:

$$\int_{a_i} \omega_j = \delta_{ij} \quad . \tag{2.7}$$

Then the periods over the b-cycles are completely determined; we name them

$$\int_{b_j} \omega_j \equiv \Omega_{ij} \tag{2.8}$$

Ω_{ij} is known as the "period matrix" of the Riemann surface; it is a symmetric matrix with positive definite imaginary part, depending on the chosen homology basis. These two properties are a simple consequence of (2.5). For instance, if we take $\theta = \omega_i$, $\eta = \omega_j$, then $\theta \wedge \eta = \omega_i \wedge \omega_j = 0$, since both ω_i, ω_j are proportional to dz. Thus

$$0 = \int_\Sigma \omega_i \wedge \omega_j = \sum_{k=1}^{g} \left(\int_{a_k} \omega_i \int_{b_k} \omega_j - \int_{a_k} \omega_j \int_{b_k} \omega_i \right) =$$

$$= \Omega_{ij} - \Omega_{ij} \quad .$$

Similarly let $\xi = u_i \omega_i$, $u_i \in \mathbb{C}$; then the following quadratic form

$$(\xi,\xi) \equiv \frac{i}{2} \int \xi \wedge \bar{\xi} \tag{2.9}$$

is positive definite. Using (2.5) and the normalization conditions (2.7-8):

$$(\xi,\xi) = (\mathrm{Im}\,\Omega)_{ij} u_i \bar{u}_j > 0 \quad , \tag{2.10}$$

i.e. $\mathrm{Im}\,\Omega$ is a positive matrix.

The space of all period matrices $\mathscr{H}_g = \{\Omega;\ \Omega_{ij} = \Omega_{ji}, \operatorname{Im}\Omega > 0\}$ is a complex $g(g+1)/2$ - dimensional space known as Siegel's upper half plane. We have seen how every Riemann surface with homology basis gives rise to an Ω. In fact it can be shown that no two inequivalent Riemann surfaces have the same Ω (Torelli's theorem), so we can use $\mathscr{H}_g$ to parametrize surfaces. This is a highly redundant description, however, since the same surface with two bases will in general have two different matrices Ω. Suppose the two canonical bases are related by

$$\begin{pmatrix} a' \\ b' \end{pmatrix} = \begin{pmatrix} D & C \\ B & A \end{pmatrix} \begin{pmatrix} a \\ b \end{pmatrix} \quad , \tag{2.11}$$

where A, B, C, D are $g \times g$ matrices. To preserve the convention (2.1), the matrix in (2.11) must leave the symplectic form (2.2-3) invariant. The matrix in (2.11) is therefore a symplectic modular matrix with integer coefficients, i.e. an element of $Sp(2g, \mathbb{Z})$.

It is easy to compute the change of ω_i and Ω_{ij} under a change of basis: We want the new abelian differentials ω_i' to be normalized as in (2.7) with respect to the new homology basis:

$$\int_{a_i'} \omega_j' = \delta_{ij} \ , \quad a_i' = D_{ij} a_j + C_{ij} b_j \quad .$$

Then

$$\omega' = \left[(C\Omega + D)^{-1} \right]^t \omega \tag{2.12}$$

and the new period matrix is given by

$$\tilde{\Omega} = (A\Omega + B)(C\Omega + D)^{-1} \quad . \tag{2.13}$$

Using the defining properties of a symplectic matrix

$$A^tC - C^tA = 0 \quad , \quad A^tD - C^tB = 1$$
$$B^tD - D^tB = 0 \quad , \quad B^tC - D^tA = -1 \tag{2.14}$$

one can directly check that $\tilde{\Omega}$ is again an element of $\mathscr{H}_g$. Since period matrices related by (2.13) refer to the same Riemann surface, we can define $\mathscr{A}_g \equiv \mathscr{H}_g/Sp(2g, \mathbb{Z})$, with every point on $\mathscr{A}_g$ representing an orbit of $Sp(2g, \mathbb{Z})$ in $\mathscr{H}_g$. Finally a metric on $\mathscr{H}_g$ invariant under the action of the symplectic modular group is given by

$$ds^2 = Tr[(Im\ \Omega)^{-1}d\Omega, (Im\ \Omega)^{-1}d\bar{\Omega}] \quad . \tag{2.15}$$

To summarize, using the map (2.8) and the equivalence (2.13) we see that every Riemann surface is represented by a point in $\mathscr{A}_g$; thus the moduli space of Riemann surfaces $\mathscr{M}_g$ sits inside $\mathscr{A}_g$. On the other hand there is no reason to think that every matrix Ω in $\mathscr{H}_g$ corresponds to any surface Σ. Indeed the moduli space $\mathscr{M}_g$ is in general much smaller than $\mathscr{A}_g$, and it is embedded in a very complicated way [11]. Nevertheless $\mathscr{M}_g$ inherits a fairly simple complex structure from $\mathscr{A}_g$ [12]. Later we will give another description of $\mathscr{M}_g$ and its complex structure based on Bers' construction [13].

We can now describe in more detail the group of disconnected diffeomorphisms of Σ. Let $Diff^+(\Sigma)$ be the full group of orientation preserving diffeomorphisms on Σ, and $Diff_0^+(\Sigma)$ the normal subgroup of diffeomorphism connected to the identity. The quotient group

$$\Omega(\Sigma) \equiv Diff^+(\Sigma)/Diff_0^+(\Sigma) \tag{2.16}$$

is known as the mapping class group. Any non-trivial class of $\Omega(\Sigma)$

is called a modular transformation, and all are generated by the "Dehn twists". A Dehn twist around a noncontractible loop $\gamma \subset \Sigma$ is constructed as follows: Given γ, choose a neighborhood of γ that is topologically equivalent to a cylinder. We can now cut Σ along γ, and keeping one of the edges of the cut fixed, we can twist the other by 2π and glue them together again. By this construction we associate to every point of the original torus a new point, in a way which is smooth and yet clearly not continuously related to the identity map. A deep theorem (see [14] for details) says that for any class in $\Omega(\Sigma)$ we can always choose a representative given by a Dehn twist. Furthermore, we can describe a complete set of generators of $\Omega(\Sigma)$, and here we find a pleasant surprise: A useful set of generators of $\Omega(\Sigma)$ includes Dehn twists around curves which wind around a single handle or at most two handles. In other words, there is nothing qualitatively new in $\Omega(\Sigma)$ beyond genus 2 (beyond two loops). In Fig. 2.3 we have drawn a set of generators of $\Omega(\Sigma)$. (Note however that this is not a *minimal* set).

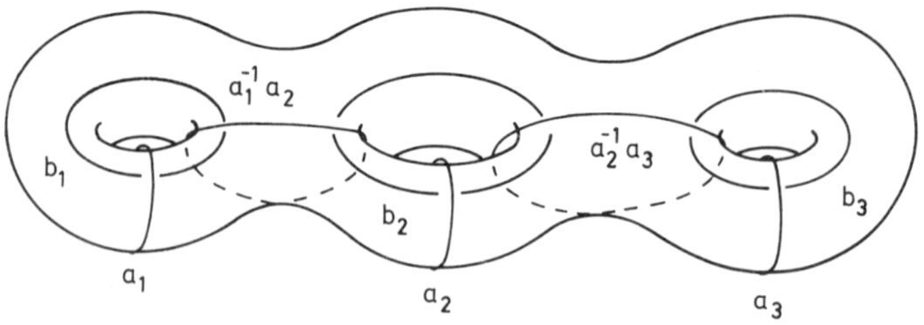

Fig. 2.3 Generators of the mapping class group for genus 3.

There is a useful way of representing the Dehn twists in terms of matrices. To do this we write the action of the twist on the homology basis. Let D_γ be the diffeomorphism defined by twisting around γ. The intersection matrix (2.1-3) is manifestly invariant under diffeomorphisms, so the action of $\Omega(\Sigma)$ on $H_1(\Sigma, \mathbb{Z})$ must necessarily preserve (2.2). Thus, the matrix $M(D_\gamma)$ representing D_γ must be an element of $Sp(2g, \mathbb{Z})$. In fact the set of matrices $M(D_\gamma)$ generate all of $Sp(2g, \mathbb{Z})$ [15]. Unfortunately a lot of information is lost when passing from D_γ to $M(D_\gamma)$, since a Dehn twist about a homologically trivial curve, while nontrivial, doesn't affect the homology class of any curve and so maps to the unit matrix. Such twists generate a subgroup $\mathscr{T}(\Sigma)$ of $\Omega(\Sigma)$ called the "Torelli group". Fortunately, as we will see later the diffeomorphisms in $\mathscr{T}(\Sigma)$ do not affect spin structures. Once we are given the proof that the heterotic string is modular invariant [16], then, we see that any new string theory obtained by modifying the sum over spin structures will automatically be $\mathscr{T}(\Sigma)$-invariant. Thus we will only need to check invariance under the quotient group $\Omega(\Sigma)/\mathscr{T}(\Sigma)$, and this is precisely $Sp(2g, \mathbb{Z})$.

Before describing the characterization of spin structures on Σ, we write down the matrices $M(D_\gamma)$ for genus 2 which implement the Dehn twists in Fig. 2.3 on $H_1(\Sigma, \mathbb{Z})$:

$$D(a_1) = \begin{bmatrix} 1 & 0 & 0 & 0 \\ 0 & 1 & 0 & 0 \\ 1 & 0 & 1 & 0 \\ 0 & 0 & 0 & 1 \end{bmatrix}, \quad D(b_1) = \begin{bmatrix} 1 & 0 & 1 & 0 \\ 0 & 1 & 0 & 0 \\ 0 & 0 & 1 & 0 \\ 0 & 0 & 0 & 1 \end{bmatrix}$$

$$D(a_2) = \begin{bmatrix} 1 & 0 & 0 & 0 \\ 0 & 1 & 0 & 0 \\ 0 & 0 & 1 & 0 \\ 0 & 1 & 0 & 1 \end{bmatrix}, \quad D(b_2) = \begin{bmatrix} 1 & 0 & 0 & 0 \\ 0 & 1 & 0 & 1 \\ 0 & 0 & 1 & 0 \\ 0 & 0 & 0 & 1 \end{bmatrix} \qquad (2.17)$$

$$D(a_1^{-1}a_2) = \begin{bmatrix} 1 & 0 & 0 & 0 \\ 0 & 1 & 0 & 0 \\ -1 & 1 & 1 & 0 \\ 1 & -1 & 0 & 1 \end{bmatrix}$$

B) Tensors and Bundles

The local differential geometry on Σ is quite simple. Suppose that a particular metric for Σ has been given. On a local coordinate patch we can always transform the metric into the form

$$ds^2 = e^{2\phi}((dx^1)^2 + (dx^2)^2) = 2\gamma_{z\bar{z}}dzd\bar{z}$$

$$z = x^1 + i\, x^2 \quad . \tag{2.18}$$

Using the notation of the first two entries in [3], we can always decompose an arbitrary tensor into its irreducible components with respect to O(2) (the frame group of the surface). These can all be written in terms of tensors of the form

$$t = t(z,\bar{z})dz^q \quad , \quad q = 0, \pm 1, \pm 2, \ldots \quad . \tag{2.19}$$

$t(z,\bar{z})$ transforms across coordinate patches in such a way that t is invariant. In complex coordinates (2.18) there is a single independent component of the Christoffel symbols:

$$\Gamma^z_{zz} = (\gamma_{z\bar{z}})^{-1}\partial_z\gamma_{z\bar{z}} \quad . \tag{2.20}$$

Γ is called the "hermitian connection" of γ.

The tensors (2.19) are sections of the line bundle K^q, where K is the holomorphic cotangent bundle i.e. $K = T^*_{(1,0)}\,\Sigma$, the set of 1-forms with no $\bar{z}$ indices. K is called the "canonical bundle" of Σ. The covariant derivative is then an operator:

$$\nabla^q_z : K^q \rightarrow K^{q+1}$$

given locally by:

$$\nabla_z^q\, t(z,\bar{z}) = (\gamma_{z\bar{z}})^{+q}\, \partial_z(\gamma_{z\bar{z}}^{-q}\, t) \tag{2.21}$$

and its adjoint is

$$\begin{aligned} &\nabla_q^z : K^q \to K^{q-1} \\ &\nabla_q^z\, t = -\gamma^{z\bar{z}}\, \partial_{\bar{z}}\, t \qquad . \end{aligned} \tag{2.22}$$

On a complex manifold the operator $\bar{\partial}$ (the Cauchy-Riemann operator) is always intrinsically defined, and ∇_q^z is essentially the Cauchy-Riemann operator acting on q-differentials. Using (2.22), we can construct two laplacians

$$\begin{aligned} \Delta_q^{(+)} &= \nabla_{q+1}^z\, \nabla_z^q \\ \Delta_q^{(-)} &= \nabla_z^{q-1}\, \nabla_q^z \end{aligned} \tag{2.23}$$

and

$$\Delta_q^{(+)} - \Delta_q^{(-)} = -\, qR/2 \tag{2.24}$$

where R is the scalar curvature of (2.18)

$$R = -\,2\, e^{-2\phi}\, \partial_z \partial_{\bar{z}} \phi \qquad . \tag{2.25}$$

Later in these lectures we will need the dimension of the space of holomorphic q-differentials. This is given by the Riemann-Roch theorem (see [10]). Using standard notation, we let $H^0(\Sigma, K^q)$ be the space of global holomorphic sections of K^q, i.e. zero modes of $\bar{\partial}$

acting on sections of K^q; which we denote by $\bar{\partial}_q$. In other words,

$$H^0(\Sigma, K^q) = \text{Ker}\,\bar{\partial}_q \quad . \tag{2.26}$$

If $q = 0$, Liouville's theorem implies that the only holomorphic function on a compact surface is a constant. Thus:

$$\dim \text{Ker}\,\bar{\partial}_0 = 1 \tag{2.27}$$

For $q = 1$, $H^0(\Sigma, K)$ is the space of abelian differentials, and we have already mentioned that there are g of them.

On a Riemann surface, one can take advantage of the complex structure to describe spinors in terms of half-differentials $q = \pm\frac{1}{2}$. Geometrically, a spinor bundle is defined as a square root of the canonical line bundle $L \cong K^{\frac{1}{2}}$, i.e. as a bundle L such that

$$L \otimes L \cong K \quad . \tag{2.28}$$

The question is whether such square roots exist; if they do one would like to know how many there are. Given a covering of Σ by coordinate patches $\{U_\alpha\}$, we can choose orthonormal frames $e^a_{(\alpha)}$ such that

$$ds^2 = \delta_{ab}\, e^a_{(\alpha)} e^b_{(\alpha)} \quad . \tag{2.29}$$

On the overlaps $U_\alpha \cap U_\beta$, the frames are related by an O(2) rotation:

$$e^a_{(\alpha)} = (R_{\alpha\beta})^a_b\, e^b_{(\beta)} \tag{2.30}$$

satisfying the "cocyle condition"

$$R_{\alpha\beta}R_{\beta\gamma}R_{\gamma\alpha} = 1 \quad . \tag{2.31}$$

When we define spinors, the transition functions $S_{\alpha\beta}$ must therefore satisfy $(S_{\alpha\beta})^2 = R_{\alpha\beta}$. However, for an arbitrary choice of square roots the cocycle condition will not be satisfied:

$$S_{\alpha\beta}S_{\beta\gamma}S_{\gamma\alpha} \equiv w_2(\alpha,\beta,\gamma) = \pm 1 \quad . \tag{2.32}$$

One says that $w_2(\alpha,\beta,\gamma)$ is a 2-cocyle with $\mathbb{Z}_2$-values; it is a representative of the second Stiefel-Whitney class $w_2 \in H^2(\Sigma, \mathbb{Z}_2)$.

We would thus like to know whether $S_{\alpha\beta}$ can be chosen to make $w_2(\alpha,\beta,\gamma) \equiv +1$. For a compact Riemann surface, however, the second Stiefel-Whitney class is the mod. 2 reduction of the Euler class [17]. Since the Euler number of Σ is always even, i.e. $2 - 2g = 0 \pmod 2$, $w_2(\alpha,\beta,\gamma)$ must always be a coboundary: $w_2(\alpha,\beta,\gamma) = \delta\eta(\alpha,\beta,\gamma)$, where η is a 1-cocycle. Replacing $S_{\alpha\beta}$ by $S'_{\alpha\beta} = S_{\alpha\beta}\eta_{\alpha\beta}$, we therefore see that we can change any set of would-be transition functions into genuine transition functions satisfying the cocycle condition.

We can think of the $S_{\alpha\beta}$ not as real 2×2 rotation matrices but as the phases of the complex transition functions $h_{\alpha\beta} = (dz_{(\alpha)}/dz_{(\beta)})^{\frac{1}{2}}$. A set of $S_{\alpha\beta}$ satisfying the cocycle condition can thus be regarded as a consistent choice of square roots of $dz_{(\alpha)}/dz_{(\beta)}$, since certainly there is no ambiguity about the square root of the modulus. Thus a spin structure is also a holomorphic bundle whose square is K, as asserted above. Given one such set of choices, any other is then given by $h'_{\alpha\beta} = \eta_{\alpha\beta}h_{\alpha\beta}$ where $\eta_{\alpha\beta}$ is the constant function ± 1 on $U_\alpha \cap U_\beta$. Thus the set of all spin bundles $\{L_\alpha\}$ is in 1-1 correspondence with $H^1(\Sigma, \mathbb{Z}_2)$; in particular the number of distinct L_α is the number of elements of $H^1(\Sigma; \mathbb{Z}_2)$, or 2^{2g}. This correspondence is not natural, however, as there is *a priori* no special "starting" spin structure.

There is another way to see that there are 2^{2g} spin structures on Σ. Given two spin bundles L_α, L_β, we can put on each one the metric inherited from some metric on K. Now consider the line bundle $\mathscr{L} = L_\alpha \otimes L_\beta^{-1}$, where L_β^{-1} is the dual bundle to L_β, with its dual metric. Then the hermitian curvature of $\mathscr{L}$ is easily seen to be $R_{\mathscr{L}} = \frac{1}{2} R_K - \frac{1}{2} R_K \equiv 0$. (See (2.25).) Even though $\mathscr{L}$ is flat and hence has zero Chern class, it may nevertheless be nontrivial as a holomorphic bundle. In fact, if we cut Σ open as before and assign transition functions across the cuts then every flat bundle on Σ is equivalent to one whose transition functions are all constant phases, one for each homology generator [18]. (These transition functions can also be regarded as the holonomy of the connection whose connection form in this trivialization is identically zero on Σ_c.) For the homology basis we have chosen (Fig. 2.1), if s is a section of a general flat line bundle $\mathscr{L}$, we identify the section s along a_i with $e^{-2\pi i\phi_i}$ times s along a_i^{-1}; and s along b_i with $e^{2\pi i\theta_i}$ times s along b_i^{-1} (see Fig. 2.2). Since $0 \le \phi_i, \theta_i \le 1$, flat holomorphic line bundles are parametrized by a torus, called the "Picard torus" of the surface $\mathrm{Pic}_0(\Sigma) = \mathbb{R}^{2g}/\mathbb{Z}^{2g}$. Since $L_\alpha \otimes L_\beta^{-1}$ is a flat holomorphic line bundle, it is represented by a point in $\mathrm{Pic}_0(\Sigma)$. However, since $L_\alpha^2 = L_\beta^2 = K$, $(L_\alpha \otimes L_\beta^{-1})^2$ is trivial, and therefore the difference between two spin structures is parametrized by a point of order 2 in $\mathrm{Pic}_0(\Sigma)$. Since $\mathbb{R}^{2g}/\mathbb{Z}^{2g}$ has 2^{2g} points of order 2, we again conclude that the number of spin structures is 2^{2g}. In this description of flat line bundles, we found it convenient to choose a homology basis of Σ to describe them. However, the group of flat holomorphic line bundles $\mathrm{Pic}_0(\Sigma)$ is an intrinsically defined object independent of any basis. The basis simply makes it easy to find an explicit parametrization.

In order to get a better understanding of spinors (and line bundles in general) on a Riemann surface, there is another equivalent way of describing line bundles in terms of "divisors". Above we regarded a holomorphic line bundle ξ over Σ as defined by its transition functions $\{h_{\alpha\beta}\}$, with $h_{\alpha\beta}$ holomorphic and nowhere vanishing on $U_\alpha \cap U_\beta$. If we denote by $\mathcal{O}^*$ the set of nowhere vanishing holomorphic functions on open sets of Σ, then the group of holomorphic line bundles is given by 1-cocycles with $\mathcal{O}^*$ values: $H^1(\Sigma, \mathcal{O}^*)$ [18]. Divisors provide an alternative description of line bundles as follows.

Let s be a meromorphic section of ξ, and let us consider the set of points $\{P_i\}$ where s vanishes and $\{Q_i\}$ where s has a pole. We define the "divisor of s" as the formal sum of points P_i of Σ with integer multiplicities attached:

$$\mathrm{div}(s) \equiv \sum n_i P_i - \sum m_i Q_i \tag{2.33}$$

where $n_i(m_i)$ is the order of the zero (pole) of s at $P_i(Q_i)$. Any other section of ξ is obtained from s via multiplication by a meromorphic function. We define an equivalence class in the group of divisors by saying that $D_1 \sim D_2$ if $D_1 - D_2$ is the divisor of a meromorphic function. Thus to every bundle we assign a divisor class.

It turns out that every distinct line bundle corresponds to exactly one divisor class. For example, given a divisor class (D), choose a representative divisor D. On any coordinate patch U_α, we can find a meromorphic function f_α whose divisor (f_α) coincides with the restriction $D|U_\alpha = (f_\alpha)$. On the overlaps $U_\alpha \cap U_\beta$, we have two meromorphic functions f_α, f_β with the same divisor. Then $h_{\alpha\beta} = f_\alpha / f_\beta$ is holomorphic and nowhere vanishing on $U_\alpha \cap U_\beta$, and the cocycle condition is trivially satisfied. Thus D defines a line bundle ξ_D.

To any divisor D, we can associate its degree

$$D = \sum n_i P_i - \sum m_i Q_i \tag{2.34}$$

$$\deg D = \sum n_i - \sum m_i \quad . \tag{2.35}$$

Since ordinary meromorphic functions have as many poles as zeros, it follows that $\deg D = \deg D'$ whenever $D \sim D'$, or in other words we can assign a degree to divisor *classes*, or to line bundles. In fact, the degree of D is the same as the first Chern class of ξ_D. Recall that in order to compute the first Chern class of a line bundle we can count the monopole singularities of any section. By construction ξ_D has a section with poles and zeros given by the divisor D. Away from P_i, Q_i this section defines a section of a U(1) bundle which winds n_i times around P_i and $-m_i$ times around the Q_i, and so has total monopole number deg D. The canonical line bundle has degree $2g-2$, i.e. minus the Euler class. Since $L_\alpha^2 = K$, we conclude that the spinor line bundles have degree $g-1$. As in the case of spinor bundles, given two holomorphic line bundles of the same degree ξ_1, ξ_2, the "difference" $\xi_1 \otimes \xi_2^{-1}$ is a flat line bundle, and it is again characterized by a point in the Picard torus $\mathrm{Pic}_0(\Sigma)$.

C) <u>ϑ - Functions</u>

When one studies complex function theory on a sphere, the basic building blocks are monomials of the form $(z-z_i)$. The analogous functions for Riemann surfaces of higher genus are "ϑ - functions". In fact given the zeros and poles of a meromorphic function f on Σ, it can be written up to a constant in terms of ϑ - functions [19]. The ϑ - function is a function of g complex variables depending parametrically on a period matrix. It is defined by the series (for more details and references see [19]).

$$\vartheta(z|\Omega) = \sum_{n\in\mathbb{Z}^g} e^{i\pi n\cdot\Omega\cdot n + 2\pi i n\cdot z} \qquad z\in\mathbb{C}^g \quad . \tag{2.36}$$

Here Ω is the period matrix in $\mathscr{H}_g$. One can also introduce the "ϑ-function with characteristics". Let a, b be vectors in $\mathbb{R}^g$. Then define

$$\vartheta\begin{bmatrix} a \\ b \end{bmatrix}(z|\Omega) = \sum_{n\in\mathbb{Z}^g} e^{i\pi(n+a)\cdot\Omega\cdot(n+a) + 2\pi i(n+a)(z+b)} \quad , \tag{2.37}$$

which is easily seen to be

$$= e^{i\pi a\cdot\Omega\cdot a + 2\pi i a\cdot(z+b)}\,\vartheta(z + \Omega a + b|\Omega) \quad .$$

We can construct a lattice in $\mathbb{C}^g$: $L_\Omega = \mathbb{Z}^g + \Omega\mathbb{Z}^g$ whose elements are of the form $\Omega n + m$, $n, m \in \mathbb{Z}^g$. The ϑ - function is completely characterized by its behavior under lattice shifts

$$\vartheta\begin{bmatrix} a \\ b \end{bmatrix}(z + \Omega n + m|\Omega) = e^{-i\pi n\cdot\Omega\cdot n - 2\pi i n\cdot(z+b) + 2\pi i a\cdot m} \cdot \vartheta\begin{bmatrix} a \\ b \end{bmatrix}(z|\Omega) \quad . \tag{2.38}$$

Notice also that

$$\vartheta\begin{bmatrix} a+n \\ b+m \end{bmatrix}(z|\Omega) = e^{2\pi i a\cdot m}\,\vartheta\begin{bmatrix} a \\ b \end{bmatrix}(z|\Omega) \quad . \tag{2.39}$$

Thus geometrically ϑ is a section of a holomorphic line bundle $\mathscr{L}$ on the complex torus $J(\Sigma) = \mathbb{C}^g/L_\Omega$. $J(\Sigma)$ is known as the "jacobian" of the marked Riemann surface Σ. $J(\Sigma)$ is similar to the Picard torus $Pic_0(\Sigma)$, but not the same thing. For one thing $J(\Sigma)$ is naturally a complex manifold. Also $Pic_0(\Sigma)$ is intrinsic, while $J(\Sigma)$ depends on the homology basis chosen. We will soon set up a correspondence between the two.

We can easily compute the first Chern class of the bundle in which ϑ defines a section. First we note that for any holomorphic line bundle we get a generalization of (2.25), expressing the curvature of the hermitian connection:

$$c_1(\mathscr{L}) = \frac{1}{2\pi i}\,\partial\bar{\partial}\,\log\| s \|^2 \quad . \tag{2.40}$$

Here $\|\cdot\|$ is the given norm and s is any nonvanishing holomorphic section. Such a section will not in general exist, but nevertheless (2.40) can be defined locally. Since any two holomorphic sections are related by a holomorphic function, $s' = fs$, we have $c_1' = \frac{1}{2\pi i} \times (\partial\bar{\partial} \log \| s \|^2 + \partial\bar{\partial} \log f + \partial\bar{\partial} \log \bar{f}) = c_1$, so (2.41) is in fact globally defined. In the case at hand we represent s by a function S on $\mathbb{C}^g$ satisfying (2.38). Then a suitable norm, continuous throughout the torus $J(\Sigma)$, is given by

$$\| s(z) \|^2 = e^{i\pi(z-\bar{z})(\Omega-\bar{\Omega})^{-1}(z-\bar{z})} \, |S(z)|^2 \quad .$$

Inserting into (2.40), the $|S(z)|^2$ factor does not contribute and we get

$$c_1(\mathscr{L}) = \sum_{i=1}^{g} d\phi^i \wedge d\theta^i \quad . \tag{2.41}$$

Using the Hirzebruch-Riemann-Roch theorem [20], one can show that (2.41) implied that $\mathscr{L}$ has only one holomorphic section. It is represented by the ϑ - function.

The relation between ϑ - functions and spin structures is provided by the Riemann vanishing theorem [19]. Choose a point P_0 on Σ, and define the Jacobi map

$$\Phi^i(P) = \int_{P_0}^{P} \omega^i \quad . \tag{2.42}$$

This is a well defined embedding of the surface Σ into the jacobian $J(\Sigma)$, independent of the path we choose to join P_0 and P. (The difference between two paths is a closed loop, and therefore it shifts Φ by an element of the lattice L_Ω.) Define the function:

$$f(P) = \vartheta(z + \int_{P_0}^{P} \omega | \Omega) \tag{2.43}$$

for fixed z. f is well defined on the cut surface Σ. The Riemann vanishing theorem states that either $f(P)$ vanishes identically, or else it has g zeros $P_1, \ldots, P_g$ satisfying the relation

$$z + \sum_{i=1}^{g} \int_{P_0}^{P_i} \omega = \Delta \quad . \tag{2.44}$$

Δ is a point of $J(\Sigma)$ called the "vector of Riemann constants". It depends on P_0 and the choice of homology basis, but not on z. In particular, if e is a point in the set Θ of zeros of $\vartheta(z|\Omega)$ ($\Theta \subseteq J(\Sigma)$ is called the "theta divisor"), (2.44) implies that there are (g-1) points $P_1, \ldots, P_{g-1}$ such that

$$e + \sum_{i=1}^{g-1} \int_{P_0}^{P_i} \omega = \Delta \quad , \tag{2.45}$$

since one of the g points must be P_0 itself. The converse is also true. This means that positive divisors of degree $g-1$ always map to points e in Θ via (2.45).

It is useful to interpret (2.42) not as a map from Σ to $J(\Sigma)$ but as a map taking the divisor $P - P_0$ to $J(\Sigma)$. More generally, given a divisor D of degree zero we can take its points and join them in pairs to get a one-cycle c satisfying $\partial c = D$. We then define

$$\vec{I}[D] = \int_c \vec{\omega} \tag{2.46}$$

which again does not depend on which c we choose. Moreover I is linear in the divisor D and its kernel is precisely the set of divisors of ordinary meromorphic functions (Abel's theorem). Thus I sends divisor *classes* to $J(\Sigma)$; it is the map $I : \mathrm{Pic}_0(\Sigma) \to J(\Sigma)$ promised before. We can thus interpret the Riemann vanishing theorem as saying that $e \in J(\Sigma)$ is in Θ, i.e. $\vartheta(e|\Omega) = 0$, if and only if there is a divisor class $[P_1 + \ldots + P_{g-1}]$ such that $I[P_1 + \ldots + P_{g-1} - (g-1)P_0] = \Delta - e$.

We next wish to show that once a homology basis is chosen then there exists a preferred spin structure L_0 on Σ defined with no further choices. This is a consequence of the Riemann vanishing theorem. We begin by noting the Riemann-Roch theorem implies that for any holomorphic bundle ξ, then [10]

$$\dim H^0(\Sigma,\xi) - \dim H^0(\Sigma,K\otimes\xi^{-1}) = \deg \xi + 1 - g \quad . \tag{2.47}$$

When $\deg \xi = g-1$, this implies that ξ has a holomorphic section whenever $K\otimes\xi^{-1}$ does. In this case the divisor class of ξ is of the form $[P_1 + \ldots + P_{g-1}]$, since there exists a section with no poles, only zeros, and the theorem says that then $K\otimes\xi^{-1}$ has a divisor class of the form $[Q_1 + \ldots + Q_{g-1}]$.

Let L_α be *any* spin bundle, with divisor class $[D_\alpha]$. Consider the set s_α in $J(\Sigma)$ given by

$$s_\alpha = \{I[P_1 + \ldots + P_{g-1} - D_\alpha]\} \subseteq J(\Sigma) \tag{2.48}$$

where $\{P_i\}$ runs over all (g-1)-tuples of points. Then from the above paragraph s_α is symmetric under reflection through the origin of $J(\Sigma)$. For, if $z = I[P_1 + \ldots + P_{g-1} - D_\alpha]$ then we choose $\{Q_i\}$ such that the divisor class of K is $[P_1 + \ldots + Q_{g-1}]$. Letting $w = I[Q_1 + \ldots + Q_{g-1} - D_\alpha]$ and remembering that $(L_\alpha)^2 \cong K$ we see that $w = -z$, so s_α is symmetric.

Rewrite s_α as the set of all z of the form $I[P_1 + \ldots + P_{g-1} - (g-1)P_0] - I[D_\alpha - (g-1)P_0]$. Then as $\{P_i\}$ vary, the first term sweeps out the theta divisor:

$$\begin{aligned} s_\alpha &= -\Theta + \Delta - I[D_\alpha - (g-1)P_0] \\ &\equiv -\Theta + e_\alpha \quad . \end{aligned} \tag{2.49}$$

Since $\vartheta(z) = \vartheta(-z)$ we have that $\Theta = -\Theta$. We already had that $s_\alpha = -s_\alpha$, and so we conclude that $\Theta + 2e_\alpha = \Theta$. Thus $\vartheta(z + 2e_\alpha|\Omega)/\vartheta(z|\Omega)$ never vanishes or blows up on the compact space $J(\Sigma)$, and so it is constant. We conclude that $2e_\alpha \in L_\Omega$, i.e. that e_α is a half-point of $J(\Sigma)$. There are 2^{2g} such points: every spin structure yields one of them, and by Abel's theorem different spin structures yield different e_α. Hence every half-point is the e_α of some L_α. In particular some L_0 maps to $e_0 = \vec{0} \in J(\Sigma)$. This the preferred spin structure which we seek.

To summarize, we have a remarkable relationship between the zeros of the complicated function (2.36) and the geometry of spin bundles. Given a Riemann surface Σ and a homology basis, we substitute the period matrix of Σ into (2.36) and define the zero set Θ in $J(\Sigma)$. Then there exists a spin bundle L_0 with divisor D_0 such that Θ is precisely the locus $\{I(D - D_0)\}$ as D ranges over all twisted spin bundles L for which $\bar{\partial}_L$ has a zero mode. From this characterization it is clear that D_0 depends on the chosen homology basis but not on the point P_0 chosen in the intermediate steps above.

If we now recall the definition of ϑ-functions with characteristics (2.37), there is a 1-1 correspondence between spin structures and characteristics α, β where $\alpha_i = 0, \frac{1}{2}$, $\beta_i = 0, \frac{1}{2}$. We can divide the spin structures into even and odd, depending on whether the corresponding ϑ-functions are even or odd functions of z. Using (2.37) and $\alpha_i, \beta_i = 0, \frac{1}{2}$:

$$\vartheta\begin{bmatrix} \alpha \\ \beta \end{bmatrix}(-z|\Omega) = (-1)^{4\alpha\cdot\beta}\, \vartheta\begin{bmatrix} \alpha \\ \beta \end{bmatrix}(z|\Omega) \quad . \tag{2.50}$$

Therefore on the torus, there are four spin structures corresponding to $(\alpha,\beta) = (0,0),\ (0,\frac{1}{2}),\ (\frac{1}{2},0),\ (\frac{1}{2},\frac{1}{2})$ and we conclude that the first three are even and the last one is odd. A simple inductive argument shows that there are $2^{g-1}(2^g+1)$ even and $2^{g-1}(2^g-1)$ odd spin structures. The splitting between even and odd can be understood in a different way. If we consider the Weyl operator for a given spin structure, the number of zero modes is either 1(mod 2), or 0(mod 2) for even or odd spin structures respectively.

Once we have used a choice of marking to identify the spin structures with the characteristics of ϑ - functions, we can determine the behavior of the spin structures under modular transformations (elements of $\Omega(\Sigma)$). The action of a non-trivial diffeomorphism on the homology basis can be represented by a symplectic matrix $\Lambda \in Sp(2g, \mathbb{Z})$:

$$\begin{bmatrix} a' \\ b' \end{bmatrix} = \begin{bmatrix} D & C \\ B & A \end{bmatrix} \begin{bmatrix} a \\ b \end{bmatrix} \equiv \Lambda \begin{bmatrix} a \\ b \end{bmatrix}$$

and the new period matrix is given by (2.13). Under this change of basis the ϑ - function with characteristics transforms according to the rule [22]:

$$\vartheta\begin{bmatrix} \tilde{\alpha} \\ \tilde{\beta} \end{bmatrix}(0|\tilde{\Omega}) = \varepsilon(\Lambda)e^{-i\pi\phi(\alpha,\beta)}\det{}^{\frac{1}{2}}(C\Omega + D)\,\vartheta\begin{bmatrix} \alpha \\ \beta \end{bmatrix}(0|\Omega) \tag{2.51}$$

with

$$\begin{pmatrix} \tilde{\alpha} \\ \tilde{\beta} \end{pmatrix} = \begin{bmatrix} D & -C \\ -B & A \end{bmatrix} \begin{bmatrix} \alpha \\ \beta \end{bmatrix} + \frac{1}{2}\begin{bmatrix} (CD^t)_d \\ (AB^t)_d \end{bmatrix} \tag{2.52}$$

$$\phi(\alpha,\beta) = [\alpha\, D^tB\alpha + \beta\, C^tA\beta] - [2\alpha\, B^tC\beta + (\alpha D^t - \beta C^t)(AB^t)_d]$$

and M_d is the vector constructed with the diagonal entries of the matrix M. ε is an eighth root of unity depending on Λ but not on α,β.

D) Uniformization Theory

Finally, we would like to discuss the general structure of Riemann surfaces from a different point of view. The basic result we will use in this subsection is the general uniformization theorem for Riemann surfaces (see [10] for details). This will also give us an easy way of counting the dimension of the moduli space $\mathcal{M}_g$. This theorem of Klein, Poincaré, and Koebe implies that there are just three distinct simply-connected Riemann surfaces:

a) the sphere S^2. We can represent its unique complex structure by the usual metric of constant positive curvature.

b) the complex plane $\mathbb{C}$ with the standard flat metric.

c) the upper half plane H with its Poincaré metric $ds^2 = dzd\bar{z}/(\mathrm{Im}\ z)^2$.

Given any Riemann surface Σ, its universal cover is one of these. That is, any Σ is the quotient of either a), b) or c) by $\pi_1(\Sigma)$ acting without fixed points. To preserve the complex structure, the group action must be by analytic isomorphisms of the simply connected surface. Therefore what we have to do is to classify the fixed point free analytic automorphisms of S^2, $\mathbb{C}$ and H.

For S^2 the answer is very easy. The group of automorphisms is $SL(2,\mathbb{C})$ acting by Möbius transformations

$$z \to \frac{az+b}{cz+d} \quad , \qquad ad-bc = 1 \tag{2.53}$$

on the extended complex plane. Since any of these transformations has three fixed points, we conclude that the sphere S^2 has a "rigid" complex structure, i.e. there is a unique Riemann surface of genus zero.

For the plane $\mathbb{C}$, only translations act without fixed points. Since translations commute, to get a compact Riemann surface we have to divide by a group of translations generated by two elements: $\mathbb{Z}\omega_1 + \mathbb{Z}\omega_2$, $\omega_1, \omega_2 \in \mathbb{C}$, $\omega_1 \neq \lambda\omega_2$ for λ real. This gives a two-dimensional torus.

To count the number of parameters describing inequivalent tori, we first note that by an automorphism (rotation) we can choose axes for the lattice $\mathbb{Z}\omega_1 + \mathbb{Z}\omega_2$ so that ω_1 is along the positive real axis. Furthermore, by another automorphism (rescaling) we can choose $\omega_1 = 1$. Finally by a reflection we can arrange that $\operatorname{Im}\tau > 0$ with $\tau = \omega_2/\omega_1$. Thus in $g = 1$ the "Siegel upper half plane" really is the upper half plane, with coordinate τ. Thus we have a family of tori realized as regions in $\mathbb{C}$ with corners $0, \tau, 1, \tau + 1$.

To determine completely the set of inequivalent tori, we have to understand the action of the mapping class group $\Omega(\Sigma)$ on τ. Since in the case of the torus $\pi_1(T^2) = H_1(T^2, \mathbb{Z})$, we do not have to worry about the Torelli group. The full mapping class group then is $SL(2, \mathbb{Z})$, the classical modular group. Indeed it turns out that the tori specified by τ and

$$\tau' = \frac{a\tau + b}{c\tau + d} \quad , \quad \begin{pmatrix} a & b \\ c & d \end{pmatrix} \in SL(2, \mathbb{Z}) \tag{2.54}$$

can be mapped to each other diffeomorphically, and that this map preserves the standard metric on $\mathbb{C}$ up to rescaling. For example, sending $\tau = i$ to $\tau' = i + 1$ can be undone by pushing the resulting rhombus in $\mathbb{C}$ back to its original square shape. This diffeomorphism is a Dehn twist about the loop represented by the edge $[0,1]$ of the square.

The set of inequivalent tori is therefore obtained as a fundamental region of $SL(2, \mathbb{Z})$ on the upper half plane, which we can choose to be $-\frac{1}{2} \leq \operatorname{Re}\tau \leq +\frac{1}{2}$ and $|\tau| \geq 1$ (see Fig. 2.4). When we identify the points on the boundary of the region drawn in Fig. 2.4 we get the genus-1 moduli space $\mathcal{M}_1$. $\mathcal{M}_1$ has two especially interesting points. This is because $SL(2, \mathbb{Z})$ acts with fixed points. The point $\tau = i$ is a fixed point of $\tau \to -1/\tau$; it corresponds to a "square torus". Similarly the point $e^{2\pi i/3}$ is a fixed point of $\tau \to 1 - 1/\tau$. Since $\tau \to (1 - 1/\tau)$ is a transformation of order 3, and $\tau \to -1/\tau$ is of order 2, we conclude that $\mathcal{M}_1$ is an "orbifold" with orbifold points of

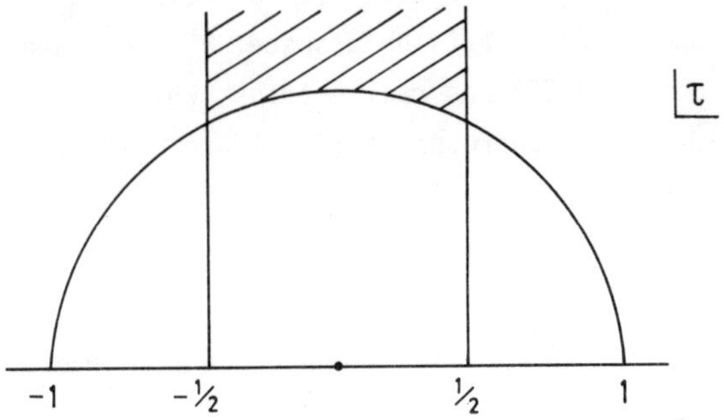

Fig. 2.4 The shaded area is a fundamental region for $SL(2,\mathbb{Z})$.

orders 2 and 3. Finally we can compactify $\mathcal{M}_1$ by adding a single point at infinity $(\tau_2 \to \infty)$ corresponding to a pinched torus. This gives a complete picture of the moduli space in genus one.

For surfaces with genus $g > 1$, the covering Riemann surface is not $\mathbb{C}$ but the upper half plane H, and its conformal structure can be represented by the Poincaré metric

$$ds^2 = dz d\bar{z}/y^2 \quad ; \quad y > 0 \quad ; \quad z = x + iy \quad . \tag{2.55}$$

This metric has constant negative curvature, implying that for any compact Riemann surface of genus higher than one we can always choose a metric of constant curvature. The group of conformal automorphisms of (2.55) is now not the abelian group of translations but the group of Möbius transformations which leave the real axis invariant i.e.

$$z \to \frac{az+b}{cz+d} \quad , \quad ad-bc = 1 \quad ,$$
$$a,b,c,d \in \mathbb{R} \quad . \tag{2.56}$$

Do not confuse this action with (2.53)! (2.56) acts on a half-plane representing Σ, while (2.53) acts on a space $\mathscr{H}_1$ representing all surfaces of genus one. (2.56) has real entries, while (2.53) has integer entries. Discrete subgroups of (2.56) are used to say when the points of H are the same point of some Σ, while (2.53) tells when two points of $\mathscr{H}_1$ describe the same surface Σ with different homology bases.

Since in (2.56) the common sign of a,b,c,d is irrelevant, these transformations form the group $PSL(2, \mathbb{R}) = SL(2, \mathbb{R})/\{+1,-1\}$. Writing an element of $SL(2,\mathbb{R})$ as a matrix $\begin{pmatrix} a & b \\ c & d \end{pmatrix}$, we can divide them into elliptic, parabolic or hyperbolic as follows:

i) T is elliptic if by a similarity transformation it can be brought into the form

$$\begin{pmatrix} \cos\theta & \sin\theta \\ -\sin\theta & \cos\theta \end{pmatrix}$$

ii) T is parabolic if it is similar to a translation

$$\begin{pmatrix} 1 & a \\ 0 & 1 \end{pmatrix}$$

iii) T is a hyperbolic if it can be brought to the form

$$\begin{pmatrix} e^{\lambda/2} & 0 \\ 0 & e^{-\lambda/2} \end{pmatrix}$$

This classification follows from the value of $|\mathrm{Tr}T|$: $|a+d| < 2$, we have an elliptic element, if $|a+d| = 2$, we have a parabolic element, and if $|a+d| > 2$ we have a hyperbolic element. Clearly elliptic elements have fixed points on H, and cannot be used to construct compact Riemann surfaces. Similarly, parabolic elements would generate punctures if they were used to generate $\pi_1(\Sigma)$.

Thus we are reduced to hyperbolic elements. It is left as an exercise for the reader to show that any hyperbolic transformation has two fixed points on the real axis: a,b; and that it can be written in the form:

$$\frac{T(z) - a}{T(z) - b} = e^{\lambda}\left(\frac{z - a}{z - b}\right) \tag{2.57}$$

λ is also known as the normalizer of T. To understand the geometrical significance of $\lambda(T)$, we have to look in more detail into the hyperbolic geometry of H. The geodesics of the hyperbolic metric are half-circles centered on the real axis and straight lines parallel to the imaginary axis (see Fig. 2.5). Given two points $z,w \in H$, the hyperbolic distance is

$$\cosh d(z,w) = 1 + \frac{1}{2}\frac{|z - w|^2}{\text{Im } z \cdot \text{Im } w} \tag{2.58}$$

which is $SL(2,\mathbb{R})$ invariant. If we now consider two points z and Tz, where T is hyperbolic, the minimum hyperbolic distance for any z is equal to $\lambda(T)$, and it is attained for z in the geodesic intersecting the real axis at the fixed points of T. We can construct the Riemann surface by choosing 2g hyperbolic elements of SL(2, R): $T_1,\ldots,T_{2g}$ satisfying the relation

$$\prod_{i=1}^{g} T_{2i-1}T_{2i}T_{2i-1}^{-1}T_{2i}^{-1} = 1 \tag{2.59}$$

(this is just (2.4)). The hyperbolic subgroups of SL(2, R) satisfying this condition are known as Fuchsian groups. Since every T_i depends on 3 real parameters, $T_1,\ldots,T_{2g}$ contain 6g real parameters. The constraint (2.59) eliminates one set of three parameters. Furthermore, since (2.59) is invariant under conjugation by an element of $SL(2,\mathbb{R})$,

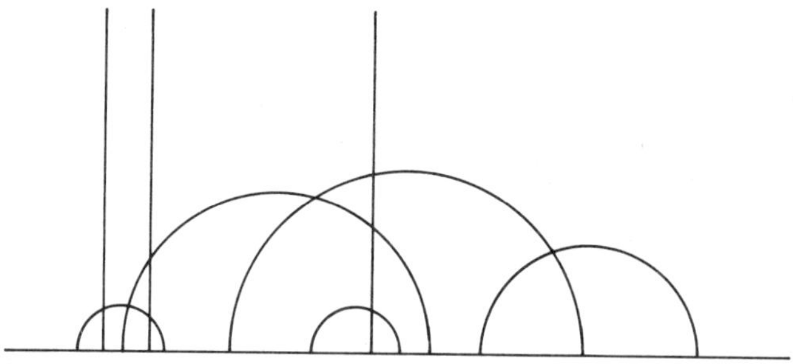

Fig. 2.5 Geodesics on the upper half plane with the Poincaré metric

we can choose the fixed points of T_{2g} to be 0 and ∞, and constrain the fixed points a_{2g-1}, b_{2g-1} of T_{2g-1} to satisfy $|a_{2g-1}b_{2g-1}| = 1$. In total we have 6g-6 real independent parameters, and we conclude that the moduli space of Riemann surfaces of genus $g \geq 2$ has real dimension 6g-6. From the geometrical interpretation of the normalizer of T, we also obtain that the normalizers of all the elements of $\pi_1(\Sigma)$ represent the lengths of all closed geodesics on Σ which start and end at the same point.

In this representation of the surface, the local formulas of part B) can be made particularly explicit. A q-differential $q \in \mathbb{Z}$ is now a modular form under the fuchsian group, i.e. it satisfies

$$\psi\left(\frac{az+b}{cz+d}\right)\left[d\left(\frac{az+b}{cz+d}\right)\right]^q = \psi(z)dz^q \qquad (2.60)$$

or

$$\psi\left(\frac{az+b}{cz+d}\right) = (cz+d)^{2q}\ \psi(z) \quad .$$

The scalar laplacian becomes

$$\Delta_0 = -y^2(\partial_x^2 + \partial_y^2) \tag{2.61}$$

and the laplacians $\Delta_n^{(\pm)}$ in (2.23) take a simple form. If we rescale $\psi(z)$ by y^{-q} for convenience, we have:

$$y^q \nabla_z^q y^{-q} = y^{-q} \partial_z y^q = \frac{1}{2}(\partial_x - i\partial_y - iqy^{-1})$$

$$y^q \nabla_q^z y^{-q} = -y^2(\partial_x + i\partial_y - iqy^{-1})$$

$$\gamma_{z\bar{z}} = 1/2y^2 \quad , \quad \gamma^{z\bar{z}} = 2y^2 \tag{2.62}$$

and

$$2y^q \Delta_q^{(+)} y^{-q} = -y^2(\partial_x^2 + \partial_y^2) + 2iqy \frac{\partial}{\partial x} + q(1 + q)$$

$$2y^q \Delta_q^{(-)} y^{-q} = -y^2(\partial_x^2 + \partial_y^2) + 2iqy \frac{\partial}{\partial x} + q(q - 1) \tag{2.63}$$

so that

$$\Delta_q^{(+)} - \Delta_q^{(-)} = + q \tag{2.64}$$

as expected (cf. 2.24). Since both $\Delta_q^{(\pm)}$ are positive semidefinite, we see that for $q > 0$, $\Delta_q^{(+)} > 0$ and for $q < 0$, $\Delta_q^{(-)} > 0$.

The heat kernels and determinants of these operators can be written in an elegant way in terms of Selberg's zeta function (see Ref. 23 for a thorough exposition of Selberg trace formulas). Divide the closed geodesics on Σ into "primitive" and "non-primitive". A geodesic is primitive if when we follow it, we go over each point on it only once. If $\ell(\gamma)$ denotes the length of a geodesic γ, we can construct the function:

$$Z(s) = \sum_{k=0}^{\infty} \prod_{\gamma \text{ primitive}} (1 - e^{-\ell(\gamma)(s+k)}) \quad . \tag{2.65}$$

For Re s large enough, Z(s) converges. Then it is defined as a meromorphic function of s by anlytic continuation. It can be shown (see [24] for details and references) that

$$\det{}' - \Delta_0 = \text{const. } Z'(1) \tag{2.66}$$

$$\det{}' - \Delta_q^{(+)} = \text{const. } Z(1+q) \quad , \quad q = 1,2,\ldots \tag{2.67}$$

As we will see later on, these formulas can be used to obtain the behavior of the string partition function when we approach a degenerate surface, i.e. a surface with node singularities. This information is important for understanding divergences on string theory. In Sections IV and V we will see that there are alternative and more efficient ways of obtaining the same (and more) information using complex algebraic geometry at the expense of making the formulation of the theory more abstract. The basic ingredient for the algebraic-geometric approach is the fact that the moduli space $\mathcal{M}_g$ is a complex space with rather remarkable properties. This will be explained in Sections IV and V. For the moment we leave the theory of Riemann surfaces, and return to string theory, to explain why modular invariance plays such a central role in the formulation of superstring theories.

III. THE MODULAR GROUP AND THE SPECTRUM OF FERMIONIC STRING THEORIES

We would like to show now that modular invariance places very stringent constraints on the allowed spectra of fermionic string theories. We will present in detail the analysis for the O(32) heterotic string [25] in the fermionic formulation, and outline the construction of the $E_8 \times E_8$ theory and the tachyon-free non-supersymmetric heterotic string with gauge group $O(16) \times O(16)$ [26].

In the light cone gauge formulation of the heterotic string [25], the spectrum of two dimensional fields on the string world sheet is as follows: The bosonic fields are the transverse coordinates of the string $X^i(\sigma,\tau)$, $i = 1,\ldots,8$, and they transform as the 8_v of $SO(8)$, the transverse part of the Lorentz group. The fermions can be divided into left movers and right movers. The right movers are eight Weyl-Majorana fermions $\psi^i(\tau-\sigma)$ transforming also as the 8_v of $SO(8)$. The left movers $\lambda^I(\tau+\sigma)$, $I = 1,\ldots,32$ are Weyl-Majorana fermions, singlets under the Lorentz group in ten dimensions, and they transform as the fundamental representation of $O(32)$. These are the fermionized form of the sixteen left moving scalars taking values in the maximal torus of $O(32)$ in the bosonic formulation of the theory [25]. For the $E_8 \times E_8$ or $O(16) \times O(16)$ theories, we would divide the $\lambda^I(\tau+\sigma)$'s into two groups of sixteen, transforming in the $(16,1) + (1,16)$ representation of the $O(16) \times O(16)$ subgroup of $E_8 \times E_8$.

In the NSR formulation, (see Ref. 1 for details and references) we consider two sectors of the theory for each fermion, depending on whether we consider fermions with antiperiodic (NS) or periodic (R) boundary conditions on σ; $0 \leq \sigma \leq 1$. Thus in the first quantized formulation, the dynamics of the free string in the light-cone gauge is described by the two dimensional action (still in Minkowski signature)

$$S = \frac{1}{2}\int d\sigma d\tau(\partial_+ X^i \partial_- X^i + i\psi^i\partial_+\psi^i + i\lambda^I\partial_-\lambda^I)$$

$$\xi^{\pm} = (\tau \pm \sigma)/\sqrt{2} \quad , \quad \partial_{\pm} = \partial/\partial\xi^{\pm} \quad . \tag{3.1}$$

To determine the spectrum of the string, we proceed to canonically quantize (3.1)

$$X^i = \frac{1}{2}(X_L^i + X_R^i)$$

$$X_L^i = q^i + p^i(\tau-\sigma) + \frac{i}{2}\sum \frac{\alpha_n^i}{n} e^{-2\pi i n(\tau-\sigma)}$$

$$X_R^i = q^i + p^i(\tau+\sigma) + \frac{i}{2}\sum \frac{\tilde{\alpha}_n^i}{n} e^{-2\pi i n(\tau+\sigma)} \qquad (3.2)$$

where

$$[\alpha_n^i, \alpha_m^j] = n\,\delta^{ij}\,\delta_{n+m,0}$$

$$[q^i, p^j] = i\,\delta^{ij} \quad . \qquad (3.3)$$

For the fermions we have to choose boundary conditions. For the right movers we have two sectors

$$\psi^i(\tau-\sigma) = \sum_{r\in\mathbb{Z}+\frac{1}{2}} b_r^i\, e^{-2\pi i r(\tau-\sigma)} \qquad \text{(N.S.)}$$

$$\psi^i(\tau-\sigma) = \sum_{r\in\mathbb{Z}} d_r^i\, e^{-2\pi i r(\tau-\sigma)} \qquad \text{(R)} \qquad (3.4)$$

and similarly for the left moving fermions

$$\lambda^I(\tau+\sigma) = \sum_{r\in\mathbb{Z}+\frac{1}{2}} \tilde{b}_r^I\, e^{-2\pi i r(\tau+\sigma)} \qquad \text{(N.S.)}$$

$$\lambda^I(\tau+\sigma) = \sum_{r\in\mathbb{Z}} \tilde{d}_r^I\, e^{-2\pi i r(\tau+\sigma)} \qquad \text{(R)} \qquad (3.5)$$

with the usual anticommutation relations

$$\{d_r^i, d_s^j\} = \delta^{ij}\delta_{r,-s}$$

$$\{b_r^i, b_s^j\} = \delta^{ij}\delta_{r,-s}$$

$$\{\tilde{b}_r^I, \tilde{b}_s^J\} = \delta^{IJ}\delta_{r,-s} \quad . \qquad (3.6)$$

When we solve the constraints in the light-cone gauge, we can express the masses of the string excitations in terms of the light-cone hamiltonian [1]. One of the many subtleties on the quantization of strings is to determine the normal ordering constant in this hamiltonian. The rigorous way to do it is by requiring the light-cone formalism to be Lorentz invariant. There is however a much quicker way of obtaining the same answer, originally due to Brink and Nielsen [27], which consists of defining the normal ordering constant by ζ-function regularization. In the cases we will treat below, we will need the following sum:

$$\eta(\alpha) = \frac{1}{2} \sum_{n=0}^{\infty} (n+\alpha)^{-s} \Big|_{s=-1} \quad . \tag{3.7}$$

$\eta(\alpha)$ is defined by analytic continuation, and it is given by:

$$\eta(\alpha) = -\frac{1}{24}(6\alpha^2 - 6\alpha + 1) \quad . \tag{3.8}$$

Then the mass operators for the right movers on the (NS) and R-sectors can be written as:

$$L_0^{(NS)} \equiv \frac{1}{4} M_R^2(NS) = \sum_{n=1}^{\infty} \alpha_{-n}^i \alpha_n^i + \sum_{r=\frac{1}{2}}^{\infty} r b_{-r}^i b_r^i + 8(\eta(1) - \eta(\tfrac{1}{2}))$$

$$L_0^{(R)} \equiv \frac{1}{4} M_R^2(R) = \sum_{n=1}^{\infty} \alpha_{-n}^i \alpha_n^i + \sum_{r=1}^{\infty} r d_{-r}^i d_r^i + 8(\eta(1) - \eta(1))$$

using (3.8),

$$L_0^{(NS)} \equiv \frac{1}{4} M_R^2(NS) = \sum_{n=1}^{\infty} \alpha_{-n}^i \alpha_n^i + \sum_{r=\frac{1}{2}}^{\infty} r b_{-r}^i b_r^i - \frac{1}{2}$$

$$L_0^{(R)} \equiv \frac{1}{4} M_R^2(R) = \sum_{n=1}^{\infty} \alpha_{-n}^i \alpha_n^i + \sum_{r=0}^{\infty} r d_{-r}^i d_r^i \quad . \tag{3.9}$$

Similarly for the left movers:

$$\tilde{L}_0(NS) = \frac{1}{4} M_L^2(NS) = \sum_{n=1}^{\infty} \tilde{\alpha}^i_{-n}\tilde{\alpha}^i_n + \sum_{r=\frac{1}{2}}^{\infty} r\tilde{b}^I_{-r}\tilde{b}^I_r - 1$$

$$\tilde{L}_0(R) = \frac{1}{4} M_L^2(R) = \sum_{n=1}^{\infty} \tilde{\alpha}^i_{-n}\tilde{\alpha}^i_n + \sum_{r=0}^{\infty} r\tilde{d}^I_{-r}\tilde{d}^I_r + 1 \quad . \tag{3.10}$$

There is one further physical constraint which must be imposed on physical states, namely that $(L_0 - \tilde{L}_0)|\mathrm{phys}> = 0$. The meaning of this condition is that $L_0 - \tilde{L}_0$ is the generator of translations of σ (the variable parametrizing the string at fixed time). Since we are dealing with closed strings, consistency requires that the theory should not depend on what point of the string we choose as its origin. Now for each set of oscillators (3.10) we can construct a Fock space, and the physical states are constructed by taking tensor products of the left and right Fock spaces subject to the constraint $L_0 = \tilde{L}_0$.

To illustrate this in more detail, we now exhibit the lowest mass states of each Fock space. The only point we need to clarify before we start is that in the Ramond sector, the vacuum is degenerate. This is because the oscillators d_0^i satisfy a Clifford algebra:

$$\{d_0^i, d_0^j\} = \delta^{ij} \tag{3.11}$$

and thus the vacuum must be a representation of this Clifford algebra, i.e. an 8-dimensional spinor. This is how space-time fermions appear in the first-quantized formulation of string theory. A similar phenomenon happens in the left sector. We will represent the group properties of the various states as (R,R') where the first entry labels the 0(8) representation, and the second the 0(32) representation.

For the right movers, the NS and R Fock spaces look as follows:

L_0			L_0		
0	$(8_v,1)$	$b^i_{-\frac{1}{2}}\|0>$	0	$(8_s \oplus 8_{s'},1)$	$\|0>$
$-\frac{1}{2}$	$(1,1)$	$\|0>$			
	(NS)			(R)	

8_s, $8_{s'}$ denote the two spinors of SO(8). In the left moving sector, we have

$\tilde{L}_0$			$\tilde{L}_0$		
			1	$(1,2^{15} \oplus 2^{15'})$	$\|0>$
0	$\tilde{\alpha}^i_{-1}\|0>$, $\tilde{b}^I_{-\frac{1}{2}}\tilde{b}^J_{-\frac{1}{2}}\|0>$	$(8_v,1) \oplus (1,496)$			
$-\frac{1}{2}$	$(1,32)$	$\tilde{b}^I_{-\frac{1}{2}}\|0>$			
-1	$(1,1)$	$\|0>$			
	(NS)			(R)	

(3.12)

The spectrum we have written down has two problems: the first is that it has tachyons, and the second is that it is not supersymmetric. This is most easily seen if we notice that the massless states have different numbers of bosons and fermions. These two problems can both be solved by "GSO projection" [28]. On each sector we project out those states with a given fermion number. This prescription instructs us to discard half the states in each sector based on the number of world-sheet fermions. For the left movers we require that $(-)^F = +1$, which at once removes the tachyonic state $\tilde{b}^I_{-\frac{1}{2}}|0>^{NS}_L \otimes |0>^{NS}_R$. For the right movers we require that $(-)^F = -1$ in the NS sector and $\Gamma(-)^F = 1$ in the R sector. Here Γ is the chirality operator acting on the spinor ground state. This projection keeps the massless states $\tilde{\alpha}^i_{-1}|0>^{NS}_L \otimes |0^\mu>^R_R$ while discarding their partners of opposite chirality. All told, the GSO projection together with the $L_0 = \tilde{L}_0$ requirement leaves a set of massless states corresponding exactly to those of $N=1$ supergravity coupled to $N=1$ super-Yang-Mills theory with group $O(32)$:

$$(8_v \otimes 8_v, 1) \oplus (8_v \otimes 8_s, 1) \oplus (8_v, 496) \oplus (8_s, 496) \tag{3.13}$$

In fact the theory is fully supersymmetric in ten dimensions, because the right moving sector obtained after the GSO projection coincides with the right moving part of the Green-Schwarz superstring [29]; this was in fact one of the original ingredients in the construction of the heterotic string [25].

To make contact with the functional integral formulation, and to see the relevance of the modular group, let us compute the partition function of the $O(32)$ heterotic string in genus 1. This is equivalent to computing

$$\mathrm{Tr}\, q^{L_0} \bar{q}^{\tilde{L}_0} \tag{3.14}$$

where $q = \exp 2\pi i\tau$, $\tau = \tau_x + i\tau_y$, $\tau_y > 0$. We can think of τ_y as a two-dimensional "temperature", and τ_x is needed in order to impose the constraint $L_0 = \tilde{L}_0$ after integration over τ. Thus the trace (3.14) splits into pieces holomorphic and antiholomorphic in τ. If we consider L_0 for the moment, there are two possible sectors $L_0(NS)$ and $L_0(R)$, and the projection operator requires that we compute:

$$\mathrm{Tr}\, \frac{1 - (-1)^F}{2}\, q^{L_0(NS)} \tag{3.15}$$

and

$$\mathrm{Tr}\ \frac{1+\Gamma(-1)^F}{2}\ q^{L_0(R)} \tag{3.16}$$

in the NS and R-sectors respectively. The relative sign between the two contributions is determined by the connection between spin and statistics in $d=10$. The sum (3.16) represents the contribution to the vacuum energy from the zero point oscillations of space-time fermions, and it must therefore have the opposite sign from (3.15), the contribution from space-time bosons. Using (3.9) and

$$(-1)^F = e^{i\pi\Sigma b^i_{-r} b^i_r}$$

we obtain for (3.15):

$$\mathrm{Tr}\ \frac{1}{2}\left(1 - e^{i\pi\Sigma b^i_{-r} b^i_r}\right) q^{\Sigma \alpha^i_{-n}\alpha^i_n + \Sigma r b^i_{-r} b^i_r - \frac{1}{2}}$$

$$= \frac{1}{2} q^{-\frac{1}{2}} \prod_{n=1}^{\infty} (1-q^n)^{-8} \left(\Pi(1+q^{n+\frac{1}{2}})^8 - \Pi(1-q^{n-\frac{1}{2}})^8\right) \tag{3.17}$$

and for the Ramond sector:

$$\mathrm{Tr}\ \frac{1}{2}\ (1+\Gamma(-1)^F) q^{\Sigma \alpha^i_{-n}\alpha^i_n + \Sigma r d^i_{-r} d^i_r} =$$

$$= 8\Pi(1-q^n)^{-8}\ \Pi(1+q^n) \quad , \tag{3.18}$$

when we account for the degeneracy of the vacuum $8_s \oplus 8_{s'}$. For the left movers, we obtain using (3.10)

$$\mathrm{Tr}\ \frac{1+(-)^F}{2}\ \bar{q}^{\tilde{L}_0(NS)} = \bar{q}^{-1} \prod_{n=1}^{\infty} (1-\bar{q}^n)^{-8}\ \frac{1}{2}\Big(\Pi(1+\bar{q}^{r+\frac{1}{2}})^{32} +$$

$$+ \Pi(1-\bar{q}^{r+\frac{1}{2}})^{32}\Big)$$

$$\mathrm{Tr}\,\frac{1+(-1)^F}{2}\,\bar{q}^{-\tilde{L}_0(R)} = 2^{15}\bar{q}\prod_{n=1}^{\infty}(1+\bar{q}^{n})^{32} \quad . \tag{3.19}$$

Incidentally, the first non-trivial check of ten-dimensional supersymmetry is that (3.17) minus (3.18) should vanish. These functions simply count (with a minus sign for fermions) the number of states at a given mass level. That this is true, is the celebrated "aequatio identica satis abstrusa" of Jacobi which is the equality of (3.17) and (3.18). In a more field theoretic language, this means that the one loop cosmological constant vanishes for the $O(32)$ and $E_8 \times E_8$ heterotic strings. The result follows for the $E_8 \times E_8$ theory because the right moving sector is the same as for the $O(32)$ string.

We can now interpret (3.17) and (3.18) geometrically if we recall the connection between Hilbert space traces and functional integrals. Let us for the moment concentrate on $\mathrm{Tr}\, q^{L_0}$. If we take $\tau = i\tau_y$ we get $\mathrm{Tr}\exp(-2\pi\tau_y L_0)$. Neglecting for the moment the contribution from the scalar modes α_n, this trace is the euclidean functional integral of a free fermion action satisfying antiperiodic boundary condition on time. Equivalently, this is the euclidean vacuum amplitude for a set of free fermions on a torus $0 \le \sigma_1 \le 1$, $0 \le \sigma_2 \le \tau_y$. The boundary conditions on (σ_1, σ_2) will be (P,A) and (A,A) for the R- and NS-sectors respectively. The notation means the following. In genus one the degree of a spin bundle is $g-1=0$, so that there is a preferred spin structure, namely the trivial bundle. (It is odd.) Sections are given by functions periodic along both the axes of the fundamental region. Other spin bundles have sections represented by functions antiperiodic along one or both of the axes given by 1 and τ, and so are denoted (P,A), etc. For example, $\mathrm{Tr}(-1)^F\exp(-2\pi\tau_y L_0)$ is the euclidean vacuum amplitude on the torus for the same fermions but now with boundary conditions (P,P) and (A,P) in the R and NS sectors respectively.

If we now take $\tau_x \ne 0$, since $(L_0 - \tilde{L}_0)$ is the generator of translations in the σ_1-direction, $\mathrm{Tr}\, q^{L_0}\bar{q}^{\tilde{L}_0}$ is equivalent to computing the free euclidean functional integral over a tilted torus (see Fig. 3.1) with axes given by 1 and τ. We conclude that (3.15,16) is simply a

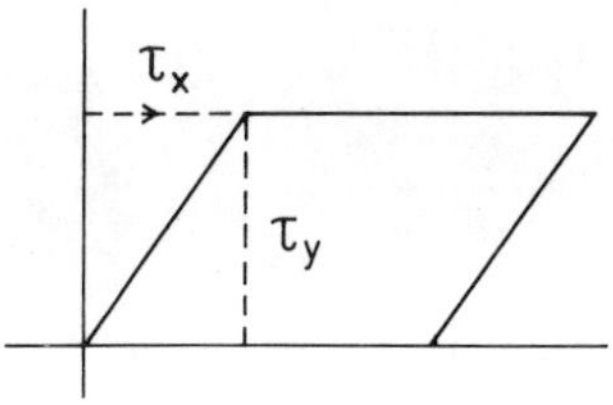

Fig. 3.1 Tilted torus.

sum of over all possible boundary conditions (periodic or antiperiodic) of the fermionic functional integral over a torus with modular parameter τ. These different boundary conditions represent geometrically the spin structures on the torus.

Now we can understand why the modular group $SL(2, \mathbb{Z})$ will play a very important role in determining the spectrum of string theories. We know from the previous section that the non-trivial diffeomorphisms act on the spin structures. Hence the possible projection operators generalizing the GSO projection are strongly constrained by the requirement of modular invariance. Let us imagine that we start with a fermion with (A,A) boundary conditions. If we now make a modular transformation $\tau \to \tau + 1$, the boundary conditions change to (A,P). If we further perform a transformation $\tau \to -1/\tau$ (exchanging σ_1 and σ_2) we obtain the boundary conditions (P,A). Thus the relative signs of the contribution of determinants with (A,A), (A,P) and (P,A) boundary conditions are completely determined by modular invariance. These three spin structures are even. The (P,P) spin structure is odd, the Dirac operator in this case has a zero mode and therefore its determinant equals zero. From the physical point of view this reflects the fact that we are free to choose what is meant by ten dimensional chirality.

Using the same arguments, we also conclude that the trace over the bosonic oscillators is related to the determinant of the scalar laplacian on the torus. In fact if we carry out the same procedure for the bosonic string in the light-cone gauge, we obtain (after account is taken of the zero mode) that

$$\det{}'\Delta_0^{(+)} = \text{const. } \tau_2 |\eta(\tau)|^4 \tag{3.20}$$

where

$$\eta(\tau) \equiv q^{1/24} \prod_{n=1}^{\infty} (1 - q^n) \tag{3.21}$$

is the "Dedekind eta-function". The τ_2 factor in (3.20) can also be obtained from the requirement of modular invariance. Under $\tau \to \tau + 1$ $\eta(\tau+1) = \eta(\tau)$ and $\tau_2 \to \tau_2$; but under $\tau \to -1/\tau$, $\eta(-1/\tau) = (-i\tau)^{\frac{1}{2}}\eta(\tau)$ and $\tau_2 \to \tau_2/|\tau|^2$ so that (3.20) is invariant. We expect modular invariance of (3.20) on general grounds because the quantum mechanics of a free scalar on a surface is free of local and global anomalies.

Let us now outline what happens for the $E_8 \times E_8$ and $O(16) \times O(16)$ theories. In this case the left moving fermions are divided into two groups of sixteen. In the partition function we will now have a sum over three sets of spin structures, one for the right moving fermions, and one each for each of the groups of sixteen fermions. Performing computations entirely analogous to (3.9,10) we again present a table of the states of the theory on the different sectors before the GSO projections. Now there are two sectors: A,P for the right movers, and four sectors for the left movers AA, PP, PA, AP. Since we are again thinking in terms of canonical quantization, the boundary conditions we give refer to the σ direction only. For the right movers we again have (in terms of representations of $O(8) \times O(16) \times O(16)$) for the lowest-lying states

L_0	A	L_0	P
	⋮		⋮
0	(8,1,1)	0	$(8_s \oplus 8_{s'},1,1)$
$-\frac{1}{2}$	(1,1,1)		

For the left movers there are four possibilities

$\tilde{L}_0$	PP	PA	AP	AA
1	(256,256)			
0		$(1,128,1)\oplus(1,128',1)$	$(1,128',1)\oplus(1,128,1)$	$(16,16)\oplus(1,120,1)\oplus(1,1,120)\oplus(8,1,1)$
$-\frac{1}{2}$				$(16,1)\oplus(1,16)$
-1				$(1,1,1)$

(3.22)

The $(-1)^F$ projections can now be chosen so that we get the full $_8 \times E_8$ super Yang-Mills theory coupled to supergravity. The detailed orm of the GSO projections [25] is such that for massless states the P;·,·) gives a Majorana-Weyl spinor 8_s (the first entry refers to he right moving fermions, and the other two, to the left moving ectors). The (A;·,·) gives a vector of O(8), 8_v. The sector ·;A,A) gives $(8_v,1,1)+(1,120,1)+(1,1,120)$, while from (·;P,A) e get (1,128,1) and from (·;A,P) we obtain (1;1,128). The sector ·;P,P) gives no massless states. Collecting results, we obtain for he massless states

$$(8_v \oplus 8_s,1,1)\otimes((8_v,1,1)\oplus(1,248,1)\oplus(1,1,248)) \tag{3.23}$$

here we have used that under the embedding $SO(16)\subset E_8$, the adjoint $\underline{48} = \underline{120} + \underline{128}$.

Schematically, any quantity computed with the heterotic string hould formally look like

$$\sum_{a,b,c} S_a^{(1)} S_b^{(2)} S_c^{(3)} \tag{3.24}$$

where the S_a's represent the contribution of each type of fermion, and the sum is over spin structures with the signs implied by the choice of GSO projections mentioned. What one would like to ask now is whether there are other possible GSO projections consistent with modular invariance and absence of tachyons. The answer is yes [26]. There is a non-supersymmetric tachyon-free theory with gauge group $O(16)\times O(16)$. Since the argument for fixing the new projection operators requires the use of all the generators of the mapping class group, we will go through the argument in some detail.

As we have seen in Section II, a spin structure on a general Riemann surface is more than just a set of boundary conditions on the generators of the homology. However we can compare two spin structures $L_a \otimes L_b^{-1}$. Since $L_a \otimes L_b^{-1}$ has degree zero, it is a flat line bundle, and so it is characterized (once we choose a homology basis) by periodic or antiperiodic boundary conditions around the loops in the chosen basis. Since we have sets of three spin structures, in order to clarify all possible signs in the sum (3.24) we will abstractly think of operators α,β which change the second and third spin structure respectively relative to the first. Since the change of a spin structure can also be viewed as a point of order two in the jacobian, we have that $\alpha^2=1$, $\beta^2=1$. Thus we have a $\mathbb{Z}_2\times\mathbb{Z}_2$ group with elements $1, \alpha, \beta, \alpha\beta$.

Given a surface Σ of genus g, we can compare the second and third spin structures with respect to the first along all the homology cycles a_i, b_i, $1\le i\le g$; thus the relative change can be written concisely as $\alpha^{k_{a_i}}, \alpha^{k_{b_i}}, \beta^{l_{a_i}}, \beta^{l_{b_i}}$ where $k, l = 0,1$. $\alpha^{k_{a_i}}$ indicates the relative sign of the second spin structure with respect to the first along the a_i-cycle. If $k_{a_i}=0$, there is no change, and if $k_{a_i}=1$, we change the relative sign between the two spin structures. Thus we modify the heterotic string by changing the sum (3.24) into:

$$\sum_{a,b,c} S_a^{(1)}\, S_b^{(2)} S_c^{(3)}\ \varepsilon(\alpha^{k_{a_i}}\beta^{l_{a_i}}, \alpha^{k_{b_i}}\beta^{l_{b_i}}) \quad . \tag{3.25}$$

Since the original sum is assumed to be modular invariant (with the sign assignments of the $E_8 \times E_8$ heterotic string), we want to determine the constraints which modular invariance and factorization place on the phases ε. The factorization condition means that if we shrink a homologically trivial (but homotopically non-trivial) cycle of Σ to create a node, the ε's should factorize into a factor for each of the surfaces to the left and right of the node.

We can carry out the analysis now for a general abelian group instead of simply $\mathbb{Z}_2 \times \mathbb{Z}_2$. Here we follow the arguments of Ref. 30. At the one-loop level we have two homology cycles a, b. If the abelian group of twisted boundary conditions is G, then we have in general a twist $g \in G$ around a and a twist $h \in G$ around b. Thus the phase in (3.25) is a function $\varepsilon(g;h)$. For a genus p surface, we have $\varepsilon(g_1,\ldots,g_p;\ h_1,\ldots,h_p)$. The factorization condition then implies that

$$\varepsilon(g_1,\ldots,g_p;\ h_1,\ldots,h_p) = \varepsilon(g_i;h_1)\ldots\ \varepsilon(g_p,h_p) \qquad (3.26)$$

since we can pinch any or all of the necks between the handles in Fig. 2.1. Under a general 1-loop modular transformation $\tau \to \frac{a\tau + b}{c\tau + d}$, the new boundary conditions are easily seen to be $g^a h^b$, $g^c h^d$. Thus 1-loop modular invariance implies

$$\varepsilon(g^a h^b, g^c h^d) = \varepsilon(g,h) \qquad . \qquad (3.27)$$

Since the modular group has no "new" generators beyond genus two, we can now consider the two-loop generator (see Fig. 2.3) corresponding to a Dehn twist around the curve $a_1^{-1}a_2$. We obtain:

$$\varepsilon(g_1h_2h_1^{-1},h_1)\varepsilon(g_2h_1h_2^{-1},h_2) = \varepsilon(g_1,h_1)\varepsilon(g_2,h_2) \tag{3.28}$$

Finally we can impose the convention $\varepsilon(1,1) = 1$.

We thus get three conditions on ε:

$$\varepsilon(g^ah^b,\ g^ch^d) = \varepsilon(g,h) \quad , \qquad ad - bc = 1 \tag{3.29a}$$

$$\varepsilon(g_1h_2h_1^{-1},h_1)\varepsilon(g_2h_1h_2^{-1},h_2) = \varepsilon(g_1,h_1)\varepsilon(g_2,h_2) \tag{3.29b}$$

$$\varepsilon(1,1) = 1 \quad . \tag{3.29c}$$

In (3.29a) choose $\begin{pmatrix} a & b \\ c & d \end{pmatrix} = \begin{pmatrix} 1 & -1 \\ 0 & 1 \end{pmatrix}$. Then

$$\varepsilon(gh^{-1},\ h) = \varepsilon(g,\ h) \quad , \text{ so}$$

$$\varepsilon(1,h) = \varepsilon(h,h) \quad . \tag{3.30}$$

Similarly if we choose $\begin{pmatrix} a & b \\ c & d \end{pmatrix} = \begin{pmatrix} 1 & 0 \\ -1 & 1 \end{pmatrix}$ we obtain:

$$\varepsilon(g,\ g^{-1}h) = \varepsilon(g,h) \quad . \tag{3.31}$$

If $g = h$ this says

$$\varepsilon(h,1) = \varepsilon(h,h)$$

or in other words

$$\varepsilon(g,g) = \varepsilon(1,g) = \varepsilon(g,1) \quad .$$

Now in (3.29b) first use (3.30) to obtain

$$\varepsilon(g_1h_2,h_1)\varepsilon(g_2h_1,h_2) = \varepsilon(g_1,h_1)\varepsilon(g_2,h_2) \tag{3.32}$$

Choosing $g_1 = g_2 = h_1 = 1$ we get

$$\varepsilon(1,h_2) = 1 \quad . \tag{3.33}$$

Choosing $g_1 = g_2 = 1$ yields

$$\varepsilon(1,h)\epsilon(h,1) = 1 \tag{3.34}$$

and setting $g_2 = 1$:

$$\varepsilon(g_1h_2,h_1) = \varepsilon(g_1,h_1)\varepsilon(h_2,h_1) \quad . \tag{3.35}$$

Summarizing, the conditions (3.29) lead to

$$\varepsilon(g_1g_2;g_3) = \varepsilon(g_1,g_3)\varepsilon(g_2,g_3) \tag{3.36a}$$

$$\varepsilon(g_1,g_2)\varepsilon(g_2,g_1) = 1 \tag{3.36b}$$

$$\varepsilon(g_1,g_1) = 1 \quad . \tag{3.36c}$$

This set of conditions is in fact equivalent to (3.29). To recover (3.29a) it is enough to consider the two generators of the modular group $\begin{pmatrix}1 & 1\\ 0 & 1\end{pmatrix}$, $\begin{pmatrix}1 & 0\\ 1 & 1\end{pmatrix}$ i.e. $\varepsilon(gh,h) = \varepsilon(g,h)$ and $\varepsilon(g,gh) = \varepsilon(g,h)$. These follow from (3.36a, b) by setting $g_1 = g$, $g_2 = g_3 = h$ and then using (3.36c). Finally (3.32) (equivalently (3.29b)) is a trivial consequence of (3.36a, b).

If we now concentrate on the case of interest to us, $G = \mathbb{Z}_2 \times \mathbb{Z}_2$, we find that $\varepsilon(\alpha,1) = \varepsilon(1,\beta) = \varepsilon(\alpha,\alpha) = \varepsilon(\beta,\beta) = \varepsilon(\alpha\beta,1) = 1$. The only non-trivial element is $\varepsilon(\alpha,\beta)$. Thus:

$$\varepsilon(\alpha^{k_{a_i}}{}_{\beta}^{l_{a_i}};\alpha^{k_{b_i}}{}_{\beta}^{l_{b_i}}) = \varepsilon(\alpha,\beta)^{\sum_i k_{a_i} l_{b_i} - \sum_i k_{b_i} l_{a_i}} \tag{3.37}$$

and there are two possibilities: If $\varepsilon(\alpha,\beta) = +1$ we get the standard heterotic $E_8 \times E_8$ string. If $\varepsilon(\alpha,\beta) = -1$ we get a new theory, the $O(16) \times O(16)$ string. To determine the changes in the projection operators, we consider the one loop case. If we have a given sector, for instance (NS;NS,NS) i.e. (A;A,A)-boundary conditions on σ_1, we consider in the time direction of the torus three boundary conditions: (P;A,A), (A;P,A), (A;A,P). Since periodic boundary conditions in time imply physically that we are computing $\mathrm{Tr}(-1)^F q^{L_0}$; we can easily determine the changes in $(-1)^{F_1}$, $(-1)^{F_2}$, $(-1)^{F_3}$ in the eight sectors of the theory. The sector (P;P,P) will have no change in $(-1)^{F_i}$, $i = 1, 2, 3$ because the ε factor is of the form $\varepsilon(\cdot,1) = 1$. Similarly in the (A;A,A) sector we again have $\varepsilon(\cdot,1) = 1$ and there are no changes. This means that in this theory we get from the (A;A,A) sector the same states as in the heterotic string i.e.

$$(8_v,1,1) \otimes [(1,120,1) + (1,1,120) + (8_v,1,1)] \quad . \tag{3.38}$$

For the (P;A,A) sector there are changes. We illustrate this by the following diagrams:

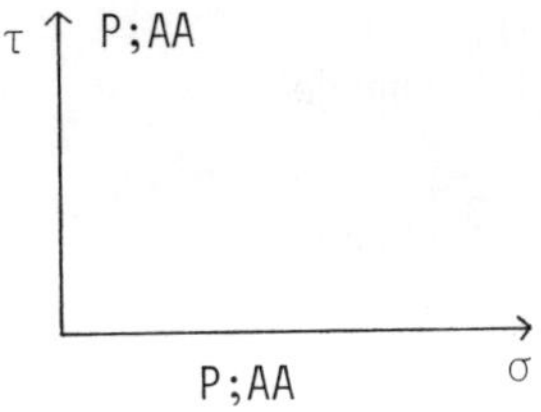

$\varepsilon(\alpha\beta,\alpha\beta) = 1$

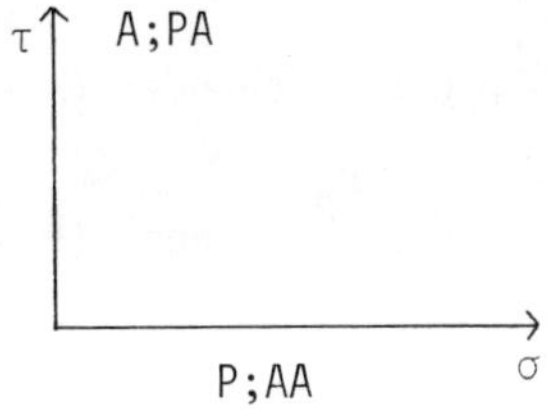

$\varepsilon(\alpha,\alpha\beta) = -1$

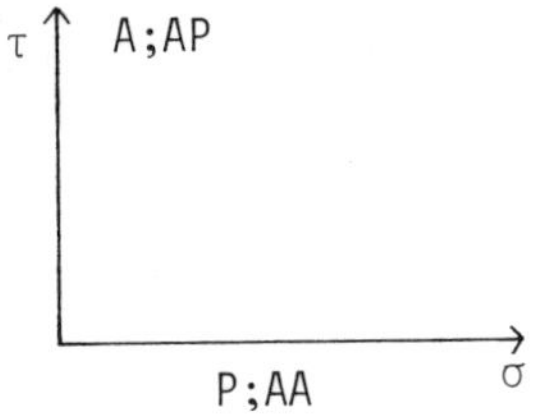

$\varepsilon(\beta,\alpha\beta) = -1$

(3.39)

hence in the (P;A,A) sector $(-1)^{F_2}$ and $(-1)^{F_3}$ change with respect to the heterotic string. In this sector we previously required $(-1)^{F_1} = 1$, $(-1)^{F_2} = 1$, $(-1)^{F_3} = 1$. Now we have to project onto $(-1)^{F_1} = 1$, $(-1)^{F_2} = -1$, $(-1)^{F_3} = -1$. This eliminates the states $(8_v,1,1) \oplus (1,120,1) \oplus (1,1,120)$ and leaves only (16,16) so we only get a spinor in the (16,16) representation of $O(16) \times O(16)$ i.e. $(8_s,16,16)$. Note that we have projected out the gravitino of the heterotic string which came from this sector. We can similarly check the other sectors.

In total we get the following list of changes:

(A;AA): no change ; (P,AA): $(-1)^{F_2}$, $(-1)^{F_3}$ change

(P;PP): no change ; (A,PP): $(-1)^{F_2}$, $(-1)^{F_3}$ change

(A;PA): $(-1)^{F_1}$, $(-1)^{F_3}$ change; (A;AP): $(-1)^{F_1}$, $(-1)^{F_2}$ change

(P;PA): $(-1)^{F_1}$, $(-1)^{F_2}$ change; (P;AP): $(-1)^{F_1}$, $(-1)^{F_3}$ change

The massless spectrum is therefore:

$$(8_v,1,1) \otimes [(1,120,1) + (1,1,120) + (8_v,1,1)] \tag{3.40a}$$

for bosons, and

$$(8_s,16,16) \oplus (8_{s'},128',1) \oplus (8_{s'},1,128') \tag{3.40b}$$

for fermions, and it is easy to check that there are no tachyons on the spectrum. We have obtained a chiral spectrum of massless fermions in $d = 10$. We might want to ask if this theory is free of gauge and gravitational anomalies. The answer is affirmative, and the easiest way to prove it is to consider the fermion content of the standard $E_8 \times E_8$ theory with positive chirality, say. If we now consider the massless fermion content of the $O(32)$ theory but with negative chirality, we find after splitting the representations with respect to $O(16) \times O(16)$ and removing the non-chiral representations that we are left with (3.40b). Since the $E_8 \times E_8$ and $O(32)$ theories are free of anomalies, then so is the $O(16) \times O(16)$ theory. For the conscientious reader who wants to check this in detail we will make a small technical remark: In the decomposition of the adjoint of E_8 with respect to $O(16)$, one gets $\underline{120} + \underline{128}$. In (3.40b) we have however $\underline{128}'$. This does not present a problem, because for an arbitrary matrix F in the spinor representation of the Lie algebra of $O(16)$, $\mathrm{Tr}\, F^n$ is the same for both spinor representations for $n < 8$. Since the anomaly only involves $n = 6$, the proof of anomaly cancellation goes through.

This concludes our illustration of the central importance of modular invariance in string theory.

IV. MULTILOOP DIAGRAMS. A HOLOMORPHIC FACTORIZATION THEOREM

A) The Measure

We now start the study of multiloop diagrams and a general analysis of the type of divergences appearing in string theory. For simplicity, we concentrate on the bosonic strings where the ideas can be illustrated more easily. In Section V we will consider in more detail the contribution of fermions to the Polyakov functional integral.

In a multiloop closed string diagram we have to come to grips with the measure (1.1). We have to make explicit the integration volume. We will follow closely the presentations of Ref. 3. Given a fixed topological surface Σ of genus $g>1$, let Γ be the space of metrics on Σ, $\mathrm{Diff}^+(\Sigma)$ the set of orientation preserving diffeomorphisms, and $W(\Sigma)$ the group of Weyl rescalings. We want to find first the jacobian for the change of variables from $[d\gamma]$ into a measure along $\mathrm{Diff}^+(\Sigma)$, $W(\Sigma)$ and the remaining moduli parameters.

Let us consider a given metric $\gamma_{ab}(\tau,\sigma)\varepsilon\Gamma$. In order to define a measure of integration we first have to define a metric on $T_\gamma\Gamma$, the tangent space to Γ at the point γ_{ab}. Since we want to keep two-dimensional gravity non-dynamical, we choose an ultralocal measure on Γ, i.e. one defined by the following metric on Γ:

$$\begin{aligned} ||\delta g||^2 &= \int d^2\sigma\sqrt{\gamma}\, G^{abcd}\delta\gamma_{ab}\delta\gamma_{cd} \\ &\equiv (\delta\gamma,\ \delta\gamma) \end{aligned} \tag{4.1}$$

((4.1) is not the most general metric; see [3].) Here G^{abcd} is the projection operator onto traceless deformations with respect to γ:

$$G^{abcd} = \gamma^{ac}\gamma^{bd} + \gamma^{ad}\gamma^{bc} - \gamma^{ab}\gamma^{cd} \quad . \tag{4.2}$$

A measure on the tangent space $T_\gamma\Gamma$ is defined by requiring

$$\int [d\delta_\gamma]\, e^{-\frac{1}{2}||\delta_\gamma||^2} = 1 \quad . \tag{4.3}$$

(As explained in [3] we only need measures on the various $T_\gamma\Gamma$ in order to find the jacobian for measures on Γ itself.) Similarly, the set of diffeomorphisms at γ are generated by vector fields ξ_a. We define a measure similar to (4.3)

$$\|\xi\|^2 = \int d^2\sigma \sqrt{\gamma}\, \gamma^{ab}\xi_a\xi_b \quad ,$$

$$\int [d\xi]\, e^{-\frac{1}{2}\|\xi\|^2} = 1 \tag{4.4}$$

and also:

$$\int [d\delta X]\, e^{-\frac{1}{2}\|\delta X\|^2} = 1 \quad ,$$

$$\|\delta X\|^2 = \int \sqrt{\gamma}\, d\sigma\; \delta X_\mu \delta X_\nu \eta^{\mu\nu} \quad . \tag{4.5}$$

The metric (4.1) is invariant under diffeomorphisms, but it is not invariant under Weyl rescalings of γ. Thus the measure will contribute to the conformal anomaly. We can split the variations of the metric into two pieces; a trace, and a trace-free part. In turn the trace free part is subdivided into two pieces: The image of coordinate transformations and the moduli deformations:

$$\delta\gamma_{ab} = \delta\sigma\gamma_{ab} + \delta h_{ab}$$

$$\delta h_{ab} = (P_1\xi)_{ab} + \delta t^i \frac{\partial}{\partial t^i}\gamma_{ab}(t)\Big|_{\text{traceless}}$$

$$(P_1\xi)_{ab} = \nabla_a\xi_b + \nabla_b\xi_a - \gamma_{ab}\nabla_c\xi^c \quad . \tag{4.6}$$

For surfaces with genus $g > 1$, P_1 has no zero modes. The zero modes of P_1 are called "conformal Killing vectors" of the surface. In genus zero, there are six (corresponding to the generators of $SL(2,\mathbb{C})$), while for genus one there are two (generating $U(1)\times U(1)$). For higher genus compact surfaces there are no conformal Killing vectors. The arguments we will present assume that P_1 has no kernel, but they can be readily modified to include conformal Killing vectors (see Ref. 3 for more details).

P_1 is an elliptic operator which takes vectors into traceless symmetric tensors. Its adjoint with respect to the metric (4.1) is simply:

$$(P_1^\dagger h)_b = -\nabla^a h_{ab} \quad . \tag{4.7}$$

This implies that the space of metric deformations $\delta\gamma_{ab}$ can be divided into orthogonal sets according to the metric (4.1) and the eigenspaces of $P_1 P_1^\dagger$:

$$\{\delta\gamma\} = \{\delta\phi\cdot\gamma_{ab}\} \oplus \{P_1\xi\} \oplus \mathrm{Ker}\, P_1^\dagger \quad . \tag{4.8}$$

To get a better idea of $\mathrm{Ker}\, P_1^\dagger$, we can choose local complex coordinates $ds^2 = 2g_{z\bar{z}}dzd\bar{z}$. Then the traceless variations have components of the form δh_{zz}, $\delta h_{\bar{z}\bar{z}}$, and the elements in $\mathrm{Ker}\, P_1^\dagger$, satisfy:

$$\nabla^z\, \delta h_{zz} = 0 \quad , \qquad \text{i.e.}$$

$$\partial_{\bar{z}}\, \delta h_{zz} = 0 \quad . \tag{4.9}$$

That is, the kernel of $P_1^\dagger$ consists of the holomorphic quadratic differentials. Using the notation of Section II and the Riemann-Roch theorem (2.47), we can compute the number of linearly independent quadratic differentials. A quadratic differential is a global holomorphic section of K^2. Since $\deg K^2 = 4g-4$ we have from (2.47):

$$\dim H^0(\Sigma,K^2) - \dim H^0(\Sigma,K^{-1}) = 3g-3 \quad . \tag{4.10}$$

In addition we have that $H^0(\Sigma,K^{-1})=0$ for $g>1$. This follows because K^{-1} is a line bundle of degree $-(2g-2)<0$. Any meromorphic section of K^{-1} must have a divisor with this same degree. But a *holomorphic* section would have to have a nonnegative divisor, and hence a nonnegative degree. Hence the space of holomorphic sections is zero and we conclude $\dim H^0(\Sigma,K^2)=3g-3$. This in turn says that the real dimension of $\mathrm{Ker}\, P^\dagger$ is $6g-6$, in agreement with our earlier counting of the moduli parameters.

Thus to parametrize all metrics we choose a slice $\hat{\gamma}(t^i)$ depending on $6g-6$ moduli parameters t and define

$$\tilde{\gamma} = f^*(\gamma) \quad , \quad \text{where} \quad \gamma \equiv e^{\phi}\hat{\gamma}(t) \quad . \tag{4.11}$$

Here ϕ is a Weyl rescaling and f is a diffeomorphism. It is convenient to write an arbitrary fluctuation $\delta\tilde{\gamma}$ of $\tilde{\gamma}$ as $\delta\tilde{\gamma} = f^*(\delta\gamma)$. This way $\delta\gamma$ involves among other things a small diffeomorphism at the origin of $\mathrm{Diff}(\Sigma)$ (instead of at f), which we represent as usual by a vector field ξ.

We now want to express the measure $[d\tilde{\gamma}]$ near $\tilde{\gamma}$ in terms of a measure over $\mathrm{Diff}(\Sigma)\times W(\xi)$ and moduli deformations. Thus we use the decomposition (4.8) and the metrics (4.1-5) to do this. A simplifying feature is that in (4.1) $\mathrm{Diff}(\Sigma)$ acts as an isometry, so that in $\|\delta\tilde{\gamma}\|^2_{\tilde{\gamma}}$ we can forget about f^*. (In order to avoid confusion, we will sometimes indicate the metric on Σ with respect to which we compute the norms, e.g. $\|\cdot\|^2_{\tilde{\gamma}}$.) More precisely,

$$\|\delta\tilde{\gamma}\|^2_{f^*(e^{\phi}\hat{\gamma})} = \|\delta\gamma\|^2_{e^{\phi}\hat{\gamma}} \quad . \tag{4.12}$$

Then we have:

$$\delta\gamma_{ab} = \delta\phi\gamma_{ab} + (P_1\xi)_{ab} + \delta t^i T^i_{ab} \tag{4.13}$$

where

$$T^i_{ab} = \frac{\partial}{\partial t^i}\gamma_{ab}(t) - \text{trace} \equiv e^{\phi}\hat{T}^i_{ab} \qquad (4.14)$$

and $\delta\phi$ has been redefined to absorb the trace parts of T and $\nabla_a\xi_b + \nabla_b\xi_a$.

Inserting (4.13) into the definition of the metric on Γ, (4.3), we get

$$\begin{aligned} \|\delta\gamma\|_\gamma &= \|\delta\phi\|^2_\gamma + \|P_1\xi + T^i\delta t^i\|^2_\gamma \\ &= \|\delta\phi\|^2_\gamma + (P_1\xi, P_1\xi) + (T^i, T^j)\delta t^i\delta t^j \\ &\quad + 2(P_1\xi, T^i)\delta t^i \quad . \end{aligned} \qquad (4.15)$$

Let ψ^a, $a = 1,\ldots,6g-6$ be an orthonormal basis of $\text{Ker}\,P_1^\dagger$. The ψ^a are orthogonal to $P_1\xi$. In order to simplify (4.15), we decompose T^i along the basis $\{\psi^a, P_1\xi\}$

$$T^i = \left(1 - P_1\frac{1}{P_1^\dagger P_1}P_1^\dagger\right)T^i + P_1\frac{1}{P_1^\dagger P_1}P_1^\dagger T^i \qquad (4.16)$$

i.e.

$$T^i = \psi^a(\psi^a, T^i) + P_1 v^i \quad . \qquad (4.17)$$

Substituting (4.17) into (4.15) and redefining $\xi \to \xi + v^i\delta t^i$ (allowed inside a gaussian integral) we obtain

$$\|\delta\gamma\|^2_\gamma = \|\delta\phi\|^2_\gamma + (T^i, \psi^a)(\psi^a, T^j)\delta t^i\delta t^j + \|P_1\xi\|^2_\gamma \quad . \qquad (4.18)$$

If we choose an arbitrary basis for Ker $P_1^\dagger$, then $\|\delta\gamma\|_\gamma^2$ becomes instead

$$\|\delta\gamma\|_\gamma^2 = \|\delta\phi\|_\gamma^2 + (T^i,\psi^a)(\psi_a,\psi_b)^{-1}(\psi^b,T^j)\delta t^i\delta t^j$$
$$+ \|P_1\xi\|_\gamma^2 \tag{4.19}$$

where the inverse refers to the inverse of a matrix of inner products.

Thus, the jacobian we were looking for is:

$$(dt^i)[d\phi][d\xi]\ \det{}_\gamma^{\frac{1}{2}}P_1^\dagger P_1\ \frac{\det_\gamma(\psi_a,T_b)}{\det_\gamma^{\frac{1}{2}}(\psi_a,\psi_b)} \tag{4.20}$$

where the determinant of $P_1^\dagger P_1$ is computed using for example ζ-function regularization. Since we are dealing with the bosonic string and there are no chiral fermions, the factors in (4.20) are diffeomorphism invariant, and we can proceed to cancel the factor $\mathrm{Vol}_\gamma(\mathrm{Diff}\ \Sigma)$ in (1.1) with the integration over $[d\xi]$.

The Weyl volume is more delicate. Notice that in (4.20) everything is computed with the metric $\gamma = e^\phi\hat{\gamma}(t)$. We would like to have an integral only over the slice $\hat{\gamma}(t)$. To see if this is possible, we have to compute the variation of the integrand under changes in ϕ. Before we do that, however, we also have to include the scalars X^μ which give the embedding of the string in space-time. Using (4.5), and removing the zero mode corresponding to translations $X^\mu \to X^\mu + X_0^\mu$, (this leads to an overall factor of the volume of space-time which we drop) we get:

$$\left(\frac{2\pi}{\int\sqrt{\gamma}}\det{}' - \nabla^2\right)^{-d/2} \quad . \tag{4.21}$$

Putting everything together, we have to analyze the behavior of

$$\det{}_\gamma^{\frac{1}{2}} P_1^\dagger P_1 \left(\frac{2\pi}{\int \sqrt{\gamma}} \det{}'_\gamma - \nabla^2 \right)^{-d/2} \frac{\det(\psi^a, T^i)}{\det^{\frac{1}{2}}(\psi^a, \psi^b)} \tag{4.22}$$

under Weyl transformations. The term $\det(T,\psi)$ is already Weyl invariant. This follows since ψ is defined by an equation depending only on the conformal class of $\tilde{\gamma}$ (4.9), so it does not change under Weyl transformations. On the other hand, $T^i = e^\phi \hat{T}^i$, Eq. (4.14). Thus

$$(T^i, \psi^a)_\gamma = \int d^2\sigma \sqrt{\gamma}\, \gamma^{\alpha\beta} \gamma^{\lambda\mu}\, T^i_{\alpha\lambda}\, \psi^a_{\beta\mu} = (\hat{T}^i, \psi^a)_{\hat{\gamma}} \quad . \tag{4.23}$$

which is indeed independent of ϕ.

Hence we only have to worry about:

$$\left[\frac{\det' P_1^\dagger P_1}{\det(\psi^a,\psi^b)} \right]_\gamma^{\frac{1}{2}} \left[\frac{\det' - \nabla^2}{\int \sqrt{\gamma}} \right]_\gamma^{-d/2} \quad . \tag{4.24}$$

The conformal variation of (4.24) can be studied using complex coordinates (see Section IIB). A vector has components ξ^z, $\xi^{\bar{z}}$, and a traceless symmetric tensor h_{zz}, $h_{\bar{z}\bar{z}}$ i.e. a vector is a section of $K \oplus \bar{K}$, and a traceless symmetric tensor a section of $K^2 \oplus (\bar{K})^2$. P_1 decomposes into

$$P_1 = \nabla^1_z \oplus \nabla^z_{-1} : K \oplus \bar{K} \to K^2 \oplus \bar{K}^2$$

$$P_1^\dagger = \nabla^z_2 \oplus \nabla^{-2}_z : K^2 \oplus \bar{K}^2 \to K \oplus \bar{K} \tag{4.25}$$

thus:

$$P_1^\dagger P_1 = \begin{pmatrix} \nabla^z_2\, \nabla^1_z & 0 \\ 0 & \nabla^{-2}_z\, \nabla^z_{-1} \end{pmatrix} \tag{4.26}$$

and

$$\det P_1^\dagger P_1 = \det \nabla_2^z \nabla_z^1 \det \nabla_z^{-2} \nabla_{-1}^z$$

$$= \det \Delta_1^{(+)} \det \Delta_{-1}^{(-)} \quad . \tag{4.27}$$

By complex conjugation it is easy to see that the spectrum of $\Delta_1^{(+)}$ is the same as the spectrum of $\Delta_{-1}^{(-)}$. If ξ_z is an eigenfunction of $\Delta_1^{(+)}$ with eigenvalue λ, then $\gamma^{z\bar{z}}(\xi_z)^*$ is an eigenfunction of $\Delta_{-1}^{(-)}$ with the same eigenvalue, and vice versa. Hence:

$$(\det P_1^\dagger P_1)^{\frac{1}{2}} = \det \Delta_1^{(+)} = \det \nabla_2^z \nabla_z^1 =$$

$$= \det{}' \nabla_z^1 \nabla_2^z \quad . \tag{4.28}$$

In the third equality we have to remove the zero modes of ∇_2^z. In more intrinsic terminology, ∇_2^z is the Cauchy-Riemann operator coupled to sections of K^2 : $\bar{\partial}_{K^2}$, and (4.28) can be written as:

$$(\det P_1^\dagger P_1)^{\frac{1}{2}} = \det{}' \bar{\partial}_{K^2}^\dagger \bar{\partial}_{K^2} \quad . \tag{4.29}$$

Turning to the zero modes, if we write ψ^a in a complex basis $\psi^a = (iS_{zz}^a, -iS_{\bar{z}\bar{z}}^a)$, then:

$$\det{}^{\frac{1}{2}}(\psi^a,\psi^b) = \det (S^a,S^b)$$

$$(S^a,S^b) \equiv \int \sqrt{\gamma}\, d\sigma (\gamma^{z\bar{z}})^2 (S_{zz}^a)^* S_{zz}^b \quad . \tag{4.30}$$

We can do something entirely similar for the scalar laplacian:

$$\Delta_0^{(+)} = \nabla_1^z \nabla_z^0 \quad , \quad \Delta_1^{(-)} = \nabla_z^0 \nabla_1^z$$

and:

$$\det \Delta_0^{(+)} = \det' \nabla_1^z \nabla_z^0 = \det' \nabla_z^0 \nabla_1^z \quad . \tag{4.31}$$

In the second equality, the prime removes the zero modes of ∇_1^z. These are just the abelian differentials ω_i, $1 \le i \le g$. A useful feature is that the inner product of abelian differentials is naturally conformally invariant:

$$(\omega_i, \omega_j) = \frac{i}{2} \int \omega_i \wedge \bar{\omega}_j = \mathrm{Im}\Omega_{ij} \quad . \tag{4.32}$$

B) The Conformal Anomaly

In the computation of the conformal anomaly we have to be careful to include the finite dimensional determinants of zero modes appearing in (4.24). Using the heat kernel of the determinant:

$$\log \det H = - \int_\varepsilon^\infty \frac{dt}{t} \mathrm{Tr}' \, e^{-tH} \quad , \tag{4.33}$$

we can now compute the variation of (4.24) under infinitesimal conformal rescalings. Since this computation is very clearly explained by O. Alvarez in his paper in Ref. 3, we simply quote the result. For the operator $\Delta_q^{(+)}$, $q \ge 0$ one gets:

$$\delta \log \frac{\det \Delta_q^{(+)}}{\det(S^a|S^b)} = \frac{1 + 6q(1+q)}{12\pi} \int d^2\sigma \sqrt{\gamma}\, R\delta\phi \quad . \tag{4.34}$$

Thus for $q = 1$ the conformal anomaly is 13 times that for $q = 0$. Thus:

$$\delta \log \left[\frac{\det^{\frac{1}{2}} P_1^\dagger P_1}{\det^{\frac{1}{2}}(\psi_a|\psi_b)} \left(\frac{\det' - \Delta_0}{\int \sqrt{\gamma}} \right)^{-d/2} \right]$$

$$= \frac{13 - d/2}{12\pi} \int d^2\sigma \sqrt{\gamma}\, R\delta\phi \tag{4.35}$$

which vanishes for $d=26$, the critical dimension of the bosonic string theory. Thus if we take $d=26$ the integration $[d\phi]$ cancels the volume of the Weyl group.

The final expression is then:

$$Z = \int_{\mathcal{M}_g} (dt)\frac{\det(\psi,T)}{(\det(\psi,\psi))^{\frac{1}{2}}} \det{}^{\frac{1}{2}}_{\hat{\gamma}} P_1^\dagger P_1 \left(\frac{\det' - \Delta_0}{\int\sqrt{\hat{\gamma}}}\right)^{-13} \quad . \qquad (4.36)$$

In addition to fixing infinitesimal gauge symmetries we have passed to $\mathcal{M}_g$, fixing the mapping class group. This is allowed in this case without any further ado because the theory is non-chiral and there are no global gravitational anomalies. What we have shown is that for $d=26$ the procedure is independent of how we choose the slice $\hat{g}(t)$.

There are several slices which are useful for explicit computations. For example we can choose a slice of the action of the conformal group which in each orbit selects the metric with constant negative curvature. We are guaranteed by the uniformization theorem (Section IIIC) that this metric is unique up to diffeomorphisms. If Γ_{const} denotes this slice, the metric induced on $\mathcal{M}_g = \Gamma_{const}/\mathrm{Diff}(\Sigma)$ by (4.1) is known as the Weil-Petersson metric, and it turns out that we can write (4.36) as:

$$Z = \int_{\mathcal{M}_g} d(\text{Weil-Petersson}) \det{}^{\frac{1}{2}} P_1^\dagger P_1 \left(\frac{\det' - \Delta_0}{\int\sqrt{\gamma}}\right)^{-13} \quad . \qquad (4.37)$$

We can use this form of the measure to understand some of the divergences of the bosonic string theory. Later on when we use the complex geometry on $\mathcal{M}_g$, we will see that there are other ways of expressing (4.36) in terms of more intrinsic quantities in $\mathcal{M}_g$.

As another application of the conformal anomaly (4.34) we can check that the field content of the heterotic string leads to a conformally invariant theory in ten dimensions. Since the heterotic string is chiral, we have to be careful in the application of (4.34). Since we work with fields whose kinetic terms involve ∇_z or ∇^z but not $\Delta^{(\pm)}$, it seems that we need not $\det \Delta$ but $\det \nabla$ in the string integrand. This however does not make literal sense, since the ∇ operator maps a tensor space to a different tensor space and so does not have a determinant in the usual sense. We will see later the sense in which this determinant is meaningful. For now, however, we can take a short cut by noticing that the Weyl anomaly is indifferent to chirality. Thus if we like we can double all the fields in the heterotic string's 2d field theory, compute the total anomaly and divide by two. In the process each chiral determinant $\det \nabla$ gets replaced by a well-defined factor of $\det \Delta$.

In the right moving sector of the heterotic string in the covariant formulation we have [25] X^μ_R, ψ^μ_R, where ψ^μ_R are a set of d right moving Weyl Majorana fermions. They contribute a determinant which is the square root of the corresponding Weyl determinant. We have the right moving part of the ghosts which reproduce the determinant of $P_1^\dagger P_1$ and we also have commuting spinor ghosts corresponding to gauge fixing of the right moving world sheet supergravity of the heterotic string. Then, in the doubled representation we have:

$$\left(\frac{\det \nabla^z_1 \nabla^0_z}{\int \sqrt{\gamma}}\right)^{-d/2} (\det \nabla^z_{\frac{1}{2}} \nabla^{-\frac{1}{2}}_z)^{d/2} \frac{\det \nabla^z_2 \nabla^1_z}{\det(S^a, S^b)} \left(\frac{\det \nabla^z_{3/2} \nabla^{\frac{1}{2}}_z}{\det\langle\nu|\nu\rangle}\right)^{-1} \tag{4.38}$$

where the last factor comes from the supersymmetry ghost, and the ν's are the zero modes of $\nabla^z_{3/2}$. Since $\nabla^z_{3/2}$ acts on sections of $K^{3/2}$, using the Riemann-Roch theorem again we find that there are $2g-2$ zero modes, representing the supermoduli of the Riemann surface [4]; for simplicity those have all been set to zero in (4.38). Adding up the conformal anomalies (4.34) we have for the right sector (let $f(q) \equiv 1 + 6q(1+q)$):

$$\text{anomaly coefficient} = -\frac{d}{2}f(0) + \frac{d}{2}f(-\tfrac{1}{2}) + f(1) - f(\tfrac{1}{2}) = \frac{30-3d}{4} \tag{4.39}$$

which vanishes for $d = 10$. Similarly, the left moving sector contains ten right moving scalars, the coordinate ghost, and 32 Weyl-Majorana fermions combined into ten Weyl fermions. We find:

$$\text{anomaly} = -\frac{d}{2}f(0) + 16\ f(-\tfrac{1}{2}) + f(1) = \frac{10-d}{2} \tag{4.40}$$

again leading to $d = 10$ as the critical dimension.

C) Complex Structure of $\mathcal{M}_g$ and Holomorphic Factorization

In the genus one case, we know that the moduli space of tori is given by a fundamental region of $SL(2,\mathbb{Z})$ in the upper half plane. From Section IID we know that the upper half plane has a natural complex structure, and therefore the moduli space of tori inherits a complex structure. The moduli space $\mathcal{M}_g$ $g > 1$ is also a complex space. In this subsection we will describe its complex structure and also prove the factorization theorem alluded to in the previous subsection.

As in the case of the torus we can consider two spaces of Riemann surfaces: moduli space $\mathcal{M}_g$, and its covering space $\mathcal{T}_g$, the "Teichmüller space". We obtain $\mathcal{M}_g$ if we divide $\mathcal{T}_g$ by the action of the modular group $\Omega(\Sigma)$, $\mathcal{M}_g = \mathcal{T}_g/\Omega(\Sigma)$. In the torus case, $\mathcal{T}_1 = H$, and it is contractible, i.e. it has no topology. This is also true for higher genus (see Ref. 13). Thus all the topological properties of $\mathcal{M}_g$ come from the quotient by $\Omega(\Sigma)$.

For the bosonic string we already know that the theory is invariant under $\Omega(\Sigma)$ because there are no chiral fields. For the heterotic string, we know (see Section III) that the theory is modular invariant at the one loop level. Modular invariance for higher genus surfaces is however a non-trivial issue in this case. In Ref. 31, Witten showed, using techniques extending those of (Ref. 32), that all the usual fermionic strings are indeed modular invariant. We will comment further on this point in the next section.

To examine the question of whether $\mathcal{M}_g$ really has a complex structure, we can ask instead whether $\mathcal{T}_g$ is a complex manifold with an analytic action of $\Omega(\Sigma)$. The proof of this fact is due to Ahlfors[11]. We will follow the treatment of Bers[13]. Bers mapped all of $\mathcal{T}_g$ isomorphically to a bounded domain of $\mathbb{C}^{3g-3}$, and then showed that $\Omega(\Sigma)$ acted analytically on this domain. Since $\mathbb{C}^{3g-3}$ is obviously a complex manifold, we can pull back the complex structure of $\mathbb{C}^{3g-3}$ to $\mathcal{T}_g$ to endow it with a complex structure. Even though the full construction of this map is complicated, for our purposes it is enough to know how to set up local complex coordinates on $\mathcal{M}_g$, and this is relatively easy to describe. The theorems in Refs. 11, 13 guarantee that this local construction gives a global complex structure.

From Section II, we know that given a Riemann surface Σ we can choose a metric compatible with the complex structure and set up coordinate patches such that the metric is everywhere proportional to $|dz|^2$. For instance we can choose gaussian coordinates, or use the local parameters induced by the upper half plane. We can describe any other Riemann surface Σ' with the same topology by choosing a tensor field $\mu_{\bar{z}}^{z}$, and writing $(ds^2)' \propto |dz + \mu_{\bar{z}}^{z}\, d\bar{z}|^2$. The tensor $\mu_{\bar{z}}^{z}$ is known as a Beltrami coefficient, and $\mu_{\bar{z}}^{z}\, d\bar{z}$ as a Beltrami differential. Notice that $|\mu_{\bar{z}}^{z}(z)|^2$ transforms as a scalar on the surface. We can solve patch by patch the Beltrami equation:

$$\partial_{\bar{z}} w = \mu_{\bar{z}}^{z}(z,\bar{z})\partial_z w \quad . \tag{4.41}$$

$w(\mu)$ defines what is known as a quasiconformal mapping from Σ to Σ', and it provides the relation between the complex structures on Σ and Σ'. In this description we still have a lot of redundancy. If we consider the original metric $|dz|^2$ and we make a diffeomorphism $z \to z + v(z,\bar{z})$, $\bar{z} \to \bar{z} + \bar{v}(z,\bar{z})$, then the metric changes in a way which we can also describe using a μ of the form $\partial_{\bar{z}} v^z/(1 + \partial_z v^z)$; thus deformations of this form are not really distinct from the original conformal structure.

Suppose μ is infinitesimal. Then it describes a tangent to the space of metrics modulo conformal rescalings. In this case the trivial Beltrami differentials can be written as $\mu = \bar{\partial}v$. These are the gauge directions corresponding to diffeomorphisms at the point γ in the space Γ of metrics. Thus the cotangent space to moduli space at Σ is the space of linear functionals of μ which annihilate those of the form $\bar{\partial}v$. (Recall that the cotangent space is the dual of the tangent space). Let $B_2(\Sigma)$ be this cotangent space. The linear functionals in $B_2(\Sigma)$ can be written in terms of integral kernels:

$$L_\varphi(\mu) = \int_\Sigma \mu_{\bar{z}}^{z} \varphi_{zz} \, dzd\bar{z} \tag{4.42}$$

and the kernel is a quadratic differential. Note that in (4.42) we do not need to include any metric factor, because $\mu_{\bar{z}}^{z} \varphi_{zz}$ is already a (1,1) form on Σ and can be integrated without any problem. If we now take $\mu = \bar{\partial}v$ and integrate by parts we find

$$L_\varphi(\partial v) = -\int_\Sigma v \, \bar{\partial}\varphi \tag{4.43}$$

which vanishes identically iff $\bar{\partial}\varphi = 0$. We reach the important conclusion that the cotangent space to $\mathscr{M}_g$ is isomorphic to the space of holomorphic quadratic differentials:

$$B_2(\Sigma) \equiv T^{*}_{(1,0)} \mathscr{M}_g \Big|_\Sigma \simeq H^0(\Sigma, K^2) \quad . \tag{4.44}$$

Since $B_2(\Sigma)$ is a (3g-3)-dimensional complex vector space, the isomorphism (4.44) gives a complex structure to the tangent spaces of $\mathscr{M}_g$. We can now give complex coordinates for $\mathscr{M}_g$ near a particular metric γ_{ab}. From the uniformization theorem, (Section IID), in the conformal class of γ_{ab} there is exactly one metric (up to diffeomorphisms) of constant negative curvature. Call it γ_0. We can choose complex coordinates z so that $\gamma_0 \propto dzd\bar{z}$. Now choose a basis $\varphi_0^{(i)}{}_{zz}$ of B_2. This also gives a system of coordinates for the space $\bar{B}_2$. Any $\bar{\varphi} \in \bar{B}_2$ can be written as $u_i \bar{\varphi}_0^{(i)}{}_{\bar{z}\bar{z}}$, where $u_i \in \mathbb{C}^{3g-3}$. Given $\vec{u}$, we build the conformal class

$$\gamma(\vec{u}) \propto |dz + \gamma_0^{z\bar{z}} \, \bar{\varphi}_0^{(i)}{}_{\bar{z}\bar{z}} \, u_i \, d\bar{z}|^2 \qquad . \tag{4.45}$$

Locally, for small enough u_i no two of these metrics are gauge-equivalent, so we have mapped a region in $\mathbb{C}^{3g-3}$ into a region of $\mathscr{M}_g$. We now declare this mapping to be holomorphic. Even though we have made many choices to reach this complex coordinate system, the general theorem of Refs. 11, 13 guarantees that the complex structure obtained is independent of choices and modular invariant.

The picture we have described so far also allows us to understand the Weil-Petersson metric from a different point of view. Namely, if we choose an orthonormal basis of $\bar{\varphi}_0^{(i)}{}_{\bar{z}\bar{z}}$ with respect to γ_0, then we can pull back from $\mathbb{C}^{3g-3}$ the natural hermitian metric on $\mathbb{C}^{3g-3}$ to get a metric on $T\mathscr{M}|_{\gamma_0}$. This is the Weil-Petersson metric.

Now we can analyze the holomorphic factorization of determinants of operators such as $\Delta_q^{(+)}$, $\Delta_q^{(-)}$. First we notice that under the metric deformation in (4.45) the covariant derivative varies to first order as

$$\delta\nabla_z = -\frac{1}{2}\,\delta\gamma_{zz}\,\nabla^z \tag{4.46}$$

since $\nabla^z \delta\gamma_{zz} = 0$ when $\delta\gamma$ is built from quadratic differentials, (4.45). Thus $\delta\nabla_z$ is proportional to $\bar{u}_i$, not to u_i. Naively one would expect then that $\det \nabla_z$ is an antiholomorphic function on $\mathscr{M}_g$, and thus that $\det \Delta$ is the absolute square of a holomorphic function. Indeed this is precisely what we find on the torus. We will now check it in general.

Let us consider in general a family of elliptic operators $D_{\bar{y}}$ parametrized by some complex space Y such that $D_{\bar{y}}$ varies antiholomorphically with y, and $D_y^\dagger$ varies holomorphically. For simplicity, let us assume that Ker $D_{\bar{y}} = 0$. This is indeed the case for ∇_z^q $q > 0$. The case when Ker $D_{\bar{y}} \neq 0$ is a straightforward extension of the computation we are going to present[34]. To determine the possible holomorphic factorization of det $D_y^\dagger D_{\bar{y}}$, we compute

$$\bar{\delta}\delta \log \det D_y^\dagger D_{\bar{y}} \equiv - \bar{\delta}\delta \int_\varepsilon^\infty \frac{dt}{t} \mathrm{Tr}\, e^{-tD_y^\dagger D_{\bar{y}}} .$$

If the determinant is the absolute value squared of a holomorphic function, then $\bar{\delta}\delta$ log det $D_y^\dagger D_{\bar{y}}$ should vanish. Thus $\delta\bar{\delta}$ log checks for anomalies with respect to holomorphic factorization[6]. (The holomorphic properties of families of Cauchy-Riemann operators on a *fixed* Riemann surface were first studied by Quillen[33].) We follow the presentation of Ref. 34.

Performing the variations (4.46) we obtain:

$$\bar{\delta}\delta \log \det D_y^\dagger D_{\bar{y}} = \bar{\delta} \int_\varepsilon^\infty \mathrm{Tr}\, \delta\, D^\dagger D\, e^{-tD^\dagger D}\, dt .$$

Integrating by parts in t (again $D^\dagger D$ has no kernel)

$$\bar{\delta}\delta \log \det D_y^\dagger D_{\bar{y}} = \mathrm{Tr}\, \delta D^\dagger (1 - D \frac{1}{D^\dagger D} D^\dagger)\, \bar{\delta} D \frac{1}{D^\dagger D} e^{-\varepsilon D^\dagger D}$$
$$- \varepsilon \int_0^1 ds\, \mathrm{Tr}\, \delta D^\dagger D \frac{1}{D^\dagger D} e^{-s\varepsilon D^\dagger D} D^\dagger \bar{\delta} D\, e^{-(1-s)\varepsilon D^\dagger D} . \tag{4.47}$$

Note that $P = 1 - D(D^\dagger D)^{-1} D^\dagger$ is the projection operator onto the kernel of $D^\dagger$. Thus, the first trace in (4.47) is finite dimensional, and we can take the $\varepsilon \to 0$ limit. Since $D_y^\dagger$ varies holomorphically with y, we can choose a basis for its kernel which also varies holomorphically with y. Let this basis be χ_i. Using ordinary degenerate perturbation theory we get:

$$\bar{\delta}\delta \log \det < \chi_i | \chi_j > = \operatorname{Tr} P \bar{\delta} D \frac{1}{D^\dagger D} \delta D^\dagger$$

$$P = 1 - D \frac{1}{D^\dagger D} D^\dagger = \sum |\chi_i > A_{ij} < \chi_j| \tag{4.48}$$

$$(A^{-1})_{ij} = < \chi_i | \chi_j > \quad .$$

In the second trace in (4.47) we first note that $(\exp - s\varepsilon D^\dagger D)D^\dagger = D \exp - s\varepsilon DD^\dagger$, and use the form of P in (4.48). Then in the trace we get an insertion of $(1 - P)$. Since the term coming from the P involves a finite dimensional trace, it will disappear in the $\varepsilon \to 0$ limit. Hence

$$\bar{\delta}_y \delta_y \log \frac{\det D_y^\dagger D_{\bar{y}}}{\det < \chi_i | \chi_j >} = - \varepsilon \int_0^1 ds \operatorname{Tr} \delta D^\dagger e^{-s\varepsilon DD^\dagger} \bar{\delta} D e^{-(1-s)\varepsilon D^\dagger D} \tag{4.49}$$

which can be computed by using heat kernel methods to obtain a local expression on the Riemann surface.

To apply these results to our case, we now specialize to the operators ∇_z^q with variation (4.46) and compute (4.49).

Since this computation is manifestly local, it can be carried out using Feynman diagrams. We are looking at metric deformations which look like

$$e^{2\phi} |dz + \mu d\bar{z}|^2 \cdot \tag{4.50}$$

where μ is a small Beltrami differential. If we call

$$W_q[\gamma] \equiv \log \frac{\det \Delta_q^{(\pm)}}{\det < \chi_i | \chi_j >} \tag{4.51}$$

we can first use the conformal anomaly to eliminate ϕ. Since by using the heat kernel in (4.49) we know that the obstruction to holomorphic factorization is local, it suffices to consider Beltrami differentials with support in a coordinate patch $U \subset \Sigma$. Even though these are not moduli deformations, the local formula is necessarily the same; later we can patch together several such expressions to get the full holomorphic anomaly. Integrating the conformal anomaly (4.34) to eliminate ϕ we obtain

$$W_q[\gamma] = W_q[\hat{\gamma}] + \frac{6q(1+q)+1}{12\pi} S_L[\phi,\hat{\gamma}]$$

$$S_L = \int_\Sigma d^2\sigma \sqrt{\hat{\gamma}} \, (\hat{\gamma}^{ab}\partial_a\phi\partial_b\phi - \hat{R}\phi) \tag{4.52}$$

(S_L is known as the "Liouville action"), and $\hat{\gamma}$ is just $|dz + \mu d\bar{z}|^2$ in the patch U.

The effective action $W[\hat{\gamma}]$ is to second order in μ simply given by a one loop diagram in two dimensions with two external graviton lines. This was computed in Ref. 35. Combining results, we get (see Ref. 34 for more details):

$$\delta\bar{\delta} W = \frac{1+6q(1+q)}{48\pi} \left[\int d^2\sigma \sqrt{\gamma} \, (\gamma^{z\bar{z}})^3 \, \nabla_z^{-2}\delta\gamma_{\bar{z}\bar{z}}\nabla_2^z\delta\gamma_{zz} \right.$$

$$\left. - \frac{3}{2}\int d^2\sigma \sqrt{\gamma} \, (\gamma^{z\bar{z}})^2 \delta\gamma_{zz}\delta\gamma_{\bar{z}\bar{z}}R \right] \quad , \tag{4.53}$$

expressed in terms of the original metric deformations $\delta\gamma_{zz}$, $\delta\gamma_{\bar{z}\bar{z}}$.

The crucial point to note in (4.53) is that the obstruction to holomorphic factorization has the same numerical coefficient as the conformal anomaly. This therefore means that the combination of determinants appearing in string theories in their critical dimensions is equal to the square of the absolute value of a holomorphic function.

In particular the quantity:

$$G \equiv \left(\frac{\det' \bar{\partial}^{\dagger}_{K} \bar{\partial}_{K}}{\det \mathrm{Im}\,\Omega \int \sqrt{g}} \right)^{-13} \frac{\det' \bar{\partial}^{\dagger}_{K^2} \bar{\partial}_{K^2}}{\det(\varphi^i, \varphi^j)} \tag{4.54}$$

is the absolute value square of a holomorphic function. Note that we have included the determinant of inner products of the zero modes of $\bar{\partial}_{K^1} = \nabla^z_1$ $(\bar{\partial}_{K^2} = \nabla^z_2)$, and we have rewritten it as $\det \mathrm{Im}\,\Omega$.

D) Degeneration of Surfaces

Now that we know string theories involves a locally holomorphic function on $\mathcal{M}_g$, we can try to understand any infinities that arise in terms of poles in this function. There is an elementary way to understand these infinities which we now explain. For example see Refs. 36, 39. In the next section we will mention how the computation to be done now has a very nice interpretation in terms of complex algebraic geometry.

Let us consider the bosonic string integrand once more in the form:

$$\int_{\mathcal{M}_g} d(\text{Weil-Petersson}) \det' \Delta_2^{(+)} \left(\frac{\det' - \Delta_0^{(+)}}{\int \sqrt{g}} \right)^{-13} \quad . \tag{4.55}$$

So far we have considered a complex coordinate system in $\mathcal{M}_g$. There is another coordinate system in terms of real coordinates that is also convenient. These are Fenchel-Nielsen coordinates (see for example Ref. 37). First choose the slice of metrics with constant negative curvature. We parametrize $\mathcal{M}_g$ by first choosing $3g-3$ closed geodesics which cut the surface into a set of "pants" (see Fig. 4.1). To each such geodesic c_j we can associate its length l_j and a twist τ_j $-\infty < \tau_j < +\infty$ defined as follows: we cut Σ along c_j, and reattach the two sides after making a relative twist of τ_j. The coordinates

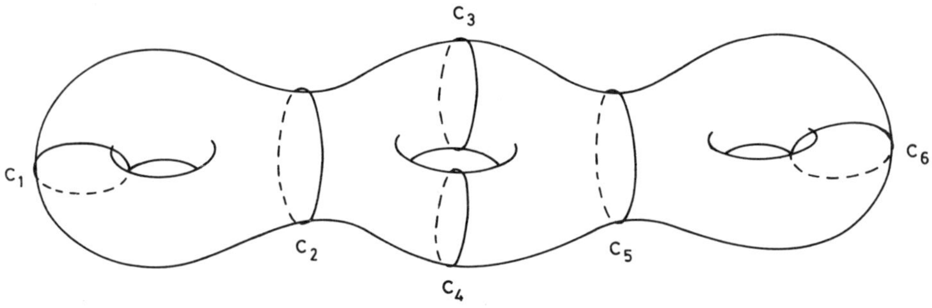

Fig. 4.1 Cutting Σ into four pairs of pants.

(l_j, τ_j) $1 \leq j \leq 3g-3$ provide the Fenchel-Nielsen coordinates of $\mathscr{M}_g$. In these coordinates the Weil-Petersson volume element is simply [38]:

$$d(\text{Weil-Petersson}) = \prod_{j=1}^{3g-3} dl_j \wedge d\tau_j \quad . \tag{4.56}$$

Since the determinants in (4.55) are defined using ζ-function regularization, we find that as long as the Riemann surface is smooth they are non-vanishing and well behaved. The only possibility for (4.55) to blow up is when we approach the boundary of moduli space and the surface degenerates (see Fig. 4.2). By taking the length l_γ in Fig. 4.2 to zero, we approach a surface with a node. Σ then has two components Σ_1, Σ_2 to the left and to the right of the node*. We can obviously shrink to zero more than one geodesic. To evaluate the divergencies of (4.55), we can write the determinant using Selberg's zeta function (2.65). We outline now the analysis of Ref. 36.

---

* We will not consider here the case where γ is homologically non-trivial. In this case the pinched surface still has one component.

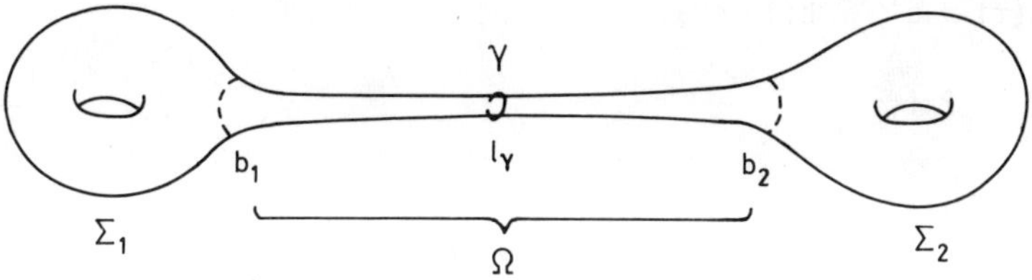

Fig. 4.2 A degenerating surface on the constant curvature slice.

Note that in Fig. 4.2 we are shrinking a primitive geodesic. Since we are always assuming that we work on the constant curvature slice, sending $l_\gamma \to 0$ implies that the neck joining Σ_1, Σ_2 gets long and thin, so that its Gaussian curvature stays constant. All other primitive geodesics different from γ fall into two classes: those that go through the neck (which will be stretched to infinite length as $l_\gamma \to 0$) and those that do not. For the ghost determinant we have:

$$(\det P_1^\dagger P_1)^{\frac{1}{2}} = \text{const.} \prod_{k=0}^{\infty} \prod_{\gamma \text{ primitive}} (1 - e^{-l_\gamma (k+2)}) \,. \tag{4.57}$$

In the limit $l_\gamma \to 0$ (4.57) becomes:

$$(\det{}' P_1^\dagger P_1)^{\frac{1}{2}}_{\Sigma_1} (\det{}' P_1^\dagger P_1)^{\frac{1}{2}}_{\Sigma_2} \prod_{k=0}^{\infty} (1 - e^{-l(\gamma)(k+2)})^2 \quad . \tag{4.58}$$

The squared factor appears because there are two primitive geodesics corresponding to γ; they both traverse γ, but in opposite directions. Using

$$\eta(-1/\tau) = (-i\tau)^{\frac{1}{2}} \eta(\tau)$$

where

$$\eta(\tau) = q^{1/24}\, \Pi(1-q^n)$$

$$q = e^{2\pi i\tau}$$

with $\tau = i l(\gamma)/2\pi$, we get as $l_\gamma \to 0$ that

$$\det{}'\, P_1^\dagger P_1 \sim l(\gamma)^{-3}\, e^{-\pi^2/3l_\gamma} \prod_{n=1}^{\infty} (1-e^{-4\pi^2 n/l_\gamma})^2 \quad . \tag{4.59}$$

The behavior of $[Z'(1)]^{-13}$ is more tricky. The strategy of Ref. 36 is to compute $Z'(1)^{-13}$ by first splitting Σ into three parts: Σ_1, Σ_2, and the neck Ω joining Σ_1, Σ_2. Keeping the lengths of the two boundaries $b_i = \partial\Sigma_i$, $i = 1, 2$ fixed, one can represent the determinant of the laplacian as a functional integral by first fixing the value of $X^\mu(\sigma)$ on b_1, b_2 to be $\bar{X}_{1,2}(\sigma)$, and integrating over $\bar{X}_{1,2}$ at the end. Thus:

$$[Z'(1)]^{-13} = \int D\bar{X}_1 D\bar{X}_2 \int_{\Sigma_1} DX\, e^{-X^\mu \Delta_0^{(+)} X_\mu} \int_{\text{neck}} DX\, e^{-X^\mu \Delta_0^{(+)} X_\mu}$$

$$\times \int_{\Sigma_2} DX\, e^{-X^\mu \Delta_0^{(+)} X_\mu} \quad . \tag{4.60}$$

We now perform the integral over the neck. This can be done as in Ref. 40 to give in the limit of degeneration at γ

$$\frac{\det{}'\,(-\Delta_0^{(+)})}{\int \sqrt{\gamma}} \sim e^{-\pi^2/3l} \quad . \tag{4.61}$$

Putting (4.59) and (4.61) together gives:

$$(\det{}' \, P_1^\dagger P_1)^{\frac{1}{2}} \left(\frac{\det{}' \, (-\Delta_0^{(+)})}{\int \sqrt{\gamma}} \right)^{-13} \sim l_\gamma^{-3} \, e^{4\pi^2/l_\gamma} \quad . \tag{4.62}$$

The term blowing up exponentially in (4.62) is a signature of the tachyon of the bosonic string. The tachyon propagates along the neck Ω in Fig. 4.2, which has length π/l_γ. The power divergence is due to the massless dilaton of the bosonic string. We would like to use the information in (4.62) to gain some understanding of the analytic behavior of (4.54) as we approach the boundary of moduli space. Since the Fenchel-Nielsen coordinates are not adapted to the complex structure of $\mathscr{M}_g$, we have to find first the correct complex coordinates in $\mathscr{M}_g$ describing a neighborhood of the node as $l_\gamma \to 0$, and also the relation between the complex coordinate and the coordinates (l_γ, τ_γ). This is done in terms of the so-called "plumbing fixture" (for details and references see Refs. 9, 41).

The neighborhood of a node is described as follows. Let Σ_1, Σ_2 be the two components of the Riemann surface Σ after we pinch γ to a point, and let $z_0 \in \Sigma_1$, $w_0 \in \Sigma_2$ be the two points on Σ_1, Σ_2 corresponding to the node. By choosing suitable coordinates, we can consider two disks $D_i \subseteq \Sigma_i$, $i = 1, 2, \ldots$ such that z_0 (resp. w_0) is at the center of D_1 (resp. D_2). Let us fix our attention now on the disks D_1, D_2 and forget the rest of Σ_1, Σ_2. A local complex neighborhood of the Riemann surface with a node at $z_0 = w_0$ is obtained by taking a complex parameter $t \in D_t$, $D_t = \{t \in \mathbb{C}, \ |t| \leq 1\}$, and considering a complex 1-parameter family of surfaces defined by

$$\mathscr{C}_t = \{(z,w), \ z \in D_1, \ w \in D_2 \ , \ zw = t, \ t \in D_t\} \quad . \tag{4.63}$$

The collection of all the $\mathscr{C}_t$ gives a subset of $\mathbb{C}^3$ that describes the desired neighborhood of Σ with a node.

If $t \neq 0$, we have a smooth surface joining Σ_1 and Σ_2. When $t = 0$, we obtain the node. For fixed $t \neq 0$, the plumbing fixture is shown in Fig. 4.3. The neighborhood (4.63) corresponds to cutting out of D_1, D_2, two small disks of radius $|t|$, and glueing the two annular regions onto each other as prescribed in (4.63) and as drawn in Fig. 4.3. The points A_i, B_i, $1 \leq i \leq 4$ have been drawn to help visualize the prescription (4.63).

In order to relate the plumbing fixture and its complex parameter t to the hyperbolic neck Ω (Fig. 4.2) and its Fenchel-Nielsen coordinates, we conformally map the plumbing fixture into the upper half plane. If we denote by u the complex coordinate in the upper half plane and w the complex coordinate on the plumbing fixture, the conformal mapping is (Fig. 4.4)

$$u = \exp\left[-i\pi \frac{\log w}{\log |t|^{-1}}\right] \tag{4.64}$$

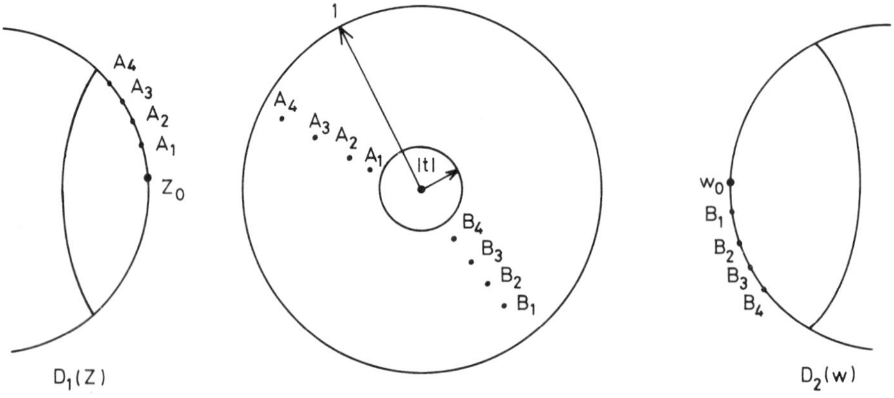

Fig. 4.3 The plumbing fixture.

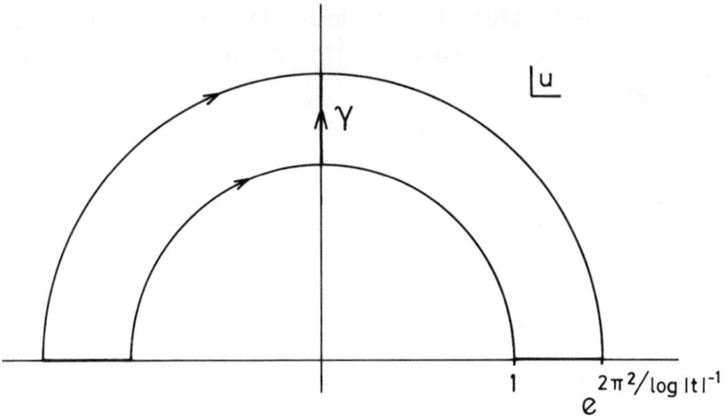

Fig. 4.4 The conformal image of the plumbing fixture on the upper half plane.

so that the annulus in w, Fig. 4.3, corresponds to the region shown in Fig. 4.4, where we identify the two semicircles. In this coordinate it's easy to see that the metric of constant negative curvature is just the Poincaré metric on the u half-plane. In this metric the circumference l_γ of the neck is the length of the contour in Fig. 4.4:

$$l_\gamma = \int_i^{i\,\exp(2\pi^2/\log|t|^{-1})} du\left(\frac{1}{u-\bar{u}}\right) \sim 1/\log|t|^{-1} \tag{4.65}$$

Also the twist τ_γ in the limit as $l_\gamma \to 0$ can be written as l_γ times the phase of t. Thus the degenerating part of the Weil-Petersson measure is

$$\begin{aligned} dl_\gamma \wedge d\tau_\gamma &\sim (\log|t|^{-1})^{-3}\left(\frac{dt}{t} + \frac{d\bar{t}}{\bar{t}}\right) \wedge \left(\frac{dt}{t} - \frac{\overline{dt}}{\bar{t}}\right) \\ &\sim (\log|t|^{-1})^{3}\,\frac{dt\wedge d\bar{t}}{t\,\bar{t}} \qquad . \end{aligned} \tag{4.66}$$

(Even though this derivation is a bit heuristic, it is in fact correct. For the rigorous derivation see Ref. 41.) Combining (4.62, 4.65, 4.66) we finally obtain:

$$d(\text{Weil-Petersson}) \times (\det{}' \, P_1^{\dagger} P_1)^{\frac{1}{2}} \left(\frac{\det{}' \, \Delta_0^{(+)}}{\int \sqrt{\gamma}} \right)^{-13}$$

$$\sim dt \wedge d\bar{t} \, \frac{1}{t^2 \bar{t}^2} \qquad (4.67)$$

i.e. we get a fourth order pole. This as argued after (4.62) is a signature of the presence of the tachyon.

The presence of this fourth order pole in the string integrand and its interpretation as a signature of the tachyon was first given by Belavin and Knizhnik. (4.67) is indeed a holomorphic square, even though we did not assume this in the derivation (4.55-67). Belavin and Knizhnik also noted that when we do make use of the general holomorphic factorization theory it turns out that (4.67) can be derived from existing results in algebraic topology. We will briefly describe this approach in Section V.

E) The Dirac Determinant

As another application of the ideas of holomorphic factorization, we will consider one more example. In the one loop case analyzed in Section III we found that in the partition function for fermionic strings one has to sum over the spin structures of all the fermions involved. Furthermore, to construct a modular invariant sum we have to understand how the spin structures behave under modular transformations. Since the spin structures in higher genus are more complicated objects than in the torus, it is an interesting question to determine the explicit spin structure dependence of the Dirac determinant on an arbitrary Riemann surface. This was one of the purposes of Ref. 34. The basic argument involves a combination of Quillen's theorem[33] (holomorphic factorization) and the Riemann vanishing theorem explained in Section IIC.

The chiral Dirac operator $\nabla_z^{-\frac{1}{2}}$ takes sections of some fixed spin bundle L_α^{-1} to sections of $L_\alpha^{-1} \otimes K \cong L_\alpha$. We will want to consider not one fixed operator, but the entire family of operators on a given Σ coupled to every twisted spin bundle, i.e. every bundle L of degree $d = g-1$, regardless of whether $L \otimes L \cong K$. That is, we want a family parametrized by all of the Picard torus Pic_{g-1}. In turn, each Picard group Pic_α in degree d can itself be parametrized by the Jacobian $J(\Sigma)$ as we have seen earlier. Since $J(\Sigma)$ is a complex torus, we can ask whether the associated family of Dirac operators has any holomorphy property.

Choose one fixed spin bundle L_α, and realize any other twisted spin bundle as $L = L_\alpha^{-1} \otimes V(\vec{\theta},\vec{\varphi})$. Here $V(\vec{\theta},\vec{\varphi})$ is a bundle of degree zero. $V(\vec{\theta},\vec{\varphi})$ will not have a nonvanishing holomorphic section if $(\theta,\varphi) \neq (0,0)$, but it will have a nonvanishing holomorphic section e on the cut Riemann surface Σ_c of Fig. 2.2. As we cross the curves a_i (resp. b_i), e can be chosen to jump by the constant phases $e^{2\pi i\varphi_i}$ (resp. $e^{-2\pi i\theta_i}$). The θ_i, φ_i are defined mod 1, so we do have a real 2g-torus of possible twists. We would like to replace it by a complex g-torus.

We can describe a smooth section of $V(\theta,\varphi)$ as $s = f \cdot e$, where f is a function on Σ_c which jumps by $e^{-2\pi i\varphi_i}$, $e^{2\pi i\theta_i}$ across a_i, b_i. The Cauchy-Riemann operator is then $\bar{\partial}s \equiv (\partial_{\bar{z}}f) \cdot e$ since e is holomorphic. We will now represent this operator in terms of a new basis section $\hat{e}$ which is periodic but *not* holomorphic.

Consider the function $U(P)$ on Σ_c given by

$$U(P) = \exp \int_{P_1}^{P} A \quad , \qquad A = 2\pi i(\theta \cdot \alpha - \varphi \cdot \beta)$$

where P_1 is some fixed chosen point. α^i, β^i are the real harmonic one-forms dual to a_i, b_i (Eq. 2.6). Hence $U(P)$ is path-independent on Σ_c. Also $U(P)$ jumps in such a way that $\hat{e} = U \cdot e$ does not, and we have

$$\bar{\partial} s = [(\partial_{\bar{z}} + A_{\bar{z}})\hat{f}]\hat{e} \tag{4.68}$$

where $s = \hat{f} \cdot \hat{e}$ is the same section as before. Thus for every twisted bundle $V(\theta, \varphi)$ we represent its Cauchy-Riemann operator by a gauged operator on the space of periodic functions $\hat{f}$. We can rewrite this family of Dirac operators in terms of the abelian differentials using [34]

$$A_z = - 2\pi i(\varphi + \Omega\theta)^t \cdot (\Omega - \bar{\Omega})^{-1} \cdot \omega \tag{4.69}$$

to get

$$D(\vec{u}) = \nabla_z^{-\frac{1}{2}} - 2\pi i \; u \cdot (\Omega - \bar{\Omega})^{-1} \cdot \omega \quad . \tag{4.70}$$

Here $u = \varphi + \Omega\theta \in \mathbb{C}^g$. When we shift u by a vector in the lattice L_Ω, θ and φ do not change mod 1 and we get essentially the same operator $D(u)$ back. Hence we have a family of differential operators depending holomorphically on the jacobian $J(\Sigma)$, as desired. We can now apply the machinery of holomorphic factorization to this family.

Since the Dirac operator has no zero modes generically, and $\delta D(\vec{u}) = \delta A(\vec{u})$ does not involve derivatives, the computation of (4.49) is much simpler than when we were considering moduli variations. By a heat kernel expansion, or a Feynman graph computation one gets [32, 34]

$$\det D^\dagger(\vec{u})D(\vec{u}) = e^{i\pi(u-\bar{u})(\Omega-\bar{\Omega})^{-1}(u-\bar{u})}|g(\vec{u})|^2 \tag{4.71}$$

where $g(\vec{u})$ is a holomorphic function of $\vec{u}$. It can be almost uniquely fixed by requiring that (4.71) be invariant under shifts of the jacobian lattice $\vec{u} \to \vec{u} + \vec{n} + \Omega\vec{m}$. This has to be true, because such shifts amount to well-defined $U(1)$ gauge transformations

$$U(z) = \exp -2\pi i\left(\vec{m} \cdot \int_{P_0}^{z} \vec{\alpha} - \vec{n} \cdot \int_{P_0}^{z} \vec{\beta} \right) \quad . \tag{4.72}$$

Since the determinant (4.71) was implicitly computed using a gauge invariant ζ-function regulator, (4.73) must be gauge invariant. Using the fact that $g(\vec{u})$ is holomorphic, we can derive its transformation rules under lattice shifts:

$$g(\vec{u} + \vec{n}) = e^{i\phi(\vec{n})} g(\vec{n})$$

$$g(\vec{u} + \Omega\vec{m}) = e^{i\psi(\vec{m})} e^{-i\pi\vec{m}\cdot\Omega\cdot\vec{m} - 2\pi i\vec{m}\cdot\vec{u}} g(\vec{u}) \quad . \tag{4.73}$$

Since $\phi(\vec{n} + \vec{m}) = \phi(\vec{n}) + \phi(\vec{m})$ and the same for $\psi(\vec{m})$, we must have $\phi(n) = \vec{a}\cdot\vec{n}$, $\psi(\vec{m}) = -\vec{m}\cdot\vec{b}$. Recalling the definition of ϑ functions with characteristics (2.37, 2.38) we obtain

$$g(\vec{u}) = \text{const.}\ \vartheta\begin{bmatrix}\vec{a}\\ \vec{b}\end{bmatrix}(\vec{u}|\Omega) \quad \cdot \tag{4.74}$$

for some real characteristics $\vec{a}$, $\vec{b}$.

To fix $\vec{a}$, $\vec{b}$ we use Riemann's theorem, stated below (2.49). The operator (4.70) has a zero mode, and hence $g(u) = 0$, whenever $\bar{\partial}$ coupled to $L_\alpha \otimes V(-u)$ does. Suppose we choose L_α to be L_0, the preferred spin structure for the given homology basis. Then we have a zero mode precisely when $-u$ (or u) lies on the theta divisor, or vanishing set of ϑ . Hence for this parametrization the characteristics $a = b = 0$ in (4.76), so

$$\det_{L_0} D^\dagger(\vec{u})D(\vec{u}) = |c|^2\, e^{i\pi(u-\bar{u})(\Omega-\bar{\Omega})^{-1}(u-\bar{u})} |\vartheta(\vec{u}|\Omega)|^2 \quad . \tag{4.75}$$

Here the subscript on det means that spin structures are parametrized relative to L_0 by $\vec{u}$. $|c|^2$ is a real function of the moduli which can be determined by matching conformal anomalies and using bosonization [34] or by more rigorous methods using algebraic geometry[43] to be

$$|c|^2 = \left(\frac{\det' - \Delta_0}{\int \sqrt{g}\, \det \mathrm{Im}\, \Omega} \right)^{-\frac{1}{2}} \tag{4.76}$$

up to an overall constant. In the convenient notation (2.37),

$$\det D^{\dagger}\left[\begin{matrix} \alpha \\ \beta \end{matrix} \right] D\left[\begin{matrix} \alpha \\ \beta \end{matrix} \right] = \text{const.} \left(\frac{\det' - \Delta_0}{\int \sqrt{g}\, \det \mathrm{Im}\, \Omega} \right)^{-\frac{1}{2}}$$

$$\times \left| \vartheta\left[\begin{matrix} \alpha \\ \beta \end{matrix} \right] (0|\Omega) \right|^2 \tag{4.77}$$

where the left hand side means we parametrize using an L_α related to L_0 by $u = \Omega\alpha + \beta$.

This formula can be used to check explicitly part of Witten's argument[31] about the modular invariance of the heterotic (and other fermionic) strings[34]. This argument proceeds in two steps. Take the chiral square root of the set of determinants (4.38) and include also the Weyl determinants for the left movers. In the first step Witten showed that, after setting the spin structures equal, the resulting expression is invariant under modular transformations preserving the chosen spin structure. This cannot be checked explicitly with (4.77) because we would also need the spin structure dependence of the Rarita-Schwinger operator. The second step however involves showing that

$$\left(\frac{\det_\alpha \nabla_z^{-\frac{1}{2}}}{\det_\beta \nabla_z^{-\frac{1}{2}}} \right)^4 \tag{4.78}$$

is invariant under diffeomorphisms preserving both α and β. This can be checked explicitly using (2.49). The interested reader is referred to Ref. 34 for the details.

In the next section we will rephrase some of the results presented in this section using more of the analytic properties of $\mathcal{M}_g$.

IV. STRINGS AND ALGEBRAIC GEOMETRY

Let us go back to the measure (4.36)

$$Z = \int_{\Xi} dt^1 \wedge \ldots \wedge dt^{6g-6} \cdot \det(\psi^i, T^j) \cdot \left(\frac{\det' \Delta_0}{\int \sqrt{\gamma}} \right)^{-13} \left(\frac{\det P_1^\dagger P_1}{\det (\psi_i, \psi_j)} \right)^{1/2} \quad . \tag{5.1}$$

The integral was derived by choosing a particular slice Ξ to represent $\mathcal{M}_g$, so that the T^i's are the tangents to the slice as we vary its coordinates t^i. We would like to express (5.1) more intrinsically on $\mathcal{M}_g$ instead of the slice Ξ. This was done in Ref. 6; see also Ref. 9. Since in the critical dimension the combination of determinants in (5.1) is gauge invariant, it should descend to some function on the quotient space $\mathcal{M}_g$. Thus we would like to rewrite (5.1) as:

$$Z = \int_{\mathcal{M}_g} \omega \, F \tag{5.2}$$

where ω is some volume form on $\mathcal{M}_g$ and F is the gauge invariant combination of determinants in (5.1) (excluding det (ψ^i, T^j) which is also gauge invariant) considered as a function on $\mathcal{M}_g$. Note that, ω and F will depend on the chosen basis ψ^i for Ker $P_1^\dagger$ in such a way that ωF does not. We can split the ψ^i's into complex conjugate pairs, involving the quadratic differentials: $(\varphi^i_{zz}, \varphi^i_{\bar{z}\bar{z}})$ and $(i\varphi^i_{zz}, -i\varphi^i_{\bar{z}\bar{z}})$. We also know that the φ^i's can be chosen to vary holomorphically with $\mathcal{M}_g$ because ∇^z_2 depends holomorphically on the moduli. The volume form ω in (5.2) is in fact

$$\omega = \psi^1 \wedge \ldots \wedge \psi^{6g-6} = \varphi^1 \wedge \ldots \wedge \varphi^{3g-3} \wedge \bar{\varphi}^1 \wedge \ldots \wedge \bar{\varphi}^{3g-3} \quad . \tag{5.3}$$

This is a volume form, because as explained in Section IVC, the φ's can be regarded as cotangent vectors to $\mathcal{M}_g$.

To see that (5.3) is the correct volume form to reproduce (5.1), note that each ψ^i is also a cotangent to the full space Γ of metrics. Since the T^i are tangents to Ξ dual to the dt^i, we have that $\sum_j (\psi^i, T^j) dt^j$ is the restriction of (5.3) to the slice Ξ. In other words, the volume form in (5.1) is the pullback to Ξ of the form (5.3); since the function being integrated is gauge-invariant we may therefore replace the integral over Ξ by the integral (5.2). See Ref. 9 for more details.

We can thus rewrite (5.1) as:

$$Z = \int_{\mathcal{M}_g} \varphi' \wedge \ldots \wedge \varphi^{3g-3} \wedge \bar{\varphi}^1 \wedge \ldots \wedge \bar{\varphi}^{3g-3} \, (\det \operatorname{Im} \Omega)^{-13} G$$

where

$$G \equiv \left(\frac{\det' \bar{\partial}_K^\dagger \bar{\partial}_K}{\det \operatorname{Im} \Omega \int \sqrt{\gamma}} \right)^{-13} \frac{\det' \bar{\partial}_{K^2}^\dagger \bar{\partial}_{K^2}}{\det(\varphi^i, \varphi^j)} \quad . \tag{5.4}$$

We have multiplied and divided by a power of $\det \operatorname{Im} \Omega$ so that, by the holomorphic factorization theorem, the function G is locally the absolute value squared of a holomorphic function.

We showed in the previous section that the spaces of zero modes of $\bar{\partial}_{K^n}$ i.e. $H^0(\Sigma, K^n)$, $n = 0, 1, \ldots$ vary holomorphically in $\mathcal{M}_g$ i.e. $H^0(\Sigma, K^n)$ is a holomorphic vector bundle over moduli space. There are two of these which are of special interest for the bosonic string, namely $H^0(\Sigma, K)$ (abelian differentials) and $H^0(\Sigma, K^2) = T^*_{(1,0)} \mathcal{M}_g$. Given a holomorphic vector bundle, its highest exterior power:

$\lambda H^0(\Sigma,K^n) \equiv \overset{\max}{\wedge} H^0(\Sigma,K^n)$ defines a holomorphic line bundle on $\mathcal{M}_g$. In particular $\lambda H^0(\Sigma,K^1) \equiv E$ is known as the Hodge line bundle, and a section of it is given by $\omega_1 \wedge \ldots \wedge \omega_g$, $\omega_i \in H^0(\Sigma,K)$; and $\lambda H^0(\Sigma,K^2) \equiv \mathcal{K}$ is the canonical line bundle of $\mathcal{M}_g$. A section of it is $\varphi^1 \wedge \ldots \wedge \varphi^{3g-3}$. A more trivial example is of course $H^0(\Sigma,K^0)$. The only section is a constant function, say 1, and it is the holomorphic section on $\mathcal{M}_g$ of the trivial line bundle $H^0(\Sigma,K^0)$.

Once we give a metric γ_{ab} on Σ, we can give a metric on $H^0(\Sigma,K^n)$. However only $H^0(\Sigma,K)$ and hence E has a natural norm independent of the specific representative metric γ in a given conformal class. What we would like to have if possible is a natural metric on $\mathcal{K}$ so that we can build a volume element on $\mathcal{M}_g$. That this is possible is a consequence of a theorem of Mumford, which asserts that $\mathcal{K}$ is actually isomorphic to E^{13}. The fact that the measure so defined actually reproduces (5.4) will turn out to be a consequence of holomorphic factorization.

Consider again the problem of giving meaning to a chiral determinant such as $\det \bar{\partial}_\xi$. To this family of operators we can assign a line bundle over $\mathcal{M}_g$ by considering its family of kernels $\ker \bar{\partial}_\xi$. We must also consider the family of vector spaces $\ker \bar{\partial}_{K\otimes\xi^{-1}}$. Physically the reason for doing this is that both sets of kernels enter into the formula (5.4) dictated by physics. Mathematically the holomorphic factorization theorem only works when both sets of kernels are included (though in our proof we assumed that one of them happened to be zero); this ultimately comes down to the fact that while the dimensions of these kernels can jump, still their difference is a constant by the Riemann-Roch theorem.

Thus given the family of operators $\bar{\partial}_\xi$ varying holomorphically on $\mathcal{M}_g$ we can construct a holomorphic line bundle $\mathrm{DET}\,\bar{\partial}_\xi$ on $\mathcal{M}_g$ as follows:

$$\mathrm{DET}\,\bar{\partial}_\xi = \lambda H^0(\Sigma,\xi)^{-1} \otimes \lambda H^0(\Sigma,K\otimes\xi^{-1})^{-1} \quad . \tag{5.5}$$

When $\deg \xi > 2g-2$ the second term on the right-hand side is absent and we are left with $\lambda H^0(\Sigma,\xi)^{-1}$.

To compute the first Chern class of (5.5), we can choose a metric for DET $\bar\partial_\xi$, take a holomorphic section $s : \mathcal{M}_g \to \mathrm{DET}\ \bar\partial_\xi$ and compute

$$c_1(\mathrm{DET}\ \bar\partial_\xi) = \frac{1}{2\pi i}\,\partial\bar\partial \log \| s \|^2 \tag{5.6}$$

where $\| s \|^2$ is the metric on the section. (Recall that (5.6) is the curvature of the hermitian connection on DET $\bar\partial_\xi$ given by the metric.) A clever choice of metric can simplify computations substantially. For determinant line bundles the natural choice was given by Quillen (see Ref. 33). Given a basis φ^i of $H^0(\Sigma,\xi)$ and a basis ψ^a of $H^0(\Sigma,K\otimes\xi^{-1})$, we can construct a section of DET $\bar\partial_\xi$ as:

$$(\varphi^1 \wedge \dots \wedge \varphi^n)^{-1} \otimes (\psi^1 \wedge \dots \wedge \psi^m)^{-1}$$

$$n = \dim H^0(\Sigma,\xi), \qquad m = \dim H^0(\Sigma,K\otimes\xi^{-1}) \quad . \tag{5.7}$$

Quillen's norm on the section (5.7) is defined by

$$\| (\varphi^1 \wedge \dots \wedge \varphi^n)^{-1} \otimes (\psi^1 \wedge \dots \wedge \psi^m)^{-1} \|_Q^2 \equiv \frac{\det'\ \bar\partial_\xi^\dagger\, \bar\partial_\xi}{\det(\varphi^i,\varphi^j)\ \det(\psi^a,\psi^b)} \quad . \tag{5.8}$$

Since (5.6) amounts to taking holomorphic and antiholomorphic variations of the logarithm of (5.8) in moduli space, we have already computed the curvature of $\bar\partial_\xi$ in Section IVC. It is precisely the obstruction to holomorphic factorization.

In order to get a natural norm on $\mathcal{K}$, we can now think of $\mathcal{K}$ and E in terms of determinant line bundles:

$$\mathcal{K} = (\mathrm{DET}\ \bar{\partial}_{K^2})^{-1} \quad , \quad E = (\mathrm{DET}\ \bar{\partial}_K)^{-1} \quad . \tag{5.9}$$

The holomorphic factorization theorem of Section IVC now says that $c_1(\mathcal{K} \otimes E^{-13}) = 0$. Thus the bundle $\mathcal{K} \otimes E^{-13}$ is flat, and with a bit more argument we can conclude that $\mathcal{K} \simeq E^{13}$. This is the result of Mumford mentioned earlier [44]. Unfortunately, even though these two bundles are isomorphic we do not yet know what isomorphism to use. There might be several possibilities. Uniqueness follows from a rather remarkable property of $\mathcal{M}_g$. Any analytic function on $\mathcal{M}_g$ is a constant, at least for genus $g \geq 3$ [45]. Now suppose that there were two global trivializations η, η' of $\mathcal{K} \otimes E^{-13}$, i.e. two holomorphic sections everywhere non-zero and finite. Then $\eta^{-1} \otimes \eta'$ would be a regular analytic function and by the result quoted, we conclude that η is unique up to a constant. As a side remark we should mention that a similar argument plays a central role in the proof of bosonization in Riemann surfaces of genus $g > 1$ for arbitrary spin[8].

Thus the section:

$$\Sigma = \varphi^1 \wedge \ldots \wedge \varphi^{3g-3} \otimes (\omega^1 \wedge \ldots \wedge \omega^g)^{-13} \tag{5.10}$$

of (5.9) is unique up to a constant. To construct a metric on $\mathcal{K}$ and a volume form on $\mathcal{M}_g$, we take $\Sigma \otimes \bar{\Sigma}$ and take the natural Hodge norm on the abelian differentials. This leaves behind a section $\|\Sigma\|_E^2$ of $\mathcal{K} \otimes \bar{\mathcal{K}}$. Hence $\|\Sigma\|_E^2$ is an object of the sort we were looking for, a volume form on moduli space. We can write:

$$\|\Sigma\|_E^2 = \frac{\zeta \wedge \bar{\zeta}}{(\det \mathrm{Im}\,\Omega)^{13}} \quad . \tag{5.11}$$

Note that while the left side of (5.11) is globally well-defined on moduli space, the holomorphic form ζ on the right is not. This had to happen because the factor $\omega' \wedge \ldots \wedge \omega^g$ in (5.10) cannot be chosen to be single-valued by itself; if it could then E would be trivial. Only the combination (5.10) is single-valued. Hence ζ is a holomorphic form on Teichmüller space, not on $\mathcal{M}_g$. Moreover, since we know explicitly the modular transformation properties of $\det \mathrm{Im}\,\Omega$ we find that in fact ζ is a modular form. See e.g. Ref. 46.

The fact that Σ lives on $\mathcal{M}_g$ now makes it clear that the bosonic string is conformal and modular invariant. Furthermore, questions concerning the infinities of the bosonic string are equivalent to questions about the divisor of the holomorphic section Σ as we approach the boundary of moduli space.

Finally we should complete the argument by showing that (5.11) coincides with the measure (5.4). Using holomorphic factorization again, G is the absolute value squared of a holomorphic function. Since $G = |f|^2$ and f is holomorphic in $\mathcal{M}_g$, then it has to be a constant. The rest of (5.4) is just $\|\Sigma\|_E^2$, as desired.

Part of the power of the analytic methods introduced in this section is that they extend naturally to the boundary of moduli space, i.e. to the pinched Riemann surfaces described earlier. We do not have room here for a detailed explanation of this subject (see Ref. 9). The point however is that when we add to $\mathcal{M}_g$ additional points corresponding to pinched surfaces, then the resulting $\bar{\mathcal{M}}_g$ is again a complex space, and it is compact. The new points sit in $\bar{\mathcal{M}}_g$ as a divisor Δ, and nothing unusual happens to the complex structure there. Thus we can refine our previous calculations, finding for example that the completed bundle $K \otimes E^{-13}$ is nontrivial on $\bar{\mathcal{M}}_g$, with a divisor given by -2Δ. This in turn implies the same singularity for the section Σ as was computed by more laborious methods in (4.67).

Similar considerations at infinity can be combined with nonexistence theorems for modular forms to show that the cosmological constant vanishes for the heterotic string in flat spacetime [46]. Finally the analytic structure of the boundary of moduli space plays a key role in the program of Friedan and Shenker [47].

Still other aspects of Riemann surface theory are finding applications in string theory. Of these we mention only the use of ideas from Arakelov and Faltings (see [7], [8], and references therein) to obtain various formulas for determinants and to relate conformal field theories of different spin. The full generalization of these analytical methods to supersymmetric systems is still just beginning.

ACKNOWLEDGEMENTS

We have benefited greatly from discussions with many colleagues. We would especially like to thank J.-B. Bost, A. Cohen, P. Ginsparg, J. Harris, D. Kazhdan, G. Moore, D. Mumford, J. Polchinski, I. Singer, C. Vafa and S. Wolpert for many hours of exposition and conversation. L.A.G. would like to thank the organizers of this meeting and in particular Prof. Salam for their kind hospitality. This work was supported in part by the Harvard Society of Fellows and by NSF grant PHY85-15249.

REFERENCES

[1] For reviews and detailed references see J. Schwarz, "Superstrings: The First Fifteen Years", World Scientific (1985).

[2] A.M. Polyakov, Phys. Lett. 103B (1981) 207; 211.

[3] D. Friedan, "Introduction to Polyakov's string theory" in "Recent Advances in Field Theory and Statistical Mechanics", J.B. Zuber and R. Stora eds., Les Houches 1982, Elsevier (Amsterdam, 1984);
O. Alvarez, Nucl. Phys. B216 (1983) 125;
J. Polchinski, Comm. Math. Phys. 104 (1986) 37;
G. Moore and P. Nelson, Nucl. Phys. B266 (1986) 58;
E. D'Hoker and D. Phong, Nucl. Phys. B269 (1986) 205.

[4] See for example G. Moore, P. Nelson and J. Polchinski, Phys. Lett. 169B (1986) 47;
E. D'Hoker and D. Phong, Nucl. Phys. B278 (1986) 225.

[5] F. Gliozzi, J. Scherk and D.I. Olive, Nucl. Phys. B122 (1977) 253.

[6] A. Belavin and V. Knizhnik, Phys. Lett. B168 (1986) 201; "Complex Geometry and Theory of Quantum Strings," Landau Inst. preprint submitted to ZETF.
J. Bost and J. Jolicoeur, Phys. Lett. B174 (1986) 273;
R. Catenacci, M. Cornalba, M. Martinelli and C. Reina, Phys. Lett. B172 (1986) 328.

[7] Yu.I. Manin, JETP Lett. 43 (1986) 204;
A.A. Beilinson and Yu.I. Manin, to appear in Comm. Math. Phys.

[8] L. Alvarez-Gaumé, J.B. Bost, G. Moore, P. Nelson and C. Vafa, Phys. Lett. B178 (1986) 41, and in preparation.

[9] P. Nelson, "Lectures on Strings and Moduli Space", to appear in Phys. Reports.

[10] See for example H. Farkas and I. Kra, "Riemann Surfaces", Springer-Verlag (1980).

[11] See for example, M. Mulase, Proc. Japan Acad. 59 (1983) 285; J. Diff. G. 19 (1984) 403.

[12] L.V. Ahlfors, "The complex analytic structure of the space of complex Riemann surfaces" eds. R. Nevanlina et al., in "Analytic Functions", Princeton U. Press (1960).

[13] L. Bers, "Quasiconformal mappings and Teichmüller's theorem" in "Analytic Functions", eds. Nevanlina et al., Princeton U. Press (1960) and in Bull. Am. Math. Soc. 5 (1981) 131 (new series).

[14] J. Birman, "Links, Braids and Mapping Class Groups", Princeton U. Press (1974).

[15] W. Magnus, A. Karras and D. Solitar, "Combinatorial Group Theory", Interscience (1966).

[16] E. Witten in "Symposium on Anomalies, Geometry and Topology", eds. W.A. Bardeen and A. White, World Scientific (1985).

[17] J. Milnor and J. Stasheff, "Characteristic Classes", Princeton U. Press (1974).

[18] See for example R. Gunning, "Lectures on Riemann Surfaces", Princeton U. Press (1966).

[19] D. Mumford, "Tata Lectures on Theta", Birkhauser (1983), (2 vols.).

[20] F. Hirzebruch, "Topological Methods in Algebraic Geometry", Springer-Verlag (1966).

[21] M.F. Atiyah, Ann. Sci. de l'Ecole N.S. 4 (1971).

[22] J. Igusa, "Theta Functions", Springer-Verlag (1972).

[23] D. Hejhal, "The Selberg Trace Formula for PSL(2, ℝ), Lecture Notes in Mathematics 47, Springer-Verlag (1974).

[24] See E. D'Hoker and D. Phong, Comm. Math. Phys. 104 (1986) 537, and references therein.

[25] D. Gross, J. Harvey, E. Martinec and R. Rohm, Phys. Rev. Lett. 54 (1985) 502; Nucl. Phys. B256 (1985) 253, B267 (1986) 75.

[26] L. Alvarez-Gaumé, P. Ginsparg, G. Moore and C. Vafa, Phys. Lett. 171B (1986) 155;
L. Dixon and J. Harvey, "String Theories in Ten Dimensions without Space-time Supersymmetry", Princeton Preprint (1986).

[27] L. Brink and H.B. Nielsen, Phys. Lett. 45B (1973) 332.

[28] See Ref. 5.

[29] M. Green and J. Schwarz, Phys. Lett. 109B (1982) 444.

[30] C. Vafa, "Modular Invariance and Discrete Torsion on Orbifolds", Nucl. Phys. B273 (1986) 592.

[31] See E. Witten in Ref. 16.

[32] E. Witten, Comm. Math. Phys. 100 (1985) 197.

[33] D. Quillen, Funct. Anal. Appl. 19 (1986) 31;
J. Bismut and D. Freed: Comm. Math. Phys. 106 (1986) 159.

[34] L. Alvarez-Gaumé, G. Moore and C. Vafa, Comm. Math. Phys. 106 (1986) 1.

[35] L. Alvarez-Gaumé and E. Witten, Nucl. Phys. B234 (1984) 269.

[36] E. Gava, R. Iengo, T. Jayaraman and R. Ramachandran, Phys. Lett. 168B (1986) 207.

[37] W. Abikoff, "The Real Analytic Theory of Teichmüller Space", Springer-Verlag (1980).

[38] S. Wolpert, Am. J. Math., 107 (1985) 969.

[39] S. Wolpert, Maryland Math. preprint MD86-10-SAW.

[40] A. Cohen, G. Moore, P. Nelson, J. Polchinski, Nucl. Phys. B267 (1986) 143.

[41] J. Fay, "Theta Functions on Riemann Surfaces", Lecture Notes in Mathematics 352, Springer-Verlag (1973).

[42] H. Masur, Duke Math. J. 43 (1976) 623.

[43] J.-B. Bost and P. Nelson, Phys. Rev. Lett. 57 (1986) 795.

[44] D. Mumford, L'Ens. Math. 23 (1977) 39.

[45] J. Harris in "Proceedings of the International Congress of Mathematicians", Warzawa, Poland, eds. C. Olech and Z. Ciesielski, Elsevier (Amsterdam, 1984).

[46] G. Moore, J. Harris, P. Nelson and I. Singer, Phys. Lett. 178B (1986) 167.

[47] D. Friedan and S. Shenker, "The Analytic Geometry of Quantum String", Phys. Lett. B175 (1986) 287; Chicago preprint EFI-86-18A.

HARMONIC SUPERSPACE IN ACTION: GENERAL N=2 MATTER SELF-COUPLINGS

A. Galperin[x], E. Ivanov[xx], V. Ogievetsky[xx] and E. Sokatchev[xxx]

[x] Institute of Nuclear Physics, Tashkent, USSR

[xx] Laboratory of Theoretical Physics, Joint Institute for Nuclear Research, Dubna, USSR

[xxx] Institute for Nuclear Research and Nuclear Energy, Sofia, Bulgaria

ABSTRACT

An introduction to the harmonic superspace approach to extended supersymmetry is presented. Its basic ideas and techniques are applied to construct the general N=2 matter hypermultiplet self-coupling $\mathcal{L}^{(+4)}$. The latter provides us with a "hyper-Kähler potential" that supposedly generates all hyper-Kähler metrics via sigma-model in a physical bosons sector. Harmonic supergraphs for hypermultiplets are developed and a simple proof of the d=2 hyper-Kähler sigma-models finiteness is given.

CONTENTS

I. INTRODUCTION

"... It is notoriously difficult to have self-interactions for the hypermultiplet " /1/. This opinion precisely reflects the 1983 status of affairs with the simplest representation of N=2 supersymmetry (SUSY) - matter hypermultiplet /2/ . The latter contains on-shell four scalars and a Dirac spinor. On these physical fields SUSY algebra is generally realized non-linearly and is closed only modulo equations of motion. Thus dealing with physical fields one has to guess simultaneously both SUSY transformations and an invariant action consistent with them. Resolution of this dilemma (common to all SUSY theories) is to separate unknowns. First, one has to find an off-shell representation for hypermultiplet. Here SUSY is manifest. It is realized linearly and does not depend on any particular action. Second, one has to look for actions invariant under SUSY and propagating the physical fields.

A high road for manifest SUSY is superspace. Here off-shell representations are realized via superfields and invariant actions are written down as integrals over superspace. An important point is that there is a variety of superspaces associated with given SUSY. Which N=2 superspace is appropriate to describe hypermultiplet we are inquiring about?

Until 1984 two well-known N=2 superspaces were tried: a real $\mathbb{R}^{4|8}$ and a chiral $\mathbb{C}^{4|4}$ ones (here $m|n$ means m bosonic and n fermionic coordinates). There were some partial successes in this direction, the N=2 tensor multiplet /3/ and the relaxed hypermultiplet /4/ theories. All these multiplets contain a finite number of auxiliary fields. Enjoying virtue of manifest N=2 SUSY these descriptions of hypermultiplet have some serious drawbacks, however. They do not allow a number of couplings that are known to be possible on-shell, e.g. their self-couplings are too restricted, the N=2 tensor multiplet cannot be coupled minimally to Yang-Mills, while the relaxed hyper-multiplet cannot be coupled to Yang-Mills in a complex representation etc. These peculiarities have analogs in N=1 SUSY. There matter can be described either by chiral multiplet or by N=1 tensor one. Unlike the former, the latter cannot be coupled minimally to N=1 Yang-Mills, its self-couplings are equivalent to a restricted class of those of the former etc. It is the chiral multiplet that provides the "ultimate" (most versatile) description of matter in N=1 SUSY. Recall that in superspace it is simply an unconstrained superfield on $\mathbb{C}^{4|2}$. What are the N=2 analogs of chiral multiplet and of $\mathbb{C}^{4|2}$?

The questions stated above were resolved in 1984 with advent of N=2 harmonic superspace /5/ . This is an extension of $\mathbb{R}^{4|8}$ by adding to it coordinates of coset SU(2)/U(1)

$$\mathbb{HR}^{4+2|8} = \mathbb{R}^{4|8} \otimes SU(2)/U(1)$$

where SU(2) is the automorphism group inherent to N=2 SUSY. There is a remarkable quotient of $\mathbb{HR}^{4+2|8}$ called analytic superspace $\mathbb{AR}^{4+2|4}$. It is real and is parametrized by 4 (Minkowski) + 2 (SU(2)/U(I)) bosonic and 4 fermionic coordinates. The most important thing to be memorized is that all N=2 supersymmetric theories (hyper-multiplet, Yang-Mills, Einstein and Weyl supergravities) have the "ultimate" formulation in $\mathbb{AR}^{4+2|4}$. For example, the hypermultiplet is described by an unconstrained superfield $q^+(\zeta,u)$ on the analytic superspace, it allows general self-couplings, it can be coupled to Yang-Mills in any representation, etc. This description of hyper-multiplet naturally involves an infinite number of auxiliary fields (due to extra coordinates of SU(2)/U(1)).

A "no-go" theorem now exists [6] that states impossibility to maintain simultaneously a finite number of auxiliary fields and an essential property of hypermultiplet to have four physical scalars as a complex doublet of SU(2). This explains why the ultimate hypermultiplet could not be invented in component or in ordinary N=1 or N=2 superspace approaches.

The present lectures do not contain presentation of N=2 Yang-Mills and supergravity theories in harmonic superspace [5,7,8]. They discuss neither off-shell unconstrained N=3 Yang-Mills theory [9] nor recent application of light-cone harmonic superspace to ten-dimensional super-Yang-Mills theory [10]. Instead we shall concentrate here on N=2 matter and its general self-couplings. According to a beautiful theorem [11] these self-couplings lead in the physical scalars sector to nonlinear σ-models with hyperKähler target manifolds. HyperKahler geometry is a branch of Kähler geometry (related to N=1 σ-models [12]) which itself is a branch of Riemannian geometry (related to N=0 σ-models)[x]. While the "prepotentials" for Riemannian and Kähler geometries are well-known (they are the metric tensor and the Kahler potential correspondingly) nothing of this sort has been found for hyperKähler case. We conjectered that the sought "hyperKähler potential" is just the Lagrangian density $\mathcal{L}^{(+4)}$ for hypermultiplets in analytic superspace [14-16].

All the N=2 matter self-couplings enjoy another interesting feature when reduced to d=2 dimensions. There they are ultraviolet finite. This fact follows almost trivially from our unconstrained manifestly N=2 supersymmetric formalism. We observe that the classical action is an integral over $\mathbb{A}\mathbb{R}^{4+2|4}$ while the quantum corrections are necessarily integrals over $\mathbb{H}\mathbb{R}^{4+2|8}$ and then apply

x) For readable reviews on supersymmetric σ-models see [13].

simple power counting /7/.

Harmonic superspace also allows one to describe succinctly the finite component formulations of the hypermultiplet mentioned above. There the hypermultiplet appears via constrained (and sometimes gauge) analytic superfields. With the aid of N=2 duality transformations we show that the corresponding self-couplings are equivalent to a restricted class of those for unconstrained formulation /16,17/ .

Such are the main themes of these lectures. Their organization is detailed in the Contents.

2. COSET SPACES FOR SUPERSYMMETRIES

2.1. Superalgebra of N-Extended Poincare Supersymmetry contains the Poincare subalgebra consisting of the 4-translations P_a and Lorentz transformations L_{ab}

$$[P_a, P_b] = 0 \quad , \quad [P_a, L_{bc}] = i(\eta_{ab}P_c - \eta_{ac}P_b)$$

$$[L_{ab}, L_{cd}] = i(\eta_{ad}L_{bc} + \eta_{bc}L_{ad} - \eta_{ac}L_{bd} - \eta_{bd}L_{ac})$$

and includes also the spinor generators Q^i_α , $\bar{Q}_{\dot{\alpha}i}$ subjected to the anticommutation relations (we use the two-component spinor formalism, $\alpha, \dot{\alpha} = 1,2$; $i = 1,\dots,N$)

$$\{Q^i_\alpha, \bar{Q}_{\dot{\alpha}j}\} = 2\delta^i_j(\sigma^a)_{\alpha\dot{\alpha}}P_a \, , \; \{Q^i_\alpha, Q^j_\beta\} = \varepsilon_{\alpha\beta}Z^{ij} \, , \; \{\bar{Q}_{\dot{\alpha}i}, \bar{Q}_{\dot{\beta}j}\} = \varepsilon_{\dot{\alpha}\dot{\beta}}\bar{Z}_{ij}$$

Here Z^{ij} , $\bar{Z}_{ij}$ are central charge operators which commute with all the other generators and among themselves. In what follows, we shall basically consider the models with zero central charges. It will be convenient to add to this superalgebra the generators of its automorphism group SU(N) acting on indices i of spinor generators.

2.2. Coset Spaces Generalities. As has been said in the Introduction, we are interested in manifestly invariant realizations of N-extended SUSY. An adequate framework for such realizations is provided by the coset space method. We begin by recalling its basics.

An action of group G (having generators X_i, $i=1,\dots,n$; Y_a, $a=1,\dots,m$) in the coset space G/H, H being a subgroup of G with generators X_i, is defined as follows. To each generators Y_a, one puts in correspondence a coordinate ξ^a and forms an exponential

$$\Omega = \exp\{i\,\xi^a Y_a\}$$

(other parameterizations of Ω are equally admissible). An arbitrary element of G in a vicinity of the identity element can be uniquely divided into the product of G/H and H-factors:

$$g = \exp\{i c^a Y_a\} \ \exp\{i\lambda^k X_k\}$$

Then the left action of G on the coset G/H is defined according to

$$G: \qquad g\cdot\exp\{i\xi^a Y_a\} = \exp\{i\,\xi'^a(\xi,g)Y_a\}\cdot h(\xi,g) \qquad (2.1)$$

where $h(\xi,g) = \exp\{i f^i(\xi,g)X_i\}$ takes values in the subgroup H and the composite parameters $f^i(\xi,g)$, depend both on the group element g and coordinates ξ^a. The representation (2.1) can easily be derived using the Baker-Campbell-Haussdorf formula [x)] and the commutation relations of G-algebra:

$$[X,X]\propto X, \quad [X,Y]\propto Y, \quad [Y,Y]\propto X+Y$$

---

x) $e^A e^B = \exp\{A+B+\frac{1}{2}[A,B]+\frac{1}{12}[A,[B,[B,A]]]+\dots\}$

Eq. (2.1) is the basic formula of group realizations in coset spaces. Apart from the transformation law of coordinates ξ^a it also defines the transformations of covariant fields given on G/H. The latter are naturally classified according to the irreducible representations of subgroup H. Some field $\psi_k(\xi)$ (k is an external index of H-representation) transforms under general G-transformations just as in H, but with composite parameters $f^i(g,\xi)$:

$$G: \ \psi'_k(\xi') = \left(e^{if^i(g,\xi)X_i}\right)_{k\ell} \psi_\ell(\xi) \tag{2.2}$$

It follows from (2.1) that

$$\begin{aligned}(\Omega^{-1}d\Omega)' &= h\,\Omega^{-1}d(\Omega h^{-1}) = \\ &= h(\Omega^{-1}d\Omega)h^{-1} + h\,dh^{-1}\end{aligned} \tag{2.3}$$

Decomposing this G-algebra valued quantity in generators of G :

$$\Omega^{-1}d\Omega = i\left(\omega^a Y_a + \omega^i X_i\right), \tag{2.4}$$

We observe that the 1-forms ω^a transform homogeneously with respect to the left action of G :

$$\omega^{a\prime} Y_a = h\,\omega^a Y_a h^{-1}$$

These 1-forms can be interpreted as covariant differntials of the coset parameters ξ^a. The inhomogeneously transforming 1-forms ω define the H-connection in the coset space G/H. Indeed, the quantity

$$(D\psi)_k = \left[(d + i\omega^i X_i)\psi\right]_k \tag{2.5}$$

transforms homogeneously in G, just as the field ψ_k itself, and can thus be called the covariant differential of ψ_K. Correspondingly, the covariant derivative of ψ_k is defined as a coefficient of $\omega^a(\xi, d\xi)$ in $(D\psi)_K$:

$$(D\psi)_K \equiv \omega^a (D_a \psi)_K \tag{2.6}$$

It is a simple exercise to check that $(D_a\psi)_K$ transforms again in accord with the general law (2.2).

In what follows, the reader will have many opportunities to apply these simple and general rules.

2.3. Coset Spaces of Poincaré and Super-Poincaré Groups. The first example is the realization of Poincare group in Minkowski space. H is now the Lorentz group. Then the coset parameters are familiar coordinates X^a, $a = 0,1,2,3$:

$$\frac{\{P_a, L_{bc}\}}{\{L_{bc}\}} = (X^a)$$

The reader can easily be convinced that (2.1) in this case yields ordinary Poincare rotations and translations of coordinates X^a, and $\omega^a = dX^a$, $\omega^i = 0$, i.e. the covariant derivatives coincide with the usual ones.

Analogously, one may realize N-extended SUSY in proper coset spaces, e.g. in

$$\mathbb{R}^{4|4N} = \frac{\{P_a, L_{bc}, Q^i_\alpha, \bar{Q}_{\dot{\alpha}i}, SU(N)\}}{\{L_{bc}, SU(N)\}} = (X^a, \theta_{\alpha i}, \bar{\theta}^i_{\dot{\alpha}})$$

where X^a and $\theta_{\alpha i}, \bar{\theta}^i_{\dot{\alpha}}$ are even and odd coordinates associated with translations and supertranslations.

This choice is not unique, there exists, e.g. the chiral superspace

$$\mathbb{C}^{4|2N} = \frac{\{P, L, Q, \bar{Q}, SU(N)\}}{\{L, SU(N), \bar{Q}\}} = (x_L^a, \theta_{\alpha i})$$

Here the stability subgroup H contains spinor generators $\bar{Q}_{\dot{\alpha} i}$ and is thus a supergroup.

We see that N-extended SUSY can be realized in different coset spaces. From a mathematical point, of view, all these realizations are equally admissible. Nevertheless, only some of them prove to be interesting for physics.

First of all, for manifest Lorentz covariance their stability subgroup should include the Lorentz group (in other words, acceptable superspaces should contain Minkowski space as an even subspace). In the case of N=1, this requirement leaves only two possibilities [x)]

$$\mathbb{R}^{4|4} = \frac{\{P_a, L_{ab}, Q_\alpha, \bar{Q}_{\dot{\alpha}}\}}{\{L_{ab}\}} = (x^a, \theta_\alpha, \bar{\theta}_{\dot{\alpha}})$$

$$\mathbb{C}^{4|2} = \frac{\{P_a, L_{ab}, Q_\alpha, \bar{Q}_{\dot{\alpha}}\}}{\{L_{ab}, \bar{Q}_{\dot{\alpha}}\}} = (x_L^a, \theta_\alpha)$$

---

x) Note that $\mathbb{R}^{4|4}$ can be regarded as a real hypersurface in $\mathbb{C}^{4|2}$. In the flat case, it can be easily seen from the possibility to rearrange the relevant coset exponential as follows

$$e^{i(x^a P_a + \theta^\alpha Q_\alpha + \bar{\theta}_{\dot{\alpha}} \bar{Q}^{\dot{\alpha}})} = e^{i(x^a P_a + \theta^\alpha Q_\alpha)} e^{i\bar{\theta}_{\dot{\alpha}} \bar{Q}^{\dot{\alpha}}}$$

(prove that $x_L^a = x^a + i\theta\sigma^a\bar{\theta}$).

Both have been utilized in N=1 SUSY: the prepotentials of N=1 Yang–Mills and supergravity theories $V(x,\theta,\bar\theta)$ and $H^m(x,\theta,\bar\theta)$ are unconstrained superfields in $\mathbb{R}^{4|4}$ while the parameters of their gauge groups and the basic N=1 matter superfields (chiral superfields) are defined in $\mathbb{C}^{4|2}$. As we shall see later, for N > 1 the number of possible superspaces increases and it becomes not so easy to single out an adequate superspace.

The N=1 chiral superfields can equally be looked upon as complex functions on $\mathbb{R}^{4|4}$ subject to the differential constraint

$$\bar D_{\dot\alpha}\,\Phi(x,\theta,\bar\theta) = 0 \tag{2.7}$$

where $D_\alpha = \frac{\partial}{\partial\theta^\alpha} + i\bar\theta^{\dot\alpha}\not\partial_{\alpha\dot\alpha}$, $\bar D_{\dot\alpha} = -\frac{\partial}{\partial\bar\theta^{\dot\alpha}} - i\theta^\alpha\not\partial_{\alpha\dot\alpha}$ $(\not\partial = \sigma^a\partial_a)$ are covariant spinor derivatives. (We suggest to the reader to derive the latter using general construction (2.5), (2.6) and the structure relations of N=1 supersymmetry algebra). The constraint (2.7) is nothing else than a kind of the Grassmann analyticity condition /18/ whose solution is an unconstrained holomorphic function over $\mathbb{C}^{4|2}$, i.e. just the chiral superfield:

$$\Phi(x,\theta,\bar\theta) = \varphi(x_L,\theta)$$

2.4. <u>N=1 Matter Self-Couplings</u>. As is well known, chiral superfield $\varphi(x_L,\theta)$ describes the matter multiplet composed of a complex scalar field $\varphi(x)$, Weyl spinor $\psi_\alpha(x)$ and, finally, a complex auxiliary field $F(x)$

$$\varphi(x_L,\theta) = \varphi(x_L) + \theta^\alpha\psi_\alpha(x_L) + \theta\theta F(x_L) \tag{2.8}$$

The free action of φ has a familiar appearance:

$$S^{free} = \int d^4x\, d^4\theta\ \varphi\bar\varphi \tag{2.9}$$

Most general action for n self-interacting superfields φ^i, $\bar\varphi^i$ which yields the second-order equations for bosons and the first-order ones for formions (supersymmetric σ-model) is as follows

$$S^{gen} = \int d^4x d^4\theta\, K(\varphi,\bar\varphi) + \int d^4x_L d^2\theta\, P(\varphi) + h.c. \tag{2.10}$$

where $K(\varphi,\bar\varphi)$, $P(\varphi)$ are general functions of their arguments. The only restriction to be fulfilled by $K(\varphi,\bar\varphi)$ is the positive definiteness of matrix $g_{ij} = \partial^2 K/\partial\varphi^i\partial\bar\varphi^j |_{\theta=\bar\theta=0}$. In eq. (2.10), $P(\varphi)$ plays the role of a potential. After passing to components and eliminating auxiliary fields the action (2.10) is represented as

$$S = \int d^4x \;\; g_{ij}(\varphi,\bar\varphi)\,\partial^a\varphi^i\partial_a\bar\varphi^j + \dots \tag{2.11}$$

Zumino /12/, Alvarez-Gaume and Freedman /11/, Bagger and Witten /19/ have discovered a remarkable correlation between a sort of SUSY and admissibly classes of metrics in the relevant σ-models. While in the case of N=0 any Riemannian metrics suits, in the case of N=1 the metrics only of the type (2.11) are possible. These are Kahler metrics. In other words, not all the σ-models can be N=1 supersymmetrized but only those constructed on an important subclass of Riemannian manifolds, the Kähler ones /12/.

N=2 SUSY imposes further more severe restrictions on a possible choice of σ-model manifolds /11,19/.

After recalling these well-known facts concerning N=1 matter we pass to the description of N=2 matter in the framework of harmonic superspace. We begin with discussing possible N=2 superspaces.

2.5. Superspaces for N=2 SUSY. The most straightforward possibilities are associated with $\mathbb{R}^{4|8}$ or $\mathbb{C}^{4|4}$. For instance, $\mathbb{R}^{4|8}$ is identified with the coset

$$\mathbb{R}^{4|8} = \frac{\{P_a, L_{ab}, Q^i_\alpha, \bar{Q}_{\dot{\alpha} i}, SU(2)\}}{\{L_{ab}, SU(2)\}} = (x^a, \theta_{\alpha i}, \bar{\theta}^i_{\dot{\alpha}}) \quad (i=1,2) \tag{2.12}$$

The corresponding covariant spinor derivatives are as follows

$$D^i_\alpha = \frac{\partial}{\partial\theta^\alpha_i} + i\bar{\theta}^{\dot{\alpha} i}\not{\partial}_{\alpha\dot{\alpha}}, \quad \bar{D}_{\dot{\alpha} i} = -\frac{\partial}{\partial\bar{\theta}^{\dot{\alpha} i}} - i\theta^\alpha_i \not{\partial}_{\alpha\dot{\alpha}} \tag{2.13}$$

(the reader can derive them by the general algorithm (2.5), (2.6)). These anticommute with the N=2 SUSY generators and, on their own, form the superalgebra coinciding with that of N=2 SUSY:

$$\{D^i_\alpha, \bar{D}_{\dot{\alpha} j}\} = -2\delta^i_j\, i\not{\partial}_{\alpha\dot{\alpha}}, \quad \{D^i_\alpha, D^j_\beta\} = \{\bar{D}_{\dot{\alpha} i}, \bar{D}_{\dot{\beta} j}\} = 0 \tag{2.14}$$

It turns out that $\mathbb{R}^{4|8}$ does not suit to adequately represent N=2 theories (see below) though, from a pure mathematical standpoint, it is still quite acceptable for realization of N=2 SUSY. The same concerns the chiral superspace $\mathbb{C}^{4|4}$:

$$\mathbb{C}^{4|4} = \frac{(P_a, L_{ab}, Q^i_\alpha, \bar{Q}_{\dot{\alpha} i}, SU(2))}{(L_{ab}, \bar{Q}_{\dot{\alpha} i}, SU(2))} = (x^a_L, \theta_{\alpha i}) \tag{2.15}$$

To obtain an adequate description of N=2 fields, a new type of superspaces should be included into play, just the harmonic ones.

2.6. The N=2 Harmonic Superspace[5] corresponds to placing in the stability subgroup H not the whole automorphism group SU(2) (as in the cases (2.12), (2.15)) but merely its subgroup U(I):

$$\mathbb{R}^{4+2|8} = \frac{(P_a, L_{ab}, Q_\alpha^i, \bar{Q}_{\dot\alpha i}, SU(2))}{(L_{ab}, U(1))} = (x^a, \theta_{\alpha i}, \bar{\theta}^i_{\dot\alpha}, u^\pm_i) \tag{2.16}$$

Here coordinates of a new sort appeared, u^{+i}, $u^-_i = \overline{(u^{+i})}$. These form a 2x2 SU(2)-matrix

$$\begin{pmatrix} u^-_1 & u^+_1 \\ u^-_2 & u^+_2 \end{pmatrix} \tag{2.17}$$

and, accordingly, satisfy the condition[x)]

$$u^{+i} u^-_i = 1 \tag{2.18}$$

Then

$$u^{+i} u^-_j - u^{-i} u^+_j = \delta^i_j \tag{2.19}$$

and, hence

$$f^\pm = f^i u^\pm_i \iff f^i = u^{+i} f^- - u^{-i} f^+ \tag{2.20}$$

Here f^i is same SU(2) doublet and $f^\pm$ are its U(I) projections. Evidently:

$$u^{+i} u^+_i = u^{-i} u^-_i = 0$$

---

x) We use the following SU(2)-conventions

$$\psi^i = \varepsilon^{ij}\psi_j, \quad \psi_j = \varepsilon_{jk}\psi^k, \quad \varepsilon^{ij}\varepsilon_{jk} = \delta^i_k$$

$$\varepsilon^{12} = -\varepsilon^{21} = -\varepsilon_{12} = 1$$

It is clear that the definitions (2.17), (2.18) allow us to consider $u^{\pm}_i$ as SU(2) doublets with respect to the index i and as carrying U(I) charges ± 1. We shall refer to them as harmonic variables according to their origin.

As is implied by U(I)-invariance, these are defined up to a U(I)-transformation

$$u^{\pm}_i \Rightarrow e^{\pm i\alpha} u^{\pm}_i \tag{2.21}$$

This property and the normalization condition (2.18) reduce the number of independent degrees of freedom in $u^{\pm}_i$ just to two. Sometimes, it is useful to know an explicit parametrization of matrix (2.17) in terms of two angles on the sphere:

$$\begin{pmatrix} \cos\theta/2 \cdot e^{-i\varphi/2} & i\sin\theta/2 \cdot e^{-i\varphi/2} \\ i\sin\theta/2 \cdot e^{i\varphi/2} & \cos\theta/2 \cdot e^{i\varphi/2} \end{pmatrix} \quad \begin{matrix} 0 \le \theta \le \pi \\ 0 \le \varphi \le 2\pi \end{matrix} \tag{2.22}$$

The ordinary complex conjugation ($\bar{\ }$) affects both the SU(2) and U(I) indices of $u^{\pm}_i$. One can define another and very important operation ($*$) which acts only on the U(I) index and is compatible with (2.18)

$$(u^+_i)^* = u^-_i \; , \quad (u^-_i)^* = -u^+_i \tag{2.23}$$

Notice a somewhat unusual property

$$(u^{\pm}_i)^{**} = -u^{\pm}_i \tag{2.24}$$

The operation $*$ has the geometric meaning of antipodal map of the sphere

$$P \to P^*$$

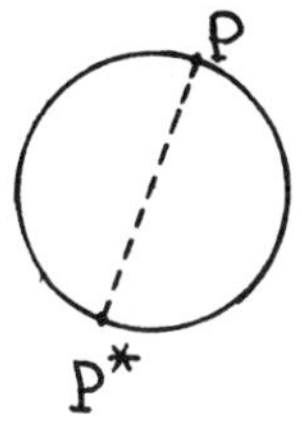

Check that in the parametrization (2.2) $\theta^* = \pi - \theta$, $\varphi^* = \varphi + \pi$

Given two involutions, $-$ and $*$, we can combine them to get an important operation $\underset{\sim}{*}$

$$(u^{\pm i})^{\underset{\sim}{*}} = -u^{\pm}_{i} \quad , \quad (u^{\pm}_{i})^{\underset{\sim}{*}} = u^{\pm i} \tag{2.25}$$

We shall consider functions on the sphere S^2. They carry a definite U(I) charge and are expanded in harmonic follows x)

$$F^{(q)}(u^{\pm}) = \sum_{n=0}^{\infty} f^{i_1 \dots i_{n+q} j_1 \dots j_n} \cdot u^{+}_{(i_1} \cdots u^{+}_{i_{n+q}} u^{-}_{j_1} \cdots u^{-}_{j_n)}$$

$F^{(q)}$ as a whole transforms as a representation of U(I) with the charge q. The coefficients $f^{i_1 \dots j_n}$ are SU(2) tensors with isospin $n + q/2$. Of course, $F^{(q)}$, $f^{(i_1 \dots j_n)}$ may also depend on some additional arguments e.g. x, θ^i_α etc.

The straightforward differentiation with respect to $u^{\pm}_{i}$ does not preserve the defining property (2.18). However, there exist three (as many as the number of independent degrees of freedom in $u^{\pm}$) differential operators which do preserve it

$$D^{++} = u^{+i} \frac{\partial}{\partial u^{-i}} \quad , \quad D^{--} = u^{-i} \frac{\partial}{\partial u^{+i}}$$
$$D^{0} = u^{+i} \frac{\partial}{\partial u^{+i}} - u^{-i} \frac{\partial}{\partial u^{-i}} \tag{2.26}$$

Exercise. Obtain (2.26) using general formulae (2.5), (2.6).

These derivatives form the algebra of SU(2)

---

x) Our symmetrization convention is

$$A^{(i_1 \dots i_n)} = \frac{1}{n!} \sum_{\text{Perm}} A^{i_1 \dots i_n}$$

$$[D^{++}, D^{--}] = D^{0}$$
$$[D^{0}, D^{++}] = 2D^{++} \;, \quad [D^{0}, D^{--}] = -2D^{--} \qquad (2.27)$$

The action of them on $u^{\pm}_{i}$ is very simple

$$D^{++} u^{+}_{i} = 0 \;, \qquad D^{++} u^{-}_{i} = u^{+}_{i} \qquad (2.28)$$

Simple tests show that the equation

$$D^{++} f^{(q)}(u) = 0 \qquad (2.29)$$

has the following solution

$$f^{(q)}(u) = 0 \;, \quad q < 0$$
$$f^{(0)}(u) = \text{const} \;, \quad q = 0 \qquad (2.30)$$
$$f^{(q)}(u) = f^{i_1 \ldots i_q} u^{+}_{i_1} u^{+}_{i_2} \cdots u^{+}_{i_q} \;, \quad q > 0$$

Finally, one can define integration over $u^{\pm}$ by the following rules

$$\int du \,.\, 1 = 1$$
$$\int du \cdot u^{+}_{(i_1} \cdots u^{+}_{i_n} u^{-}_{j_1} \cdots u^{-}_{j_m)} = 0 \qquad (2.31)$$

These rules are in complete accordance with the rules obtained from the explicit realization (2.22) with

$$du = \frac{1}{4\pi} \sin\theta \, d\theta \, d\varphi$$

In fact, this integral simply extracts the singlet component of the integrand. Integration by parts is possible due to the property that can be easily checked by the reader:

$$\int du \,.\, D^{++} f^{--}(u) = 0 \qquad (2.31')$$

To close this subsection we give a brief account of the central basis (CB) of superspace $\mathbb{R}^{4+2|8}$ as a starting point for discussing the analytic basis and analytic superspace in the next subsection. In CB we use the coordinates

$$\{ Z^M = (x^a, \theta_{\alpha i}, \bar{\theta}^i_{\dot{\alpha}}) \ , \ u^\pm_i \} \qquad i = 1,2 \qquad (2.32)$$

Here $\theta_{\alpha i}, \bar{\theta}^i_{\dot{\alpha}}$ are conjugated SU(2) doublets. The N=2 SUSY transformations are given by

$$\begin{aligned} \delta x^a &= i(\varepsilon^i \sigma^a \bar{\theta}_i - \theta^i \sigma^a \bar{\varepsilon}_i) \\ \delta \theta_{\alpha i} &= \varepsilon_{\alpha i} \ , \quad \delta \bar{\theta}^i_{\dot{\alpha}} = \bar{\varepsilon}^i_{\dot{\alpha}} \end{aligned} \qquad (2.33)$$

The spinor and harmonic covariant derivatives are defined by (2.13) and (2.26).

2.8. Analytic Basis, Superspace and Superfields.

Now we proceed to the main topic of this Section. From CB (2.32) we can pass to another basis (AB)

Analytic basis

$$\{ \zeta = (x^a_A, \theta^+_\alpha, \bar{\theta}^+_{\dot{\alpha}}) \ , u^\pm_i \ , \theta^-_\alpha \ , \bar{\theta}^-_{\dot{\alpha}} \} \qquad (2.34)$$

The new independent variables are related to the old ones as follows

$$\begin{aligned} x^a_A &= x^a - 2i\, \theta^{(i} \sigma^a \bar{\theta}^{j)} \cdot u^+_i u^-_j \\ \theta^\pm_\alpha &= \theta^i_\alpha u^\pm_i \ , \quad \bar{\theta}^\pm_{\dot{\alpha}} = \bar{\theta}^i_{\dot{\alpha}} u^\pm_i \end{aligned} \qquad (2.35)$$

In this new basis N=2 SUSY is realized as

$$\begin{aligned} \delta x^a_A &= -2i\,(\varepsilon^i \sigma^a \bar{\theta}^+ + \theta^+ \sigma^a \bar{\varepsilon}^i)\, u^-_i \ , \\ \delta \theta^+_\alpha &= \varepsilon^i_\alpha u^+_i \ , \quad \delta \bar{\theta}^+_{\dot{\alpha}} = \bar{\varepsilon}^i_{\dot{\alpha}} u^+_i \\ \delta u^\pm_i &= 0 \end{aligned} \qquad (2.36)$$

We stress that in (2.36) the SUSY parameters are the same $\varepsilon_{\alpha i}$, $\bar{\varepsilon}_{\dot\alpha}^{i}$ as in (2.33).

It is most remarkable that in (2.34) (z, u) form a subset closed under N=2 SUSY. This is due to the existence of a coset

$$\mathbb{A}\mathbb{R}^{4+2|4} = \frac{(P_a, L_{ab}, Q_\alpha^i, \bar{Q}_{\dot\alpha i}, SU(2))}{(L_{ab}, Q_\alpha^+, \bar{Q}_{\dot\alpha}^+, U(1))} = (z, u) \qquad (2.37)$$

Exercise: obtain (2.35), (2.36) using coset structure (2.16), (2.37) and general formula (2.1).

We shall refer to this coset space as to

Analytic superspace

$$(z = (x_A^a, \theta_\alpha^+, \bar{\theta}_{\dot\alpha}^+), u_i^\pm) \qquad (2.38)$$

An important property of $\mathbb{H}\mathbb{R}$ (2.32) is its reality

$$\overline{(x^a)} = x^a \ , \ \overline{(\theta_{\alpha i})} = \bar{\theta}_{\dot\alpha}^i \ , \ \overline{(u^{+i})} = u^-_i$$

It allows us to consider real SF's. Remarkably, the analytic superspace $\mathbb{A}\mathbb{R}$ is also real but with respect to the modified conjugation rule:

$$\overset{*}{\overline{(x_A^a)}} = x_A^a \ , \ \overset{*}{\overline{(\theta_\alpha^+)}} = \bar{\theta}_{\dot\alpha}^+ \ , \ \overset{*}{\overline{(\bar{\theta}_{\dot\alpha}^+)}} = -\theta_\alpha^+ \qquad (2.39)$$

This property will permit us to consider real superfields, in this superspace too.

Consider now the covariant derivatives in AB. The reader can easily establish that in this basis (by changing of coordinates (2.35) or by the coset considerations

$$D^{++} = u^{+i}\frac{\partial}{\partial u^{-i}} - 2i\theta^+\sigma^a\bar{\theta}^+\partial_a + \theta^{+\alpha}\frac{\partial}{\partial\theta^{-\alpha}} + \bar{\theta}^{+\dot\alpha}\frac{\partial}{\partial\bar{\theta}^{-\dot\alpha}}$$

and analogously for D^{--}, D^{0}. The spinor derivatives are decomposed into $D^{\pm}$ parts

$$D^{+}_{\alpha} = \frac{\partial}{\partial\theta^{-\alpha}} \quad , \quad \bar{D}^{+}_{\dot\alpha} = \frac{\partial}{\partial\bar\theta^{-\dot\alpha}} \quad , \qquad (2.40)$$
$$D^{-}_{\alpha} = -\frac{\partial}{\partial\theta^{+\alpha}} + 2i\bar\theta^{-\dot\alpha}\partial_{\alpha\dot\alpha} \; , \; \bar{D}^{-}_{\dot\alpha} = -\frac{\partial}{\partial\bar\theta^{+\dot\alpha}} - 2i\theta^{-\alpha}\partial_{\alpha\dot\alpha}$$

The property that D^{+}_{α}, $\bar{D}^{+}_{\dot\alpha}$ are reduced to partial derivatives with respect to θ^{-}_{α}, $\bar\theta^{-}_{\dot\alpha}$ is simply another expression of the fact of existence of the invariant analytic quotient superspace (2.37).

Now we can define

<u>Analytic superfields</u>

$$\phi^{(q)}(\zeta,u) \quad , \quad D^{+}_{\alpha}\phi^{(q)} = \bar{D}^{+}_{\dot\alpha}\phi^{(q)} = 0 \qquad (2.41)$$

i.e. superfields living on the coset $\mathbb{A}\mathbb{R}$ (2.37).

In general, analytic superfields carry U(I) charge q. The analytic (e.g. scalar) SF's have the decomposition

$$\begin{aligned}\phi^{(q)}(\zeta,u) = {} & F^{(q)}(x_A,u) + \theta^{+\alpha}\psi^{(q-1)}_{\alpha}(x_A,u) + \bar\theta^{+}_{\dot\alpha}\bar\varphi^{\dot\alpha(q-1)}(x_A,u) + \\ & + \theta^{+}\theta^{+}M^{(q-2)}(x_A,u) + \bar\theta^{+}\bar\theta^{+}N^{(q-2)}(x_A,u) + \theta^{+}\sigma^{a}\bar\theta^{+}A^{(q-2)}_{a}(x_A,u) + \qquad (2.42) \\ & + (\theta^{+})^{2}\bar\theta^{+}_{\dot\alpha}\bar\chi^{\dot\alpha(q-3)}(x_A,u) + (\bar\theta^{+})^{2}\theta^{+\alpha}\xi^{(q-3)}_{\alpha}(x_A,u) + (\theta^{+})^{2}(\bar\theta^{+})^{2}D^{(q-4)}(x_A,u)\end{aligned}$$

Using the rules (2.39) of combined conjugation one can impose the reality condition on $\phi^{(q)}$ (for even q):

$$\left[\phi^{(q)}(\zeta,u)\right]^{\star} = \phi^{(q)}(\zeta,u) \qquad (2.43)$$

It means, e.g.

$$\left(F^{(q)}\right)^{*} = F^{(q)} \ , \ \overset{*}{(\psi_\alpha)} = -\bar{\varphi}_{\dot\alpha} \ , \ \overset{*}{(A_a)} = -A_a \ \text{etc} \qquad (2.44)$$

The subsequent components in the u - expansion of bosonic fields become alternately real or imaginary.

Due to dependence on extra bosonic coordinates analytic SF's contain an infinite number of component fields.

It is instructive to rewrite the analytic SF's in CB

$$\phi^{(q)}(\zeta(z,u), u) = \sum_{n=0}^{\infty} \phi^{(i_1 \dots i_{n+q} j_1 \dots j_n)}(z) \cdot u^+_{i_1} \dots u^+_{i_{n+q}} \cdot u^-_{j_1} \dots u^-_{j_n} \qquad (2.45)$$

The coefficients in (2.45) are ordinary $u^\pm$ independent SF's satisfying the constraints

$$D^{(i}_{\alpha} \phi^{j_1 \dots j_{2n+q})} = \frac{n+1}{2n+q+3} D_{\alpha l} \phi^{(l i j_1 \dots j_{2n+q})} \qquad (2.46)$$

(the same is valid for $\bar{D}^i_{\dot\alpha}$). One can clearly see how the compact and unconstrained object (2.41) defined in becomes an infinite tower of constrained ordinary SF's in CB.

A final remark concerns the integration over the analytic superspace. The measure

$$d\zeta^{(-4)} du \equiv d^4 x_A \, d^2\theta^+ d^2\bar{\theta}^+ du \qquad (2.47)$$

is real and has dimension cm^2 .
Integration over $u^\pm$ was defined in (2.31). The Grassmann integration is equivalent to differentiation, so

$$\int d\zeta^{(-4)} du \iff \int d^4 x_A \, du (D^-)^2 (\bar{D}^-)^2 \qquad (2.48)$$

This explains the negative charge of the measure. Equation (2.31') now becomes

$$\int d\zeta^{(-4)} du \; D^{++} f^{++}(\zeta, u) = 0 \qquad (2.49)$$

3. GENERAL SELF-COUPLINGS OF N=2 MATTER

The basic N=2 matter multiplet (hypermultiplet) is an N=2 analogue of the N=1 chiral multiplet. Like the latter it can be described by an unconstrained superfield defined on some submanifold (on the analytic N=2 superspace). This hypermultiplet can be represented either in the first-order formalism (q^+) or in the second-order one (ω). We begin with

3.1. Free Hypermultiplet. First Order Formalism.

The Fayet-Sohnius hypermultiplet contains on-shell an SU(2) doublet of scalars and a Dirac spinor $(\varphi^i, \psi_\alpha, \bar{\chi}^{\dot\alpha})$. It is described by a complex analytic SF $q^+(\zeta, u)$, $[q^+] = cm^{-1}$. The free action for q^+ is

$$S = \frac{1}{2} \int dz^{(-4)} du \, \overset{*}{q}{}^+ \, \overleftrightarrow{D}^{++} q^+ \tag{3.1}$$

Note it close resemblance free Dirac action with D^{++} instead of $\not{\partial}$. It is real with respect to conjugation rules (2.25).

The equation of motion is obtained by straightforward variation of (3.1)

$$D^{++} q^+(\zeta, u) = 0 \tag{3.2}$$

__Exercise:__ show that the equation of motion (3.2) kills an infinite tail of fields in the decomposition of q^+ (i.e. these fields are auxiliary) and on-shell

$$q^+ = \varphi^i(x_A) u_i^+ + \theta^{+\alpha}\psi_\alpha + \bar\theta^+_{\dot\alpha} \bar{\chi}^{\dot\alpha} + 2i\,\theta^+\sigma^a\bar\theta^+ \partial_a \varphi^i \cdot u_i^-$$

with

$$\Box \varphi^i = \not{\partial}^{\dot\alpha\alpha}\psi_\alpha = \not{\partial}_{\alpha\dot\alpha}\bar{\chi}^{\dot\alpha} = 0 \tag{3.3}$$

To make a contact with the ordinary SF formulation of this theory let us rewrite (3.2) in CB (2.32)

$$D^{++} q^+ = 0 \quad \Rightarrow \quad q^+(\zeta(z,u), u) = q^i(z)\, u_i^+ \qquad (3.4)$$

Utilizing the analiticity condition we are left with the familiar constrained description of the hypermultiplet /2/

$$D_\alpha^{(i} q^{j)}(z) = \bar{D}_{\dot\alpha}^{(i} q^{j)}(z) = 0 \qquad (3.5)$$

These constraints give on-shell descriptions of free fields because the lead to the free equations of motion (Exercise: prove this statement using D-algebra (2.14)).

The free action has a lot of symmetries and it is worth to mention three of them

i) $SU(2)_A$-symmetry of automorphisms

$$\delta u^{\pm i} = \lambda^i{}_j u^{\pm j}, \quad \lambda^i{}_j \in su(2) \quad ; \quad \delta x_A^a = \delta\theta_\alpha^+ = \delta\bar\theta_{\dot\alpha}^+ = 0$$
$$\delta q^+ \equiv q^{+\prime}(\zeta', u') - q^+(\zeta, u) = 0 \qquad (3.6)$$

ii) The Pauli-Gursey x) SU(2) symmetry

$$\delta\zeta = \delta u = 0 \quad , \quad \delta q^+_a = \lambda_a{}^b q^+_b \quad , \quad \lambda_a{}^b \in SU(2) \qquad (3.7)$$

where we have introduced the notation

---

x) Indeed, the action for free physical fields

$$S = \frac{1}{2}\int d^4x \left(\partial_a \bar\varphi^i \partial^a \varphi_i + i\psi \partial\!\!\!/ \bar\psi + i\chi \partial\!\!\!/ \bar\chi \right)$$

has a Pauli-Gursey invariance under

$$\delta\psi = i\alpha\psi + \beta\chi, \quad \delta\chi = -i\alpha\chi - \bar\beta\psi$$

$$\alpha = \bar\alpha = -i\lambda^1{}_1, \quad \beta = -\lambda^2{}_1$$

and $\delta\varphi^i = i\alpha\varphi^i + \beta\bar\varphi^i$

$$q_a^+ = (q^+, \overset{*}{q}{}^+) \quad ; \quad q^{+a} = \varepsilon^{ab} q_b^+ = (q_a^+)^* \tag{3.8}$$

In this notation the free action

$$S = -\frac{1}{2} \int dz^{(-4)} du \; q^{+a} D^{++} q_a^+ \tag{3.9}$$

The transformations (3.7) evidently commute with N=2 SUSY and SU(2) group of automorphisms (3.6).

iii) Besides, the action (3.9) is invariant under N=2 superconformal group SU(2,2 | 2). We refer to /8/ for details.

3.2. Free Hypermultiplet. Second Order Formalism

In this formalism the basic hypermultiplet is described by a real analytic superfield with zero U(I) charge ω. Passing from q^+ to ω hypermultiplet goes as follows. Let us make an intertible change of variables

$$q_a^+ = u_a^+ \omega + u_a^- f^{++} \quad ; \quad \omega = \overset{*}{\omega}, \; f^{++} = \overset{*}{f}{}^{++} \tag{3.10}$$

$$\omega = u_a^- q^{+a} \; , \quad f^{++} = u^{+a} q_a^+$$

Then we obtain

$$S^{free} = \frac{1}{2} \int dz^{(-4)} du \left[(f^{++})^2 + 2 f^{++} \cdot D^{++} \omega \right] \tag{3.11}$$

The equation of motion tells that $f^{++} = - D^{++} \omega$ Substituting this back into the action (3.11) we arrive at

$$S_\omega^{free} = -\frac{1}{2} \int dz^{(-4)} du \left[D^{++} \omega \right]^2 \tag{3.12}$$

i.e. we obtain an analogue of the Klein-Gordon action. The on-shell field content of the superfield $\omega(z,u)$ differs from that of q^+ by SU(2) prescriptions only: ω contains four physical scalars as $\underline{1} + \underline{3}$ and two

Weyl fermions as $\underline{2}$ of SU(2) . Prove that on-shell they enter the decomposition of ω as follows

$$\omega(z,u) = \omega(x_A) + \omega^{(ij)}(x_A)\, u^+_i u^-_j + \theta^{+\alpha}\psi^i_\alpha \cdot u^-_i + \bar\theta^+_{\dot\alpha}\bar\Psi^{\dot\alpha}_i u^{-i} +$$
$$+ 2i\,\theta^+\sigma^a\bar\theta^+ \cdot \partial_a \omega^{(ij)} \cdot u^-_i u^-_j \tag{3.13}$$

So much for free case. Now we are going to discuss

3.3. General Self-Couplings of N=2 Hypermultiplets

Let us start with a simple example of self-coupling. If we wish to preserve $SU(2)_A$ invariance then self-couplings have not contain harmonics explicitly. In this case general form of $\mathcal{L}^{+4}_{int}$ is

$$\mathcal{L}^{(+4)}_{int} = \frac{\lambda}{2}(q^+)^2(\overset{*}{q}{}^+)^2 + \beta (q^+)^3\overset{*}{q}{}^+ + \gamma (q^+)^4 + h.c. \tag{3.14}$$

The equation of motion now becomes (consider, e.g. $\beta=\gamma=0$)

$$D^{++}q^+ + \lambda (q^+)^2\overset{*}{q}{}^+ = 0 \tag{3.15}$$

Un fact, however, we can introduce in the action harmonics explicitly without violating N=2 SUSY. These harmonics of course will break $SU(2)_A$ symmetry. E.g.

$$\mathcal{L}^{(+4)}_{int} = q^+\overset{*}{q}{}^+ u^+_1 u^+_2 \cdot \sin(q^+u^-_1) + h.c.$$

etc. Resorting to dimensionality and analyticity reasoning we can write down the most general self-interaction of any number of hypermultiplets in the form

$$S^{gen} = \frac{1}{\varkappa^2}\int d\zeta^{(-4)} du\, \mathcal{L}^{(+4)}\left(q^+_K, \overset{*}{q}{}^+_K, D^{++}q^+_K, D^{++}\overset{*}{q}{}^+_K, (D^{++})^2 q^+_K, \ldots, u^\pm_i\right) \tag{3.16}$$

Here $\mathcal{L}^{(+4)}$ is a general U(I) charge + 4 function of harmonics, q^+_K and their harmonic derivatives of any order (because these derivatives are dimensionless), $\varkappa$ is a coupling constant, $[\varkappa] = cm^1$ for d=4, $[\varkappa] = cm^0$ for d=2. In (3.16) we set q^+'s to be dimensionless. So, spatial and spinor derivatives ($[\partial/\partial x] = cm^{-1}$ and $[\partial/\partial\theta] = cm^{-1/2}$, respectively) are inadmissible for dimensionality considerations, and in fact they would lead to a higher derivative action for physical fields.

After elimination of auxiliary fields every $\mathcal{L}^{(+4)}$ results in a N=2 supersymmetric nonlinear sigma model.

3.4. Hypermultiplets and Hyper-Kähler Metrics

A coincidence of prefixes in two words of heading turns out to be not accidental. Due to a remarkable theorem by Alvarez-Gaume and Freedman /11/ any self-coupling of N=2 hypermultiplets lead in its physical boson sector to a sigma-model with a hyper-Kähler target manifold. This theorem adds new link to a known chain mentioned above: a metric of non-linear bosonic σ-model g_{mn},

$$S = \int dx\; g_{mn}(\varphi)\, \partial^a \varphi^m \cdot \partial_a \varphi^n \tag{3.17}$$

can be

i) arbitrary Riemannian in N=0 case

ii) arbitrary Kählerian in N=1 case

iii) arbitrary hyper-Kähler in N=2 case

In the N=1 case we deal with the Kähler potential /12/, from which the Kähler metric can be obtained

$g_{m\bar{n}}(\varphi,\bar{\varphi}) = \dfrac{\partial^2 K(\varphi,\bar{\varphi})}{\partial\varphi^m \partial\bar{\varphi}^n}$. Up to now no analogues of the Kähler potential were known in the hyper-Kähler case. The harmonic superspace off-shell description of N=2 matter seems to suggest such an analogue because the general self-coupling $\mathcal{L}^{(+4)}$ apparently provides us with a kind of hyper-Kähler potential. Given $\mathcal{L}^{(+4)}$ one can eliminate auxiliary fields and obtain some hyper-Kähler metric. <u>We conjecture that the converse is also true, i.e. each hyper-Kähler metric can be extended to some</u> $\mathcal{L}^{(+4)}$. However, at present we do not have a proof of this conjecture

Instead we present a series of examples.

Below we list several hypermultiplet actions leading to widely known hyper-Kähler metrics: Taub – NUT, Eguchi-Hanson, Calabi, multi-Eguchi-Hanson x)

1) Taub- NUT action /14/

$$S_{TN} = \frac{1}{2\varkappa^2}\int d\zeta^{-4} du \left[\overset{*}{q}{}^{+} \overset{\leftrightarrow}{D}{}^{++} q^{+} + \lambda (q^{+})^{2} (\overset{*}{q}{}^{+})^{2} \right] \tag{3.18}$$

2) The Eguchi-Hanson action /15/ xx)

$$S_{EH} = \frac{1}{2\varkappa^2}\int d\zeta^{(-4)} du \left[q^{+a} D^{++} q^{+}_{a} - (\xi^{++})^{2} (q^{+a} u^{-}_{a})^{-2} \right] \tag{3.19}$$

where

$$\xi^{++} = \xi^{(ij)} u^{+}_{i} u^{+}_{j} \,, \quad \overline{(\xi^{ij})} = \varepsilon_{ik} \varepsilon_{je} \xi^{ke} \tag{3.20}$$

This formula is obtained after some algebra from a Fayet-Iliopoulos type action (for details see /15/)

$$S_{EH} = \frac{1}{2\varkappa^2}\int d\zeta^{(-4)} du \left[\overset{*}{q}{}^{+} \overset{\leftrightarrow}{D}{}^{++} q^{+} + V^{++}(\zeta,u) \left(\overset{*}{q}{}^{+} T q^{+} + \xi^{++} \right) \right] \tag{3.21}$$

where q^{+} is promoted to a doublet of an extra SU(2) and T is the generator of its Cartan subalgebra. This way is extended to

3) the multi-Eguchi-Hanson action that is obtained by i) coupling of n hypermultiplets in the quark ($\underline{n}$) representation of SU(n) $(q^{+} = (q^{+}_{1}, \ldots, q^{+}_{n}))$ ii) gauging the (n-1)-dimensional Abelian group generated by the Cartan subalgebra of SU(n) and iii) adding (n-1) Fayer-Iliopoulos terms:

---

x) See, e.g., ref. /13a, 20/ .

xx) For this action be meaningful q^{+}_{a} should start with a constant:

$$q^{+}_{a} = \mathrm{const} \cdot u^{+}_{a} + \tilde{q}^{+}_{a}$$

$$S_{MEH} = \frac{1}{2\varkappa^2} \int d\zeta^{(-4)} du \left\{ \left[\overset{*}{q}{}^{+} \overleftrightarrow{D}^{++} q^{+} + \sum_{k=1}^{n-1} V^{++}_{(k)}(\zeta,u) \left[\overset{*}{q}{}^{+} T_{(k)} q^{+} + \xi^{++}_{(k)} \right] \right. \right. \tag{3.22}$$

where $T_{(k)}$ are generators of Cartan subalgebra.

4) Target manifolds above were four-dimensional. The 4n-dimensional Calabi manifold can be obtained by an analogous procedure. The action is

$$S = \frac{1}{2\varkappa^2} \int d\zeta^{(-4)} du \left\{ \overset{*}{q}{}^{+}_{A} \overleftrightarrow{D}^{++} q^{+}_{A} + V^{++} \cdot \left(\overset{*}{q}{}^{+}_{A} M_{AB} q^{+}_{B} + \xi^{++} \right) \right\} \tag{3.23}$$

where M is any constant anti-Hermitean $n \times n$ matrix.

The set of examples can be enlarged.

3.5. Comment on Central Charges

Here we shall briefly discuss an extension of the hypermultiplet description to nonzero central charge. As was shown in the Appendix to the first ref. /5/, the central charge can be included into the harmonic superspace scheme by a standard method, i.e. by adding to (ζ, u) an extra bosonic coordinate x^5_A x).

This entails the following modification of harmonic derivative (written down in the analytic basis)

$$D^{++} \to D^{++}_{cc} = D^{++} + i\left(\theta^{+}\theta^{+} - \bar{\theta}^{+}\bar{\theta}^{+}\right)\frac{\partial}{\partial x^5_A} \tag{3.24}$$

Now analytic superfields are allowed to depend in a general way on the coordinate x^5_A. To preserve the number of physical fields, we follow the dimensional reduction procedure of Scherk and Schwarz /21/. Let the action (3.16) have an U(I) -symmetry commuting with N=2 SUSY

x) Harmonic superspace with two central charges was discussed in /22/.

and possessing a Killing vector $G^+(q^+, u^\pm, \ldots)$

$$\delta q^+ = \alpha\, G^+(q^+, \ldots) \tag{3.25}$$

where α is the U(I) parameter. Then x_A^5 dependence of q^+ is restricted as (m is a parameter of dimension of the mass)

$$\frac{\partial}{\partial x_A^5} q^+ = m\, G^+ \tag{3.26}$$

After substituting (3.26) into harmonic derivative (3.24) and then (3.24) into the action (3.16) we arrive at the theory invariant under N=2 SUSY with a central charge. In such a theory the potential terms (in particular, a mass term) become possible. E.g. for the free theory (3.1) such a procedure with $G^+ = i q^+$ results in a mass term (prove this):

$$S_{cc} = \frac{1}{2} \int d\zeta^{(-4)} du \left[\overset{*}{q}{}^+ \overleftrightarrow{D}^{++} q^+ + 2m(\theta^+\theta^+ - \bar\theta^+\bar\theta^+) q^+ \overset{*}{q}{}^+ \right] \tag{3.27}$$

4. DUALITY TRANSFORMATIONS

First attempts at the off-shell formulation of N=2 matter were undertaken using N=2 multiplets with finite sets of auxiliary fields. Our aim here is to show that they are described in the harmonic superspace approach as transparently as the "ultimate" multiplet q^+ is. By means of a kind of Legendre transformation (the "duality" one) their general couplings are proven to be equivalent to some restricted class of the general q^+ coupling. We restrict ourselves here to considering the "tensor" /3/, "improved tensor" /1/ and "relaxed hypermultiplets" /4/ while a proof for other multiplets the reader can find in /16/. We begin with

4.1. <u>N=2 Tensor Multiplet.</u> In harmonic superspace it is described by a real analytic superfield $L^{++}(\zeta, u) = [L^{++}(\zeta, u)]^*$ having U(I) charge 2 and obeying the constraint

$$D^{++}L^{++}(\zeta,u) = 0 \tag{4.1}$$

In the central basis (4.1) implies

$$L^{++}(\zeta(z,u), u) = u^+_i u^+_j L^{(ij)}(z) \tag{4.2}$$

Then the analyticity conditions

$$D^+_\alpha L^{++} = u^+_i D^i_\alpha L^{k\ell}(z) u^+_k u^+_\ell = 0 \quad (\text{and h.c.})$$

are reduced to

$$D^{(i}_\alpha L^{jk)}(z) = \bar{D}^{(i}_{\dot\alpha} L^{jk)}(z) = 0 \tag{4.3}$$

where brackets mean symmetrization, i.e. we come to the familiar picture of the N=2 tensor multiplet /3/.

Let us return now to analytic superspace and give the free action. It is bilinear in L^{++}

$$S^{free} = \frac{1}{2\varkappa^2}\int d\zeta^{(-4)} du \, (L^{++})^2 \; ; \quad [\varkappa] = cm^1, \; [L^{++}] = cm^0.$$

By dimensionality and analyticity arguments the general N=2 supersymmetric self-coupling is given by

$$S'^{gen} = \frac{1}{\varkappa^2}\int d\zeta^{(-4)} du \, F^{(+4)}(L^{++}, u^\pm) \tag{4.5}$$

where $F^{(+4)}$ may have an arbitrary dependence on L^{++}, $u^\pm_i$ (provided its total U(I)-charge is four). The constraint (4.1) is implied as before, being the definition of tensor multiplet.

Superconformally-invariant version of N=2 tensor multiplet is

4.2. Improved Tensor N=2 Multiplet. It has drawn

attention mainly in connection with the possibility to use it as a compensator for N=2 conformal supergravity /1/. It has somewhat more complicated action which is written down formally like (4.4)

$$S_{impr} = \frac{1}{2\varkappa^2} \int d\zeta^{(-4)} du \, (g^{++})^2 \tag{4.6}$$

but with

$$g^{++} = \frac{2\,(L^{++} - 2i\, u_1^+ u_2^+)}{1 + (1 - 4 u_1^+ u_2^+ u_1^- u_2^- - 2i L^{++} u_1^- u_2^-)^{1/2}} \tag{4.7}$$

and is accompanied by the same constraint (4.1)

$$D^{++} L^{++} = 0$$

The form (4.6), (4.7) is dictated by N=2 superconformal invariance. We shall not give here the whole superconformal transformations in analytic superspace (these can be found, e.g., in our papers /8,17/) and restrict ourselves to dilatations (parameter ℓ) and conformal SU(2)-transformations (parameters $\lambda^{(ij)}$):

$$\delta x_A^a = \ell x_A^a - 2i\, \lambda^{(ij)} u_i^- u_j^- \, \theta^+ \sigma^a \bar\theta^+$$

$$\delta \theta^+_{\alpha(\dot\alpha)} = \frac{\ell}{2} \theta^+_{\alpha(\dot\alpha)} + \lambda^{(ij)} u_i^+ u_j^- \, \theta^+_{\alpha(\dot\alpha)} \tag{4.8}$$

$$\delta u_i^+ = \Lambda^{++} u_i^- \,, \quad \delta u_i^- = 0$$

$$\delta\, (d\zeta^{(-4)} du) = -2\Lambda \, d\zeta^{(-4)} du \,, \quad \delta L^{++} = 2\Lambda L^{++}$$

$$\Lambda = -\ell + \lambda^{(ij)} u_i^+ u_j^- \,, \quad \Lambda^{++} = D^{++}\Lambda \,, \quad D^{++}\Lambda^{++} = 0$$

We again leave to the reader to check invariance of (4.1) and (4.6) (and noninvariance of (4.4)) under (4.8) (or to consult with [17]).

The action (4.6) is nonpolynomial in L^{++} . Nevertheless, just as in the N=1 case, it contains no dynamics, being equivalent to a free action. This was proven first in terms of N=1 superfields [3b].

N=2 harmonic superspace provides us with another proof. The basic ingredient of the latter is the use of

4.3. N=2 Duality Transformations. Let us illustrate it first by the example of the free tensor multiplet action (4.4). Implementing the constraint (4.1) in the action with the help of analytic Lagrange multiplier $\omega(\zeta,u)$ we arrive at the new action

$$S=\frac{1}{2\varkappa^2}\int d\zeta^{(-4)}du\left[(L^{++})^2+\omega\cdot D^{++}L^{++}\right] \qquad (4.9)$$

which is reduced to the initial one (4.4) by using the result of varying ω . On the other hand (4.9) is none other than the free action for the ultimate multiplet (3.11) After a change of field variables (3.10) it becomes the free q^+-action (3.9) . The same is true for the general tensor multiplet action (4.5) and we arrive at its dual form as

$$S^{dual}=\frac{1}{\varkappa^2}\int d\zeta^{(-4)}du\left[F^{(+4)}(u^+q^+,u)-\frac{1}{2}(u^+q^+)^2-\frac{1}{2}q^{+i}D^{++}q^+_i\right] \qquad (4.10)$$

(the second term in the integrand cancels in fact with a similar term coming from the free part of $F^{(+4)}$: $F^{(+4)}=\frac{1}{2}(L^{++})^2+\cdots=\frac{1}{2}(u^+q^+)^2+\cdots$. Comparing (4.10) with the general q^+ -action we observe that (4.10) exhibits an invariance under the shifts

$$\delta q^+_i = \text{const}\cdot u^+_i \qquad (4.11)$$

while (3.16) in general has not any symmetries (apart from supersymmetry itself). Thus N=2 tensor multiplet may produce only a restricted class of N=2 matter action, not the general one.

After all, what's about the improved N=2 tensor multiplet action (4.6)? It is certainly not reduced to the free q^+ -one by the variable change (3.10). It turns out that in this case one needs more sophisticated change:

$$\check{q}^+_1 = -i\left(2u^+_1 + ig^{++}u^-_1\right)\cdot e^{-i\omega/2} \qquad (4.12)$$

$$\check{q}^+_2 = i\left(2u^+_2 - ig^{++}u^-_2\right) e^{i\omega/2} \qquad (4.13)$$

It is an useful exercise for the reader to verify that the action

$$S = \frac{1}{\varkappa^2}\int d\zeta^{(-4)} du \left[\frac{1}{2}(g^{++})^2 + \omega\cdot D^{++}L^{++}\right] \qquad (4.14)$$

being rewritten through $\check{q}^+_1 = q^+$, $\check{q}^+_2 = \overset{*}{q}{}^+$ is

$$S = +\frac{1}{2\varkappa^2}\int d\zeta^{(-4)} du \; \overset{*}{q}{}^+ \overleftrightarrow{D}^{++} q^+$$

This completes the proof of dual equivalence between the improved tensor multiplet and free q^+- hypermultiplet actions.

M. Roček called our attention to the following interesting fact. Summing free actions for the N=2 tensor and improved tensor multiplets leads to a non-trivial self-interaction equivalent to the Taub- NUT one. Indeed, in the harmonic superspace language, it is easy to extract from (4.12), (4.13) a beautiful identity (prove it):

$$q^+ \overset{*}{q}{}^+ = 2i\, L^{++}$$

Then

$$(4.14) + \lambda (4.12) \overset{dual}{=} + \frac{1}{2\varkappa^2} \int d\zeta^{(-4)} du \Big[\overset{*}{q}{}^+ \overset{\leftrightarrow}{D}{}^{++} q^+ + \\ + \frac{\lambda}{2} (q^+)^2 (\overset{*}{q}{}^+)^2 \Big]$$

that is familiar now (we hope) Taub- NUT action (3.18). It is interesting that SU(2) x U(I)- symmetry of this action has entirely different meaning depending on whether it is realized in terms of q^+ , $\overset{*}{q}{}^+$ or L^{++} , ω . The abelian factor U(I) produces, on the one hand, phase rotations of q^+ and, on the other hand, constant shifts of ω : $\omega \to \omega$ + const. The SU(2) coincides with the automorphism group in the q^+ -language and with the unbroken subgroup SU(2) of superconformal group SU (2,2|2) when is represented through L^{++}, ω (parameters λ^{ij}) in (4.8) x) .

The last topic we wish to discuss in this Section is

4.4. Relaxed Hypermultiplet. The main disadvantage of tensor multiplet was the impossibility of minimal Yang-Mills coupling xx) . To overcome this difficulty Howe, Stelle, and Townsend /4/ have invented the relaxed matter hypermultiplet which contains no conserved vector and so may have minimal couplings with Yang-Mills.

x) Note that the automorphism group is always defined up to an adding of isomorphic group commuting with the original automorphism group and supersymmetry.

xx) Because this multiplet contains an antisymmetric tensor field, notoph, gauge invariance of which is incompatible with the Yang-Mills one.

In harmonic superspace it is described by real analytic superfields $L^{++}(\zeta,u)$ and $V(\zeta,u)$. The former superfield is subjected to the constraint

$$(D^{++})^2 L^{++} = 0 \tag{4.15}$$

which is a relaxed form of (4.1). The superfield V is defined up to a gauge transformation

$$V' = V + D^{++}\lambda^{--}(\zeta,u) \tag{4.16}$$

where λ^{--} is an analytic real parameter.
The free action

$$S^{free}_{HST} = \frac{1}{\varkappa^2}\int d\zeta^{(-4)}du\left[(L^{++})^2 + V D^{++}L^{++}\right] \tag{4.17}$$

is compatible with gauge freedom (4.16) because of the constraint (4.15).

The dimensionality, analyticity and gauge invariance reasonings lead to the following general form of self-couplings

$$S_{HST} = \frac{1}{\varkappa^2}\int d\zeta^{(-4)}du\left[F^{(+4)}(L^{++}, D^{++}L^{++}, u) + V\cdot D^{++}L^{++}\right] \tag{4.18}$$

with $F^{(+4)}$ being an arbitrary U(I)-charge four function of its arguments. To couple L^{++} to Yang-Mills one simply should put L^{++} and V into a real representation of the Yang-Mills group and to covariantize the constraint (4.15) and the gauge transformation (4.16).

To establish a contact with the original description [4] we apply again to the central basis (2.32) where eq. (4.15) says that

$$L^{++} = u^+_i u^+_j L^{ij}(z) + 5\, u^+_{(i} u^+_j u^+_k u^-_{l)} L^{(ijkl)}(z)$$

(the factor 5 is introduced for further convenience).

Analyticity of L^{++} then implies

$$D^{(i}_{\alpha(\dot\alpha)} L^{jk)} = D_{\alpha(\dot\alpha)l} L^{ijkl} \tag{4.19}$$
$$D^{(i}_{\alpha(\dot\alpha)} L^{jklm)} = 0$$

Finally, scalar superfield $V(z)$ of /4/ is

$$V(z) = \int du\, V(\zeta(z,u),u) \tag{4.20}$$

Obviously, $V(z)$ is invariant under (4.16) and solves the constraints /4/

$$D^i_\alpha D_{\beta i} V = \bar D_{\dot\alpha i} \bar D^i_{\dot\beta} V = [D^i_\alpha, \bar D_{\dot\beta i}] V = 0 \tag{4.21}$$

To pass to the dual description one has an before, to include first of all the constraint (4.15) into the action with the help of Lagrange multiplier

$$S = \frac{1}{\kappa^2}\int d\zeta^{(-4)} du \Big[F^{(+4)}(L^{++}, D^{++}L^{++}, u) + $$
$$+ V \cdot D^{++}L^{++} + V^{--}(D^{++})^2 L^{++} \Big] \tag{4.22}$$

where V^{--} should transform under the gauge group (4.16) as

$$V^{--\prime} = V^{--} + \lambda^{--}$$

in order to maintain the gauge invariance of the action. Now one can combine V and V^{--} into one (nongauge) analytic superfield

$$\omega = V - D^{++} V^{--} \tag{4.23}$$

Making the change of variables (3.10) we obtain S (4.22) in a dual form

$$S^{dual} = \frac{1}{\varkappa^2} \int dz^{(-4)} du \Big[F^{(+4)}(u^+q^+, u^{+i} D^{++} q^+_i, u) - \\ - \frac{1}{2}(u^+q^+)^2 - \frac{1}{2} q^{+i} D^{++} q^+_i \Big] \qquad (4.24)$$

Again, the action is not of a general form. Moreover, it is guaranteed to be equivalent to the action (4.5) for the tensor multiplet (on-shell, after eliminating auxiliary fields), and it is also invariant under constant shifts (4.11).

The same reasonings can equally be applied to all other known N=2 matter multiplets with finite sets of components. The careful analysis given in our paper[16] shows that all their self-interactions are reduced by proper N=2 duality transformations to particular classes of the general q^+-hypermultiplet action (3.16). The latter is thus a most likely candidat to play a role of general off-shell N=2 matter action.

Finally, we would like to stress that all the actions considered in this Section admit an equivalent representation in ordinary N=2 superspace (see Appendix B of [6]). One may compare simple and transparent analytic superspace expressions given here with their lengthy and somewhat ugly prototypes in conventional N=2 superspace.

5. HARMONIC SUPERGRAPHS

So the harmonic superspace approach gives us an ultimate description of N=2 matter on the classical level. In this section we shall discuss briefly a corresponding quantization procedure. We introduce generalized functions (distributions) including δ -functions in the harmonic superspace and, thereafter, the Green functions of hypermultiplets. Then we formulate Feynman rules for the latter and give some examples of supergraph calculations. d=4, N=2 general self-interactions are nonrenormalizable. Reduction to two-dimensional theories gives a simple proof of finiteness of d=2, N=4 nonlinear σ-models merely by power counting.

5.1. Examples of Integration over Harmonics

We begin with reminding that any function on $SU(2)/U(1)$ has U(I) charge q and is expanded in terms of symmetrized (and, consequently, irreducible with respect to SU(2)) products of $u^{\pm}{}_i$

$$D^0 f^{(q)}(u) = q\, f^{(q)}(u) \tag{5.1}$$

$$f^{(q)}(u) = \sum_{n=0}^{\infty} f^{(i_1 \dots i_{n+q} j_1 \dots j_n)}\, u^+_{(i_1} \dots u^+_{i_{n+q}} u^-_{j_1} \dots u^-_{j_n)}$$

Coefficients in the decomposition (5.2) are defined as

$$f^{(i_1 \dots i_{n+q} j_1 \dots j_n)} = (-1)^{n+q} \frac{(2n+q+1)!}{(n+q)!\, n!} \int du\, f^{(q)}(u) \cdot u^{+(j_1} \dots u^{+j_n} u^{-i_1} \dots u^{-i_{n+q})} \tag{5.3}$$

The integration over harmonics is realized according to rules (2.31) above and these rules are consistent with the explicit parametrization (2.22).

It is worthwhile to mention that an integral does not vanish, iff its integrand has a zero U(I) charge and contains an SU(2) singlet in its decomposition over irreducible representations of SU(2), e.g.

$$\int du\; u^+_i u^-_j = \frac{1}{2} \varepsilon_{ij} \tag{5.4}$$

These reasonings are important when performing integrations. Let us consider several examples

1) Calculate the integral $\quad I_i = \int du \dfrac{u^+_i}{(u^+ v^+)}$ (5.5)

where v_i^+ is some outer harmonic and $u^+v^+ = u^{+i}v_i^+$.
Hint: I_i has one SU(2) index and U(I)-charge -1 with respect to the harmonic v. So we have

$$\int du \; \frac{u_i^+}{u^+v^+} = \text{Const}\; v_i^- \tag{5.6}$$

Multiplying both sides of (5.6) by v^{+i} and using (2.31) and (2.18) we obtain Const $=-1$ and

$$\int du \; \frac{u_i^+}{u^+v} = -\,v_i^- \tag{5.7}$$

2) Show in an analogous way that

$$\int du \; \frac{u_{(i}^+ u_j^+ u_{k)}^-}{u^+v^+} = -\frac{1}{2}\, v_{(i}^+ v_j^- v_{k)}^- \tag{5.8}$$

3) Check that

$$\int du \; \frac{u_i^+ u_j^+}{(u^+v^+)(u^+w^+)} = +\,\frac{v_{(i}^+ v_{j)}^-}{w^+v^+} - \frac{w_{(i}^+ w_{j)}^-}{w^+v^+} \tag{5.9}$$

In this case the reader has to be careful about U(I) charges of both the harmonics v, w separately.
4) Prove the formula

$$\int du \frac{1}{(u^+v^+)^n}\, u_{(i_1}^+ \cdots u_{i_{n+1}}^+ u_{i_{n+2})}^- = \frac{(-1)^n}{2}\cdot n\, v_{(i_1}^- \cdots v_{i_{n+1}}^- v_{i_{n+2})}^+ \tag{5.10}$$

In some cases one has to use a fixed parametrization, e.g. (2.22) to make integration.

5.2. δ -Functions on SU(2)/U(I)

These δ -functions have to be defined for each value of the U(I) charge:

$$\int dv\, \delta^{(q,-q)}(u,v) F^{(p)}(v) = \delta^{pq} F^{(q)}(u)$$

(5.11)

Note the separate preservation of U(I)-charges corresponding to each harmonic variable. The δ-function can be differentiated in a natural way

$$\int dv \cdot D^{++}_v \delta^{(q,-q)}(u,v) \cdot F^{(q-2)}(v) = = -\int dv \cdot \delta^{(q,-q)}(u,v)\, D^{++}_v F^{(q-2)}(v) = -D^{++} F^{(q-2)}(u) \qquad (5.12)$$

$$D^{++}_v \delta^{(q,-q)}(u,v) = -D^{++}_u \delta^{(q-2,2-q)}(u,v) \qquad (5.13)$$

The reader can obtain a number of further interesting properties of the δ-functions. These are

a) $$\delta^{(q,-q)}(u,v) = \delta^{(-q,q)}(v,u) \qquad (5.14)$$

b) $$F^{(p)}(v) \cdot \delta^{(q,-q)}(u,v) = F^{(p)}(u)\, \delta^{(q-p,p-q)}(u,v) \qquad (5.15)$$

and, consequently,

c)
$$(u^+v^+) \cdot \delta^{(q,-q)}(u,v) = (u^-v^-) \cdot \delta^{(q,-q)}(u,v) = 0 \qquad (5.16)$$

d)
$$\delta^{(q,-q)}(u,v) = (u^+v^-)\, \delta^{(q-1,1-q)}(u,v) = (u^+v^-)^q \cdot \delta^{(0,0)}(u,v) \qquad (5.17)$$

e) $$u^+v^+ \cdot (D^{--}_u)^n \delta^{(m,-m)}(u,v) = n(D^{--}_u)^{n-1} \delta^{(m-1,1-m)}(u,v) \qquad (5.17')$$

The properties a) and b) follow from (5.11) and the useful lemma: If

$$\int du\, f^{(q)}(u)\cdot g^{(-q)}(u) = 0 \tag{5.19}$$

for any $g^{(-q)}(u)$ then $f^{(q)} = 0$.

To decompose $\delta^{(q,-q)}$-function in harmonics one has to use the fact that symmetrized products of $u^{\pm}$ form the complete orthogonal set on S^2

$$\int du\, (u^+)^{(m}(u^-)^{n)}\cdot(u^+)_{(k}(u^-)_{\ell)} = \begin{cases} \frac{(-1)^n\, m!\, n!}{(m+n+1)!}\,\delta^{(i_1}_{(j_1}\cdots\delta^{i_{m+n})}_{j_{k+\ell})}\,, & \text{if } \begin{matrix} m=\ell \\ n=k \end{matrix} \\ 0\,, & \text{otherwise} \end{cases} \tag{5.20}$$

$$(u^+)^{(m}(u^-)^{n)} \equiv u^{+(i_1}\cdots u^{+i_m}\, u^{-i_{m+1}}\cdots u^{-i_{m+n})}$$

Consequently, one obtain the decomposition

$$\delta^{(q,-q)}(u,v) = \sum_{n=0}^{\infty} (-1)^{n+q}\,\frac{(2n+q+1)!}{n!\,(n+q)!}\times$$
$$\times (u^+)_{(n+q}(u^-)_{n)}\cdot(v^+)^{(n}(v^-)^{n+q)} \tag{5.21}$$

Its validity can be easily checked. However, in practice, we shall never use this decomposition. Two more comments;

i) The chargeless harmonic δ-function $\delta^{0,0}(u,v)$ (which is the basic one, see (5.17), can be written down as an ordinary δ-function with a special argument

$$\delta^{(0,0)}(u,v) = 2\,\delta\left[(u^+v^+)(u^-v^-)\right] \tag{5.22}$$

There is a useful identity (prove it)

$$(u^+v^+)(u^-v^-) = 1 + (u^+v^-)(u^-v^+) \qquad (5.23)$$

5.3. Harmonic Distributions

To find Green functions for hypermultiplets q^+, ω, we have to learn how to solve equations of the type

$$D^{++}_u X^{(-n,-n)}(u,v) = \frac{1}{(n-1)!}(D^{--}_u)^{n-1}\delta^{n,-n}(u,v), \quad n>0 \qquad (5.24)$$

for an unknown function $X^{(-n,-n)}(u,v)$ having U(I) charges $(-n)$, $(-n)$ with respect to each argument. According to (2.30) the solution to this equations is unique.

Multiplying the l.h.s. and r.h.s. of (5.24) by $(u^+v^+)^n$ we get x)

$$D^{++}_u X^{(0,0)}(u,v) = 0, \quad X^{0,0} \equiv (u^+v^+)^n X^{(-n,-n)}(u,v) \qquad (5.25)$$

where we have used (5.16, 17') in the r.h.s. and we have

$$(2.30) \Rightarrow X^{0,0} = c = const, \quad X^{(-n,-n)}(u,v) = \frac{c}{(u^+v^+)^n}$$

To define c, we multiply (5.24) by $(u^+v^+)^{n-1}$ and integrate over u. As a result, $c=1$, so

$$X^{(-n,-n)}(u,v) = \frac{1}{(u^+v^+)^n} \qquad (5.26)$$

It is a distribution on the sphere SU(2)/U(I) and has a singularity at $u^+ = v^+$. It is analogous, in a way, to the distribution x^{-n} in the ordinary analysis. Like the latter, it has the property

$$\frac{u^+v^+}{(u^+v^+)^n} = \frac{1}{(u^+v^+)^{n-1}} \tag{5.27}$$

The proof goes straightforwardly by multiplying eq. (5.27) by u^+v^+ and using (5.16).

In contradistinction to D^{++} - differentiation the D^{--} one is done in the naive way

$$D^{--}_u \frac{1}{(u^+v^+)^n} = -n \frac{u^-v^+}{(u^+v^+)^{n+1}} \tag{5.28}$$

Indeed, applying D^{++}_u to this equation one gets an identity (prove this) and (5.28) holds due to (5.24).

Finally,

$$\left[(u^+v^+)^{-n}\right]^{\overset{*}{-}} = (u^+v^+)^{-n} \tag{5.29}$$

Exercise. Using (5.24) check that

$$\int du \cdot \frac{1}{(u^+v^+)^n} \cdot u^+_{(i_1} \cdots u^+_{i_{n+k}} u^-_{j_1} \cdots u^-_{j_k)} = \tag{5.30}$$

$$= \frac{(-1)^n \cdot (n+k-1)!}{(n-1)!\,(k+1)!} \cdot v^-_{(i_1} \cdots v^-_{i_{n+k}} v^+_{j_1} \cdots v^+_{j_k)}$$

Hint. Take into account that

$$u^+_{(i_1} \cdots u^+_{i_{n+k}} u^-_{j_1} \cdots u^-_{j_k)} = \frac{1}{k+1} D^{++}\left[u^+_{(i_1} \cdots u^+_{i_{n+k-1}} u^-_{i_{n+k}} u^-_{j_1} \cdots u^-_{j_k)} \right]$$

5.4. Analytic δ -Functions

The δ -functions for the analytic subspace are defined by the equation

$$\int d\zeta_2^{(-4)} du_2\, \delta_A^{(q,4-q)}(\zeta_1,u_1|\zeta_2,u_2) \cdot \phi^{(p)}(\zeta_2,u_2) = \delta^{qp}\, \phi^{(p)}(\zeta_1,u_1) \tag{5.31}$$

Note the exact balance of the U(I) charges with respect to both arguments 1 and 2. Now we proceed to construction of such δ -functions starting from the δ -function for the harmonic superspace. It is defined as follows

$$\int d^{12}z_2 du_2 \cdot \delta^{12}(z_1-z_2) \cdot \delta^{(q,-q)}(u_1,u_2) \cdot f^{(p)}(z_2,u_2) = \\ = \delta^{qp} f^{(p)}(z_1,u_1) \tag{5.32}$$

where

$$\delta^{12}(z_1-z_2) = \delta^4(x_1-x_2) \cdot \delta^8(\theta_1-\theta_2) \tag{5.33}$$

Let us now consider an analytic superfield as a function of the coordinates (Z, U) of the full superspace, i.e.

$$\phi^{(p)} = \phi^{(p)}(\zeta(z,u),u) \tag{5.34}$$

(see (2.35)). Then one can write down

$$\int d^{12}z_2 du_2 \, \delta^{12}(z_1-z_2) \cdot \delta^{(q,-q)}(u_1,u_2) \cdot \phi^{(p)}(\zeta(z_2,u_2),u_2) = \\ = \delta^{qp} \phi^{(p)}(\zeta(z,u_1),u_1) \tag{5.35}$$

Using

$$d^{12}z\,du = d\zeta^{(-4)}du \cdot (D^+)^4 , \quad (D^+)^4 \equiv \frac{1}{16}(D^{+\alpha}D^+_\alpha)(\bar{D}^+_{\dot{\alpha}}\bar{D}^{+\dot{\alpha}}) \tag{5.36}$$

and treating (ζ,u) as independent variables one finds

$$\int d\zeta_2^{(-4)} du_2 \left[(D_2^+)^4 \delta^{12}(z_1-z_2)\right] \cdot \delta^{(q,-q)}(u_1,u_2) \cdot \phi^{(p)}(\zeta_2,u_2) = \\ = \delta^{qp} \phi^{(p)}(\zeta_1,u_1) \tag{5.37}$$

Comparing this equation with the definition (5.31) one obtains

$$\delta_A^{(q,4-q)}(\zeta_1,u_1|\zeta_2,u_2) = (D_2^+)^4 \delta^{12}(z_1-z_2) \cdot \delta^{(q,-q)}(u_1,u_2) = \\ = (D_1^+)^4 \delta^{12}(z_1-z_2) \cdot \delta^{(q-4,4-q)}(u_1,u_2) \tag{5.38}$$

The second form is derived with the help of the property

$$D_{1\alpha}^{\;i}\,\delta^{12}(z_1-z_2) = -\,D_{2\alpha}^{\;i}\,\delta^{12}(z_1-z_2)$$

and the identity (5.15) . Note that in the expressions (5.38) the analyticity of δ_A is manifest.

In fact, the analytic δ-functions are analogous of the chiral ones in N=1 supersymmetry just as analyticity is analogous to chirality. For the left-handed chiral superspace one has

$$\delta_L^6(\zeta_1^L-\zeta_2^L) = (\bar{D}_2)^2\,\delta^8(z_1-z_2) = (\bar{D}_1)^2\,\delta^8(z_1-z_2)$$

Let us emphasize, however, that δ_L^6 factorizes into $\delta(x)$ and $\delta(\theta)$

$$\delta_L^6(\zeta_1^L-\zeta_2^L) = \delta^4(x_1^L-x_2^L)\;\delta^2(\theta_1-\theta_2)$$

whereas δ_A (5.38) does not (the reason is that one cannot write down $\delta^4(\theta_1^+-\theta_2^+)$ because θ_1^+ and θ_2^+ transform under different U(I) groups).

5.5. Green Functions for Hypermultiplets

At last we are prepared to start constructing Green functions for various N=2 theories. The first case is the hypermultiplet matter in its Fayet–Sohnius form. As explained above, the FS–hypermultiplet is described by an analytic superfield $q^+(\zeta,u)$ which satisfies

$$D^{++}q^+(\zeta,u) = J^{(3)}(\zeta,u) \qquad (5.39)$$

where $J^{(3)}$ is an analytic source of U(I) charge +3. The corresponding Green function

$$G^{(1,1)}(\zeta_1,u_1|\zeta_2,u_2) \equiv \langle \overset{*}{q}{}^+(1)\;\; q^+(2)\rangle$$

obeys the equation

$$D_1^{++}\,G^{(1,1)}(1|2) = \delta_A^{(3,1)}(1|2) \qquad (5.40)$$

and gives the solution of (5.39)

$$q^+(\zeta_1,u_1) = \int d\zeta_2^{(-4)} du_2\, G^{(1,1)}(1|2)\, J^{(3)}(2)$$

It is not difficult to see that the following manifestly analytic expression $\left(u_1^+ u_2^+ \equiv u_1^{+i} u_{2i}^+ \right)$

$$G^{(1,1)}(1|2) = -\frac{1}{\Box_1}(D_1^+)^4 (D_2^+)^4 \delta^{12}(z_1-z_2)\,\frac{1}{(u_1^+u_2^+)^3} \tag{5.41}$$

is the solution of equation (5.40). Indeed, with the help of (5.24,26) one finds

$$D_1^{++} G^{(1,1)}(1|2) = -\frac{1}{2\Box_1}(D_1^+)^4 (D_2^+)^4 \delta^{12}(z_1-z_2)\cdot(D_1^{--})^2 \delta^{(3,-3)}(u_1,u_2) \tag{5.42}$$

Now, using the algebra of the derivatives one can prove a very useful identity:

$$-\frac{1}{2}(D^+)^4 (D^{--})^2\, \phi(\zeta,u) = \Box\, \phi(\zeta,u) \tag{5.43}$$

for any analytic superfield ϕ. In (5.42) such a superfield is $(D_2^+)^4 \delta^{12}(z_1-z_2)\delta^{(3,-3)}(u_1,u_2) \equiv \delta_A^{(3,1)}(1,2)$ so, one arrives at (5.40).

From (5.41) one clearly sees that the Green function is antisymmetric:

$$G^{(1,1)}(1|2) = -G^{(1,1)}(2|1)$$

and real with respect to the conjugation $\widetilde{*}$

In the second order formalism the equation of motion is

$$(D^{++})^2 \omega(\zeta,u) = J^{(4)}(\zeta,u) \tag{5.44}$$

The Green function

$$G^{(0,0)}(1|2) \equiv \langle \omega(1)\,\omega(2) \rangle \tag{5.45}$$

satisfies the equation

$$(D_1^{++})^2\, G^{(0,0)}(1|2) = \delta_A^{(4,0)}(1|2) \tag{5.46}$$

and provides the solution for (5.44)

$$\omega(z_1,u_1) = \int dz_2^{(-4)} du_2\, G^{(0,0)}(1|2)\, J^{(4)}(2)$$

Exercise. Prove that

$$G^{(0,0)}(1|2) = -\frac{1}{\Box_1}(D_1^+)^4 (D_2^+)^4 \delta^{12}(z_1-z_2)\, \frac{u_1^- u_2^-}{(u_1^+ u_2^+)^3} \tag{5.47}$$

The Green functions for the hypermultiplets derived above are the basis for the perturbation theory for N=2 quantum matter x)

5.6. Feynman Rules for Hypermultiplets

Let us consider the general action for self-interacting hypermultiplets. It can be divided into free and interaction parts

$$S = \int dz^{(-4)} du \left[\frac{1}{2} \overset{*}{q}{}^+ \overset{\leftrightarrow}{D}{}^{++} q^+ + \mathcal{L}^{(+4)}_{int}(q^+, D^{++}q^+, u, \dots) \right] \tag{5.48}$$

For simplicity we deal below with one hypermultiplet (Generalization to several hypermultiplets is obvious) Now we give a list of Feynman rules. They will be formulated in momentum space, i.e. after Fourier transforming the x -dependence of the analytic superfields $q^+(\zeta(x,\theta,u), u)$ The propagator is determined by the free part of (5.48)

$$\underset{1 \quad\; 2}{\xrightarrow{p_1 - p_2 = p}} \quad \langle \overset{*}{q}{}^+(1)\, q^+(2) \rangle = \frac{i}{p^2} \frac{(D_1^+)^4 (D_2^+)^4}{(u_1^+ u_2^+)^3} \delta^8(\theta_1 - \theta_2) \tag{5.49}$$

$$\langle \overset{*}{q}{}^+ \overset{*}{q}{}^+ \rangle = \langle q^+ q^+ \rangle = 0 \tag{5.50}$$

The vertices can be read off from the $\mathcal{L}^{(+4)}_{int}$. To this end one expands $\mathcal{L}^{(+4)}_{int}(q^+, D^{++}q^+, \dots)$ in powers of q^+ . To n^{th} power of q^+ there corresponds an n-legged vertex. For example, if

x) Some of these studied have been performed in parallel with us in /23/ (in the second paper for nonzero central charges)

$$\mathcal{L}^{(+4)}_{int} = \frac{\lambda}{2}(q^+)^2(\bar{q}^+)^2 \qquad (5.51)$$

then the only vertex is

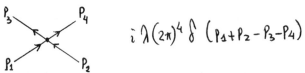

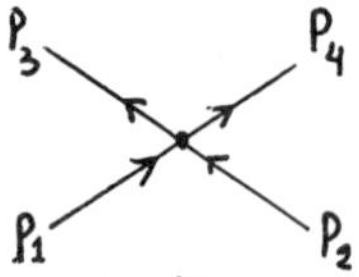

$$i\lambda(2\pi)^4\delta(p_1+p_2-p_3-p_4)$$

At each vertex one integrates over all the internal momenta (with the measure $(2\pi)^{-4}\cdot d^4p$). Besides, an integration $\int d^4\theta^+ du$ (remaining from $\int dz^{(-4)} du$ in the momentum representation) is also implied. Inspecting the propagators one can see that at each analytic vertex there are factors $(D^+)^4$ coming from the propagators which can always be used to restore the full Grassmann measure at the vertex. This is an important feature of the N=2 supergraph technique. An analogous property (restoration of full Grassman measure) is also inherent to the N=1 supersymmetry

Here we consider only the contribution of the diagram of Fig.1 to the 4-point function for a self-interacting hypermultiplet in the Taub - NUT model (3.18). The corresponding analytic expression is

$$\Gamma = \lambda^2 \int \frac{d^4p_1 \cdots d^4p_4\, d^4k}{(2\pi)^{16}}\, d^4\theta_1^+ d^4\theta_2^+\, du_1 du_2\, \delta(p_1+p_2-p_3-p_4)$$

$$\times\, q^+(p_1,\theta_1,u_1)\, q^+(p_2,\theta_1,u_1)\, \overset{*}{q}{}^+(p_3,\theta_2,u_2)\, \overset{*}{q}{}^+(p_4,\theta_2,u_2) \qquad (5.52)$$

$$\times\, \frac{(D_1^+)^4(D_2^+)^4}{(u_1^+u_2^+)^3}\,\delta^8(\theta_1-\theta_2)\; \frac{(D_1^+)^4(D_2^+)^4}{(u_1^+u_2^+)^3}\,\delta^8(\theta_1-\theta_2)\cdot\frac{1}{k^2(p_1+p_2-k)^2}$$

The general rule for handling such expressions is first to do all the θ - integrations but one using the Grassman

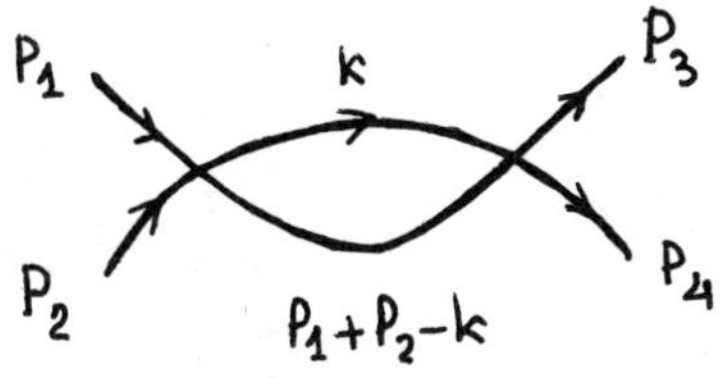

Fig.1. One-loop correction to the 4-point function for a self-interacting q^+- hypermultiplet: s-channel.

δ -functions from the propagators. For this purpose one has to restore the full measures $d^8\theta_1 d^8\theta_2$. This can be achieved by taking $(D_1^+)^4 (D_2^+)^4$ off one of the propagators and using ($d^{12}z\,du = dz^{(-4)}du\,(D^+)^4$). (Note that the other propagators and external superfields are analytic, so D_1^+, D_2^+ do not act on them). Then one applies the identity

$$\delta^8(\theta_1-\theta_2)\,(D_1^+)^4(D_2^+)^4\,\delta^8(\theta_1-\theta_2) = (u_1^+ u_2^+)^4\,\delta^8(\theta_1-\theta_2) \qquad (5.53)$$

and do the integral. The results is

$$\Gamma = \lambda^2 \int \frac{d^4p_1 \ldots d^4p_4\, d^4k}{(2\pi)^{16}}\, d^8\theta\, du_1\, du_2 \cdot \frac{\delta(p_1+p_2-p_3-p_4)}{(u_1^+u_2^+)^2 k^2 (p_1+p_2-k)^2} \times \qquad (5.54)$$

$$\times\, q^+(p_1,\theta,u_1)\cdot q^+(p_2,\theta,u_1)\, \overset{*}{q}{}^+(p_3,\theta,u_2)\, \overset{*}{q}{}^+(p_4,\theta,u_2)$$

In (5.54) we observe an important phenomenon. Although in the initial expression (5.52) there seemed to be a product of two singular harmonic distributions, in the process of doing the D-algebra one of them cancelled out. The distribution remaining in (5.54) does not lead to new, harmonic divergences. This can be most easily demonstrated if the external lines are put on-shell, i.e., $D^{++}q^+ = 0$. In this case (see (2.28) and (3.4)

$$q^+(u_1)\, q^+(u_1) = \frac{1}{2}\, D_1^{++} D_1^{--}\left[q^+(u_1)\, q^+(u_1)\right]$$

and the u_2 integral can be computed (see (5.24)):

$$\frac{1}{2}\int du_1 du_2 \left[D_1^{++} D_1^{--}(q_1^+ q_1^+)\right]\cdot \overset{*}{q}{}_2^+ \overset{*}{q}{}_2^+ \frac{1}{(u_1^+u_2^+)^2} =$$

$$= -\frac{1}{2}\int du_1 du_2\, D_1^{--}(q_1^+q_1^+)\cdot \overset{*}{q}{}_2^+ \overset{*}{q}{}_2^+ \cdot D_1^{--}\,\delta^{(2,-2)}(u_1,u_2) =$$

$$= \frac{1}{2}\int du\, (\overset{*}{q}{}^+ \overset{*}{q}{}^+)\cdot (D^{--})^2 (q^+q^+) \qquad (5.55)$$

One sees that the harmonic nonlocality present in (5.54) has disappeared and there are no harmonic divergences. The momentum integral diverges logarithmically. Its divergent part is local in x-space

$$\Gamma_\infty = C_\infty \cdot \lambda^2 \int d^{12}z\, du\, (\overset{*}{q}{}^+)^2 (D^{--})^2 (q^+)^2 =$$
$$= C_\infty \lambda^2 \int d\zeta^{(-4)} du\, (\overset{*}{q}{}^+)^2 (D^+)^4 (D^{--})^2 (q^+)^2 = \qquad (5.56)$$
$$= -2\, C_\infty \lambda^2 \int d\zeta^{(-4)} du\, (\overset{*}{q}{}^+)^2 \,\Box\, (q^+)^2$$

(see (5.43)). Obviously, Γ_∞ differs from the initial action. The same is true for the contribution of the t-channel diagram of Fig. 2 that modifies the integrand of (5.56) by a different new structure $4(q^+\overset{*}{q}{}^+)\Box(q^+\overset{*}{q}{}^+)$ (check it!) So the theory is nonrenormalizable (in $d=4$).

5.7. Finiteness of d=2, N=4 Sigma Models

It is remarkable that in a three-dimensional space-time (d=3) the graph in Fig. 1 is convergent. Moreover, in d=2 we may easily prove (along the lines presented in Ref. /26/) that the theory of the self-interacting q^+ hypermultiplet is finite off-shell for any self-interaction $\mathcal{L}^{(+4)}(q^+, \overset{*}{q}{}^+, D^{++}q^+, \ldots)$. Indeed, in d=2 $[\lambda] = cm^0$, $[q^+(Z)] = cm^0$, so $[q^+(p,\theta,u)] = cm^2$ The n-particle contribution to the effective action has the generic form (for the most general self-coupling)

$$\Gamma_n = \int d^8\theta\, du\, (d^2p)^{n-1} \, [q(p,\theta,u)]^n \, I(p) \qquad (5.57)$$

The fact that the θ integral has the full measure $d^8\theta$ follows from the Feynman rules as explained above. We see that the momentum integral $I(p)$ has dimension cm^2 and hence is convergent. Note that on-shell q^+-theory with a general coupling describes any hyper-Kähler nonlinear supersymmetric sigma-model (N=4 in d=2). The

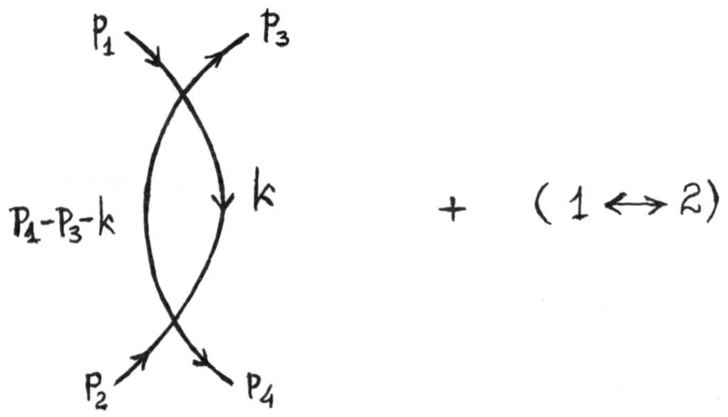

Fig. 2. One-loop correction to the 4-point function for a self-interacting q^+-hypermultiplet: t-channel.

finiteness of hyper-Kahler N=4 σ-models in d=2 was proved in [1] by completely different means [x)] .

CONCLUSION

So much for matter couplings in the framework of the harmonic superspace. There certainly remains a lot of problems, in particular,

1) To work out a simple and systematic way of extracting geometric information from the "hyper-Kähler potential" $\mathcal{L}^{(+4)}$. We mean simplifications in the construction of metrics and techniques to draw conclusions concerning general characteristics of a sought hyper-Kähler manifold, e.g. its Euler number, or compactness, etc.

2) To look for new interesting hyper-Kahler metrics. This study is stimulated also by general relativity as all the four-dimensional hyper-Kahler metrics are gravitational instantons [20] .

3) To prove the conjecture of Sect.3 that the harmonic self-interactions are the hyper-Kahler potentials [xx)] .

REFERENCES

1. de Wit B., Philippe R. and Van Proeyen A., Nucl. Phys. B219 , 143 (1983).

2. Fayet P., Nucl.Phys. B113, 135 (1976).
Salam A. and Strathdee J., Nucl. Phys. B97, 293 (1975).

3. a) Wess J., Acta Phys.Austriaca, 41, 409 (1975).
b) Lindstrom U. and Rocek M., Nucl. Phys. B222, 285 (1983).

---

x) Finiteness of a large class of the Kahler d=2, N=2 sigma-models is now questionable [27].

xx) One of the ways [24] seems to start from the on-shell superfield formulation [25] .

c) Karlhede A., Lindström U. and Rocek M., Phys.Lett. 147B,297 (1984).
d) Siegel W., Phys.Lett. 153B , 51 (1985).
4. Howe P., Stelle K. and Townsend P., Nucl. Phys. B214, 519 (1983).
5. Galperin A., Ivanov E., Kalitzin S., Ogievetsky V. and Sokatchev E., Class. Quantum Grav., 1, 469 (1984); Proc. 1984 Trieste Spring School on Supersymmetry and Supergravity, World Scient., p.449 (1984).
6. Howe P., Stelle K. and West P., Class. Quantum Grav. 2, 815 (1985).
7. Galperin A., Ivanov E., Ogievetsky V. and Sokatchev E. Class. Quantum Grav. 2, 601, 617 (1985).
8. Galperin A., Ivanov E., Ogievetsky V. and Sokatchev E. JINR preprint E2-85-363 (Dubna, 1985); In "Quantum Field Theory and Quantum Statistics. Essays in Honour of the 60th Birthday of E.S.Fradkin", C.Isham edit., Adam Hilger (1986).
9. Galperin A., Ivanov E., Kalitzin S., Ogievetsky V. and Sokatchev E., Class.Quant.Grav. 2, 155 (1985); Phys.Lett. 151B , 215 (1985).
10. Sokatchev E., Phys.Lett., 169B , 209 (1986).
11. Alvarez-Gaume L. and Freedman D.Z., Comm.Math.Phys. 80, 443 (1981).
12. Zumino B. Phys.Lett., 87B , 203 (1979).
13. a) Rocek M., Physica 15D , 75 (1985).
b) Bagger J., preprint SLAC-PUB-3461 (1984); in "Proc. of 1984 Bonn-NATO Advanced Study Institute on SUSY".
14. Galperin A., Ivanov E., Ogievetsky V. and Sokatchev E. Comm.Math.Phys. 103 , 515 (1986).
15. Galperin A., Ivanov E., Ogievetsky V. and Townsend P. JINR preprint E2-85-732 (Dubna, 1985), Class.Quantum Grav. in press.

16. Galperin A., Ivanov E. and Ogievetsky V., JINR pre - print E2-86-277 (Dubna, 1986), Nucl.Phys. in press.
17. Galperin A., Ivanov E. and Ogievetsky V., JINR pre-print E2-85-897 (Dubna, 1985), Yadern.Fyz. in press.
18. Galperin A., Ivanov E. and Ogievetsky V., JETP Pisma 33 , 176 (1981).
19. Bagger J. and Witten E., Nucl. Phys. B222 , 1 (1983).
20. Eguchi T., Gilkey P. and Hanson A. , Phys.Rep. 66, 213 (1980).
21. Scherk J. and Schwarz J., Nucl. Phys., B153 ,61 (1979).
22. Ohta N., Sugata H. and Yamaguchi H., Osaka Univ. Report No. OS-GE-85-02 (1985).
23. Kubota T. and Sawada S., Progr.Theor.Phys. 74 , 1329 (1985).
Ohta N. and Yamaguchi H., Phys.Rev. D32, 1954 (1985).
24. Rosly A. and Schwarz A. In: Proc. of III Int.Seminar "Quantum Gravity", M.A.Markov et al. ed., World Scient. p.308 (1985).
25. Sierra G. and Townsend P. , Nucl. Phys. B233 , 289 (1984), Phys.Lett. 124B , (1983).
26. Gates J., Grisaru M., Rocek M. and Siegel W., Superspace (Benjamin/Cummings, Reading (1983)).
27. Grisaru M.T., van de Ven A. and Zanon D., prepr. HUTP - 86/A026, A027.
Pope C.N., Sohnius M.F., Stelle K.S., prepr. Imperial /TP/85-86/16.